Controversial Statistical Issues in Clinical Trials

Chapman & Hall/CRC Biostatistics Series

Editor-in-Chief

Shein-Chung Chow, Ph.D.
Professor
Department of Biostatistics and Bioinformatics
Duke University School of Medicine
Durham, North Carolina

Series Editors

Byron Jones
Senior Director
Statistical Research and Consulting Centre
(IPC 193)
Pfizer Global Research and Development
Sandwich, Kent, U.K.

Jen-pei Liu
Professor
Division of Biometry
Department of Agronomy
National Taiwan University
Taipei, Taiwan

Karl E. Peace
Georgia Cancer Coalition
Distinguished Cancer Scholar
Senior Research Scientist and
Professor of Biostatistics
Jiann-Ping Hsu College of Public Health
Georgia Southern University
Statesboro, Georgia

Bruce W. Turnbull
Professor
School of Operations Research
and Industrial Engineering
Cornell University
Ithaca, New York

Chapman & Hall/CRC Biostatistics Series

Adaptive Design Theory and Implementation Using SAS and R
Mark Chang

Advanced Bayesian Methods for Medical Test Accuracy
Lyle D. Broemeling

Advances in Clinical Trial Biostatistics
Nancy L. Geller

Applied Statistical Design for the Researcher
Daryl S. Paulson

Basic Statistics and Pharmaceutical Statistical Applications, Second Edition
James E. De Muth

Bayesian Adaptive Methods for Clinical Trials
Scott M. Berry, Bradley P. Carlin,
J. Jack Lee, and Peter Muller

Bayesian Analysis Made Simple: An Excel GUI for WinBUGS
Phil Woodward

Bayesian Methods for Measures of Agreement
Lyle D. Broemeling

Bayesian Missing Data Problems: EM, Data Augmentation and Noniterative Computation
Ming T. Tan, Guo-Liang Tian,
and Kai Wang Ng

Bayesian Modeling in Bioinformatics
Dipak K. Dey, Samiran Ghosh,
and Bani K. Mallick

Causal Analysis in Biomedicine and Epidemiology: Based on Minimal Sufficient Causation
Mikel Aickin

Clinical Trial Data Analysis using R
Ding-Geng (Din) Chen and Karl E. Peace

Clinical Trial Methodology
Karl E. Peace and Ding-Geng (Din) Chen

Computational Methods in Biomedical Research
Ravindra Khattree and Dayanand N. Naik

Computational Pharmacokinetics
Anders Källén

Controversial Statistical Issues in Clinical Trials
Shein-Chung Chow

Data and Safety Monitoring Committees in Clinical Trials
Jay Herson

Design and Analysis of Animal Studies in Pharmaceutical Development
Shein-Chung Chow and Jen-pei Liu

Design and Analysis of Bioavailability and Bioequivalence Studies, Third Edition
Shein-Chung Chow and Jen-pei Liu

Design and Analysis of Clinical Trials with Time-to-Event Endpoints
Karl E. Peace

Design and Analysis of Non-Inferiority Trials
Mark D. Rothmann, Brian L. Wiens,
and Ivan S. F. Chan

Difference Equations with Public Health Applications
Lemuel A. Moyé and Asha Seth Kapadia

DNA Methylation Microarrays: Experimental Design and Statistical Analysis
Sun-Chong Wang and Arturas Petronis

DNA Microarrays and Related Genomics Techniques: Design, Analysis, and Interpretation of Experiments
David B. Allsion, Grier P. Page,
T. Mark Beasley, and Jode W. Edwards

Dose Finding by the Continual Reassessment Method
Ying Kuen Cheung

Elementary Bayesian Biostatistics
Lemuel A. Moyé

Chapman & Hall/CRC Biostatistics Series

Controversial Statistical Issues in Clinical Trials

Shein-Chung Chow

Duke University School of Medicine

Durham, North Carolina, USA

CRC Press
Taylor & Francis Group
Boca Raton London New York

CRC Press is an imprint of the
Taylor & Francis Group, an **informa** business

A CHAPMAN & HALL BOOK

CRC Press
Taylor & Francis Group
6000 Broken Sound Parkway NW, Suite 300
Boca Raton, FL 33487-2742

First issued in paperback 2020

© 2011 by Taylor & Francis Group, LLC
CRC Press is an imprint of Taylor & Francis Group, an Informa business

No claim to original U.S. Government works

ISBN-13: 978-0-367-57693-6 (pbk)
ISBN-13: 978-1-4398-4961-3 (hbk)

**Visit the Taylor & Francis Web site at
http://www.taylorandfrancis.com**

**and the CRC Press Web site at
http://www.crcpress.com**

Contents

Preface

In pharmaceutical/clinical development of a test drug or treatment, relevant clinical data are usually collected from subjects with the diseases under study in order to evaluate the safety and efficacy of the test drug or treatment under investigation. It is necessary to conduct well-controlled clinical trials under a valid study design to provide an accurate and reliable assessment. A clinical trial process is a lengthy and costly process but is nevertheless necessary to ensure a fair and reliable assessment of the test treatment under investigation. It consists of protocol development, trial conduct, data collection, statistical analysis/interpretation, and reporting. In practice, controversial issues inevitably occur regardless of the compliance with good statistical practice (GSP) and good clinical practice (GCP). Controversial issues are basically debatable issues that are commonly encountered during the conduct of clinical trials. In practice, these issues could be raised from, but are not limited to, (1) compromises between theoretical and real/common practices; (2) miscommunication and/or misunderstanding in perception/interpretation among regulatory agencies, clinical scientists, and biostatisticians; and (3) disagreement, inconsistency, miscommunication/misunderstanding, and errors in clinical practice.

In clinical trials, commonly seen controversial issues include, but are not limited to, (1) appropriateness of traditional statistical hypotheses (which primarily focus on efficacy) for the clinical evaluation of *both* efficacy and safety, (2) the instability of classical sample size calculation based on information from a small pilot study, (3) the integrity of randomization and blinding, (4) clinical strategies for selecting an appropriate endpoint from some endpoints that are derived based on data collected from the same patient population, (5) the impact of major protocol amendments that may have resulted in a population shift, (6) the feasibility/applicability of the use of adaptive design methods in clinical trials, (7) issues of multiplicity in clinical trials, (8) the independence of the independent data monitoring committee (IDMC), (9) the determination of non-inferiority margin in active control (or non-inferiority) trials, and (10) the assessment of the probability of success in clinical development. In this book, we will post these controversial issues rather than provide resolutions. Other practical and/or controversial issues are also briefly described. The impact of these issues on the evaluation of the safety and efficacy of the test treatment under investigation is discussed with examples whenever applicable. Recommendations regarding possible resolutions of these issues are also provided whenever possible. It is our goal that regulatory agencies, clinical scientists, and biostatisticians should (1) pay attention to these issues, (2) identify the possible causes, (3) resolve/correct the issues, and, consequently, (4) enhance good statistical/clinical practices for achieving the study objectives of the intended clinical trials.

This book is intended to be the first book entirely devoted to the discussion of controversial issues in clinical trials. It covers controversial issues that are commonly encountered at various stages of clinical research and development, including bench-to-bedside translational research. It is our goal to provide a useful desk reference and state-of-the art examination of controversial issues in clinical trials to (1) scientists who are engaged in clinical research and development, (2) statistical and/or medical reviewers from regulatory agencies who have to make decisions on the evaluation/approval of test treatments under investigation, and (3) biostatisticians who provide statistical support for the design and analysis of clinical trials or related projects. We hope that this book can serve as a bridge among scientists from the pharmaceutical industry, medical/statistical reviewers from government regulatory agencies, and researchers from academia.

The scope of this book is restricted to controversial issues that are commonly seen in clinical development including early-phase clinical development such as bioavailability/bioequivalence and bench-to-bedside translational research. This book consists of 27 chapters. Chapter 1 provides a background on pharmaceutical/clinical research and development and describes some commonly seen controversial issues in clinical research. Chapter 2 emphasizes the importance of GSP in clinical research and development. In Chapter 3, some controversial issues that are commonly seen in bench-to-bedside translational research are discussed. Chapter 4 discusses practical issues encountered during the assessment of bioequivalence. Chapter 5 introduces composite hypotheses for the clinical evaluation of efficacy and safety simultaneously. Chapter 6 examines the instability of sample size calculation/justification based on data obtained from a small pilot study. Chapter 7 discusses tests for the integrity of randomization/blinding while Chapter 8 attempts to provide some insight into clinical strategies for the selection of an appropriate endpoint for the assessment of treatment effect. Chapter 9 studies the impact of major protocol amendments that have caused population shifts during the conducting of clinical trials. Chapter 10 investigates the feasibility/applicability for the use of adaptive design methods in clinical trials. Chapter 11 discusses the issue of multiplicity in clinical trials. Chapter 12 challenges the independence of an IDMC. Chapter 13 studies the impact of analysis results under an incorrect model (e.g., data collected under a one-way analysis of variance model but analyzed using a two-way analysis of variance model).

Chapter 14 reviews some performance characteristics for the validation of a subjective instrument (questionnaire) to assess the clinical benefit of the test treatment under investigation such as quality-of-life assessment. Chapter 15 provides a summary of statistical methods for missing data imputation in clinical trials. Chapter 16 compares several approaches for center grouping for clinical trials with a number of small centers. Chapter 17 provides a summary of statistical methods for determining the non-inferiority margin in non-inferiority (active-control) trials. In Chapter 18, design and analysis for QT/QTc studies with recording replicates for the assessment of cardio-toxicity in terms of QT/QTc prolongation are reviewed. Chapter 19 discusses some practical issues that are commonly encountered in multiregional (multinational) clinical trials. Chapter 20 compares commonly considered dose escalation trial designs in cancer trials such as algorithm-based traditional escalation rule (TER) and model-based continual reassessment method (CRM) trial designs. Chapter 21 focuses on the enrichment process in target clinical trials. Chapter 22 discusses basic concepts and principles for conducting clinical trial simulation. Chapter 23 provides an outline of fundamental differences between Western medicine and traditional Chinese medicine. Chapter 24 discusses practical issues encountered during the assessment of biosimilarity between follow-on biologics (FOB). Chapter 25 deals with the calculations of the probabilities of generalizability and reproducibility of a given clinical trial based on the observed clinical data of the clinical trial. Chapter 26 provides a review of good regulatory practices, especially the good review practice (GRP) published by the Center for Drug Evaluation and Research (CDER) at the United States Food and Drug Administration (FDA). Chapter 27 evaluates the probability of success for the pharmaceutical and/or clinical development of a test treatment under investigation. In each chapter, examples and possible recommendations and/or resolutions are provided whenever possible.

I would like to thank David Grubbs from Taylor & Francis for providing me the opportunity to work on this book. I would also like to thank my colleagues from the Department of Biostatistics and Bioinformatics, Duke Clinical Research Institute (DCRI), Duke Clinical Research Unit (DCRU), and Center for AIDS Research (CFAR) of Duke University School of Medicine for their support during the preparation of this book. I wish to express my gratitude to the following individuals for their encouragement and support: Robert Califf, MD, Robert Harrington, MD, and Ralph Corey, MD, of DCRI; John Sundy, MD, PhD of DCRU; Ken Weinhold, MD, of CFAR; John Rush, MD, of Duke-NUS; and Liz DeLong, PhD, of the Department of Biostatistics and Bioinformatics, Duke University School of Medicine, as well as many friends from academia, the pharmaceutical industry, and regulatory agencies such as U.S. FDA and EU EMEA.

Finally, the views expressed are mine and not necessarily those of Duke University School of Medicine. I am solely responsible for the contents and errors of this book. Any comments and suggestions will be very much appreciated.

Shein-Chung Chow, PhD
Duke University School of Medicine
Durham, North Carolina

1

Introduction

1.1 Introduction

In the past several decades, it has been recognized that increasing spending for biomedical research does not reflect an increase in the success rate of pharmaceutical (clinical) development. Woodcock (2005) indicated that the low success rate of pharmaceutical development could be due to (1) a diminished margin for improvement that escalates the level of difficulty in proving drug benefits, (2) genomics and other new sciences have not yet reached their full potential, (3) mergers and other business arrangements have decreased candidates, (4) easy targets are the focus as chronic diseases are harder to study, (5) failure rates have not improved, (6) rapidly escalating costs and complexity decrease the willingness/ability to bring many candidates forward into the clinic. In the early 2000s, the U.S. Food and Drug Administration (FDA) kicked off a *Critical Path Initiative* to assist the sponsors in identifying the scientific challenges underlying the medical product pipeline problems. In its 2004 *Critical Path Report*, the FDA presented its diagnosis of the scientific challenges underlying the medical product pipeline problems.

On March 16, 2006, the FDA released a Critical Path Opportunity List that outlines six broad topic areas, which include 76 initial projects to bridge the gap between the quick pace of new biomedical discoveries and the slower pace at which those discoveries are currently being developed into therapies. These six broad topic areas include (1) better evaluation tools, (2) streamlining clinical trials, (3) harnessing bioinformatics, (4) moving manufacturing into the twenty-first century, (5) developing products to address urgent public health needs, and (6) specific at-risk populations such as pediatrics. In this book, we will focus on the second broad topic area of streamlining clinical trials, which includes (1) design of active controlled trials, (2) enrichment designs, (3) use of prior experience or accumulated information in trial design, (4) development of best practices for handling missing data, (5) development of trial protocols for specific therapeutic areas, and (6) analysis of multiple endpoints. The first topic for the design of active controlled trials has led the research for design and statistical methodology development for non-inferiority trials. The enrichment designs have stimulated research for using biomarkers in the enrichment process of target clinical trials for achieving the ultimate goal of personalized medicine. The recommendation for the use of prior experience or accumulated information in the trial design has provoked tremendous discussion for the use of adaptive methods in clinical trials and the use of the Bayesian approach in drug research and evaluation. The encouragement for the development of best practices for handling missing data has triggered (1) the study of the validity of the commonly used method of last observation carry forward (LOCF) and (2) research for the methodology development of missing data imputation (see, e.g., NRC, 2010). The last topic of analysis of multiple endpoints has attracted much attention on the issue of multiplicity in clinical trials.

In pharmaceutical/clinical research and development, clinical trials are necessarily conducted for the evaluation of the efficacy and safety of the test treatment under investigation. In practice, the clinical trial process involves (1) protocol development, (2) conduct of clinical trial, analysis, and interpretation, and (3) regulatory review and approval. For a given clinical trial, good clinical practice (GCP) and good statistics practice (GSP), which is the foundation of GCP, are key to the success of the intended clinical trial. GSP and GCP ensure the validity and integrity of the clinical data collected from the clinical trial. In clinical trials, controversial issues inevitably occur regardless of whether the clinical trial process is in compliance with both GCP and GSP. In this book, controversial issues in clinical trials are referred to as *debatable* issues that are commonly encountered while conducting clinical trials. Controversial issues could be raised from, but are not limited to, (1) compromises between theoretical and real/common practices, (2) miscommunication and/or misunderstanding in perception/interpretation among regulatory agencies, clinical scientists, and biostatisticians, and (3) disagreement, inconsistency, miscommunication/misunderstanding, and errors in clinical practice.

In Section 1.2, the process of pharmaceutical development including nonclinical, preclinical, and clinical development is briefly outlined. Some commonly seen controversial issues are briefly described in Section 1.3. The aim and structure of the book are given in the last section.

1.2 Pharmaceutical Development

As pointed out by Chow and Shao (2002) and Chow and Liu (2004), pharmaceutical development is a lengthy and costly process to ensure the safety and efficacy of the drug products under investigation before they can be approved by the regulatory agencies for use in humans. In addition, this lengthy and costly development process is necessary to assure that the approved drug product will possess some good drug characteristics such as identity, purity, quality, strength, stability, and reproducibility. A typical pharmaceutical development process involves drug discovery, formulation, laboratory development, animal studies for toxicity, clinical development, and regulatory submission/review and approval. Pharmaceutical development is a continual process that can be classified into three phases of development, namely, nonclinical development (e.g., drug discovery, formulation, laboratory development, scale-up, manufacturing process validation, stability, and quality control/assurance), preclinical development (e.g., animal studies for toxicity, bioavailability and bioequivalence studies, and pharmacokinetic and pharmacodynamic studies), and clinical development (e.g., phases I–III clinical trials for the assessment of safety and efficacy). These phases may occur in sequential order or be overlapped during the development process. To provide a better understanding of the pharmaceutical development process, these critical phases of pharmaceutical development are briefly outlined in the following sections.

1.2.1 Nonclinical Development

Nonclinical development includes drug discovery, formulation, laboratory development such as analytical method development and validation, (manufacturing) process validation, stability, statistical quality control, and quality assurance (see, e.g., Chow and Liu, 1995). Drug discovery usually consists of the phases of drug screening and drug lead optimization. In the drug screening phase, the mess compounds are screened to identify those that are active from those that are not. Lead optimization is a process of finding a compound with some advantages over related leads based on some physical, chemical, and/or pharmacological properties. In practice, the success rate for identifying a promising active compound is usually relatively low. As a result, there may be a few compounds that are identified as promising active compounds.

The purpose of formulation is to develop a dosage form (e.g., tablets or capsules) such that the drug can be delivered to the site of action efficiently. For laboratory development, an analytical method is necessarily developed to quantitate the potency (strength) of the drug product. Analytical method development and validation play an important role in quality control and quality assurance of the drug product. To ensure that a drug product will meet the U.S. Pharmacopeia and National Formulary (USP/NF, 2000) standards for the identity, strength, quality, and purity of the drug product, a number of tests such as potency testing, weight variation testing, content uniformity testing, dissolution testing, and disintegration testing are usually performed at various stages of the manufacturing process. These tests are referred to as USP/NF tests. At the same time, stability studies are usually conducted to characterize the degradation of the drug product over time under appropriate storage conditions. Stability data can then be used to determine drug expiration dating period (or drug shelf life) as it is required by the regulatory agency to be indicated in the immediate label of the container (Chow, 2007b).

After the drug product has been approved by the regulatory agency for use in humans, a scale-up program is usually carried out to ensure that a production batch can meet USP/NF standards for the identity, strength, quality, and purity of the drug before a batch of the product is released to the market. The purpose of a scale-up program is not only to identify, evaluate, and optimize critical formulation and/or (manufacturing) process factors of the drug product but also to maximize or minimize the excipient range. A successful scale-up program can result in an improvement in formulation/process or at least a recommendation on a revised procedure for formulation/process of the drug product. During nonclinical development, the manufacturing process is necessarily validated in order to produce drug products with good drug characteristics such as identity, purity, strength, quality, stability, and reproducibility (Bergum, 1988). Process validation is important in nonclinical development to ensure that the manufacturing process does what it purports to do.

1.2.2 Preclinical Development

The primary focus of preclinical development is to evaluate the safety of the drug product through *in vitro* assays and animal studies. In general, *in vitro* assays or animal toxicity studies are intended to alter the clinical investigators to the potential toxic effects associated with the investigational drugs so that those effects may be watched for during the clinical investigation. Preclinical testing involves dose selection, toxicological testing for toxicity and carcinogenicity, and animal pharmacokinetics. For selection of an appropriate dose, dose response (dose ranging) studies in animals are usually conducted to determine the effective dose, such as the median effective dose (ED_{50}). Preclinical development is critical in the pharmaceutical development process because it is not ethical to investigate certain toxicities such as the impairment of fertility, teratology, mutagenicity, and overdose in humans (Chow and Liu, 1998a). Animal models are then used as a surrogate for human testing under the assumption that they can be predictive of clinical outcomes in humans.

Following the administration of a drug, it is also important to study the rate and extent of absorption, the amount of drug in the bloodstream that hence becomes available, and the elimination of the drug. For this purpose, a comparative bioavailability study in humans is usually conducted to characterize the profile of the blood or plasma concentration–time curve by means of several pharmacokinetic parameters such as area under the blood or plasma concentration–time curve (AUC), maximum concentration (C_{max}), and time to achieve maximum concentration (t_{max}) (Chow and Liu, 2000a). It should be noted that the identified compounds will have to pass the stages of nonclinical/preclinical development before they can be used in humans.

1.2.3 Clinical Development

Clinical development in the development of a pharmaceutical entity is to scientifically evaluate benefits (e.g., efficacy) and risks (e.g., safety) of promising pharmaceutical entities at a minimum cost and within a relatively short time frame. As indicated by Chow and Liu (2004), approximately 75% of pharmaceutical development is devoted to clinical development and regulatory registration. In a set of new regulations promulgated in 1987 and known as the investigational new drug (IND) Rewrite, the phases of clinical investigation adopted by the FDA since the late 1970s is generally divided into three phases (see, e.g., Part 21 Code of Federal Regulations, 312.21). These phases of clinical investigation are usually conducted sequentially but may overlap.

The primary objective of phase I is not only to determine the metabolism and pharmacological activities of the drug in humans, the side effects associated with increasing doses, and the early evidence on effectiveness, but also to obtain sufficient information about the drug's pharmacokinetics and pharmacological effects to permit the design of well-controlled and scientifically valid phase II studies. The primary objectives of phase II studies are not only to first evaluate the effectiveness of a drug based on clinical endpoints for a particular indication or indications in patients with the disease or condition under study, but also to determine the dosing ranges and doses for phase III studies and the common short-term side effects and risks associated with the drug. Note that some pharmaceutical companies further differentiate phase II into phases IIa and IIb. For example, clinical studies designed to evaluate dosing are referred to as phase IIa studies, while studies designed to determine the effectiveness of the drug are called phase IIb. In some cases, clinical studies based on clinical endpoints are considered phase IIb studies. The primary objectives of phase III studies are (1) to gather additional information about the effectiveness and safety needed to evaluate the overall benefit–risk relationship of the drug and (2) to provide an adequate basis for physician labeling. Note that studies conducted after regulatory submission before approval are generally referred to as phase IIIb studies.

In addition to these three phases of clinical development, many pharmaceutical companies consider performing studies after a drug is approved for marketing called phase IV studies. The purpose for conducting phase IV studies is to elucidate further the incidence of adverse reactions and determine the effect of a drug on morbidity or mortality. In addition, a phase IV trial may be conducted to study a patient population not previously studied, such as children. In practice, phase IV studies are usually considered useful market-oriented comparison studies against competitors such as quality-of-life studies. As indicated by Chow and Shao (2002), in practice, it is estimated that about 1 in 8 to 10×10^3 compounds screened may finally reach the phase of clinical development for human testing. The probability of success for those compounds that reach clinical development is relatively low.

As a result, a thoughtful clinical development plan is necessary to ensure the success of the development of a promising pharmaceutical entity.

In practice, phases I and II are considered early-phase clinical development, while phases III and IV are viewed as later-phase clinical development. However, in the pharmaceutical industry, some pharmaceutical companies consider clinical studies up to phase IIa as early-phase clinical development. Phase I clinical investigation provides an initial introduction of an IND to humans. Phase I clinical investigation includes studies of drug metabolism, bioavailability, dose ranging, and multiple doses. Phase I studies usually involve 20–80 normal volunteer subjects or patients. In several therapeutic areas, patients with the diseases are subjects rather than healthy volunteers. This tradition is strongest in oncology because many cytotoxic agents cause damage to DNA. For similar reasons, many anti-AIDS drugs are not tested initially in healthy subjects. It should be noted that some categories of drugs such as neuropharmacology may have an acclimatization or tolerance aspect, which makes them difficult to study in healthy subjects. For phase I clinical investigation, FDA's review focuses on the assessment of safety. Therefore, extensive safety information such as detailed laboratory evaluations is usually collected at very extensive schedules. A typical phase I design for tolerability and safety is a dose escalation trial design in which successive groups (cohorts) of patients are given successively higher doses of the treatment until some of the patients in a cohort experience unacceptable side effects. In most phase I trials of this kind, there are—three to six patients in each cohort. The starting dose at the first cohort is usually rather low. If unacceptable side effects are not seen in the first cohort, patients in the next cohort will receive a higher dose. This continues until a dose is reached at which it is too toxic for some patients (say one out of three). Then, the previous dose level is considered to be the maximum tolerated dose (MTD). It should be noted that MTD usually is the most effective dose, which is often chosen as the optimal dose for phase II studies in practice. Also, as indicated by the FDA, phase I studies are usually less detailed and more flexible than for subsequent phases, and therefore adaptive (flexible) designs are usually considered.

Phase II studies are the first controlled clinical studies of the drug under investigation. Phase II studies usually involve not more than several hundred patients. A commonly employed study design for a phase II study is a randomized, parallel group (either a placebo-control or an active-control) study. Patients will be randomly assigned to either of the treatment groups to receive the dose determined in the prior phase I study. Many phase II trials, however, are conducted in two stages. The idea is to stop the trial as soon as it can be known that the treatment is ineffective. On the other hand, we wish to continue the trial if the treatment has shown to be effective. In a two-stage design, after a predetermined number of patients have been treated, the trial is paused and the response rate (RR) is evaluated. If the RR is less than a prespecified minimum goal (undesirable RR), it is concluded that the treatment is not worth pursuing and the trial is stopped. Otherwise, the trial continues and additional patients are enrolled to permit determination of the RR for achieving desired accuracy with certain statistical power. It should be noted that if the trial has reached the second stage, it indicates that at least some of the patients are responding to the treatment though the RR could still be low at the first stage.

1.3 Controversial Issues

In clinical development, the success of a well-controlled clinical trial relies on both clinical operation and statistical operation. Clinical operation is responsible for (1) the involvement of protocol development, (2) site management including selecting qualified study sites, and patient recruitment, (3) Institutional Review Board review, (4) conducting/monitoring of the trial, (5) protocol amendments, and (6) data management. On the other hand, statistical operation is responsible for (1) evaluation of alternative study designs for achieving the study objectives of the intended trial, (2) setting up appropriate (statistical) hypotheses according to study objectives, (3) performing a pre-study power analysis for sample size calculation based on primary study endpoint, (4) preparing statistical section for inclusion in the study protocol including randomization model/method with blinding procedure for preventing potential bias, (5) clinical strategy for endpoint selection and development of appropriate statistical methods for data analysis, (6) addressing possible statistical impact on protocol amendments, (7) providing support to an established independent data safety monitoring committee (IDMC) (if applicable) to ensure the validity, integrity, and safety of the intended clinical trials, and (8) interaction with regulatory agencies for feasibility and applicability of the use of adaptive design methods in clinical trials (if applicable). During the conduct of a clinical trial, some controversial issues are commonly encountered regardless of the compliance of GSP and GCP. These controversial issues will not only have an influence on the validity of statistical analysis for providing a fair and unbiased assessment of the treatment under investigation, but also have an impact on the probability of the success for bringing promising compounds or innovative therapies into the marketplace. In the subsequent sections, these controversial issues that are commonly encountered are briefly described.

Drug recall/withdrawal: A commonly asked question in pharmaceutical/clinical development is "Why did a newly approved drug product get recalled or withdrawn (usually due to safety concern) after a lengthy and costly process of pharmaceutical/clinical development?" Subsequent questions include the following: (1) Is the current drug review/approval process adequate? (2) Is the observed safety issue which led to the recall/withdrawal of the drug product by chance alone? (3) Are the trial design, data management, and programming and statistical methods employed for data analysis valid? (4) Are the clinical data interpreted correctly? (5) Did the regulatory submission contain all of the information regarding efficacy/safety and good drug characteristics of identity, purity, quality, strength, stability, and reproducibility? In practice, there may exist no definite answers to any of these questions. In this book, we intend to provide some insights in some chapters, which may be useful to revisit these questions.

One-fits-all criterion: For approval of generic drug products, most regulatory agencies including the FDA require that evidence of average BE in terms of the extent and rate of drug absorption, which are measured by AUC and C_{max}, be provided. The FDA requires that the 90% confidence interval for the ratio of means (e.g., AUC) be totally within the BE limits (80%, 125%) for claiming BE. This one-fits-all criterion is applicable to all drug products across therapeutic areas regardless of narrow/wide therapeutic index and/or intra-subject variability. One of the controversial issues that is frequently challenged by clinical scientists is "What if we fail to meet the BE limits by a relatively small margin?" This is similar to the question "What is the difference between a *p*-value of 0.49 (pass) and a *p*-value of 0.51 (fail) in clinical trials?" In addition, the following questions are often asked: (1) Can an approved generic drug product reach a similar therapeutic effect of the innovative drug product—what is the compromise between scientific validity and regulatory consideration? (2) Can all of the approved generic drug products be used interchangeably (drug interchangeability in terms of drug prescribability for new patients and drug switchability for current patients)? (3) What if a BE study meets the BE criterion based on the raw data but fails to meet the BE criterion based on log-transformed data (current FDA requirement) or vice versa? (4) What if AUC meets the BE criterion but C_{max} fails? More details and discussions of the above controversial issues are given in Chapter 4.

Lost-in-translation: One of the major concerns in bench-to-bedside translational research is probably the appropriateness of the one-way translational process from basic drug discovery to clinical outcome. The most commonly asked question is "Is an animal model (or *in vitro* activity) predictive of the human model (or *in vivo* activity)?" or "Does an *in vitro–in vivo* correlation exist?" Under the one-way translational process from bench to bedside, what is the potential lost-in-translation? The possibility that a significant lost-in-translation from bench to bedside could lead to the failure of the clinical trial despite the test treatment is in fact promising. As a result, it is suggested that a two-way translational process between bench (basic drug discovery) and bedside (clinical application) be considered for the improvement of the pharmaceutical/clinical development of a test treatment under investigation. More details can be found in Chapter 3.

Instability of sample size: In practice, sample size calculation/justification is usually performed based on the information obtained from previous studies or a small pilot study. It is, however, of concern whether the selected sample size can achieve statistical significance with a desired power for correctly detecting a clinically meaningful difference at a prespecified level of significance. One of the controversial issues regarding sample size calculation is why the selected sample size does not guarantee the success of the intended clinical trials? In addition, Why sample size reestimation is recommended? For a given clinical trial, can we always start with a small number of subjects and then increase the sample size later if necessary? Is this approach acceptable to regulatory agencies? It should be noted that sample size calculation is usually performed under certain assumptions that are closely related to the uncertainties of the target patient population. Thus, the formula or procedure for sample size calculation is very sensitive to assumptions of the study parameters. Any deviations to the assumption could lead to instability of the estimated sample size. The instability of the sample size in clinical trials is examined in Chapter 6.

Integrity of randomization/blinding: The purpose of randomization and blinding in a double-blind randomized clinical trial is to prevent possible biases that may be introduced during the conduct of the clinical trial. However, because of human nature, both patients and investigators may guess which treatment a patient receives. Thus, "Does the randomization/blinding work in randomized double-blind studies?" is an interesting question to clinical scientists. Chow and Shao (2004) proposed a method for testing the integrity of blinding. This, however, raises the following controversial issues. First, should a test for the integrity of blinding be performed at the end of the study? Second, what action should be taken for those positive trials which fail to pass the test for the integrity of blinding? Similarly, can the sponsor appeal if a negative trial fails to pass the test for integrity of blinding? Finally, should the clinical data that fail to pass the test for integrity of blinding be rejected for clinical evaluation of the test treatment under investigation? For randomization, the integrity of randomization can be tested in terms of the probability of correctly guessing the treatment codes. For comparative clinical trials, a blocking size of 2 or 4 is usually employed for the generation of randomization schedules in order to maintain treatment balance. As a result, which blocking size will give a higher probability of correctly guessing the treatment codes right has become an interesting question in clinical trials. More details regarding the integrity of randomization/blinding can be found in Chapter 7.

Clinical strategy for endpoint selection: In clinical trials, the sponsor always seeks an appropriate study endpoint that can lead to or increase the probability of success of the intended clinical trial. As a result, two major controversial issues are raised. As an example, for cancer trials, the following study endpoints are often considered: RR, time to disease progression (TTP), and survival. Different study endpoints may exhibit different effect sizes, which relate to *overall* clinical evaluation of the efficacy of the test treatment. Williams et al. (2004) indicate that a cancer drug product could be approved based either on RR, TTP, and survival alone or combinations of RR, TTP, and survival. One of the controversial issues is that there exists no gold standard for the assessment of cancer drugs. As another example, for a given study endpoint, when data are collected from clinical trials, the following *derived* study endpoints are usually considered: (1) absolute changes from baseline, (2) relative change from baseline, (3) responder defined based on absolute change, (4) responder based on relative change, and (5) any combinations of the above. Different derived study endpoints may lead to different conclusions regarding the treatment effect, which has led to the controversial issue "Which (derived) endpoint is telling the truth?" and "How these (derived) endpoints translate one another?" In practice, it should be noted that regulatory agencies may prefer one derived endpoint over the other without scientific justification. More discussions are given in Chapter 8.

Protocol amendments: Protocol amendments are commonly issued during the conduct of the clinical trials for various reasons such as change in eligibility criteria due to slow enrollment or modification of dose/dose regimen due to safety. For a given clinical trial, it is not uncommon to have—three to five protocol amendments during the conduct of the clinical trial. It is a concern that frequent protocol amendments may cause a shift in the target patient population. A clinical trial with frequent protocol amendments (with significant changes) could result in a totally different trial that is unable to address the scientific/medical questions the original trial is intended to address. Thus, one of the controversial issues is "How many protocol amendments are allowed for a given clinical trial?" Since, currently, there are no regulations on the protocol amendment, it is suggested that regulatory guidelines/guidance on protocol amendment be developed in order to maintain the integrity of the clinical trial. The impact of protocol amendments on clinical outcomes is studied in Chapter 9.

Independence of IDMC: In recent years, an IDMC is often established for clinical trials conducted in the later phases (e.g., phases IIb and III) of clinical development. The intention of IDMC is good. However, the *independence* of IDMC has been challenged. As a result, "Is an established IDMC really independent?" has become a controversial issue in clinical trials. In practice, most IDMCs do not communicate with regulatory agencies directly, while the sponsor makes every attempt to influence the IDMC. The other controversial issue is then whether the IDMC should have the authority to communicate with regulatory agencies regarding serious misconduct or wrongdoing of the clinical trial. Some observations that are commonly seen in the function/ activity of an IDMC are described in Chapter 12.

Multiplicity: One of the controversial issues in clinical trials that has attracted much attention is probably the issue of multiplicity in clinical trials. It is not clear to clinical scientists at to when and how adjustment for multiplicity in clinical trials should be done for controlling the overall type I error rate at a prespecified level of significance. It should be noted that the purpose of a clinical trial is to detect a clinically meaningful difference for achieving statistical significance (i.e., the observed difference is not by chance alone and is reproducible). Multiplicity refers to simultaneous statistical inference. Thus, one should always refer to the null hypothesis of interest (i.e., scientific/medical question) that one wishes to answer since the test statistic should be derived under the null hypothesis. The derived test statistic is then evaluated under the alternative hypothesis for achieving the desired power. Thus, the impact on power after adjustment for multiplicity is also a great concern in practice. Westfall and Bretz (2010) pointed out that the commonly encountered controversial issues regarding multiplicity in clinical trials include (1) penalizing for doing more or good job (i.e., performing additional test), (2) adjusting α for all possible tests conducted in the trial, and (3) the family of hypotheses to be tested. These controversial issues will be further discussed in Chapter 11.

Feasibility of seamless adaptive design: The use of adaptive design methods in clinical trials has become very popular in recent years due to their flexibility and efficiency for identifying any signals of safety and/or efficacy (preferably optimal clinical benefit) of a test treatment under investigation. As indicated by Chow and Chang (2006), there are several different types of adaptive designs depending upon the nature of adaptations applied either before, during, or after the conduct of a clinical trial. Among these adaptive designs is a two-stage seamless adaptive design that combines two separate (independent) studies (e.g., a phase IIb study and a phase III study) into a single study. Although the application of a seamless adaptive design enjoys the advantages of (1) reducing lead time between trials, (2) potential saving of the cost and resources, and (3) increasing the efficiency and consequently the probability of success, there are a few issues that remain unsolved. First, it is not clear how the overall type I error can be controlled, especially when the study objectives and study endpoints at different stages are different. Second, it is not clear whether the classic O'Brien-Fleming type of boundary is appropriate. Third, it is not clear how the data collected from both stages can be combined for a final analysis. Even if the above questions can be addressed, it is still a controversial issue whether the two-stage seamless adaptive design is feasible, especially when there is a population shift due to protocol amendments as described above.

Missing values imputation: In the past decade or two, when there were missing values, subjects with missing values were often excluded from the analysis. In recent years, patients with missing values are included in the analysis with imputed data in order to (1) fully utilize all information (even it is incomplete) collected from the trial and (2) increase power by imputing the missing values based on some valid statistical methods. In clinical trials, the method of LOCF is often considered. The validity of LOCF, however, has been challenged by many researchers. Although the validity of LOCF is questionable, it is still widely accepted in practice. Alternatively, many other methods for missing data imputation are available, which include (1) mean imputation, (2) median imputation, and (3) the method of regression analysis. One of the controversial issues is "Can missing data imputation be applied if there is a large proportion of subjects with missing values?" Another controversial issue is the potential impact on power when applying missing data imputation in clinical trials.

Non-inferiority margin: For clinical trials with life-threatening diseases such as cancer, it is unethical to use a placebo-control when approved and effective therapies are available. In this case, an active-control trial is often considered. The purpose of such an active-control trial is to show that the test treatment is at least as effective as the active-control agent or that it is not worse than the active-control agent within a prespecified margin, which is usually referred to as a non-inferiority margin. One of the controversial issues in active-control trials (or non-inferiority trials) is the determination of the non-inferiority margin. A different choice of non-inferiority margin could alter the conclusion of the clinical study. As indicated in the International Conference Harmonization (ICH) guideline, the selection of non-inferiority margin should be based on both clinical justification and statistical reasoning. Since the selection of the non-inferiority margin could be based on either absolute change or relative change, both of which have a significant impact on sample size calculation and the probability for achieving study objectives, it is very controversial as to whether the non-inferiority margin based on absolute change or the non-inferiority margin based on relative change should be used. More discussions in this regard can be found in Chapters 8 and 17.

Reproducibility/generalizability probability: For marketing approval of a new drug product, the FDA requires that at least *two* adequate and well-controlled clinical trials be conducted to provide substantial evidence regarding the effectiveness of the drug product under investigation. The purpose of conducting the second trial is to study whether the observed clinical result from the first trial is reproducible on the same target patient population. One of the controversial issues is "Can a large trial serve as two adequate and well-controlled clinical trials?" Shao and Chow (2002) studied the reproducibility probability of a future study based on observed data from a given clinical trial. The result indicates that a positive trial with a p-value less than 0.001 will have approximately 90% reproducibility probability. Under certain circumstances, the FDA Modernization Act (FDAMA) of 1997 includes a provision (Section 115 of FDAMA) to allow data from *one* adequate and well-controlled clinical trial investigation and confirmatory evidence to establish effectiveness for the risk–benefit assessment of drug and biological candidates for approval. More details regarding the application of reproducibility and generalizability probabilities are given in Chapter 25.

Probability of success: In the past several decades, it has been recognized that increasing spending of biomedical research does not reflect an increase in the success rate of pharmaceutical/clinical research and development. The low success rate of pharmaceutical/clinical development could be because (1) a diminished margin for improvement that escalates the level of difficulty in proving drug benefits, (2) genomics and other new sciences have not yet reached their full potential, (3) mergers and other business arrangements have decreased candidates, (4) easy targets are the focus as chronic diseases are harder to study, (5) failure rates have not improved, (6) rapidly escalating costs and complexity decrease the willingness/ability to bring many candidates forward into the clinic (Woodcock, 2005). One of the controversial issues is "How to correctly assess the probability of success based on available data?" Other controversial issues are "How to identify the possible causes of failure?" and "What actions should be taken for improving the failure rate?" More discussions are given in the last chapter of this book.

Other controversial issues: In addition to the controversial issues described above, there are other controversial issues such as (1) validation of subjective instruments—do we ask the *right* questions? (2) center grouping—how to group small centers into a reasonable size of *dummy* center? (3) QT/QTc studies with recording replicates—is a recording replicate a *real* replicate? (4) multi-regional trials—how many subjects should be included at a specific region in order to produce *consistent* results? (5) dose escalation trials—what is the probability of correctly identifying the MTD? (6) enrichment process in target clinical trials—how to *estimate* the proportion of patients with positive diagnostic test results? (7) clinical trial simulation—is clinical trial simulation *a* solution or *the* solution? (8) traditional Chinese medicine—how to *calibrate* Chinese diagnostic procedures against well-established clinical endpoints used in Western medicines? (9) follow-on biologics (FOB)—how similar is similar? (10) good regulatory (review) practices—do gold standards for drug evaluation exist? These controversial issues have an impact on the clinical evaluation of the treatment effect under investigation. These controversial issues will be discussed in subsequent chapters of this book.

In clinical development, randomized clinical trials are usually conducted to collect data for the evaluation of the efficacy and safety of a test treatment (e.g., a drug product or a therapy). To provide an accurate and fair assessment of the test treatment under investigation, well-controlled clinical trials following GCP at different phases of clinical development are necessarily conducted. In practice, a clinical trial process consists of protocol development, trial conduct, data collection, statistical analysis/interpretation, and reporting. A clinical trial is a lengthy but costly process, which is necessary to ensure the quality, identity, purity, strength, and stability of the test treatment under investigation. However, some controversial issues evitably occur regardless of whether the intended clinical trial is well planned. Basically, these controversial issues present conceptual differences in perspectives of clinicians (investigators/sponsors), biostatisticians, and reviewers for the evaluation of the test treatment under investigation. The major concern of the clinicians is whether the observed difference is of clinical significance, while the biostatisticians are interested in demonstrating whether the observed difference is of any statistical significance (i.e., whether the observed difference is not by chance alone and it is reproducible). The reviewers from the regulatory agencies would like to make sure whether the observed clinically meaningful difference (clinical benefits) has a statistical significance before they can approve the test treatment under investigation. A clinical trial is considered successful if it can meet the expectations of clinicians, biostatisticians, and regulatory reviewers.

1.4 Aim and Structure of the Book

In this book, we will post commonly seen controversial issues rather than provide resolutions. It is our goal that regulatory agencies, clinical scientists, and biostatisticians will pay much attention to these issues, identify the possible causes, resolve/correct the issues, and consequently enhance good clinical/statistical practices for achieving the study objectives of the intended clinical trials. This book is intended to be the first book entirely devoted to the discussion of controversial issues in clinical research and development. It covers controversial issues that are commonly encountered at various stages of clinical research and development including from bench-to-bedside translational research. It is our goal to provide a useful desk reference and state-of-the art examination of controversial issues in clinical trials to scientists engaged in clinical research and development, those in government regulatory agencies who have to make decisions on the evaluation/approval of test treatments under investigation, and to biostatisticians who provide the statistical support for the design and analysis of clinical trials or related projects. We hope that this book will serve as a bridge between scientists from the pharmaceutical industry, medical/statistical reviewers from government regulatory agencies, and researchers from the academia.

In this chapter, the background of pharmaceutical/clinical research and development, critical path initiatives, and some commonly seen controversial issues in clinical research have been discussed. In Chapter 2, GSP, which is the foundation of GCP for ensuring the success of the conduct of clinical trials, including some general concepts for statistics such as type I error versus type II error, one-sided test versus two-sided test, p-value versus confidence interval, and statistical difference versus clinical difference are described. In Chapter 3, some controversial issues such as one-way translational process versus two-way translational process, animal model versus human model, and the impact of lost-in-translation from bench to bedside on the probability of success for pharmaceutical/clinical development that are commonly encountered in bench-to-bedside translational research are discussed. Practical issues for the assessment of BE for generic approval under a standard 2×2 crossover design will be discussed in Chapter 4. Unlike the traditional approach for clinical evaluation of the effectiveness and safety by first demonstrating efficacy and then assessing the tolerability of the safety, Chapter 5 describes the possibility of evaluating composite hypotheses that include both efficacy and safety at the same time. Also included in this chapter is a recommended approach of significant digits for reporting the observed clinical results.

Chapter 6 examines the instability of sample size calculation/justification based on data obtained from previous studies and/or a small pilot study. The instability of sample size calculation has led to the justification of sample size reestimation at interim analysis, which has an impact on the success of the intended clinical trial. As a result, a more robust method such as a Bayesian-bootstrap median approach is recommended. As is well known, randomization and/or blinding are often employed in clinical trials in order to prevent potential biases that might be introduced during the conduct of the intended clinical trial. However, it is not clear whether the randomization and/or blinding will achieve the objective of preventing biases. Chapter 7 discusses the integrity of randomization/blinding based on post-study patients and/or investigators' guesses of the treatment codes that the patients receive. In clinical trials, it is debatable whether the absolute change from baseline to endpoint, the relative change from baseline to endpoint, or responder that is defined based on either absolute change or relative change should be used for the assessment of treatment effect. Chapter 8 attempts to provide some insight regarding the clinical strategy for the selection of an appropriate endpoint for the assessment of treatment effect. As it is a common practice to issue protocol amendments due to various reasons, it is a major concern that frequent protocol amendments may lead to a shift of target patient population; consequently, the original clinical trial may become a totally different trial that is unable to address the scientific/medical questions the original clinical trial intended to answer. Chapter 9 studies the impact of protocol amendments in data collections and consequently statistical inference at the end of the study. Chapter 10 investigates the feasibility/applicability for the use of adaptive design methods in clinical trials, which has become very popular and widely accepted by the pharmaceutical/biotechnology industry although the regulatory agencies still have some reservation in terms of its validity and integrity. The chapter will only focus on the most commonly employed seamless adaptive trial designs that combine two separate (independent) studies into a single trial.

In clinical trials, the issue of multiplicity often occurs due to multiple doses, multiple endpoints, multiple testing, and/or multiple comparisons. It is a concern as to when and how the overall type I error rate should be controlled due to multiplicity. Chapter 11 discusses controversial issues regarding multiplicity in clinical trials. Chapter 12 challenges the independence of an IDMC, which is often established to maintain the integrity of the trial, monitor ongoing safety data, and/or perform interim analysis for efficacy. In clinical research, data collected from a one-way analysis of variance model with repeated measures is often wrongly analyzed under a two-way analysis of variance model, which may lead to a wrong conclusion of the treatment effect. Chapter 13 studies the impact of analysis results under an incorrect model. Chapter 14 reviews some performance characteristics for validation of a subjective instrument (questionnaire) for the assessment of the clinical benefit of the test treatment under investigation such as quality-of-life assessment. Missing values are commonly encountered due to various reasons regardless of missing at random or not. Chapter 15 provides a summary of statistical methods for missing data imputation in clinical trials.

To expedite patient recruitments in clinical trials, a multicenter trial is often considered. One of the disadvantages is that we may end up with a few large sites and a number of small centers. In addition, it is likely to increase the probability of observing treatment-by-site interaction, which makes the overall assessment of the treatment effect almost impossible. Chapter 16 compares several approaches for center grouping in clinical trials with a number of small centers. Statistical methods for determining the non-inferiority margin in non-inferiority (active-control) trials are summarized in Chapter 17. The design and analysis for QT/QTc studies with recording replicates for assessment of cardio-toxicity in terms of QT/QTc prolongation are reviewed in Chapter 18. Chapter 19 discusses some practical issues that are commonly encountered in multiregional (multinational) clinical trials. Also included in this chapter is the determination of sample size at specific regions as compared to the entire multiregional trial. Algorithm-based traditional escalation rule trial design and model-based continual reassessment method trial design for dose escalation trials in cancer clinical trials are compared in Chapter 20.

Chapter 21 focuses on the enrichment process in target clinical trials, which will identify patient populations who are most likely to respond to the test treatment under study and consequently may lead to personalized medicine. Chapter 22 provides basic concepts and principles for conducting clinical trial simulation, which are useful for evaluating clinical performance under an assumed model with certain assumptions. Fundamental differences in terms of dose/dose regimen, culture, and medical theory/practice between Western medicine and traditional Chinese medicine are outlined in Chapter 23. Also included in this chapter are some statistical methods for testing consistency and stability analysis. Practical issues for assessment of biosimilarity between FOB are described in Chapter 24. Also included in this chapter are some statistical considerations regarding the design and analysis and current regulatory position for assessment of biosimilarity.

Chapter 25 deals with the calculations of the probabilities of generalizability and reproducibility of a given clinical trial based on the observed clinical data of the clinical trial. Good regulatory (or review) practices (GRP), especially good review practices published by the Center for Drug Evaluation and Research at the FDA, are reviewed in Chapter 26. Also included in this chapter are some observations of inconsistencies that are commonly seen during regulatory submissions. The probability of success for a pharmaceutical and/or clinical development of a test treatment under investigation is evaluated in the last chapter of this book. In each chapter, examples and possible recommendations and/or resolutions are provided whenever possible.

2

Good Statistical Practices

2.1 Introduction

Good statistical practice (GSP) in pharmaceutical/clinical research and development is defined as a set of statistical principles and/or standard operating procedures for the best biopharmaceutical practices in design, conduct, analysis, evaluation, reporting, and interpretation of studies at various stages of pharmaceutical research and development (see, e.g., Spriet and Dupin-Spriet, 1992; Wiles et al., 1994; Chow, 1997). The purpose of GSP is not only to minimize bias but also to minimize variability that may occur before, during, and after the conduct of the studies. More importantly, GSP provides a valid and fair assessment of the drug product under study. The concept of GSP in pharmaceutical/clinical research and development can be seen in many regulatory requirements, standards/specifications, and guidelines/guidances set by most health authorities, such as the U.S. Food and Drug Administration (FDA) and the Committee for Proprietary Medicinal Products (CPMP) in the European Community (CPMP, 1990). For example, the U.S. regulatory requirements for pharmaceutical/clinical research and development are codified in the U.S. Code of Federal Regulations (CFR), while the U.S. Pharmacopeia and National Formulary (USP/NF) and National Committee for Clinical Laboratory Standards (NCCLS) include standard procedures, test and sampling plans, and acceptance criteria and specifications of many pharmaceutical compounds (USP/NF, 2000; NCCLS, 2001). In addition, the FDA also develops a number of guidelines and guidances to assist the sponsors in drug research and development. These guidelines and guidances are considered gold standards for achieving good laboratory practice (GLP), good clinical practice (GCP), current good manufacturing practice (cGMP), and good regulatory (review) practice (GRP). The concept of GSP is well outlined in the guideline on *Statistical Principles for Clinical Trials* issued by the International Conference on Harmonization (ICH, 1997). As a result, GSP not only provides accuracy and reliability of the results derived from the studies but also ensures the validity and integrity of the studies.

In pharmaceutical/clinical research and development, statistics are necessarily applied at various critical stages of development to meet regulatory requirements for the effectiveness, safety, identity, strength, quality, purity, and stability of the drug product under investigation. These critical stages include pre-IND (investigational new drug application), IND, new drug application (NDA), and post NDA. At the very early stages of pre-IND, pharmaceutical scientists may have to screen thousands of potential compounds in order to identify a few promising compounds. An appropriate use of statistics with efficient screening and/or optimal designs will assist pharmaceutical scientists to identify the promising compounds within a relatively short time frame and cost effectively.

As indicated by the FDA, an IND should contain information regarding chemistry, manufacturing, and controls (CMC) of the drug substance and drug product to ensure the drug identity, strength, quality, and purity of the investigational drug. In addition, the sponsors are required to provide adequate information about pharmacological studies for absorption, distribution, metabolism, and excretion (ADME) and acute, subacute, and chronic toxicological studies and reproductive tests in various animal species to show that the investigational drug is reasonably safe to be evaluated in clinical trials in humans. At this stage, statistics are usually applied to (1) validate a developed analytical method, (2) establish a drug expiration dating period through stability studies, and (3) assess toxicity through animal studies. Statistics are necessarily applied to meet standards of accuracy and reliability.

Before the drug can be approved, the FDA requires that substantial evidence of the effectiveness and safety of the drug be provided in the Technical Section of Statistics of an NDA submission. Since the validity of statistical inference regarding the effectiveness and safety of the drug is always a concern, it is suggested that a careful review be performed to ensure an accurate and reliable assessment of the drug product. In addition, to have a fair assessment, the FDA also establishes advisory committees, each consisting of clinical, pharmacological, and statistical experts and one advocate (not employed by the FDA) in designated drug classes and subspecialties, to provide a second but independent review of the submission. The responsibility of the statistical expert is not only to ensure that a valid design is used but also to evaluate whether statistical methods used are appropriate for addressing the scientific and medical questions regarding the effectiveness and safety of the drug.

After the drug is approved, the FDA also requires that the drug product be tested for its identity, strength, quality, and purity before it can be released for use. For this purpose, the cGMP is necessarily implemented to (1) validate the manufacturing process, (2) monitor the performance of the manufacturing process, and (3) provide quality assurance of the final product. At each stage of the manufacturing process, the FDA requires that sampling plans, acceptance criteria, and valid statistical analyses be performed for the intended tests, such as potency, content uniformity, and dissolution. For each test, sampling plan, acceptance criteria, and valid statistical analysis are crucial for determining whether the drug product passes the test based on the results from a representative sample.

In the next section, some key statistical principles for GSP are briefly described. GSPs that are commonly employed in the European Community are reviewed in Section 2.3. Some recommendations for the implementation of GSP are given in Section 2.4. Brief concluding remarks are presented in the last section of this chapter.

2.2 Statistical Principles

In this section, we discuss some key statistical principles in the design and analysis of studies that may be encountered at various stages of drug development. These key statistical principles include bias/variability, confounding/interaction, hypothesis testing, type I error and power, randomization, sample size calculation/justification, statistical difference versus clinical difference, and one-sided test versus two-sided test.

2.2.1 Bias and Variability

For the approval of a drug product, regulatory agencies usually require that the results of the studies conducted at various stages of drug research and development be accurate and reliable to provide a valid and fair assessment of the treatment effect. The accuracy and reliability are usually referred to as the closeness and the degree of closeness of the results to the true value (i.e., true treatment effect). Any deviation from the true value is considered a bias, which may be due to selection, observation, or statistical procedures. Pharmaceutical scientists would make any attempt to avoid bias, whenever possible, to ensure that the collected results are accurate.

The reliability of a study is an assessment of the precision of the study, which measures the degree of the closeness of the results to the true value. The reliability reflects the ability to repeat or reproduce similar outcomes in the targeted population. The more precise a study is, the more likely it is that the results would be reproducible. The precision of a study can be characterized by the variability incurred during the conduct of the study.

In practice, since studies are usually planned, designed, executed, analyzed, and reported by a team that consists of pharmaceutical scientists from different disciplines, bias and variability inevitably occur. It is suggested that possible sources of bias and variability be identified at the planning stage of the study, not only to reduce the bias but also to minimize the variability.

2.2.2 Confounding and Interaction

In pharmaceutical/clinical research and development, there are many sources of variation that have an impact on the evaluation of the treatment. If these variations are not identified and properly controlled, then they may be mixed up with the treatment effects that the studies are intended to demonstrate. In this case, the treatment is said to be confounded with the effects due to these variations. To provide a better understanding, consider the following example. Last winter, Dr. Smith noticed that the temperature in the emergency room was relatively low, which had caused some discomfort among medical personnel and patients. Dr. Smith suspected that the heating system might not function properly and decided to improve it. As a result, the temperature of the emergency room has been raised to a comfortable level this winter. However, this winter is not as cold as last winter. Therefore, it is not clear whether the improvement of emergency room temperature was due to the improvement of the heating system or the effect of a warmer winter.

The statistical interaction is to investigate whether the joint contribution of two or more factors is the same as the sum of the contributions from each factor when considered alone. If an interaction between factors exists, an overall assessment cannot be made. For example, suppose that a placebo-controlled clinical trial was conducted at two study centers to assess the effectiveness and safety of a newly developed drug product. Suppose that the results turned out that the drug is efficacious (better than placebo) at one study center and inefficacious (worse than placebo) at the other study center. As a result, a significant interaction between treatment and study center occurred. In this case, an overall assessment of the effectiveness of the drug product can be made.

In practice, it is suggested that possible confounding factors be identified and properly controlled at the planning stage of the studies. When significant interactions among factors are observed, subgroup analyses may be necessary for a careful evaluation of the treatment effect.

2.2.3 Hypotheses Testing

In clinical trials, a hypothesis is a postulation, assumption, or statement that is made about the population relative to a test treatment under investigation. As an example, the statement that there is a difference between the test treatment and a placebo control is a hypothesis for the treatment effect. A random sample is usually drawn through a bioavailability study to evaluate hypotheses about the test treatment. To perform a hypothesis testing, the following steps are essential:

Step 1: Choose the hypothesis that is to be questioned, denoted by H_0, where H_0 is usually referred to as the null hypothesis.

Step 2: Choose an alternative hypothesis, denoted by H_a, where H_a is usually the hypothesis of particular interest to the investigators.

Step 3: Derive a test statistic under the null hypothesis and define the rejection region (or a rule) for decision making about when to reject the null hypothesis and when to fail to reject it.

Step 4: Draw a random sample by conducting a clinical trial.

Step 5: Calculate test statistic(s).

Step 6: Draw conclusion(s) according to the predetermined rule as specified in Step 3.

In practice, we would reject the null hypothesis at a prespecified level of significance and favor the alternative hypothesis. Basically, two kinds of errors occur when testing hypotheses. If the null hypothesis is rejected when it is true, then a type I error has occurred. If the null hypothesis is not rejected when it is false, then a type II error has been made. The probabilities of making type I and type II errors are given as

$$\alpha = P(\text{type I error})$$
$$= P(\text{reject, } H_0 \text{ given that } H_0 \text{ is true}).$$
$$\beta = P(\text{type II error})$$
$$= P(\text{fail to reject } H_0 \text{ when } H_0 \text{ is false}).$$

The probability of makings a type I error α is called the level of significance. In practice, α is also known as the consumer's risk, while β is sometimes referred to as the producer's risk. Table 2.1 summarizes the relation between type I and type II errors when testing hypotheses.

The power of the test is defined as the probability of correctly rejecting H_0 when H_0 is false; that is,

$$\text{Power} = 1 - \beta$$
$$= P(\text{reject } H_0 \text{ when } H_0 \text{ is false}).$$

Note that α decreases as β increases and α increases as β decreases. The only way to decrease both α and β is to increase the sample size. In practice, because a type I error is usually considered to be a more important or serious error, which one would like to avoid, a typical approach in hypothesis testing is to control α at an acceptable level and try to minimize β by choosing an appropriate sample size. In other words, the null hypothesis can be tested

TABLE 2.1

Relationship between Type I and Type II Errors

		If H_0 Is	
		True	False
When	Fail to reject	No error	Type II error
	Reject	Type I error	No error

at a predetermined level (or nominal level) of significance with a desired power. For a fixed α, β increases when H_a moves toward H_0. This means that we will not have sufficient power to detect a small difference between H_0 and H_a. On the other hand, β decreases when H_a moves away from H_0, increasing the test power.

In practice, the null hypothesis H_0 and the alternative hypothesis H_a are sometimes reversed and evaluated for different interests. However, a test for H_0 versus H_a is not equivalent to a test for $H_0' = H_a$ versus $H_a' = H_0$. Two tests under different null hypotheses may lead to a totally different conclusion. For example, a test for H_0 versus H_a may lead to the rejection of H_0 in favor of H_a. However, a test for $H_0' = H_a$ versus $H_a' = H_0$ may reject the alternative hypothesis. Thus, the choice of the null hypothesis and the alternative hypothesis may have some influence on the parameter to be tested. The following criteria are commonly used as a rule of thumb for choosing the null hypothesis.

Rule 1: Choose H_0 based on the importance of a type I error. Under this rule, we believe that a type I error is more important and serious than a type II error. We would like to control the chance of making a type I error at a tolerable limit (i.e., α). Thus, H_0 is chosen so that the maximum probability of making a type I error (i.e., P [reject H_0 when H_0 is true]) will not exceed α level.

Rule 2: Choose the hypothesis we wish to reject as H_0 (Colton, 1974; Ott, 1984; Ware et al., 1986). The purpose of this rule is to establish H_a by rejecting H_0. Note that we will never be able to prove that H_0 is true even though the data fail to reject it.

Occasionally, for a given set of hypotheses, it may be easy to determine whether a type I error is more important or serious than a type II error. If a type II error appears to be more important or serious than a type I error, rule 1 suggests that the null hypothesis and the alternative hypothesis be reversed. Frequently, however, the relative importance of the type I error and the type II error is usually very subjective. In this case, rule 2 is useful in choosing H_0 and H_a. To illustrate the use of these two criteria, consider the following example given in Chow and Liu (2008).

Example Effectiveness/Ineffectiveness

In practice, the following two errors occur in the assessment of effectiveness of a test treatment under investigation when comparing the test treatment with a placebo control:

Hypothesis 1: We conclude that the test treatment is effective when, in fact, the test treatment is not effective as compared to the placebo control.

Hypothesis 2: We conclude that the test treatment is ineffective when, in fact, the test treatment is effective as compared to the placebo control.

In the interest of controlling the chance of making a type I error, the FDA may consider hypothesis 1 more important than hypothesis 2 and, consequently, prefer the following hypothesis:

$$H_0 : \text{Not effectiveness versus } H_a : \text{Effectiveness.} \tag{2.1}$$

On the other hand, pharmaceutical companies may want to eliminate the probability of wrongly rejecting the null hypothesis of bioequivalence (BE). Thus, the following hypotheses are used:

$$H_0 : \text{Effectiveness versus } H_a : \text{Not effectiveness.} \tag{2.2}$$

It is very subjective whether hypothesis 1 is more important than hypothesis 2 or hypothesis 2 is more important than hypothesis 1 when comparing two drug products for the same indication. In clinical trials, rule 2 is usually applied to choose H_0. For example, when a test treatment is newly developed, the sponsor will want to show effectiveness by disproving the hypothesis of ineffectiveness. In this case, hypothesis (2.1) may be considered.

2.2.4 Type I Error and Power

In statistical analysis, two different kinds of mistakes are commonly encountered when performing hypotheses testing. For example, suppose that a physician is to determine whether or not one of his/her patients is still alive. If the patient is dead, then the physician may remove his/her life-support equipment for other patients who need it. Therefore, the null hypothesis of interest is that the patient is still alive, while the alternative hypothesis is that the patient is dead. Under these hypotheses, the physician may make two mistakes, which are: (1) he/she concludes that the patient is dead when in fact the patient is still alive and (2) he/she claims that the patient is still alive when in fact the patient is dead. The first kind of mistake is usually referred to as a type I error; the latter is the so-called type II error. Since a type I error is usually considered more important or serious, we would like to limit the probability of committing this kind of error to an acceptable level. This acceptable level of probability of committing a type I error is known as the significance level. As a result, if the probability of observing a type I error based on the data is less than the significance level, we conclude that a statistically significant result is observed. A statistically significant result suggests that the null hypothesis be rejected in favor of the alternative hypothesis. The probability of observing a type I error is usually referred to as the p-value of the test. On the other hand, the probability of committing a type II error subtracted from 1 is called the power of the test. In our example, the power of the test is the probability of correctly concluding the death of the patient when the patient is dead.

For the pharmaceutical application, suppose that a pharmaceutical company is interested in demonstrating that the newly developed drug is efficacious. The null hypothesis that the drug is inefficacious is often chosen versus the alternative hypothesis that the drug is efficacious. The objective is to reject the null hypothesis in favor of the alternative hypothesis and consequently to conclude that the drug is efficacious. Under the null hypothesis, a type I error is made if we conclude that the drug is efficacious when in fact it is not. This error is also known as the consumer's risk. Similarly, a type II error is committed if we conclude that the drug is inefficacious. This error is sometimes called the producer's risk. The power is then considered to be the probability of correctly concluding that the drug is efficacious, when in fact it is. For the assessment of drug effectiveness and safety, a sufficient sample size is often selected to have a desired power with a prespecified significance level. The purpose is to control both type I (significance level) and type II (power) errors.

2.2.5 Randomization

Statistical inference on a parameter of interest of a population under study is usually derived under the probability structure of the parameter. The probability structure depends upon the randomization method employed in sampling. The failure of the randomization will have a negative impact on the validity of the probability structure. Consequently, the validity, accuracy, and reliability of the resulting statistical inference of the parameter are questionable. Therefore, it is suggested that randomization be performed using an appropriate randomization method under a valid randomization model according to the study design to ensure the validity, accuracy, and reliability of the derived statistical inference.

2.2.6 Sample Size Determination/Justification

One of the major objectives of most studies during drug research and development is to determine whether the drug is effective and safe. During the planning stage of a study, the following questions are of particular interest to pharmaceutical scientists: (1) How many subjects are needed in order to have a desired power for detecting a meaningful difference? (2) What is the trade-off if only a small number of subjects are available for the study due to a limited budget and/or some scientific considerations? To address these questions, a statistical evaluation for sample size determination/justification is often employed. Sample size determination usually involves the calculation of sample size for some desired statistical properties, such as power or precision; sample size justification is to provide statistical justification for a selected sample size, which is often a small number.

For a given study, sample size can be determined/justified based on some criteria of a type I error (a desired precision) or a type II error (a desired power). The disadvantage for sample size, determination/justification based on the criteria of precision is that it may have a small chance of detecting a true difference. As a result, sample size determination/justification based on the criteria of power becomes the most commonly used method. Sample size is selected to have a desired power for detection of a meaningful difference at a prespecified level of significance.

In practice, however, it is not uncommon to observe discrepancies among study objective (hypotheses), study design, statistical analysis (test statistic), and sample size calculation. These inconsistencies often result in (1) the wrong test for the right hypotheses, (2) the right test for the wrong hypotheses, (3) the wrong test for the wrong hypotheses, or (4) the right test for the right hypotheses with insufficient power. Therefore, before the sample size can be determined, it is suggested that the following be carefully considered: (1) the study objective or the hypotheses of interest should be clearly stated, (2) a valid design with appropriate statistical tests should be used, and (3) sample size should be determined based on the test for the hypotheses of interest.

2.2.7 Statistical Difference and Scientific Difference

A statistical difference is defined as a difference that is unlikely to occur by chance alone, while a scientific difference is a difference that is considered to be of scientific importance. A statistical difference is also referred to as a statistically significant difference. The difference between the concepts of statistical difference and scientific difference is that statistical difference involves chance (probability) while scientific difference does not. When we claim that there is a statistical difference, the difference is reproducible with a high probability.

When conducting a study, there are basically four possible outcomes. The result may show that (1) the difference is both statistically and scientifically significant, (2) there is a statistically significant difference yet the difference is not scientifically significant, (3) the difference is scientifically significant yet it is not statistically significant, and (4) the difference is neither statistically significant nor scientifically significant. If the difference is both statistically and scientifically significant or if it is neither statistically nor scientifically significant, then there is no confusion. However, in many cases, a statistically significant difference does not agree with the scientifically significant difference. This inconsistency has created confusion/arguments among pharmaceutical scientists and biostatisticians. The inconsistency may be due to large variability and/or insufficient sample size.

2.2.8 One-Sided Test versus Two-Sided Test

For the evaluation of a drug product, the null hypothesis of interest is often that there is no difference. The alternative hypothesis is usually that there is a difference. The statistical test for this setting is called a two-sided test. In some cases, the pharmaceutical scientist may test the null hypothesis of no difference against the alternative hypothesis that the drug is superior to the placebo. The statistical test for this setting is known as a one-sided test. For a given study, if a two-sided test is employed at the significance level of 5%, then the level of proof required is 1 out of 40. In other words, at the 5% level of significance, there is 2.5% chance (or 1 out of 40) that we may reject the null hypothesis of no difference in the positive direction and conclude that the drug is effective on one side. On the other hand, if a one-sided test is used, the level of proof required is 1 out of 20. It turns out that a one-sided test allows more ineffective drugs to be approved because of chance as compared to the two-sided test. It should be noted that when testing at the 5% level of significance with 80% power, the sample size required increases by 27% for a two-sided test as compared to a one-sided test. As a result, there is a substantial cost saving if a one-sided test is used.

However, agreement is not universal among the regulatory, the academia, and the pharmaceutical industry as to whether a one-sided test or a two-sided test should be used. The FDA tends to oppose the use of a one-sided test, though this position has been challenged by several pharmaceutical companies on Drug Efficacy Study Implementation (DESI) drugs at the Administrative Hearing. Dubey (1991) pointed out that several viewpoints that favor the use of one-sided tests were discussed in an administrative hearing. These points indicated that a one-sided test is appropriate in the following situations: (1) where there is truly concern with outcomes in one tail only and (2) where it is completely inconceivable that the results could go in the opposite direction.

2.3 Good Statistical Practices in Europe

In February 2005, the Statistical Program Committee (SPC) adopted the European statistics code of good practice and undertook to observe the 15 principles established therein, as well as to periodically review their application using the good practice indicators corresponding to each of the 15 principles (see also http://www.ine.es/en/ine/codigobp/codigobp_en.htm). This code has been embraced by *Instituto Natcional de Estadistica* (INE) by way of a resolution by the Board of Directors, which thus undertakes to comply with the aforementioned when establishing the general principles regulating the generating of statistics for State purposes. In this way, INE endeavors to guarantee an improvement in the service it provides to society, which will undoubtedly reinforce its image as an institution. In May 2005, the SPC agreed a formula for monitoring the implementation of the code, for a duration of 3 years. During that period, the various countries must carry out quality self-assessment, taking as a reference the aforementioned good practice indicators, which in turn must be contrasted and checked via so-called peer reviews. The end result was submitted to the Board and to the European Parliament in 2008. The 15 principles are briefly described as follows:

Principle 1: Professional independence—The professional independence of statistical authorities from other policy, regulatory, or administrative departments and bodies, as well as from private sector operators, ensures the credibility of European statistics.

Principle 2: Mandate for data collection—Statistical authorities must have a clear legal mandate to collect information for European statistical purposes. Administrations, enterprises, and households, and the public at large may be compelled by law to allow access to or deliver data for European statistical purposes at the request of statistical authorities.

Principle 3: Adequacy of resources—The resources available to statistical authorities must be sufficient to meet European statistics requirements.

Principle 4: Quality commitment—All European Statistical System (ESS) members commit themselves to work and cooperate according to the principles fixed in the "Quality declaration of the European statistical system."

Principle 5: Statistical confidentiality—The privacy of data providers (households, enterprises, administrations, and other respondents), the confidentiality of the information they provide, and its use only for statistical purposes must be absolutely guaranteed.

Principle 6: Impartiality and objectivity—Statistical authorities must produce and disseminate European statistics respecting scientific independence and in an objective, professional, and transparent manner in which all users are treated equitably.

Principle 7: Sound methodology—Sound methodology must underpin quality statistics. This requires adequate tools, procedures, and expertise.

Principle 8: Appropriate statistical procedures—Appropriate statistical procedures, implemented from data collection to data validation, must underpin quality statistics.

Principle 9: Non-Excessive burden on respondents—The reporting burden should be proportionate to the needs of the users and should not be excessive for respondents. The statistical authority monitors the response burden and sets targets for its reduction over time.

Principle 10: Cost Effectiveness—Resources must be effectively used.

Principle 11: Relevance—European statistics must meet the needs of users.

Principle 12: Accuracy and Reliability—European statistics must accurately and reliably portray reality.

Principle 13: Timeliness and Punctuality—European statistics must be disseminated in a timely and punctual manner.

Principle 14: Coherence and Comparability—European statistics should be consistent internally, over time and comparable between regions and countries; it should be possible to combine and make joint use of related data from different sources.

Principle 15: Accessibility and clarity—European Statistics should be presented in a clear and understandable form, disseminated in a suitable and convenient manner, available and accessible on an impartial basis with supporting metadata and guidance.

2.4 Implementation of GSP

The implementation of GSP in drug research and development is a team project that requires mutual communication, confidence, respect, and cooperation among statisticians, pharmaceutical scientists in the related areas, and regulatory agents. The implementation of GSP involves some key factors that have an impact on the success of GSP. These factors include (1) regulatory requirements for statistics, (2) dissemination of the concept of statistics, (3) appropriate use of statistics, (4) effective communication and flexibility, and (5) statistical training. These factors are briefly described next.

In the drug development and approval process, regulatory requirements for statistics are the key to the implementation of GSP. They not only enforce the use of statistics but also establish standards for statistical evaluation of the drug products under investigation. An unbiased statistical evaluation helps pharmaceutical scientists and regulatory agents in determining (1) whether the drug product has the claimed effectiveness and safety for the intended disease and (2) whether the drug product possesses good drug characteristics, such as proper identity, strength, quality, purity, and stability. A set of guideline standard operating procedures is often developed to fulfill regulatory requirements for good statistics practice. For example, Spriet and Dupin-Spriet (1992) proposed a set of procedures to fulfill quality requirements set by company policy according to regulatory requirements of GCP. Wiles et al. (1994) indicated that the Professional Standards Working Party of the Statisticians in the Pharmaceutical Industry (PSI) in the United Kingdom has developed a set of guideline standard operating procedures for GSP. These guideline standard operating procedures cover clinical development plan, clinical trial protocol, statistical analysis plan, determination of evaluability of subjects for analysis, randomization and blinding procedures, data management, interim analysis plan, statistical report, archiving and documentation, data overview, and quality assurance and quality control.

In addition to regulatory requirements, it is always helpful to disseminate the concept of statistical principles described earlier whenever possible. It is important for pharmaceutical scientists and regulatory agents to recognize that (1) a valid statistical inference is necessary to provide a fair assessment with certain assurance regarding the uncertainty of the drug product under investigation, (2) an invalid design and analysis may result in a misleading or wrong conclusion about the drug product, and (3) a larger sample size is often required to increase the statistical power and precision of the studies. The dissemination of the concept of statistics is critical to establish the pharmaceutical scientists' and regulatory agents' brief in statistics for scientific excellence.

One of the commonly encountered problems in drug research and development is the misuse or sometimes abuse of statistics in some studies. The misuse or abuse of statistics is critical, which may result in either having the right question with the wrong answer or having the right answer for the wrong question. For example, for a given study, suppose that a right set of hypotheses (the right question) is established to reflect the study objective. A misused statistical test may provide a misleading or wrong answer to the right question. On the other hand, in many clinical trials, point hypotheses for equality (the wrong question) are often wrongly used for the establishment of equivalency. In this case, we have the right answer (for equality) for the wrong question. As a result, it is recommended that appropriate statistical methods be chosen to reflect the design that should be able to address the scientific or medical questions regarding the intended study objectives for the implementation of GSP.

Communication and flexibility are important factors for the success of GSP. Inefficient communication between statisticians and pharmaceutical scientists or regulatory agents may result in a misunderstanding of the intended study objectives and consequently in an invalid design and/or inappropriate statistical methods. Thus, effective communication among statisticians, pharmaceutical scientists, and regulatory agents is essential for the implementation of GSP. In addition, in many studies, the assumption of a statistical design or model may not be met due to the nature of the drug product under investigation, the experimental environment, and/or other causes related/unrelated to the studies. In this case, the traditional approach of doing everything by the book does not help. In practice, since a concern from a pharmaceutical scientist or the regulatory agent may translate into a constraint for a valid statistical design and appropriate statistical analysis, it is suggested that a flexible yet innovative solution be developed under the constraints for the implementation of GSP.

Since regulatory requirements for the drug development and approval process vary from drug to drug and from country to country, various designs and/or statistical methods are often required for a valid assessment of a drug product. Therefore, it is suggested that statistical continued/advanced education and training programs be routinely held for both statisticians and nonstatisticians, including pharmaceutical scientists and regulatory agents. The purpose of such a continued/advanced education and/or training program is threefold. First, it enhances communications within the statistical community. Statisticians can certainly benefit from such a training and/or educational program by acquiring more practical experience and knowledge. In addition, it provides the opportunity to share/exchange information, ideas, and/or concepts regarding drug development between professional societies. Finally, it identifies critical practical and/or regulatory issues that are commonly encountered in the drug development and regulatory approval process. A panel discussion from different disciplines may result in some consensus to resolve the issues, which helps in establishing standards of statistical principles for the implementation of GSP.

2.5 Concluding Remarks

During the development and regulatory approval process, good pharmaceutical practices are necessarily implemented to ensure (1) the effectiveness and safety of the drug product under investigation before approval and (2) that the drug product possesses good drug characteristics, such as proper identity, strength, quality, purity, and stability, in compliance with the standards as specified in the USP/NF after regulatory approval. These good pharmaceutical practices include GLP for animal studies, GCP for clinical development, cGMP for CMC, and GRP for the regulatory review and approval process. In essence, GSP is the foundation of GLP, GCP, cGMP, and GRP. The implementation of GSP is a team project that involves statisticians, pharmaceutical scientists, and regulatory agents. The success of GSP depends upon mutual communication, confidence, respect, and cooperation among statisticians, pharmaceutical scientists, and regulatory agents.

In recent years, the use of adaptive design methods in clinical trials has become very popular due to its flexibility and efficiency in identifying any potential signals of safety and efficacy for the test treatment under investigation. In practice, however, while enjoying the flexibility of adaptive design methods, the quality, integrity, and validity of the trial may be at a greater risk. From a regulatory perspective, it is always a concern whether the p-value or confidence interval regarding the treatment effect under an adaptive trial design is reliable or correct. In addition, the misuse or abuse of statistical methods under a specific adaptive design may be biased and misleading, and therefore unable to address medical questions that the trial intends to answer. GSP plays an extremely important role for clinical trials utilizing adaptive designs, especially for those less-well-understood adaptive designs as described in the 2010 FDA draft guidance on adaptive clinical trial designs (FDA, 2010b).

3

Bench-to-Bedside Translational Research

3.1 Introduction

As pointed out in Chapter 2, the United States Food and Drug Administration (FDA) kicked off the *Critical Path Initiative* in the early 2000s to assist the sponsors to identify possible causes of the scientific challenges underlying the medical product pipeline problems. The *Critical Path Opportunities List* released by the FDA on March 16, 2006, identified (1) better evaluation tools and (2) streamlining clinical trials as the top two topic areas to bridge the gap between the quick pace of new biomedical discoveries and the slower pace at which those discoveries are currently developed into therapies. This has led to the consideration of the use of adaptive design methods in clinical development and the focus of translational science/research, which attempt not only to identify the best clinical benefit of a drug product under investigation but also to increase the probability of success. Statistical methods for the use of adaptive trial designs in clinical development can be found in Chow and Chang (2006), Chang (2007), Pong and Chow (2010). In this chapter, we will focus on statistical methods that are commonly employed in translational science/research.

Chow (2007a) and Cosmatos and Chow (2008) classified translational science/research into three areas, namely, translation in language, translation in information, and translation in (medical) technology. Translation in language refers to possible lost in the translation of the informed consent form and/or case report forms in multinational clinical trials. Lost in translation is commonly encountered due to not only difference in language but also differences in perception, culture, medical practices, etc. A typical approach for the assessment of the possible lost in translation is to first translate the informed consent form and/or the case report forms by an experienced expert and then translate it back by a different experienced but independent expert. The back-translated version is then compared with the original version for consistency. If the back-translated version passes the test for consistency, then the translated version is validated through a small-scale pilot study before it is applied to the intended multinational clinical trial. Translation in information is referred to as bench-to-bedside translational science/research, which is also known as translational medicine. Translation in technology includes biomarker development and translation in diagnostic procedures between traditional Chinese medicine and Western medicine. In this chapter, we focus on statistical methods for translation in information and translation in technology. Note that, in practice, translational medicine is often divided into two areas, namely, discovery translational medicine and clinical translational medicine. Discovery translational medicine refers to biomarker development, bench-to-bedside, and animal model versus human model, while clinical translational medicine includes translation among study endpoints, translation in technology, and generalization from a target patient population to another.

In the next section, a statistical method for optimal variable screening in microarray analysis is outlined. Also included in this section is a cross-validation method for model selection and validation. Sections 3.3 and 3.4 discusses statistical methods for the assessment of one-way/two-way translation and lost in translation in the bench-to-bedside translational process in pharmaceutical development, respectively. Whether or not an established animal model is predictive of a human model is examined in Section 3.5. Some concluding remarks are provided in the last section of this chapter.

3.2 Biomarker Development

Biomarker is a characteristic that is objectively measured and evaluated as an indicator of normal biological processes, pathogenic processes, or pharmacologic responses to a therapeutic intervention. Biomarkers can be classified into classifier marker, prognostic marker, and predictive marker. A classifier marker usually does not change over the course of the study and can be used to identify the patient population who would benefit from the treatment from those who would not. A typical example is a DNA marker for population selection in the enrichment process of clinical trials. A prognostic marker informs the clinical outcomes, which is independent of the treatment. A predictive marker informs the treatment effect on the clinical endpoint, which could be population specific. That is, a predictive marker could be predictive for population A but not for population B. It should be noted that the correlation between biomarker and true endpoint makes a prognostic marker. However, the correlation between biomarker and true endpoint does not make a predictive biomarker.

In clinical development, a biomarker could be used to select the right population, to identify the natural course of the disease, for early detection of the disease, and to develop personalized medicine. The utilization of a biomarker could lead to a better target population, detection of a larger effect size with a smaller sample size, and timely decision making. As indicated in the FDA *Critical Path Initiative Opportunity List*, better evaluation tools call for biomarker qualification and standards. Statistical methods for early-stage biomarker qualification include, but are not limited to, (1) distance-dependent K-nearest neighbors, (2) K means clustering, (3) single/average/complete linkage clustering, and (4) distance-dependent Jarvis–Patrick clustering. More information can be found at the following Web site: http://www.ncifcrf.gov/human_studies.shtml.

In what follows, we will review statistical methods that are commonly used in biomarker development for optimal variable screening. The selected variables will then be used to establish a predictive model through a model selection/validation process.

3.2.1 Optimal Variable Screening

DNA microarrays have been used extensively in medicinal practice. Microarrays identify a set of candidate genes that are possibly related to a clinical outcome of a disease (in disease diagnoses) or a medical treatment. However, there are many more candidate genes than the number of available samples (the sample size) in almost all studies, which leads to an irregular statistical problem in disease diagnoses or treatment outcome prediction. Some available statistical methods deal with a single gene at a time (e.g., Chen and Chen, 2003), which clearly do not provide the best solution for polygenic diseases. In practice, meta-analysis and/or combining several similar studies is often considered to increase sample size. These approaches, however, may not be appropriate due to the fact that (1) the combined data set may still be much too small and (2) there may be heteroscedasticity among the data from different studies. Alternatively, Shao and Chow (2007) proposed an optimal variable screening approach for dealing with the situation where the number of variables (genes) is much larger than the sample size.

Let y be a clinical outcome of interest and x be a vector of p candidate genes that are possibly related to y. Shao and Chow (2007) simply considered inference on the population of y conditional on x and noted that their proposed method can be applied to the unconditional analysis (i.e., both y and x are random). Consider the following model:

$$y = \beta'x + \varepsilon, \tag{3.1}$$

where β is a p-dimensional vector and the distribution of ε is independent of x with $E(\varepsilon) = 0$ and $E(\varepsilon^2) = \sigma^2$. Under the model (3.1), assume that there is a positive integer p_0 (which does not depend on n) such that only p_0 components of β are nonzero. Furthermore, β is in the linear space generated by the rows of $X'X$ for sufficiently large n, where X is the $n \times p_n$ matrix whose ith row is x'_i. In addition, assume that there is a sequence $\{\xi_n\}$ of positive numbers such that $\xi_n \to \infty$ and $\lambda_{in} = b_i\xi_n$, where λ_{in} is the ith nonzero eigenvalue of $X'X$,

$i = 1, \ldots, n$ and $\{b_i\}$ is a sequence of bounded positive numbers. Note that in many problems $\xi_n = n$. Furthermore, there exists a constant $c > 0$ such that $p_n / \xi_n^c \to 0$. For the estimation of β, Shao and Chow (2007) considered the following ridge regression estimator:

$$\hat{\beta} = (X'X + h_n I_{p_n})^{-1} X'Y, \tag{3.2}$$

where
　$Y = (y_1, \ldots, y_n)'$
　I_{p_n} is the identity matrix of order p_n
　$h_n > 0$ is the ridge parameter

The bias and variance of $\hat{\beta}$ are given by

$$\text{bias}(\hat{\beta}) = E(\hat{\beta}) - \beta = -(h_n^{-1} X'X + I_{p_n})^{-1}\beta$$

and

$$\text{var}(\hat{\beta}) = \sigma^2 (X'X + h_n I_{p_n})^{-1} X'X (X'X + h_n I_{p_n})^{-1}.$$

Let β_i and $\hat{\beta}_i$ be the ith component of β and $\hat{\beta}$, respectively. Under the assumptions as described earlier, we have $E(\hat{\beta}_i - \beta_i)^2 \to 0$ (i.e., $\hat{\beta}_i$ is consistent for β_i in mean squared error) if h_n is suitably chosen. Thus, we have

$$\text{var}(\hat{\beta}) = \frac{\sigma^2}{h_n} \left(\frac{X'X}{h_n} + I_{p_n} \right)^{-1} \frac{X'X}{h_n} \left(\frac{X'X}{h_n} + I_{p_n} \right)^{-1}. \tag{3.3}$$

Hence, $\text{var}(\hat{\beta}_i) \to 0$ for all i as long as $h_n \to \infty$. Note that the analysis of the bias of $\hat{\beta}_i$ is more complicated. Let Γ be an orthogonal matrix such that

$$\Gamma' X'X \Gamma = \begin{pmatrix} \Lambda_n & 0_{n \times (p_n - n)} \\ 0_{(n - p_n) \times n} & 0_{(p_n - n) \times (p_n - n)} \end{pmatrix},$$

where
　Λ_n is a diagonal matrix whose ith diagonal element is λ_{in}
　$0_{l \times k}$ is the $l \times k$ matrix of 0's

Then, it follows that

$$\text{bias}(\hat{\beta}) = -\left[\Gamma \left(\frac{\Gamma' X'X \Gamma}{h_n} + I_{p_n} \right) \Gamma' \right]^{-1} \beta = -\Gamma A \Gamma' \beta, \tag{3.4}$$

where A is a $p_n \times p_n$ diagonal matrix whose first n diagonal elements are

$$\frac{h_n}{h_n + \lambda_{in}}, \quad i = 1, \ldots, n,$$

and the last diagonal elements are all equal to 1. Under the above-mentioned assumptions, combining the results for variance and bias of $\hat{\beta}_i$, that is, (3.3) and (3.4), it can be shown that for all i

$$E(\hat{\beta}_i - \beta_i)^2 = \text{var}(\hat{\beta}_i) + \left[\text{bias}(\hat{\beta}_i)\right]^2 \to 0$$

if h_n is chosen so that $h_n \to \infty$ at a rate slower than ξ_n (e.g., $h_n = \xi_n^{2/3}$). Based on this result, Shao and Chow (2007) proposed the following optimal variable screening procedure:

Let $\{a_n\}$ be a sequence of positive numbers satisfying $a_n \to 0$. For each fixed n, we screen out the ith variable if and only if $|\hat{\beta}_i| \le a_n$.

Note that, after screening, only variables associated with $|\hat{\beta}_i| > a_n$ are retained in the model as predictors. The idea behind this variable screening procedure is similar to that in the Lasso method (Tibshirani, 1996). Under certain conditions, Shao and Chow (2007) showed that their proposed optimal variable screening method is consistent in the sense that the probability that all variables (genes) unrelated to y, which will be screened out, and all variables (genes) related to y, which will be retained, are 1 as n tends to infinity.

3.2.2 Model Selection and Validation

Suppose that n data points are available for selecting a model from a class of models. Several methods for model selection are available in the literature. These methods include, but are not limited to, Akaike information criterion (AIC) (Akaike, 1974; Shibata, 1981), the C_p (Mallows, 1973), the jackknife and the bootstrap (Efron, 1983, 1986). These methods, however, are not asymptotically consistent in the sense that the probability of selecting the model with the best predictive ability does not converge to 1 as the total number of observations $n \to \infty$. Alternatively, Shao (1993) proposed a method for model selection and validation using the method of cross-validation. The idea of cross-validation is to split the data set into two parts. The first part contains n_c data points which will be used for fitting a model (model construction), whereas the second part contains $n_v = n - n_c$ data points which are reserved for assessing the predictive ability of the model (model validation). It should be noted that all of the $n = n_v + n_c$ data, not just n_v are used for model validation. Shao (1993) showed that all of the methods of AIC, C_p, jackknife and bootstrap are asymptotically equivalent to the cross-validation with $n_v = 1$, denoted by CV(1), although they share the same deficiency of inconsistency. Shao (1993) indicated that the inconsistency of the leave-one-out cross-validation can be rectified by using leave-n_v-out cross-validation with n_v satisfying $n_v/n \to 1$ as $n \to \infty$.

In addition to the cross-validation with $n_v = 1$, denoted by CV(1), Shao (1993) also considered the other two cross-validation methods, namely, a Monte Carlo cross-validation with $n_v (n_v \neq 1)$, denoted by MCCV(n_v), and an analytic approximate CV(n_v), denoted by APCV(n_v). MCCV(n_v) is a simple and easy method utilizing the method of Monte Carlo by randomly drawing (with or without replacement) a collection $\mathfrak{R}$ of b subsets of $\{1, 2, \ldots, n\}$ that have size n_v and select a model by minimizing

$$\hat{\Gamma}_{\alpha,n} = \frac{1}{n_v b} \sum_{s \in \mathfrak{R}} \left\| y_s - \hat{y}_{\alpha,s^c} \right\|^2.$$

On the other hand, APCV(n_v) selects the optimal model based on the asymptotic leading term of balance incomplete CV(n_v), which treats each subset as a block and each i as a treatment. Shao (1993) compared these three cross-validation methods through a simulation study under the following model with five variables with $n = 40$:

$$y_i = \beta_1 x_{1i} + \beta_2 x_{2i} + \beta_3 x_{3i} + \beta_4 x_{4i} + \beta_5 x_{5i} + e_i,$$

where
   e_i are independent and identically distributed from $N(0,1)$
   x_{ki} is the i th value of the k th prediction variable x_k, $x_{1i} = 1$

and the values of x_{ki}, $k = 2, \ldots, 5$, $i = 1, \ldots, 40$, are taken from an example in Gunst and Mason (1980). Note that there are 31 possible models, and each model was denoted by a subset $\{1, \ldots, 5\}$ that contains the indices of the variable x_k in the model. Shao (1993) indicated that MCCV(n_v) has the best performance among the three methods under study except for the case where the largest model is the optimal model. APCV(n_v) is slightly better than CV(1) in all cases. CV(1) tends to select unnecessarily large models. The probability of selecting the optimal model by using CV(1) could be very low (e.g., less than 0.5).

3.2.3 Remarks

In practice, it is suggested that the optimal variable screening method proposed by Shao and Chow (2007) be applied to select a few relevant variables, say 5–10 variables. Then, apply the cross-validation method to select the optimal model based on linear model selection (Shao, 1993) or non-linear model selection (Li, Chow, and Smith, 2004). The selected model can then be validated based on the cross-validation methods as described in the previous subsection.

3.3 One-Way/Two-Way Translational Process

Pizzo (2006) defines translational medicine as *bench-to-bedside* research wherein a basic laboratory discovery becomes applicable to the diagnosis, treatment, or prevention of a specific disease and is brought forth by either a physician-scientist who works at the interface between the research laboratory and patient care or by a team of basic and clinical science investigators. Thus, translational medicine refers to the translation of basic research discoveries into clinical applications. More specifically, translational medicine is to take the discoveries from basic research to a patient and measures an endpoint in a patient. Scientists are becoming increasingly aware that this bench-to-bedside approach to translational research is a two-way street. Basic scientists provide clinicians with new tools for use in patients and for assessment of their impact, and clinical researchers make novel observations about the nature and progression of diseases that often stimulate basic investigations. As indicated by Pizzo (2006), translational medicine can also have a much broader definition, referring to the development and application of new technologies, biomedical devices, and therapies in a patient-driven environment such as clinical trials, where the emphasis is on early patient testing and evaluation. Thus, translational medicine also includes epidemiological and health-outcomes research and behavioral studies that can be brought to the bedside or ambulatory setting.

Mankoff et al. (2004) pointed out that there are three major obstacles to effective translational medicine in practice. The first is the challenge of translating basic science discoveries into clinical studies. The second hurdle is the translation of clinical studies into medical practice and health-care policy. A third obstacle is philosophical. It may be a mistake to think that basic science (without observations from the clinic and without epidemiological findings of possible associations between different diseases) will efficiently produce novel therapies for human testing. Pilot studies such as nonhuman and nonclinical studies are often used to transition therapies developed using animal models to a clinical setting. The statistical process plays an important role in translational medicine. In this chapter, we define a statistical process of translational medicine as a translational process for (1) determining the association between some independent parameters observed in basic research discoveries and a dependent variable observed from clinical application, (2) establishing a predictive model between the independent parameters and the dependent response variable, and (3) validating the established predictive model. As an example, in animal studies, the independent variables may include *in vitro* assay results, pharmacological activities such as pharmacokinetics and pharmacodynamics, and dose toxicities, and the dependent variable could be a clinical outcome (e.g., a safety parameter).

3.3.1 One-Way Translational Process

Let x and y be the observed values from basic research discoveries and clinical application, respectively. In practice, it is important to ensure that the translational process is accurate and reliable with some statistical assurance. One of the statistical criteria is to examine the closeness between the observed response y and the predicted response $\hat{y}$ via a translational process. To study this, we will first study the association between x and y and build up a model. Then, we will validate the model based on some criteria. For simplicity, we assume that x and y can be described by the following linear model

$$y = \beta_0 + \beta_1 x + \varepsilon, \tag{3.5}$$

where ε follows a normal distribution with mean 0 and variance σ_e^2.

Suppose that n pairs of observations $(x_1, y_1), \ldots, (x_n, y_n)$ are observed in a translational process. To define notation, let

$$X^T = \begin{pmatrix} 1 & 1 & \cdots & 1 \\ x_1 & x_2 & \cdots & x_n \end{pmatrix}$$

and

$$Y^T = \begin{pmatrix} y_1 & y_2 & \cdots & y_n \end{pmatrix}.$$

Then, under model (3.5), the maximum likelihood estimates (MLE) of the parameters β_0 and β_1 are given as

$$\begin{pmatrix} \hat{\beta}_0 \\ \hat{\beta}_1 \end{pmatrix} = (X^T X)^{-1} X^T Y$$

with

$$\mathrm{var}\begin{pmatrix} \hat{\beta}_0 \\ \hat{\beta}_1 \end{pmatrix} = (X^T X)^{-1} \sigma_e^2.$$

Thus, we have established the following relationship:

$$\hat{y} = \hat{\beta}_0 + \hat{\beta}_1 x. \tag{3.6}$$

Given x_i, from (3.6), the corresponding fitted value $\hat{y}_i$ is $\hat{y}_i = \hat{\beta}_0 + \hat{\beta}_1 x_i$. Furthermore, the corresponding MLE of σ_e^2 is give by

$$\hat{\sigma}_e^2 = \frac{1}{n}\sum_{i=1}^{n}(y_i - \hat{y}_i)^2 = \frac{n-2}{n}\,\text{MSE},$$

where MSE is the mean squared errors of the fitted model.

For a given $x = x_0$, suppose that the corresponding observed value is given by y; using (3.6), the corresponding fitted value is $\hat{y} = \hat{\beta}_0 + \hat{\beta}_1 x_0$. Note that $E(\hat{y}) = \beta_0 + \beta_1 x_0 = \mu_0$ and

$$\text{var}(\hat{y}) = (1 \quad x_0)(X^T X)^{-1}\begin{pmatrix} 1 \\ x_0 \end{pmatrix}\sigma_e^2 = c\sigma_e^2,$$

where

$$c = (1 \quad x_0)(X^T X)^{-1}\begin{pmatrix} 1 \\ x_0 \end{pmatrix}.$$

Furthermore, $\hat{y}$ is normally distributed with mean μ_0 and variance $c\sigma_e^2$, that is, $\hat{y} \sim N(\mu_0, c\sigma_e^2)$.

We may validate the translation model by considering how close an observed y and its predicted value $\hat{y}$ obtained based on the fitted regression model given by (3.6) are. To assess the closeness, we propose the following two measures, which are based either on the absolute difference or the relative difference between y and $\hat{y}$:

Criterion I: $p_1 = P\{|y - \hat{y}| < \delta\}$

Criterion II: $p_2 = P\left\{\left|\dfrac{y - \hat{y}}{y}\right| < \delta\right\}$

In other words, it is desirable to have a high probability that the difference or the relative difference between y and $\hat{y}$, given by p_1 and p_2, respectively, is less than a clinically or scientifically meaningful difference d. Then, for either $i = 1$ or 2, it is of interest to test the following hypotheses:

$$H_0: p_i \le p_0 \quad \text{versus} \quad H_a: p_i > p_0, \tag{3.7}$$

where p_0 is some prespecified constant. If the conclusion is to reject H_0 in favor of H_a, this would imply that the established model is considered validated. The technical details of the test of hypothesis corresponding to the two criteria are outlined in the following sections.

3.3.1.1 Test of Hypothesis for the Measures of Closeness

Case 1: Measure of Closeness Based on Absolute Difference
Since y and $\hat{y}$ are independent, we have

$$(y - \hat{y}) \sim N(0, (1+c)\sigma_\varepsilon^2).$$

It can be verified that

$$p_1 = \Phi\left(\frac{\delta}{\sqrt{(1+c)\sigma_e^2}}\right) - \Phi\left(\frac{-\delta}{\sqrt{(1+c)\sigma_e^2}}\right).$$

Thus, the MLE of p_1 is given by

$$\hat{p}_1 = \Phi\left(\frac{\delta}{\sqrt{(1+c)\hat{\sigma}_e^2}}\right) - \Phi\left(\frac{-\delta}{\sqrt{(1+c)\hat{\sigma}_e^2}}\right).$$

Using the delta rule, for a sufficiently large sample size n,

$$\text{var}(\hat{p}_1) = \left(\phi\left(\frac{\delta}{\sqrt{(1+c)\hat{\sigma}_e^2}}\right) + \phi\left(\frac{-\delta}{\sqrt{(1+c)\hat{\sigma}_e^2}}\right)\right)^2 \frac{\delta^2}{2(1+c)n\sigma_e^2} + o\left(\frac{1}{n}\right),$$

where $\phi(z)$ is the probability density function of a standard normal distribution. Furthermore, $\text{var}(\hat{p}_1)$ can be estimated by V_1, which is given by

$$V_1 = \frac{2\delta^2}{(1+c)n\hat{\sigma}_e^2} \phi^2\left(\frac{\delta}{\sqrt{(1+c)\hat{\sigma}_e^2}}\right).$$

Using the Sluksty theorem, $(\hat{p}_1 - p_0)/\sqrt{V_1}$ can be approximated by a standard normal distribution. For the testing of the hypotheses $H_0: p_1 \leq p_0$ versus $H_a: p_1 > p_0$, H_0 is rejected if

$$\frac{\hat{p}_1 - p_0}{\sqrt{V_1}} > z_{1-\alpha},$$

where $z_{1-\alpha}$ is the $100(1-\alpha)$th percentile of a standard normal distribution.

Case 2: Measure of Closeness Based on the Absolute Relative Difference
Note that y^2/σ_e^2 and $\hat{y}^2/c\sigma_e^2$ follow a noncentral χ_1^2 distribution with noncentrality parameter μ_0^2/σ_e^2 and $\mu_0^2/c\sigma_e^2$, respectively, where $\mu_0 = \beta_0 + \beta_1 x_0$. Hence, $\hat{y}^2/cy^2$ is doubly noncentral F distributed with $\upsilon_1 = 1$ and $\upsilon_2 = 1$ degrees of freedom and noncentrality parameters $\lambda_1 = \mu_0^2/c\sigma_e^2$ and $\lambda_2 = \mu_0^2/\sigma_e^2$. According to Johnson and Kotz (1970), a noncentral F distribution with υ_1 and υ_2 degrees of freedom can be approximated by

$$\frac{1+\lambda_1 \upsilon_1^{-1}}{1+\lambda_2 \upsilon_2^{-1}} \, F_{\upsilon,\upsilon'}$$

where $F_{\upsilon,\upsilon'}$ is a central F distribution with degrees of freedom

$$\upsilon = \frac{(\upsilon_1 + \lambda_1)^2}{\upsilon_1 + 2\lambda_1} = \frac{(1+\mu_0^2/c\sigma_e^2)^2}{1+2\mu_0^2/c\sigma_e^2}$$

and

$$\upsilon' = \frac{(\upsilon_1 + \lambda_2)^2}{\upsilon_2 + 2\lambda_2} = \frac{(1+\mu_0^2/\sigma_e^2)^2}{1+2\mu_0^2/\sigma_e^2}.$$

Thus,

$$\begin{aligned}
p_2 &= P\left\{\left|\frac{y-\hat{y}}{y}\right| < \delta\right\} \\[2mm]
&= P\left\{\frac{(1-\delta)^2}{c} < \frac{\hat{y}^2}{cy^2} < \frac{(1+\delta)^2}{c}\right\} \\[2mm]
&\sim P\left\{\frac{(1-\delta)^2}{c} < \frac{1+\lambda_1}{1+\lambda_2} F_{\upsilon,\upsilon'} < \frac{(1+\delta)^2}{c}\right\} \\[2mm]
&= P\left\{\frac{(1-\delta)^2}{c}\frac{1+\lambda_2}{1+\lambda_1} < F_{\upsilon,\upsilon'} < \frac{1+\lambda_2}{1+\lambda_1}\frac{(1+\delta)^2}{c}\right\}.
\end{aligned}$$

Thus, p_1 can be estimated by

$$\hat{p}_1 = P\left\{\frac{(1-\delta)^2}{c}\frac{1+\hat{\lambda}_2}{1+\hat{\lambda}_1} < F_{\hat{\upsilon},\hat{\upsilon}'} < \frac{(1+\delta)^2}{c}\frac{1+\hat{\lambda}_2}{1+\hat{\lambda}_1}\right\} = P\{u_1 < F_{\hat{\upsilon},\hat{\upsilon}'} < u_2\},$$

where

$$u_1 = \frac{(1+\hat{\lambda}_2)}{c(1+\hat{\lambda}_1)}(1-\delta)^2 \quad u_2 = \frac{(1+\hat{\lambda}_2)}{c(1+\hat{\lambda}_1)}(1+\delta)^2$$

and $(\hat{\lambda}_1, \hat{\lambda}_2, \hat{\upsilon}, \tilde{\upsilon}')$ are the corresponding MLE of $(\lambda_1, \lambda_2, \upsilon, \upsilon')$.

For a sufficiently large sample size, using the Sluksty theorem, $\hat{p}_2$ can be approximated by a normal distribution with mean p_2 and variance V_2, where

$$V_2 = \left(\frac{\partial \hat{p}_2}{\partial \hat{\beta}_0}, \frac{\partial \hat{p}_2}{\partial \hat{\beta}_1}, \frac{\partial \hat{p}_2}{\partial \hat{\sigma}_e^2} \right) \begin{pmatrix} (X^T X)^{-1}\hat{\sigma}_e^2 & 0 \\ 0' & \dfrac{2\hat{\sigma}_e^4}{n-2} \end{pmatrix} \begin{pmatrix} \dfrac{\partial \hat{p}_2}{\partial \hat{\beta}_0} \\[2mm] \dfrac{\partial \hat{p}_2}{\partial \hat{\beta}_1} \\[2mm] \dfrac{\partial \hat{p}_2}{\partial \hat{\sigma}_e^2} \end{pmatrix};$$

with

$$\frac{\partial \hat{p}_2}{\partial \hat{\beta}_0} = \frac{2(c-1)\hat{\mu}_0}{c^2 \hat{\sigma}_e^2 (1+\hat{\lambda}_1)^2}[(1+\delta)^2 f_{\hat{\upsilon},\hat{\upsilon}'}(u_2) - (1-\delta)^2 f_{\hat{\upsilon},\hat{\upsilon}'}(u_1)]$$

$$+ \frac{4\hat{\lambda}_1^2(1+\hat{\lambda}_1)}{\hat{\mu}_0(1+2\hat{\lambda}_1)^2} \int_{u_1}^{u_2} \frac{\partial f_{\hat{\upsilon},\hat{\upsilon}'}(x)}{\partial \hat{\upsilon}} \, dx + \frac{4\hat{\lambda}_2^2(1+\hat{\lambda}_2)}{\hat{\mu}_0(1+2\hat{\lambda}_2)^2} \int_{u_1}^{u_2} \frac{\partial f_{\hat{\upsilon},\hat{\upsilon}'}(x)}{\partial \hat{\upsilon}'} \, dx;$$

$$\frac{\partial \hat{p}_2}{\partial \hat{\beta}_1} = x_0 \frac{\partial \hat{p}_2}{\partial \hat{\beta}_0};$$

$$\frac{\partial \hat{p}_2}{\partial \hat{\sigma}_e^2} = \frac{\hat{\lambda}_1 - \hat{\lambda}_2}{c\hat{\sigma}_e^2(1+\hat{\lambda}_1)^2}[(1+\delta)^2 f_{\hat{\upsilon},\hat{\upsilon}'}(u_2) - (1-\delta)^2 f_{\hat{\upsilon},\hat{\upsilon}'}(u_1)]$$

$$- \frac{2\hat{\lambda}_1^2(1+\hat{\lambda}_1)}{\hat{\sigma}_e^2(1+2\hat{\lambda}_1)^2} \int_{u_1}^{u_2} \frac{\partial f_{\hat{\upsilon},\hat{\upsilon}'}(x)}{\partial \hat{\upsilon}} \, dx - \frac{2\hat{\lambda}_2^2(1+\hat{\lambda}_2)}{\hat{\sigma}_e^2(1+2\hat{\lambda}_2)^2} \int_{u_1}^{u_2} \frac{\partial f_{\hat{\upsilon},\hat{\upsilon}'}(x)}{\partial \hat{\upsilon}'} \, dx;$$

$$\frac{\partial f_{\hat{\upsilon},\hat{\upsilon}'}(x)}{\partial \hat{\upsilon}} = \frac{1}{2} f_{\hat{\upsilon},\hat{\upsilon}'}(x)\left[(\log \Gamma(\hat{\upsilon}+\hat{\upsilon}'))^{(1)} - (\log \Gamma(\hat{\upsilon}))^{(1)} + \log\left(\frac{\hat{\upsilon}x}{\hat{\upsilon}x + \hat{\upsilon}'} \right) + \frac{\hat{\upsilon}'(1-x)}{\hat{\upsilon}x + \hat{\upsilon}'} \right],$$

where,

$$\frac{\partial f_{\hat{v},\hat{v}'}(x)}{\partial \hat{v}'} = \frac{1}{2} f_{\hat{v},\hat{v}'}(x) \left[(\log \Gamma(\hat{v}+\hat{v}'))^{(1)} - (\log \Gamma(\hat{v}'))^{(1)} + \log\left(\frac{\hat{v}'}{\hat{v}x+\hat{v}'}\right) + \frac{\hat{v}'(1-x)}{\hat{v}x+\hat{v}'} \right],$$

and $(\log \Gamma(s))^{(1)}$ is the first-order derivative of the natural logarithm of the gamma function with respect to s. Thus, the hypotheses given in (3.7) for one-way translation based on the probability of relative difference can be tested. In particular, H_0 is rejected if

$$Z = \frac{\hat{p}_2 - p_0}{\sqrt{V_2}} > z_{1-\alpha},$$

where $z_{1-\alpha}$ is the $100(1 - \alpha)$th percentile of a standard normal distribution. Note that V_2 is an estimate of $\text{var}(\hat{p}_2)$ which is obtained by simply replacing the parameters with their corresponding estimates of the parameters.

3.3.1.2 An Example

For the two measures proposed in Section 3.1, p_1 is based on the absolute difference between y and $\hat{y}$. Given a p_0 and the selected observation (x_0, y_0), the hypothesis $H_0: p_1 \leq p_0$ is rejected in favor of $H_a: p_1 > p_0$ when

$$Z = \frac{\hat{p}_1 - p_0}{\sqrt{V_1}} > z_{1-\alpha}.$$

Equivalently, H_0 is rejected when

$$\left(\hat{p}_1 - p_0 - z_{1-\alpha}\sqrt{V_1}\right) > 0.$$

Note that the value of $\hat{p}_1$ depends on the value of δ and it can be shown that $\left(\hat{p}_1 - p_0 - z_{1-\alpha}\sqrt{V_1}\right)$ is an increasing function of δ over $(0, \infty)$. Thus, $\hat{p}_1 - p_0 - z_{1-\alpha}\sqrt{V_1} > 0$ if and only if $\delta > \delta_0$. Thus, the hypothesis H_0 can be rejected based on δ_0 instead of $\hat{p}_1$ as long as we can find the value of δ_0 for the given x_0. On the other hand, from a practical point of view, p_2 is more intuitive to understand because it is based on the relative difference, which is equivalent to measuring the percentage difference relative to the observed y and d can be viewed as the upper bound of the percentage error.

For illustration purpose, suppose that the following data are observed in a translational study, where x is a given dose level and y is the associated toxicity measure:

x	0.9	1.1	1.3	1.5	2.2	2.0	3.1	4.0	4.9	5.6
y	0.9	0.8	1.9	2.1	2.3	4.1	5.6	6.5	8.8	9.2

When this set of data is fitted to model (3.5), the estimates of the model parameters are given by $\hat{\beta}_0 = -0.704$, $\hat{\beta}_1 = 1.851$, and $\hat{\sigma}^2 = 0.431$. Thus, based on the fitted results, given $x = x_0$, the proposed translational model is given by $\hat{y} = -0.704 + 1.851x_0$.

In this study, choose $\alpha = 0.05$ and $p_0 = 0.8$. In particular, two dose levels $x_0 = 1.0$ and 5.2 are considered. Based on the study, the corresponding toxicity measures y_0 are 1.2 and 9.0, respectively. However, based on the translational model, the predicted toxicity measures are 1.147 and 8.921, respectively. In the following, the validity of the translational model is assessed by the two proposed closeness measures p_1 and p_2, respectively. Without loss of generality, choose $\alpha = 0.05$ and $p_0 = 0.8$.

Case 1: Testing of $H_0: p_1 \leq p_0$ versus $H_a: p_1 > p_0$
Using the above results, for $x_0 = 1.0$, δ is 1.112, since $|y_0 - \hat{y}| = |9.0 - 8.921| = 0.079$, which is less than $\delta = 1.112$, therefore H_0 is rejected.

Case 2: Testing of $H_0: p_2 \leq p_0$ versus $H_a: p_2 > p_0$
Suppose that $\delta = 1$, for the given two values of x, estimates of p_2 and the corresponding values of the test statistic Z are given in the following table.

x_0	y_0	$\hat{y}$	$\hat{p}_2$	Z	
1.0	1.2	1.147	0.870	1.183	Do not reject H_0
5.2	9.0	8.921	0.809	1.164	Do not reject H_0

3.3.2 Two-Way Translational Process

3.3.2.1 Process Validation

The above translational process is usually referred to as a one-way translation in translational medicine. That is, the information observed in basic research discoveries is translated to clinic. As indicated by Pizzo (2006), the translational process should be a two-way translation. In other words, we can exchange x and y in (3.5)

$$x = \gamma_0 + \gamma_1 y + \varepsilon$$

and come up with another predictive model $\hat{x} = \hat{\gamma}_0 + \hat{\gamma}_1 y$.

Following similar ideas, using either one of the measures p_i, the validation of a two-way translational process can be summarized by the following steps:

Step 1: For a given set of data (x, y), establish a predictive model, say, $y = f(x)$.

Step 2: Select the bound δ_{yi} for the difference between y and $\hat{y}$. Evaluate $\hat{p}_{yi} = P\{|y - \hat{y}| < \delta_{yi}\}$. Assess the one-way closeness between y and $\hat{y}$ by testing the hypotheses (3.7). Proceed to the next step if the one-way translational process is validated.

Step 3: Consider x as the dependent variable and y as the independent variable. Set up the regression model. Predict x at the selected observation y_0, denoted by $\hat{x}$, based on the established model between x and y (i.e., $x = g(y)$), that is, $\hat{x} = g(y) = \hat{\gamma}_0 + \hat{\gamma}_1 y$.

Step 4: Select the bound δ_{xi} for the difference between x and $\hat{x}$. Evaluate the closeness between x and $\hat{x}$ based on a test for the following hypotheses:

$$H_0 : p_i \le p_0 \quad \text{versus} \quad H_a : p_i > p_0,$$

where

$$p_i = P\left\{ \left| \frac{y - \hat{y}}{y} \right| < \delta_{yi} \quad \text{and} \quad \left| \frac{x - \hat{x}}{x} \right| < \delta_{xi} \right\}.$$

The above test can be referred to as a test for two-way translation. If, in Step 4, H_0 is rejected in favor of H_a, this would imply that there is a two-way translation between x and y (i.e., the established predictive model is validated). However, the evaluation of p involves the joint distribution of $(x - \hat{x})/x$ and $(y - \hat{y})/y$. An exact expression is not readily available. Thus, an alternative approach is to modify Step 4 of the above procedure and proceed with a conditional approach instead. In particular, Step 4 is modified as follows:

Step 4 (modified): Select the bound δ_{xi} for the difference between x and $\hat{x}$. Evaluate the closeness between x and $\hat{x}$ based on a test for the following hypotheses:

$$H_0 : p_{xi} \le p_0 \quad \text{versus} \quad H_a : p_{xi} > p_0, \tag{3.8}$$

where

$$p_{xi} = P\left\{ \left| x - \hat{x} \right| < \delta_{xi} \right\}.$$

Note that the evaluation of p_{xi} is much easier and can be computed in a similar way by interchanging the role of x and y for the results given in Section 3.3.1.1.

3.3.2.2 An Example

Using the data set given in Section 3.3.1.2, we set up the regression model $x = \gamma_0 + \gamma_1 y + \varepsilon$ with y as the independent variable and x as the dependent variable. The estimates of the model parameters are $\hat{\gamma}_0 = 0.468$, $\hat{\gamma}_1 = 0.519$, and $\hat{\sigma}^2 = 0.121$. Based on this model, for the same α and p_0, given $(x_0, y_0) = (1.0, 1.2)$ and $(5.2, 9.0)$, the fitted values are given by $\hat{x} = 0.468 + 0.519 y_0$.

Case 1: Testing of $H_0: p_{x1} \leq p_0$ versus $H_a: p_{x1} > p_0$
Using the above results, for $y_0 = 1.2$, δ is 0.587, since $|x_0 - \hat{x}| = |1.0 - 1.09| =$
0.09, which is less than $\delta_x = 0.587$, therefore H_0 is rejected. Similarly, for $y_0 =$
9.0, the corresponding δ is 0.624; then $|x_0 - \hat{x}| = |5.2 - 5.139| = 0.061$, which is
again smaller than $\delta = 0.624$, thus H_0 is rejected.

Case 2: Testing of $H_0: p_{x2} \leq p_0$ versus $H_a: p_{x2} > p_0$
Suppose that $\delta = 1$, for the given two values of y, estimates of p_{x2} and the corresponding values of the test statistic Z are given in the following table.

x_0	y_0	$\hat{x}_0$	$\hat{p}_{x2}$	Z	
1.0	1.2	1.090	0.809	1.300	Do not reject H_0
5.2	9.0	5.139	0.845	16.53	Do not reject H_0

3.4 Lost in Translation

It can be noted that δ_y and δ_x can be viewed as the maximum bias (or possible
lost in translation) from the one-way translation (e.g., from basic research
discovery to clinic) and from the other way of translation (e.g., from clinic
to basic research discovery), respectively. If δ_y and δ_x given in Steps 2 and 4
of the previous subsection are close to 0 with a relatively high probability,
then we conclude that the information from the basic research discoveries
(clinic) is fully translated to the clinic (basic research discoveries). Thus,
one may consider the following parameter to measure the degree of lost in
translation:

$$\zeta = 1 - p_{xy}p_{yx},$$

where
  p_{xy} is the measure of closeness from x to y
  p_{yx} is the measure of closeness from y to x

When $\varsigma \approx 0$, we consider that there is no lost in translation. Overall lost in
translation could be significant even if lost in translation from the one-way
translation is negligible. For illustration purpose, if there is a 10% lost in
translation in the one-way translation and 20% lost in translation in the
other way, there would be up to 28% loss in overall translation. In practice,
an estimate of ς can be obtained for a given set of data (x, y). In particular,
$\hat{\varsigma} = 1 - \hat{p}_{xy}\hat{p}_{yx}.$

   As an illustration, consider the example discussed in Section 3.3.1.2.
Suppose that the measure of closeness based on relative difference is used,

given $(x_0, y_0) = (1.0, 1.2)$ and $(5.2, 9.0)$, the corresponding lost in translation for the two-way translation with $\delta = 1$ is tabulated in the following table:

x_0	y_0	$\hat{y}$	$\hat{p}_{xy}$	$\hat{x}$	$\hat{p}_{yx}$	$\hat{z}$
1.0	1.2	1.147	0.870	1.090	0.809	0.296
5.2	9.0	8.921	0.809	5.139	0.845	0.316

3.5 Animal Model versus Human Model

In translational medicine, a commonly asked question is whether an animal model is predictive of a human model. To address this question, we may assess the similarity between an animal model (population) and a human model (population). For this purpose, we first establish an animal model to bridge the basic research discovery (x) and clinic (y). For illustration purpose, consider a one-way translation. Let $\hat{y} = \hat{\beta}_0 + \hat{\beta}_1 x$ be the predictive model obtained from the one-way translation based on data from an animal population. Thus, for a given x_0, $\hat{y}_0 = \hat{\beta}_0 + \hat{\beta}_1 x_0$ follows a distribution with mean μ_y and σ_y^2. Under the predictive model $\hat{y} = \hat{\beta}_0 + \hat{\beta}_1 x$, denote by (μ_y, σ_y) the target population. Assume that the predictive model works for the target population. Thus, for an animal population, $\mu_y = \mu_{\text{animal}}$ and $\sigma_y = \sigma_{\text{animal}}$, while for a human population, $\mu_y = \mu_{\text{human}}$ and $\sigma_y = \sigma_{\text{human}}$. Assuming that the linear predictive model can be applied to both animal population and human population, we can link the animal and human model by the following:

$$\mu_{\text{human}} = \mu_{\text{animal}} + \varepsilon,$$

and

$$\mu_{\text{human}} = C\mu_{\text{animal}}.$$

In other words, we expect differences in population mean and population standard deviation under the predictive model due to the possible difference in response between animals and humans. As a result, the effect size adjusted for standard deviation under the human population can be obtained as follows:

$$\left| \frac{\mu_{\text{human}}}{\sigma_{\text{human}}} \right| = \left| \frac{\mu_{\text{animal}} + \varepsilon}{C\sigma_{\text{animal}}} \right| = |\Delta| \left| \frac{\mu_{\text{animal}}}{\sigma_{\text{animal}}} \right|$$

where $\Delta = (1 + \varepsilon/\mu_{animal})/C$. Chow et al. (2002a) refer to Δ as a sensitivity index when changing from one target population to another. As can be seen, the effect size under the human population is inflated (or reduced) by the factor of Δ. If $\varepsilon = 0$ and $C = 1$, we then claim that there is no difference between the animal population and the human population. Thus, the animal model is predictive of the human model. Note that the shift and scale parameters (i.e., ε and C) can be estimated by

$$\hat{\varepsilon} = \hat{\mu}_{human} - \hat{\mu}_{animal}$$

and

$$\hat{C} = \frac{\hat{\sigma}_{human}}{\hat{\sigma}_{animal}},$$

respectively, in which $(\hat{\mu}_{animal}, \hat{\sigma}_{animal})$ and $(\hat{\mu}_{human}, \hat{\sigma}_{human})$ are estimates of $(\mu_{animal}, \sigma_{animal})$ and $(\mu_{human}, \sigma_{human})$, respectively. Thus, the sensitivity index can be assessed as follows:

$$\hat{\Delta} = \frac{(1 + \hat{\varepsilon}/\hat{\mu}_{animal})}{\hat{C}}.$$

In practice, there may be a shift in population mean (i.e., ε) and/or in population standard deviation (i.e., C), Chow et al. (2005) indicated that shifts in population mean and population standard deviation can be classified into the following four cases where (1) both ε and C are fixed, (2) ε is random and C is fixed, (3) ε is fixed and C is random, and (4) both ε and C are random. For the case where both ε and C are fixed, (5) can be used for the estimation of Δ. Chow et al. (2005) derived statistical inference of Δ for the case where ε is random and C is fixed by assuming that y conditional on μ follows a normal distribution $N(\mu, \sigma^2)$. That is,

$$y\,|_{\mu=\mu_{human}} \sim N(\mu, \sigma^2),$$

where
 μ is distributed as $N(\mu_{\mu}, \sigma_{\mu}^2)$
 σ, μ_{μ}, and σ_{μ}^2 are some unknown constants

It can be verified that y follows a mixed normal distribution with mean μ_{μ} and variance $\sigma^2 + \sigma_{\mu}^2$. That is, $y \sim N(\mu_{\mu}, \sigma^2 + \sigma_{\mu}^2)$. As a result, the sensitivity index can be assessed based on data collected from both animal and human populations under the predictive model.

Note that for other cases where C is random, the above method can also be derived similarly. The assessment of sensitivity index can be used to adjust the treatment effect to be detected under a human model when applying an animal model to a human model, especially when there is a significant or major shift between an animal population and the human population. In practice, it is of interest to assess the impact of the sensitivity index on both lost in translation and the probability of success. This, however, requires further research.

3.6 Concluding Remarks

Translational medicine is a multidisciplinary entity that bridges basic scientific research with clinical development. As the expense in developing therapeutic pharmaceutical compounds continues to increase and the success rates for getting such compounds approved for marketing and to the patients needing these treatments continues to decrease, a focused effort has emerged in improving the communication and planning between basic and clinical science. This will likely lead to more therapeutic insights being derived from new scientific ideas, and more feedback being provided back to research so that their approaches are better targeted. Translational medicine spans all the disciplines and activities that lead to making key scientific decisions as a compound traverses across the difficult preclinical–clinical divide. Many argue that improvement in making correct decisions on what dose and regimen should be pursued in the clinic, likely human safety risks of a compound, likely drug interactions, and pharmacologic behavior of the compound are likely the most important decisions made in the entire development process. Many of these decisions and the path for uncovering this information within later development are defined at this specific time within the drug development process. Improving these decisions will likely lead to a substantial increase in the number of safe and effective compounds available to combat human diseases.

In clinical research and development, before the first-in-human study, one of the controversial issues is whether the established animal model (e.g., mice) is predictive of the human model. For the first-in-human study, the starting dose is usually selected as 1/10 of LD_{10} in animals. The selected initial dose, however, may be too low to be effective or too high to have toxic effect. The other controversial issue is the potential lost in translation from bench (basic discoveries) to bedside (first-in-human) translational research. In current practice, it is recognized that bench-to-bedside translational research is a one-way translational process, which is not efficient due to potential lost in translation. Significant lost in translation will decrease the probability of success of the pharmaceutical/clinical development. Thus, it is suggested that a two-way translational process be considered.

4

Bioavailability and Bioequivalence

4.1 Introduction

According to Saul (2007), the United States spends about $275 billion annually on prescription drug products. In addition, Saul (2007) also pointed out that, in the next 5 years, a series of innovative drug products with a total combined annual sale of $60 billion are going off patents. This opens the door for a tidal wave of generic drug products that are 30%–80% cheaper than the innovative drug products. In 1984, the United States Congress passed the *Drug Price Competition and Patent Term Restoration Act*, which allows a regulatory framework for a low-cost pathway for generic drug products to enter the market (Frank, 2007). As a result, when an innovative (brand-name) drug product is going off a patent, pharmaceutical or generic companies can file an abbreviated new drug application (ANDA) for generic approval. For the approval of a generic drug product, most regulatory agencies require that evidence of average bioavailability (in terms of drug absorption) be provided through the conduct of bioequivalence (BE) studies. However, as pointed out by Saul (2007), a survey conducted in 2002 by the Association of American Retire People (AARP) indicated that 22% of the responders considered that generic drug products are less effective or of poor quality than the innovator drug products. This shows that a sizable portion of the public in the United States still lacks confidence in generic drug products even if they are approved by the United States Food and Drug Administration (FDA). Therefore, in May 2007, the FDA added generic drugs in the *Critical Path Opportunities* to use latest breakthroughs in technique to assure that the efficacy and safety of the generic drug products are the same as those of the innovator drug products. However, the FDA critical path opportunities for generic drugs do not cover all important emerging challenges for generic drugs.

For the assessment of average bioequivalence (ABE), a standard two-sequence, two-period (or 2 × 2) crossover design is usually employed. A BE study is often conducted on healthy volunteers for characterizing drug absorption in the bloodstream. Qualified subjects are randomly assigned to receive either a test (generic or new formulation) drug or a reference (brand-name or innovator) drug first and then be crossed over to receive the other drug after a sufficient length of washout. A commonly used statistical method is a confidence interval approach (or equivalently a two one-sided tests procedure) which is derived under the standard 2 × 2 crossover design. Note that the FDA requires that log transformation be performed before data analysis. The test product is then claimed bioequivalent to the reference product if the obtained 90% confidence interval for the ratio of means of the primary study endpoint such as area under the blood or plasma concentration time curve (AUC) or the peak or maximum concentration (C_{max}) is totally within the BE limit of (80%, 125%).

In the next section, the design and analysis for the assessment of BE are briefly outlined. Drug interchangeability in terms of drug prescribability and drug switchability are discussed in Section 4.3. Section 4.4 presents some controversial issues that are commonly encountered when conducting BE studies for the assessment of ABE. These controversial issues include, but are not limited to, (1) challenge of the *Fundamental Bioequivalence Assumption*, (2) adequacy of one-fits-all criterion, and (3) appropriateness of log transformation. Some frequently asked questions during the ANDA submission for generic approval are given in Section 4.5. Section 4.6 provides some concluding remarks to end the chapter.

4.2 Bioequivalence Assessment

For the approval of generic drug products, the FDA requires that the evidence of ABE in drug absorption in terms of some pharmacokinetic (PK) parameters such as AUC and C_{max} be provided through the conduct of BE studies. We claim that a test drug product is bioequivalent to a reference (innovative) drug product if the 90% confidence interval for the ratio of means of the primary PK parameter is totally within the BE limit of (80%, 125%). The confidence interval for the ratio of means of the primary PK parameter is obtained based on log-transformed data. In what follows, study designs that are commonly considered in BE studies are briefly introduced.

4.2.1 Study Design

As indicated in the *Federal Register* [Vol. 42 No. 5 Sec. 320.26(b) and Sec. 320.27(b), 1977], a bioavailability study (single dose or multidose) should be crossover in design, unless a parallel or other design is more appropriate for valid scientific reasons. Thus, in practice, a standard 2 × 2 crossover design is often considered for a bioavailability/BE study. Denote T and R by the test product and the reference product, respectively. The 2 × 2 crossover design can be expressed as (TR, RT), where TR is the first sequence of treatments and RT denotes the second sequence of treatments. Under the (TR, TR) design, qualified subjects who are randomly assigned to sequence 1 (*TR*) will receive the test product (*T*) first and then receive the reference product (*R*) after a sufficient length of washout period. Similarly, subjects who are randomly assigned to sequence 2 (*RT*) will receive the reference product (*R*) first and then receive the test product (*T*) after a sufficient length of washout period.

Satistically, one of the limitations of the standard 2 × 2 crossover design is that it does not provide independent estimates of intra-subject variability (ISV) since each subject only receives the same treatment once. In the interest of assessing ISV, the following alternative designs for comparing two drug products are often considered:

1. Balaam's design: (*TT, RR, RT, TR*)
2. Two-sequence, three-period dual design: (*TRR, RTT*)
3. Four-sequence, four-period design: (*TTRR, RRTT, TRT\RT, RTTR*)

Note that the above study designs are also referred to as higher-order crossover designs. A higher-order crossover design is defined as a design with the number of sequences or the number of periods greater than the number of treatments to be compared.

For comparing more than two drug products, a Williams' design is often considered. For example, for comparing three drug products, a six-sequence, three-period (6 × 3) Williams' design is usually considered, while a 4 × 4 Williams' design is employed for comparing four drug products. Williams' design is a variance stabilizing design. More information regarding the construction and good design characteristics of Williams' designs can be found in Chow and Liu (2008).

In the interest of assessing population bioequivalence (PBE) and/or individual bioequivalence (IBE), the FDA recommends that a replicated design be considered for obtaining independent estimates of ISV and variability due to subject-by-drug product interaction. A commonly considered replicated crossover design is the replicate of a 2 × 2 crossover design, which is given by (*TRTR, RTRT*).

In some cases, an incomplete block design or an extra-reference design such as (*TRR, RTR*) may be considered depending upon the study objectives of the bioavailability/BE studies. Under a given design, sample size calculation for achieving a desired power at the 5% level of significance can then be obtained (see, e.g., Chow and Wangm 2001; Chow, Shao and Wang, 2008; Chow and Liu, 2008).

4.2.2 Statistical Methods

As indicated earlier, BE is claimed if the ratio of average bioavailabilities between a test product and a reference product is within the BE limit of (80%, 125%) with 90% assurance based on log-transformed data. Along this line, commonly employed statistical methods are the confidence interval approach and the method of interval hypotheses testing.

For the confidence interval approach, a 90% confidence interval for the ratio of means of the primary PK response such as AUC or C_{max} is obtained under an analysis of the variance model. We claim BE if the obtained 90% confidence interval is totally within the BE limit of (80%, 125%). For the method of interval hypotheses testing, the interval hypothesis

$$H_0 : \text{Bioinequivalence} \quad \text{versus} \quad H_a : \text{Bioequivalence}$$

was decomposed into two sets of one-sided hypotheses. The first set of hypotheses is to verify that the average bioavailability of the test product is not too low (efficacy), whereas the second set of hypotheses is to verify that average bioavailability of the test product is not too high (safety). Schuirmann's two one-sided tests procedure is commonly employed for the interval hypotheses testing for ABE (Schuirmann, 1987).

In practice, other statistical methods such as Westlake's symmetric confidence interval approach, exact confidence interval based on Fieller's theorem, Chow and Shao's joint confidence region approach, Bayesian methods (e.g., Rodda and Davis' method and Mandallaz and Mau's method), and nonparametric methods (e.g., Wilcoxon–Mann–Whitney two one-sided tests procedure, distribution-free confidence interval based on the Hodges–Lehmann estimator, and bootstrap confidence interval) are sometimes considered.

4.2.3 Remarks

Although the assessment of ABE for generic approval has been in practice for years, it has the following limitations: (1) it focuses only on population average; (2) it ignores distribution of the metric; (3) it does not provide independent estimates of ISV; and (4) it ignores subject-by-formulation interaction. Many authors criticize that the assessment of ABE does not address the question of drug interchangeability and it may penalize drug products with less variability.

4.3 Drug Interchangeability

As indicated by the regulatory agencies, a generic drug can be used as a substitution of the brand-name drug if it has been shown to be bioequivalent to the brand-name drug. Current regulations do not indicate that two generic copies of the same brand-name drug can be used interchangeably, even though they are bioequivalent to the same brand-name drug. BE between generic copies of a brand-name drug is not required. Thus, one of

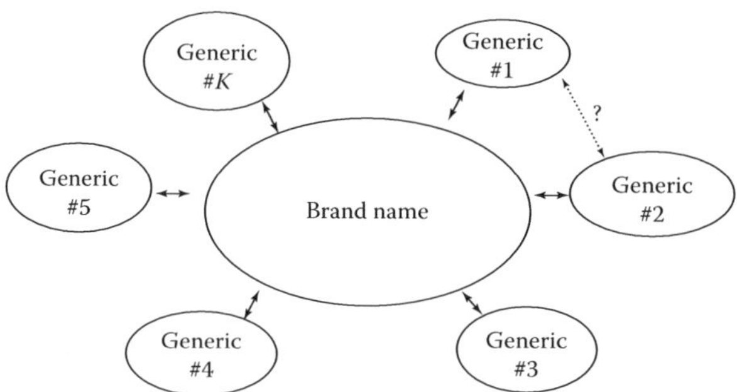

FIGURE 4.1
Safety concern of drug interchangeability.

the controversial issues is that whether these approved generic drug products can be used safely and interchangeably (see Figure 4.1).

4.3.1 Drug Prescribability and Drug Switchability

Basically, drug interchangeability can be classified as drug prescribability or drug switchability (Liu, 1998; Chow and Liu, 2008). Drug prescribability is referred to as the physician's choice for prescribing an appropriate drug product for his/her new patients between a brand-name drug product and a number of generic drug products of the brand-name drug product that have been shown to be bioequivalent to the brand-name drug product. The underlying assumption of drug prescribability is that the brand-name drug product and its generic copies can be used interchangeably in terms of the efficacy and safety of the drug product. Drug prescribability, therefore, is the interchangeability for the new patient.

Drug switchability, on the other hand, is related to the switch from a drug product (e.g., a brand-name drug product) to an alternative drug product (e.g., a generic copy of the brand-name drug product) within the same subject, whose concentration of the drug product has been titrated to a steady, efficacious, and safe level. As a result, drug switchability is considered more critical than drug prescribability in the study of drug interchangeability for patients who have been on medication for a while. Drug switchability, therefore, is interchangeability within the same subject.

4.3.2 Population and Individual Bioequivalence

As indicated by Chow and Liu (2008), ABE can guarantee neither drug prescribability nor drug switchability. Therefore, it is suggested that the assessment of BE should take into consideration drug prescribability and drug switchability for drug interchangeability. To address drug interchangeability, it is recommended that PBE and IBE be considered for testing drug prescribability and drug switchability, respectively. More specifically, the FDA recommends that PBE be applied to new formulations, additional strengths, or new dosage forms in new drug applications (NDAs), while IBE should be considered for ANDA or abbreviated antibiotic drug application (AADA) for generic drugs.

To address drug prescribability, the FDA proposed the following aggregated, scaled, moment-based, one-sided population bioequivalence criterion (PBC):

$$\text{PBC} = \frac{(\mu_T - \mu_R)^2 + (\sigma_{TT}^2 - \sigma_{TR}^2)}{\max(\sigma_{TR}^2, \sigma_{T0}^2)} \leq \theta_P,$$

where

μ_T and μ_R are the mean of the test drug product and the reference drug
  product, respectively

σ_{TT}^2 and σ_{TR}^2 are the total variance of the test drug product and the reference drug product, respectively

σ_{T0}^2 is a constant that can be adjusted to control the probability of passing
  PBE

θ_P is the BE limit for PBE

The numerator on the left-hand side of the criterion is the sum of the squared difference of the population averages and the difference in total variance between the test and reference drug products, which measures the similarity for the marginal population distribution between the test and reference drug products. The denominator on the left-hand side of the criterion is a scaled factor that depends upon the variability of the drug class of the reference drug product. The FDA guidance suggests that θ_P be chosen as

$$\theta_P = \frac{(\log 1.25)^2 + \varepsilon_P}{\sigma_{T0}^2},$$

where ε_P is guided by the consideration of the variability term $\sigma_{TT}^2 - \sigma_{TR}^2$ added to the ABE criterion. As suggested by the FDA guidance, it may be appropriate that ε_P is chosen to be 0.02. For the determination of σ_{T0}^2, the guidance suggests the use of the so-called population difference ratio (PDR), which is defined as

$$\text{PDR} = \left[\frac{E(T-R)^2}{E(R-R')^2}\right]^{1/2} = \left[\frac{(\mu_T - \mu_R)^2 + \sigma_{TT}^2 + \sigma_{TR}^2}{2\sigma_{TR}^2}\right]^{1/2} = \left[\frac{\text{PBC}}{2} + 1\right]^{1/2}.$$

Therefore, assuming that the maximum allowable PDR is 1.25, substitution of $(\log 1.25)^2/\sigma_{T0}^2$ for PBC without adjustment of the variance term approximately yields $\sigma_{T0} = 0.2$.

Similarly, to address drug switchability, the FDA recommended the following aggregated, scaled, moment-based, one-sided individual bioequivalence criterion (IBC):

$$\text{IBC} = \frac{(\mu_T - \mu_R)^2 + \sigma_D^2 + (\sigma_{WT}^2 - \sigma_{WR}^2)}{\max(\sigma_{WR}^2, \sigma_{W0}^2)} \leq \theta_I,$$

where

σ_{WT}^2 and σ_{WR}^2 are within the subject variance of the test drug product and the reference drug product, respectively

σ_D^2 is the variance due to subject-by-drug interaction

σ_{W0}^2 is a constant that can be adjusted to control the probability of passing IBE

θ_I is the BE limit for IBE. The FDA guidance suggests that θ_I be chosen as

$$\theta_I = \frac{(\log 1.25)^2 + \varepsilon_I}{\sigma_{W0}^2},$$

where ε_I is the variance allowance factor, which can be adjusted for sample size control. As indicated in the FDA guidance, ε_I may be fixed between 0.04 and 0. For the determination of σ_{W0}^2, the guidance suggests the use of individual difference ratio (IDR), which is defined as

$$\text{IDR} = \left[\frac{E(T-R)^2}{E(R-R')^2} \right]^{1/2} = \left[\frac{(\mu_T - \mu_R)^2 + \sigma_D^2 + (\sigma_{WT}^2 + \sigma_{WR}^2)}{2\sigma_{WR}^2} \right]^{1/2} = \left[\frac{\text{IBC}}{2} + 1 \right]^{1/2}.$$

Therefore, assuming that the maximum allowable IDR is 1.25, substitution of $(\log 1.25)^2 / \sigma_{W0}^2$ for IBC without adjustment of the variance term approximately yields $\sigma_{W0} = 0.2$.

4.4 Controversial Issues

In this section, we will focus on controversial issues related to Fundamental Bioequivalence Assumption, one-fits-all criterion, and issues related to log transformation of PK data prior to analysis. These controversial issues are briefly described in the following sections.

4.4.1 Fundamental Bioequivalence Assumption

As indicated by Chow and Liu (2008), BE studies are performed under so-called Fundamental Bioequivalence Assumption, which constitutes the legal basis for regulatory approval of generic drug products. Fundamental Bioequivalence Assumption states:

> If two drug products are shown to be bioequivalent, it is assumed that they will reach the same therapeutic effect or they are therapeutically equivalent and hence can be used interchangeably.

To protect the exclusivity of a brand-name drug product, the sponsors of the innovator drug products will make every attempt to prevent generic drug products from being approved by regulatory agencies such as the FDA. One of the strategies is to challenge the Fundamental Bioequivalence Assumption by filing a *citizen petition* with scientific/clinical justification. Upon the receipt of a citizen petition, the FDA has legal obligation to respond within 180 days. It, however, should be noted that the FDA will not suspend the review/approval process of generic submission of a given brand-name drug even if a citizen petition is under review within the FDA.

Under the Fundamental Bioequivalence Assumption, one of the controversial issues is that BE may not necessarily imply therapeutic equivalence and therapeutic equivalence does not guarantee BE either. The assessment of ABE for generic approval has been criticized that it is based on legal/political consideration rather than scientific consideration. In the past several decades, many sponsors/researchers have made an attempt to challenge this assumption with no success.

In practice, the verification of the Fundamental Bioequivalence Assumption is often difficult, if not impossible, without the conduct of clinical trials. For some drug products, the Fundamental Bioequivalence Assumption may be verified through the study of *in vitro–in vivo* correlation (IVIVC). It should be noted that the Fundamental Bioequivalence Assumption is for drug products with identical active ingredient(s). Whether the Fundamental Bioequivalence Assumption is applicable to (1) drug products with similar but different active ingredient(s) and (2) biological products which are made of living cells then become an interesting but controversial question.

4.4.2 One-Fits-All Criterion

For the assessment of ABE, the FDA adopted a one-fits-all criterion. That is, a test drug product is said to be bioequivalent to a reference drug product if the obtained 90% confidence interval for the ratio of means of the primary study endpoint such as AUC or C_{max} is totally within the BE limit of (80%, 125%) based on log-transformed data. The one-fits-all criterion does not take into consideration of individual therapeutic window (ITW) and ISV, which have been identified to have nonnegligible impact on the safety and efficacy of generic drug products as compared to innovative drug products.

In the past several decades, this one-fits-all criterion has been challenged and criticized by many researchers. It is suggested that flexible criteria in

TABLE 4.1

Classification of Drugs

Class	ITW	ISV	Example
A	Narrow	High	Cyclosporine
B	Narrow	Low	Theophylline
C	Wide	Low to moderate	Most drugs
D	Wide	High	Chlorpromazine or topical corticosteroids

Source: Patnaik, R.N. et al., *Clinical Pharmacokinetics*, 33, 1, 1997. With permission.

Note: ITW, individual therapeutic window; ISV, intra-subject variability.

terms of safety (upper BE limit) and efficacy (lower BE limit) be developed based on ITW and ISV according to the nature of drug class under study (Table 4.1). However, the one-fits-all criterion is still considered by most regulatory agencies until a recent proposal that based on reference-scaled average bioequivalence (RSAB) criterion for highly variable drug products proposed by Haider et al. (2008). This is probably because no (documented) evidence of safety issues are raised for those generic drug products approved based on the one-fits-all criterion. More discussions regarding the one-fits-all criterion can be found in Section 26.4.2 of Chapter 26.

4.4.3 Issues Related to Log Transformation

In practice, BE is assessed either based on raw data or log-transformed data depending upon whether the data are normally distributed. This has raised a controversial issue regarding which model should be used for a fair assessment of BE. The sponsors often choose the model that can serve their purposes (e.g., demonstration of BE). In many cases, the raw data model may reach a different conclusion regarding BE than the log-transformation model. This controversial issue has been discussed excessively that a guidance on BE published by the FDA recommends that a log transformation be performed prior to the assessment of BE (FDA, 2001). For the assessment of BE, in practice, the 2001 FDA guidance provides a rationale for the use of logarithmic transformation of exposure measures. The guidance emphasizes that the limited sample size in a typical BE study precludes a reliable determination of the distribution of the data. For some unknown reasons, the guidance does not encourage the sponsors to test for normality of error distribution after log transformation or to use normality of error distribution as a reason for carrying out the statistical analysis on the original scale.

With respect to the (PK) rationale, deterministic multiplicative PK models are used to justify the routine use of logarithmic transformation for $AUC(0-\infty)$ and C_{max}. However, the deterministic PK models are theoretical derivations of $AUC(0-\infty)$ and C_{max} for a single object. The guidance suggests that $AUC(0-\infty)$ be calculated from the observed plasma–blood concentration–time curve using the trapezoidal rule, and that C_{max} be obtained directly from the curve, without interpolation. It is not known whether the observed AUC(0–∞) and C_{max} can provide good approximations to those under the theoretical models if the models are correct.

It should be noted that the AUC(0–∞) and C_{max} are calculated from the observed plasma–blood concentrations. Therefore, the distributions of the observed AUC(0–∞) and C_{max} depend on the distributions of plasma–blood concentrations. Liu and Weng (1994) showed that the log-transformed AUC(0–∞) and C_{max} do not generally follow a normal distribution, even when either the plasma concentrations or log-plasma concentrations are normally distributed. This argues against the routine use of the logarithmic transformation in the assessment of BE. Moreover, Patel (1994) also pointed out that performing a routine log transformation of data and then applying normal, theory-based methods is not a scientific approach. In addition, the sample size of a typical BE study is generally too small to allow an adequate large-sample normal approximation.

Because current statistical methods for the evaluation of BE are based on the normality assumption of the inter-subject and intra-subject variabilities, the examination of the normal probability plots for the studentized inter-subject and intra-subject residuals should always be carried out for the scale intended to be used in the analysis. In addition, formal statistical tests for normality of the inter-subject and intra-subject variabilities can also be carried out through Shapiro–Wilk's method. Contrary to the misconception of many people, Shapiro–Wilk's method is an appropriate method for small samples, such as BE studies. It is then scientifically imperative that tests for normality be routinely performed for the scale used in the analysis, such as log scale, suggested in the guidance. If normality cannot be satisfied by both original scale and log scale, nonparametric methods should be employed.

Other issues concerning the routine use of the logarithmic transformation of exposure responses are the equivalence limits and presentation of the results on the original scale. The guidance recommends that the BE limits of (80%, 125%) on the original scale for the assessment of ABE be used. On the log scale, they are [log(0.8), log(1.25)] = (−0.2331, 0.2331), where log denotes the natural logarithm. This set of limits is symmetrical about zero on the log scale, but it is not symmetrical on the original scale. It should be noted that the rejection region of Schuirmann's two one-sided tests procedure associated with the new limits of (80%, 125%) is larger than that with the limits of (80%, 120%). As a result, a 90% confidence interval of (82%, 122%), for the ratio of averages of AUC(0–∞) between the test and reference formulations, will pass the BE test by the new limits, but not by the old limits. The new BE limits are 12.5% wider and 25% more liberal in the upper limit than the old limits. A new, wider upper BE limit may have an influence on the safety of the test formulation, which should be carefully examined if the new BE limits are adopted.

The FDA guidance requires that the results of analyses be presented on the log scale as well as on the original scale, which can be obtained by taking the inverse transformation. Because the logarithmic transformation is not linear, the inverse transformation of the results to the original scale is not straightforward (Liu and Weng, 1992). For example, the point estimator of the ratio of averages on the original scale obtained from the antilog of the estimator of difference in averages on the log scale is biased and is always overestimated. Furthermore, the antilog of the standard deviation of the difference in averages on the log scale is not the standard deviation for the point estimator of the ratio of the averages on the original scale. Further research is needed for the presentation of the results on the original scale, especially the estimation of variability after the analyses are performed on the log scale.

For the limitation of ABE, the consideration of ITWs, and the objective of interchangeability, Chen (1995) summarized the merits of individual BE as follows:

1. Comparison of both averages and variances
2. Considerations of subject-by-formulation interaction
3. Assurance of switchability
4. Provision of flexible BE criteria for different drugs based on their therapeutic windows
5. Provision of reasonable BE criteria for drugs with highly ISV
6. Encouragement or reward of pharmaceutical companies to manufacture a better formulation

To achieve the objective of exchangeability among bioequivalent pharmaceutical products, the criteria for assessment of BE must possess certain important properties. Chen (1995, 1997) outlined the desirable characteristics of BE criteria proposed by the FDA which is provided in Table 4.2. In

TABLE 4.2

Desirable Features of Bioequivalence Criteria

Comparison of both averages and variances
Assurance of switchability
Encouragement or reward of pharmaceutical companies to manufacture a better formulation
Control of type I error rate (consumer's risk) at 5%
Allowance for determination of sample size
Admission of the possibility of sequence and period effects as well as missing values
User-friendly software application for statistical methods
Provision of easy interpretation for scientists and clinicians
Minimization of increased cost for conducting bioequivalence studies

Source: Chen, M.L., Individual bioequivalence. Invited presentation at *International Workshop: Statistical and Regulatory Issues on the Assessment of Bioequivalence.* Dusseldorf, Germany, October 19–20, 1995.

addition, to address the issues of ISV and subject-by-formulation interaction and to ensure drug switchability, valid statistical procedures, both estimation and hypothesis testing, should be developed from the criteria to control the consumer's risk at the prespecified nominal level (e.g., 5%). In addition, the statistical methods developed from the criteria should be able to provide sample size determination; to take into consideration the nuisance design parameters, such as period or sequence effects; and to develop user-friendly computer software. The most critical characteristics for any proposed criteria will be their interpretation to scientists and clinicians and the cost of conducting BE studies to provide inference for the criteria.

4.5 Frequently Asked Questions

Although the concepts of PBE and IBE for addressing drug prescribability and drug switchability have been discussed vastly since the early 1990s, FDA's current position regarding the assessment of BE is:

> Average bioequivalence is required and individual/population bioequivalence may be considered.

However, the FDA encourages that medical/statistical reviewers be consulted if IBE/PBE is to be used. For the assessment of BE, some questions are frequently asked during the regulatory submission and review. In what follows, frequently asked questions in BE assessment are briefly described.

4.5.1 What If We Pass Raw Data Model but Fail Log-Transformed Data Model?

Most regulatory agencies including FDA, European Medicines Agency (EMEA), and the World Health Organization (WHO) recommend that a log transformation of PK parameters of $AUC(0-t)$, $AUC(0-\infty)$, and C_{max} be performed before analysis. *No assumption checking or verification of the log-transformed data is encouraged.* However, the sponsors often conduct analyses based on both raw data and log-transformed data and submit the one that passes BE testing. If the sponsor passes BE testing under the log-transformed data model, then there is no problem because it meets regulatory requirements. In practice, however, the sponsor may fail BE testing under the log-transformed data model but pass under the raw data model. In this case, the sponsor often provides scientific/statistical justification for the use of the raw data model. One of the most commonly seen scientific/statistical justifications is that the raw data model is a more appropriate statistical model than the log-transformed data model because all of the assumptions for the raw data model are met. However, for the raw data model, the BE limit is often expressed in terms of the ratio of the population means between the test and reference formulations, and then the equivalence limit is expressed as a percentage of the population reference average which has to be estimated from the data. Therefore, the variability of the estimated reference average is not considered in the equivalence limit. Hence, the false positive rate for claiming ABE for the two one-sided tests procedure can be inflated to 50%. As a result, one should apply the modified two one-sided tests procedure using the raw data proposed by Liu and Weng (1995) to control the size at the nominal level.

Many researchers have criticized that the use of log-transformed data is not scientifically/statistically justifiable. Liu and Weng (1992) studied the distribution of log-transformed PK data assuming that the hourly concentrations are normally distributed. The results indicated that the log-transformed data are not normally distributed. Their findings argue against the use of log-transformed data since the primary normality assumption is not met and consequently the assurance of the obtained statistical inference is questionable. In this case, it is suggested that either other transformations such as the Box–Cox transformation or a nonparametric method be considered. However, the interpretation of such a transformation is challenging to both pharmacokineticists and biostatisticians.

4.5.2 What If We Pass AUC but Fail C_{max}?

Based on the log-transformed data, the FDA requires that both AUC and C_{max} meet the (80%, 125%) BE limit for the establishment of ABE. In practice, however, it is not uncommon to pass AUC (the extent of absorption) but fail C_{max} (the rate of absorption). In this case, ABE cannot be claimed according to the FDA guidance on BE. However, for C_{max}, the EMEA and WHO guidelines use a more relaxed equivalence margin of (70%, 143%). Thus, the sponsors often argue with the FDA based on the EMEA and WHO guidelines.

In the case where we pass AUC but fail C_{max}, Endrenyi et al. (1991) suggested considering C_{max}/AUC as an alternative BE measure for the rate of absorption. However, C_{max}/AUC is not currently selected as the required PK responses for the approval of generic drug products by any of the regulatory authorities in the world including the FDA, EMEA, and WHO. On the other hand, it is very likely that we may pass C_{max} but fail AUC. In this case, it is suggested that we may look at partial AUC as an alternative measure of BE (see, e.g., Chen et al., 2001) if we fail to pass BE testing based on AUC from 0 to the last time point or AUC from 0 to infinity.

4.5.3 What If We Fail by a Relatively Small Margin?

In practice, it is very possible that we fail BE testing for either AUC or C_{max} by a relatively small margin. For example, suppose the 90% confidence interval for AUC is given by (79.5%, 120%), which is slightly outside the lower limit of (80%, 125%). In this case, the FDA's position is very clear that *Rule is rule and you fail*. With respect to regulatory review and approval, the FDA is very strict about this rule that the 90% confidence interval has to be totally within the BE limit of (80%, 125%) as described in the 2003 FDA guidance. However, the sponsor usually performs either an outlier detection analysis or a sensitivity analysis to resolve the issue. In other words, if a subject is found to be an outlier statistically, it may be excluded from the analysis with appropriate clinical justification. Once the identified outlier is excluded from the analysis, a 90% confidence interval is recalculated. If the 90% confidence interval after excluding the identified outlier is totally within the BE limit of (80%, 125%), the sponsor then argues to claim BE.

4.5.4 Can We Still Assess Bioequivalence If There Is a Significant Sequence Effect?

As indicated by Chow and Liu (2008), under a standard 2×2 crossover design, significant sequence effect is an indication of possible (1) failure of randomization, (2) true sequence effect, (3) true carryover effect, and/or (4) true formulation-by-period effect. Under the standard 2×2 crossover design, the sequence effect is confounded with the carryover effect. Therefore, if a significant sequence effect is found, the treatment effect and its corresponding 90% confidence interval cannot be estimated in an unbiased way due to possible unequal carryover effects. However, in the 2001 FDA guidance, the following list of conditions is provided to rule out the possibility of unequal carryover effects:

1. It is a single-dose study.
2. The drug is not an endogenous entity.
3. More than an adequate washout period has been allowed between periods of the study and in the subsequent periods the predose biological matrix samples do not exhibit a detectable drug level in any of the subjects.
4. The study meets all scientific criteria (e.g., it is based on an acceptable study protocol and it contains a validated assay methodology).

The 2001 FDA guidance also recommends that sponsors conduct a BE study with parallel designs if unequal carryover effects become an issue.

4.5.5 What Should We Do When We Have Almost Identical Means but Still Fail to Meet the Bioequivalence Criterion?

It is not uncommon to run into the situation that we have almost identical means but still fail to meet the BE criterion. This may indicate that (1) the variation of the reference product is too large to establish BE between the test product and the reference product, (2) the BE study was poorly conducted, and (3) the analytical assay methodology is inadequate and not fully validated. The concept of IBE and/or PBE is an attempt to overcome this problem. As a result, it is suggested that either PBE or IBE be considered to establish BE. However, in our experience, unless the variability of the test formulation is much smaller than that of the reference formulation, it is still unlikely to pass either PBE or IBE. In addition, to avoid masking the effect of PBE or IBE, the 2001 FDA guidance requires that the geometric test/reference averages be within 80%–125% too.

4.5.6 Power and Sample Size Calculation Based on Raw Data Model and Log-Transformed Model Are Different

Power analysis calculation and sample size based on the raw data model are different from those based on the log-transformed model due to the fact that they are different models. Under different models, means, standard deviations, and coefficients of variation are different. As mentioned earlier, for the assessment of BE, all regulatory authorities including the FDA, EMEA, WHO, and Japan require that log transformation of $AUC(0-t)$, $AUC(0-\infty)$, and C_{max} be done before the analysis and evaluation of BE. As a result, one should use differences in mean and standard deviation or coefficient of variation for power analysis and sample size calculation based on the method for the log-transformed model (see, e.g., Chapter 5 of Chow and Liu, 2008).

Note that sponsors should make the decision as to which model (the raw data model or the log-transformed data model) will be used for BE assessment. Once the model is chosen, appropriate formulas can be used to determine the sample size. Fishing around for obtaining the smallest sample size is not a good clinical practice.

4.5.7 Adjustment for Multiplicity

The 2003 FDA guidance for general considerations requires that for $AUC(0-t)$, $AUC(0-\infty)$, and C_{max}, the following information be provided:

1. Geometric means
2. Arithmetic means
3. Ratio of means
4. Ninety percent confidence interval

In addition, the 2003 FDA guidance recommends that logarithmic transformation be provided for measures for BE demonstration using a BE limit of 80%–125%. Therefore, to pass the ABE, each 90% confidence interval of $AUC(0-t)$, $AUC(0-\infty)$, and C_{max} must fall within 80% and 125%. It follows that according to the intersection–union principle (Berger, 1982), the type I error rate of ABE is still controlled under the nominal level of 5%. Therefore, there is no need for adjustment due to multiple PK measures.

4.6 Concluding Remarks

As indicated in Chapter 1, the FDA kicked off a critical path initiative to assist the sponsors in identifying the scientific challenges underlying the medical product pipeline problems. A critical path opportunities list was released in 2006 to bridge the gap between the quick pace of new biomedical discoveries and the slower pace at which those discoveries are currently developed into therapies. However, the assessment of BE for generic approval was not included until a year later. In May 2007, the FDA issued the critical path opportunities for generic drugs which lay out the opportunities as well as the challenges that are unique to the generic drug products. Note that the critical path opportunities for generic drugs were issued by the Office of Generic Drugs, Center for Drug Evaluation and Research. Consequently, the critical path opportunities for generic drugs are only confined to the traditional chemical drug products.

In pharmaceutical development, the concept of equivalence should not be limited to BE for the approval of generic drug products. The concept of equivalence can be applied to substantial equivalence for medical devices and biosimilarity for follow-on biologics (FOB). For medical devices, based on the risk of medical devices posed to the patient and/or user, the FDA categorized medical devices into three classes. Regulations for Class I devices require the general controls while the Class II devices require both general controls and special controls. On the other hand, because of higher risks, in addition to the general controls and special controls, the FDA requests that Class III devices require a premarket approval (PMA) to obtain marketing clearance. However, for Class I and II devices, the sponsor can make a premarket notification through a 510 (k) submission to the FDA. Under 510 (k), the new device must demonstrate that it is at least safe and effective as a legal U.S. market device or a predicate device. This concept of equivalence for the approval of medical devices under 510 (k) is referred to as substantial equivalence. According to the FDA, a device is considered substantially equivalent if it has either (1) the same intended use as the predicate and (2) the same technological characteristics as the predicate or (1) the same intended use as the predicate and (2) different technological characteristics and the information submitted to the FDA. Therefore, according to the submissions under 510 (k), as compared to the predicate, a device must demonstrate a two-sided equivalence in technological characteristics or a one-sided equivalence or non-inferiority in safety and effectiveness.

For the approval of biosimilars in the European Union (EU) community, the EMEA has issued a new guideline describing general principles for the approval of similar biological medicinal products, or biosimilars. The guideline is accompanied by several concept papers that outline areas in which the agency intends to provide more targeted guidance. Specifically, the concept papers discuss approval requirements for four classes of human recombinant products containing erythropoietin, human growth hormone, granulocyte-colony stimulating factor, and insulin. The guideline consists of a checklist of documents published to date relevant to data requirements for biological pharmaceuticals. It is not clear what specific scientific requirements will be applied to biosimilar applications. In addition, it is not clear how the agency will treat innovator data contained in the reference product dossiers. The guideline provides a useful summary of the biosimilar legislation and previous EU publications, and it also provides a few answers to the issues.

Note that very little literature on statistical methods for the assessment of (1) substantial equivalence for the approval of medical products and (2) biosimilarity of FOB can be found. In addition, even the selection of equivalence limits for the evaluation of substantial equivalence and biosimilarity of FOB has not been fully investigated or mentioned in the regulatory guidelines. More research in these areas is urgently needed. More details regarding the assessment of follow-on biologics can be found in Chapter 24 of this book.

5

Hypotheses for Clinical Evaluation and Significant Digits

5.1 Introduction

In clinical trials, a typical approach for clinical evaluation of the safety and efficacy of a test treatment is to first test for the null hypothesis of no treatment difference in efficacy based on clinical data collected under a valid trial design. The investigator would reject the null hypothesis of no treatment difference and then conclude the alternative hypothesis that there is a difference in favor of the test treatment under investigation. As a result, if there is a sufficient power for correctly detecting a clinically meaningful difference if such a difference truly exists, we claim that the test treatment is efficacious. The test treatment will be reviewed and approved by the regulatory agency if the recommended dose is well tolerated and there appear no safety concerns. In some cases, the regulatory agencies such as the United States FDA will issue a letter of approval pending a commitment for conducting large-scale long-term safety surveillance.

In practice, the intended clinical trial is always powered to achieve the study objective with a desired power (say 80%) at a prespecified level of significance (say 5%). However, the study based on a single primary endpoint (usually efficacy endpoint) may not be appropriate because one single primary efficacy endpoint may not be able to fully describe the performance of the treatment with respect to both the efficacy and safety under study. Statistically, the traditional approach based on a single primary efficacy endpoint for the clinical evaluation of both safety and efficacy is a conditional approach (i.e., conditional on safety performance). It should be noted that under the traditional (conditional) approach, the observed safety profile may not be of any statistical meaning (i.e., the observed safety profile could be by chance alone and is not reproducible). As a result, the traditional approach for the clinical evaluation of both efficacy and safety may have inflated the false positive rate of the test treatment in treating the disease under investigation.

In the past several decades, the traditional approach has been found to be inefficient as many drug products have been withdrawn from the market because of the risks to patients. Table 5.1 (reproduced from http://en.wikipedia.org/wiki/List_of_withdrawn_drugs) provides a list of (significant) withdrawn drugs between 1950 and 2010. As can be seen from Table 5.1, most drugs withdrawn from the market are due to safety concerns (risks to the patients). Usually this is prompted by unexpected adverse effects that were not detected during phase III clinical trials and were only apparent in the postmarketing surveillance data from the wider patient population. Note that the list of withdrawn drugs given in Table 5.1 was approved by the regulatory agencies such as the U.S. FDA and EMEA in European Community. Note that some of the drug products on the list were approved to be marketed in Europe but had not yet been approved by the FDA for marketing in the United States.

In addition to drug withdrawals, drug products may be recalled due to lack of good drug characteristics such as quality and stability. Table 5.2 summarizes the number of prescription and over-the-counter drugs that were recalled between the fiscal years of 2004 and 2005 for illustration purpose. Most of the drug recalls are due to or related to safety issues although some of the causes for recalls are due to failing to pass FDA inspection for stability testing and/or dissolution testing, which have an impact on the safety of the drug products currently on the marketplace. Thus, one of the controversial issues is whether the traditional (conditional) hypotheses testing approach (based on efficacy alone) for the evaluation of the safety and efficacy of a test treatment under investigation is appropriate.

In clinical trials, clinical results are often reported by rounding up the number to certain decimal places. Statistical inference obtained based on data with different decimal places may lead to different conclusions. Therefore, the selection of the number of decimal places could be critical if the treatment effect is of marginal significance. Thus, how many decimal places should be used for reporting the clinical results has become an interesting question to the investigators who conduct clinical trials at various phases of the clinical development. Chow (2000) introduced the concept of signal-noise for determining the number of decimal places for results obtained from clinical trials. The idea is to select the minimum number of decimal places in such a way that there is no statistically significant difference between the data set presented by using the minimum decimal places and any other data sets with more decimal places.

In the next section, several composite hypotheses which will take both efficacy and safety into consideration are proposed. In Section 5.3, for illustration purpose, statistical methods for testing the composite hypothesis that H_0: not NS versus H_a: NS are derived, where N represents testing for non-inferiority of the efficacy endpoint and S stands for superiority testing of the safety endpoint. Section 5.4 studies the impact on power and sample size calculation when switching from testing for a single hypothesis to

TABLE 5.1

Significant Withdrawals of Drug Products between 1950 and 2010

Drug Name	Withdrawn	Remarks
Thalidomide	1950s–1960s	Withdrawn because of risk of teratogenicity; returned to market for use in leprosy and multiple myeloma under FDA orphan drug rules
Lysergic acid diethylamide	1950s–1960s	Marketed as a psychiatric cure-all; withdrawn after it became widely used recreationally
Diethylstilbestrol	1970s	Withdrawn because of risk of teratogenicity
Phenformin and Buformin	1978	Withdrawn because of risk of lactic acidosis
Ticrynafen	1982	Withdrawn because of risk of hepatitis
Zimelidine	1983	Withdrawn worldwide because of risk of Guillain–Barré syndrome
Phenacetin	1983	An ingredient in "APC" tablet; withdrawn because of risk of cancer and kidney disease
Methaqualone	1984	Withdrawn because of risk of addiction and overdose
Nomifensine (Merital)	1986	Withdrawn because of risk of hemolytic anemia
Triazolam	1991	Withdrawn in the United Kingdom because of risk of psychiatric adverse drug reactions. This drug continues to be available in the United States
Temafloxacin	1992	Withdrawn in the United States because of allergic reactions and cases of hemolytic anemia, leading to three patient deaths
Flosequinan (Manoplax)	1993	Withdrawn in the United States because of an increased risk of hospitalization or death
Alpidem (Ananxyl)	1996	Withdrawn because of rare but serious hepatotoxicity
Fen-phen (popular combination of fenfluramine and phentermine)	1997	Phentermine remains on the market, dexfenfluramine and fenfluramine—later withdrawn as caused heart valve disorder
Tolrestat (Alredase)	1997	Withdrawn because of risk of severe hepatotoxicity
Terfenadine (Seldane)	1998	Withdrawn because of risk of cardiac arrhythmias; superseded by fexofenadine
Mibefradil (Posicor)	1998	Withdrawn because of dangerous interactions with other drugs
Etretinate	1990s	Risk of birth defects; narrow therapeutic index

(continued)

TABLE 5.1 (continued)

Significant Withdrawals of Drug Products between 1950 and 2010

Drug Name	Withdrawn	Remarks
Temazepam (Restoril, Euhypnos, Normison, Remestan, Tenox, Norkotral)	1999	Withdrawn in Sweden and Norway because of diversion, abuse, and a relatively high rate of overdose deaths in comparison to other drugs of its group. This drug continues to be available in most of the world including the United States, but under strict controls
Astemizole (Hismanal)	1999	Arrhythmias because of interactions with other drugs
Troglitazone (Rezulin)	2000	Withdrawn because of risk of hepatotoxicity; superseded by pioglitazone and rosiglitazone
Alosetron (Lotronex)	2000	Withdrawn because of risk of fatal complications of constipation; reintroduced in 2002 on a restricted basis
Cisapride (Propulsid)	2000s	Withdrawn in many countries because of risk of cardiac arrhythmias
Amineptine (Survector)	2000	Withdrawn because of hepatotoxicity, dermatological side effects, and abuse potential
Phenylpropanolamine (Propagest, Dexatrim)	2000	Withdrawn because of risk of stroke in women under 50 years of age when taken at high doses (75 mg twice daily) for weight loss
Trovafloxacin (Trovan)	2001	Withdrawn because of risk of liver failure
Cerivastatin (Baycol, Lipobay)	2001	Withdrawn because of risk of rhabdomyolysis
Rapacuronium (Raplon)	2001	Withdrawn in many countries because of risk of fatal bronchospasm
Rofecoxib (Vioxx)	2004	Withdrawn because of risk of myocardial infarction
Mixed amphetamine salts (Adderall XR)	2005	Withdrawn in Canada because of risk of stroke. See Health Canada press release. The ban was later lifted because the death rate among those taking Adderall XR was determined to be no greater than those not taking Adderall
Hydromorphone extended-release (Palladone)	2005	Withdrawn because of a high risk of accidental overdose when administered with alcohol
Pemoline (Cylert)	2005	Withdrawn from the U.S. market because of hepatotoxicity
Natalizumab (Tysabri)	2005–2006	Voluntarily withdrawn from the U.S. market because of risk of progressive multifocal leukoencephalopathy (PML). Returned to market in July 2006

TABLE 5.1 (continued)

Significant Withdrawals of Drug Products between 1950 and 2010

Drug Name	Withdrawn	Remarks
Ximelagatran (Exanta)	2006	Withdrawn because of risk of hepatotoxicity (liver damage)
Pergolide (Permax)	2007	Voluntarily withdrawn in the United States because of the risk of heart valve damage. Still available elsewhere
Tegaserod (Zelnorm)	2007	Withdrawn because of imbalance of cardiovascular ischemic events, including heart attack and stroke. Was available through a restricted access program until April 2008
Aprotinin (Trasylol)	2007	Withdrawn because of increased risk of complications or death; permanently withdrawn in 2008 except for research use
Lumiracoxib	2007–2008	Progressively withdrawn around the world because of serious side effects, mainly liver damage
Rimonabant (Accomplia)	2008	Withdrawn around the world because of risk of severe depression and suicide
Efalizumab (Raptiva)	2009	Withdrawn because of increased risk of PML; to be completely withdrawn from market by June 2009
Sibutramine (Reductil)	2010	Withdrawn in Europe because of increased cardiovascular risk

Source: Wikipedia, List of withdrawn drugs, http://en.wikipedia.org/wiki/List_of_withdrawn_drugs, 2010.

TABLE 5.2

Summary of Drug Recalls between 2004 and 2005

Fiscal Year	Prescription Drug	Over-the-Counter Drug
2004	215	71
2005	401	101

Source: Report to the Nation issued by CDER/FDA.

testing for a composite hypothesis. In clinical trials, clinical results are often reported by rounding up the number to certain decimal places. Statistical inference obtained based on data with different decimal places may lead to different conclusions. In Section 5.5, some statistical justification for Chow's proposal for determination of appropriate decimal places in observations obtained from clinical research is provided.

5.2 Hypotheses for Clinical Evaluation

In clinical trials, for the clinical evaluation of efficacy, commonly considered approaches include tests for hypotheses of superiority (S), non-inferiority (N), and (therapeutic) equivalence (E). For safety assessment, the investigator usually examines the safety profile in terms of adverse events and other safety parameters to determine whether the test treatment is either better (superiority), non-inferior (non-inferiority), or similar (equivalence) as compared to the control. As an alternative to the traditional approach, Chow and Shao (2002) suggest testing composite hypotheses that take both safety and efficacy into consideration. For illustration purpose, Table 5.3 provides a summary of all possible scenarios of composite hypotheses for the clinical evaluation of safety and efficacy of a test treatment under investigation.

Statistically, we would reject the null hypothesis at a prespecified level of significance and conclude the alternative hypothesis with a desired power. For example, the investigator may be interested in testing non-inferiority in efficacy and superiority in safety of a test treatment as compared to a control. In this case, we can consider testing the null hypothesis that H_0: not NS, where N denotes non-inferiority in efficacy and S represents superiority of safety. We would reject the null hypothesis and conclude the alternative hypothesis that H_a: NS, i.e., the test treatment is non-inferior to the active control agent and its safety is superior to the active control agent. To test the null hypothesis that H_0: not NS, appropriate statistical tests should be derived under the null hypothesis. The derived test statistics can then be evaluated for achieving the desired power under the alternative hypothesis. The selected sample size will ensure that the intended trial will achieve the study objectives of (1) establishing non-inferiority of the test treatment in efficacy and (2) showing superiority of the safety profile of the test treatment at a prespecified level of significance.

Note that the composite hypothesis problem described above is different from multiple comparisons. Multiple comparisons usually consist of a set of null hypotheses. The overall hypothesis is that all individual null hypotheses

TABLE 5.3

Composite Hypotheses for Clinical Evaluation

	Safety		
Efficacy	N	S	E
N	NN	NS	NE
S	SN	SS	SE
E	EN	ES	EE

Note: N, Non-inferiority; S, Superiority; E, Equivalence.

are true, and the alternative hypothesis is that at least one of the null hypotheses is not true. In contrast, when it comes to the composite hypothesis problem, the alternative hypothesis is that the test drug is non-inferior (N) in efficacy and superior (S) in safety. Then, the null hypothesis is not NS, i.e., the test drug is inferior in efficacy *or* the test drug is *not* superior in safety. In other words, the null hypothesis consists of three subsets of null hypothesis: first, the test drug is inferior in efficacy and superior in safety; second, the test drug is non-inferior in efficacy and *not* superior in safety; third, the test drug is inferior in efficacy *and not* superior in safety. It would be complicated to consider all these three subsets of null hypothesis. If the third subset of null hypothesis is considered, naturally the alternative hypothesis is that the test drug is either non-inferior in efficacy *or* superior in safety, which is different from the hypothesis that the test drug is non-inferior in efficacy *and* superior in safety.

It also should be noted that in the interest of controlling the overall type I error rate at the α level, appropriate α levels (say α_1 for efficacy and α_2 for safety) should be chosen. When switching from a single hypothesis testing to a composite hypothesis testing, an increase in sample size is expected.

5.3 Statistical Methods for Testing Composite Hypotheses of *NS*

For illustration purpose, consider the composite hypotheses that H_0: not NS versus H_a: NS in the clinical evaluation of a test treatment under investigation, where N represents the hypothesis for testing non-inferiority in efficacy and S stands for the hypothesis for testing superiority in safety (Chow and Lu, 2011). Let X and Y be the efficacy and safety endpoints, respectively. Assume that (X, Y) follows a bivariate normal distribution with mean (μ_X, μ_Y) and variance–covariance matrix Σ, i.e., where

$$\Sigma = \begin{pmatrix} \sigma_X^2 & \rho\sigma_X\sigma_Y \\ \rho\sigma_X\sigma_Y & \sigma_Y^2 \end{pmatrix}.$$

Suppose that the investigator is interested in testing non-inferiority in efficacy and superiority in safety of a test treatment as compared to a control (e.g., an active control agent). The corresponding composite hypotheses may be considered:

$$H_0 : \mu_{X1} - \mu_{X2} \leq -\delta_X \text{ or } \mu_{Y1} - \mu_{Y2} \leq \delta_Y \quad \text{versus}$$

$$H_1 : \mu_{X1} - \mu_{X2} > -\delta_X \text{ and } \mu_{Y1} - \mu_{Y2} > \delta_Y,$$

where

(μ_{X1}, μ_{Y1}) and (μ_{X2}, μ_{Y2}) are the means of (X, Y) for the test treatment and the control, respectively

δ_X and δ_Y are the corresponding non-inferiority margin and superiority margin

Note that δ_X and δ_Y are positive constants. If the null hypothesis is rejected based on a statistical test, we conclude that the test treatment is non-inferior to the control in the efficacy endpoint X, and is superior over the control in the safety endpoint Y.

To test the above composite hypotheses, suppose that a random sample of (X, Y) is collected from each treatment arm. In particular, $(X_{11}, Y_{11}), \ldots, (X_{1n_1}, Y_{1n_1})$ are i.i.d. $N((\mu_{X1}, \mu_{Y1}), \Sigma)$, which is the random sample from the test treatment, and $(X_{21}, Y_{21}), \ldots, (X_{2n_2}, Y_{2n_2})$ are i.i.d. $N((\mu_{X2}, \mu_{Y2}), \Sigma)$, which is the random sample from the control treatment. Let $\bar{X}_1$ and $\bar{X}_2$ be the sample means of X in the test treatment and the control, respectively. Similarly, $\bar{Y}_1$ and $\bar{Y}_2$ are the sample means of Y in the test treatment and the control, respectively. It can be verified that the sample mean vector $(\bar{X}_i, \bar{Y}_i)$ follows a bivariate normal distribution. In particular, $(\bar{X}_i, \bar{Y}_i)$ follows $N((\mu_{Xi}, \mu_{Yi}), n_i^{-1}\Sigma)$. Since $(\bar{X}_1, \bar{Y}_1)$ and $(\bar{X}_2, \bar{Y}_2)$ are independent bivariate normal vectors, it follows that $(\bar{X}_1 - \bar{X}_2, \bar{Y}_1 - \bar{Y}_2)$ is also normally distributed as $N((\mu_{X1} - \mu_{X2}, \mu_{Y1} - \mu_{Y2}), (n_1^{-1} + n_2^{-1})\Sigma)$. For simplicity, we assume that Σ is known, i.e., the values of parameters σ_X^2, σ_Y^2, and ρ are known. To test the composite hypothesis H_0 for both efficacy and safety, we may consider the following test statistics:

$$T_X = \frac{\bar{X}_1 - \bar{X}_2 + \delta_X}{\sqrt{(n_1^{-1} + n_2^{-1})\sigma_X^2}}, \quad T_Y = \frac{\bar{Y}_1 - \bar{Y}_2 - \delta_Y}{\sqrt{(n_1^{-1} + n_2^{-1})\sigma_Y^2}}.$$

Thus, we would reject the null hypothesis H_0 for large values of T_X and T_Y. Let C_1 and C_2 be the critical values for T_X and T_Y, respectively. Then, we have

$$P(T_X > C_1, \ T_Y > C_2) = P\left(U_X > C_1 - \frac{\mu_{X1} - \mu_{X2} + \delta_X}{\sqrt{\left(n_1^{-1} + n_2^{-1}\right)\sigma_X^2}}, \ U_Y > C_2 - \frac{\mu_{Y1} - \mu_{Y2} - \delta_Y}{\sqrt{\left(n_1^{-1} + n_2^{-1}\right)\sigma_Y^2}} \right),$$

$$(5.1)$$

where (U_X, U_Y) is the standard bivariate normal random vector, i.e., a bivariate normal random vector with zero means, unit variances, and a correlation coefficient of ρ.

Under the null hypothesis H_0 that $\mu_{X1} - \mu_{X2} \leq -\delta_X$ or $\mu_{Y1} - \mu_{Y2} \leq \delta_Y$, it can be shown that the upper limit of $P(T_X > C_1, T_Y > C_2)$ is the maximum of the two probabilities, i.e., $\max\{1 - \Phi(C_1), 1 - \Phi(C_2)\}$, where Φ is the cumulative distribution function of the standard normal distribution. A brief proof is as follows.

For given constants a_1 and a_2 and a standard bivariate normal vector $(U_X, U_Y) \sim N\left((0,\ 0),\ \begin{pmatrix} 1 & \rho \\ \rho & 1 \end{pmatrix}\right)$, we have

$$P(U_X > a_1,\ U_Y > a_2) = \frac{1}{2\pi\sqrt{1-\rho^2}} \int_{a_1}^{+\infty} \int_{a_2}^{+\infty} \exp\left\{-\frac{x^2 + y^2 - 2\rho xy}{2(1-\rho^2)}\right\} dy\, dx$$

$$= \frac{1}{\sqrt{2\pi}} \int_{a_1}^{+\infty} \exp\left\{-\frac{x^2}{2}\right\} \int_{a_2}^{+\infty} \frac{1}{\sqrt{2\pi(1-\rho^2)}} \exp\left\{-\frac{(y-\rho x)^2}{2(1-\rho^2)}\right\} dy\, dx$$

$$= 1 - \Phi(a_1) - \frac{1}{\sqrt{2\pi}} \int_{a_1}^{+\infty} \Phi\left(\frac{a_2 - \rho x}{\sqrt{1-\rho^2}}\right) \exp\left\{-\frac{x^2}{2}\right\} dx. \tag{5.2}$$

Since the joint distribution of (U_X, U_Y) is symmetric, (5.2) is also equal to

$$1 - \Phi(a_2) - \frac{1}{\sqrt{2\pi}} \int_{a_2}^{+\infty} \Phi\left(\frac{a_1 - \rho y}{\sqrt{1-\rho^2}}\right) \exp\left\{-\frac{y^2}{2}\right\} dy. \tag{5.3}$$

Based on (5.1), $P(T_X > C_1, T_Y > C_2)$ can be expressed by (5.2) and (5.3) with a_1 and a_2 replaced by

$$D_1 = C_1 - \frac{\mu_{X1} - \mu_{X2} + \delta_X}{\sqrt{\left(n_1^{-1} + n_2^{-1}\right)\sigma_X^2}} \quad \text{and} \quad D_2 = C_2 - \frac{\mu_{Y1} - \mu_{Y2} - \delta_Y}{\sqrt{\left(n_1^{-1} + n_2^{-1}\right)\sigma_Y^2}}$$

respectively. Under the null hypothesis H_0 that $\mu_{X1} - \mu_{X2} \leq -\delta_X$ or $\mu_{Y1} - \mu_{Y2} \leq \delta_Y$, it is true that either $D_1 \geq C_1$ or $D_2 \geq C_2$. Since the integrals in (5.2) and (5.3) are positive, it follows that $P(T_X > C_1, T_Y > C_2 \mid H_0) < \max(1 - \Phi(C_1), 1 - \Phi(C_2))$.

To complete the proof, we need to show for any $\varepsilon > 0, \delta_X$, and δ_Y (>0), and given values of other parameters, there exist values of $\mu_{X1} - \mu_{X2}$ and $\mu_{Y1} - \mu_{Y2}$ such that (5.2) is larger than $1 - \Phi(C_1) - \varepsilon$ and $1 - \Phi(C_2) - \varepsilon$. Let $\mu_{X1} - \mu_{X2} = -\delta_X$. Then (5.2) becomes

$$1 - \Phi(C_1) - \frac{1}{\sqrt{2\pi}} \int_{C_1}^{+\infty} \Phi\left(\frac{D_2 - \rho x}{\sqrt{1-\rho^2}}\right) \exp\left\{-\frac{x^2}{2}\right\} dx. \tag{5.4}$$

For $\rho > 0$, there exists a negative value K such that when $D_2 < K$, for any x in $[C_1, +\infty)$,

$$\Phi\left(\frac{D_2 - \rho x}{\sqrt{1 - \rho^2}}\right) < \varepsilon.$$

For sufficiently large $\mu_{Y1} - \mu_{Y2}$, it can happen that $D_2 < K$. Therefore, for sufficiently large $\mu_{Y1} - \mu_{Y2}$, (5.4) $> 1 - \Phi(C_1) - \varepsilon$. For $\rho \leq 0$, express the integral in (5.4) as $I_1 + I_2$, where

$$I_1 = \int_{C_1}^{E} \Phi\left(\frac{D_2 - \rho x}{\sqrt{1 - \rho^2}}\right) \exp\left\{-\frac{x^2}{2}\right\} dx \quad \text{and} \quad I_2 = \int_{E}^{+\infty} \Phi\left(\frac{D_2 - \rho x}{\sqrt{1 - \rho^2}}\right) \exp\left\{-\frac{x^2}{2}\right\} dx.$$

ε is chosen such that $I_2 \leq \int_{E}^{+\infty} \exp\{-x^2/2\} dx < 0.5\varepsilon$. The first inequality holds as the cumulative distribution is always ≤ 1. For a chosen value of ε, the argument for $\rho > 0$ can be applied to prove $I_1 < 0.5\varepsilon$ for sufficiently large $\mu_{Y1} - \mu_{Y2}$. Hence, $P(T_X > C_1, T_Y > C_2 | H_0)$ is greater than $1 - \Phi(C_1) - \varepsilon$ for $\mu_{X1} - \mu_{X2} = -\delta_X$ and sufficiently large $\mu_{Y1} - \mu_{Y2}$. Similarly, it can be proven that $P(T_X > C_1, T_Y > C_2 | H_0)$ is greater than $1 - \Phi(C_2) - \varepsilon$ for $\mu_{Y1} - \mu_{Y2} = \delta_Y$ and sufficiently large $\mu_{X1} - \mu_{X2}$. This completes the proof.

Therefore, the type I error of the test based on T_X and T_Y can be controlled at the level of α by appropriately choosing corresponding critical values of C_1 and C_2. Denote by z_α the upper α-percentile of the standard normal distribution. Then, the power function of the above test is $P(T_X > z_{\alpha_1}, T_Y > z_{\alpha_2})$, which can be calculated from (5.1) and the cumulative distribution function of the standard bivariate distribution.

5.4　Impact on Power and Sample Size Calculation

5.4.1　Fixed Power Approach

As indicated earlier, when switching from testing a single hypothesis (i.e., based on a single study endpoint such as the efficacy endpoint in clinical trials) to testing a composite hypothesis (i.e., based on two study endpoints such as both efficacy and safety endpoints in clinical trials), an increase in sample size is expected. Let X be the efficacy endpoint in clinical trials. Consider testing the following single non-inferiority hypothesis with a non-inferiority margin of δ_X:

$$H_{01}: \mu_{X1} - \mu_{X2} \leq -\delta_X \quad \text{versus} \quad H_{11}: \mu_{X1} - \mu_{X2} > -\delta_X.$$

Then, a commonly used test is to reject the null hypothesis H_{01} at the α level of significance if $T_X > z_\alpha$. The total sample size for concluding the test treatment is non-inferior to the control with $1 - \beta$ power if the difference of mean $\mu_{X1} - \mu_{X2} > - \delta_X$ is

$$N_X = \frac{(1+r)^2 (z_\alpha + z_\beta)^2 \sigma_X^2}{r(\mu_{X1} - \mu_{X2} + \delta_X)^2},$$

where $r = n_2/n_1$ is the sample size allocation ratio between the control and test treatment. Table 5.4 gives total sample size (N_X) for the test of non-inferiority based on the efficacy endpoint X and total sample size (N) for testing the composite hypothesis based on both efficacy endpoint X and safety endpoint Y, for various scenarios. In particular, we calculated sample sizes for $\alpha = 0.05$, $\beta = 0.20$, $\mu_{Y1} - \mu_{Y2} - \delta_Y = 0.3$, $r = 1$, and several values of $\Delta = \mu_{X1} - \mu_{X2} + \delta_X$ and other parameters. For a hypothesis of superiority of the test treatment in safety, i.e., the component with respect to safety in the composite hypothesis, the preceding specified values of type I error rate, power, and $\mu_{Y1} - \mu_{Y2} - \delta_Y$ and σ_Y require a total sample size $N_Y = 275$.

For many scenarios in Table 5.4, the total sample size N for testing the composite hypothesis is much larger than the sample size for testing non-inferiority in efficacy (N_X). However, it happens in some cases that they are the same or their difference is quite small. Actually N is associated with

TABLE 5.4

Comparison of Sample Size between Tests for Multiple Endpoints and Single Endpoint

σ_X	ρ	N_X	N	N/N_X	N_X	N	N/N_X	N_X	N	N/N_X
		$\Delta = 0.2$			$\Delta = 0.3$			$\Delta = 0.4$		
0.5	−1.0	155	304	1.96	69	276	4.00	39	275	7.05
	−0.5	155	303	1.95	69	276	4.00	39	275	7.05
	0.0	155	300	1.94	69	276	4.00	39	275	7.05
	0.5	155	289	1.86	69	275	3.99	39	275	7.05
	1.0	155	275	1.77	69	275	3.99	39	275	7.05
1.0	−1.0	619	647	1.05	275	381	1.39	155	304	1.96
	−0.5	619	646	1.04	275	381	1.39	155	303	1.95
	0.0	619	642	1.04	275	373	1.36	155	300	1.94
	0.5	619	629	1.02	275	352	1.28	155	289	1.86
	1.0	619	619	1.00	275	275	1.00	155	275	1.77
1.5	−1.0	1392	1392	1.00	619	647	1.05	348	433	1.24
	−0.5	1392	1392	1.00	619	646	1.04	348	432	1.24
	0.0	1392	1392	1.00	619	642	1.04	348	424	1.22
	0.5	1392	1392	1.00	619	629	1.02	348	402	1.16
	1.0	1392	1392	1.00	619	619	1.00	348	348	1.00

the sample sizes for individual testing of non-inferiority in efficacy (N_X) and of superiority in safety (N_Y), and the correlation coefficient (ρ) between X and Y. When a large difference exists between N_X and N_Y, N is quite close to the larger of N_X and N_Y, and changes little along with changes in ρ. In this numerical study, for $N_X = 69$ and 39 ($\ll 275$), N is mostly equal to 275; for $N_X = 1392$ and 619 ($\gg 275$), the difference between N and N_X is 0 or negligible compared with the size of N. In the preceding four scenarios, a change in correlation coefficient between X and Y has little impact on N. On the other hand, the larger of N_X and N_Y is not always close to N, especially when N_X and N_Y are close to each other. For example, in Table 5.4, when both values of N_X are equal to 275 ($=N_Y$), N is 352 for $\rho = 0.5$, and 373 for $\rho = 0$. In addition, the results in Table 5.4 suggest that the correlation coefficient between X and Y is unlikely to have great influence on N, especially when the difference between N_X and N_Y is quite substantial. The above findings are consistent with the following underlying "rule": when the two sample sizes are substantially different, taking N as the larger of N_X and N_Y will ensure that the powers of two individual tests for efficacy and safety are essentially 1 and $1 - \beta$, "resulting" in a power of $1 - \beta$ for testing the composite hypotheses; when N_X and N_Y are close to each other, taking N as the larger of N_X and N_Y will power the test of composite hypotheses at about $(1 - \beta)^2$. Therefore, a significant increment in N is required for achieving a power of $1 - \beta$.

5.4.2 Fixed Sample Size Approach

Based on the sample size in Table 5.4, the power of the test of the composite hypothesis H_0 was calculated with results presented in Table 5.5, where P is the power of the test of the composite hypothesis with N_X in Table 5.4. P_M is the power of the same test with max $(N_X, 275)$. With the sample size N_X, the power of the test of the composite hypothesis is always not greater than the target value 80% as N_X is always not larger than N in Table 5.4. In some cases where $\sigma_X = 1.5 > \sigma_Y = 1.0$, $N_X = N$. Hence the corresponding $P = 80\%$. However, P is less than 60% for many cases in our numerical study. The worst scenario is $P = 4.3\%$ when $N_X = 39$ for $\sigma_X = 0.5$, $\rho = -1$, and $\Delta = 0.4$. Therefore, the test for the composite hypothesis of both efficacy and safety using a sample size N_X for achieving a certain power in testing the hypothesis of efficacy only may not have enough power to reject the null hypothesis. Interestingly, testing the composite hypothesis with max(N_X, 275), the power P_M is close to the target value 80% in most scenarios. Some exceptions happen when N_X is close to 275 (corresponding to ($\Delta = 0.3, \sigma_X = 1.0$) and ($\Delta = 0.4$, $\sigma_X = 1.5$), such that a significant increment in sample size from max(N_X, 275) to N is required. This suggests taking N as the larger of the two sample sizes N_X and N_Y for testing the hypothesis of individual endpoints when one of the two is much larger, say, twofold larger than the other.

TABLE 5.5

Power (%) of Test of Composite Hypothesis

σ_X	ρ	$\Delta = 0.2$		$\Delta = 0.3$		$\Delta = 0.4$	
		P	P_M	P	P_M	P	P_M
0.5	−1.0	38.9	75.3	14.7	80.0	4.3	80.0
	−0.5	41.9	75.4	22.0	80.0	14.2	80.0
	0.0	47.1	76.2	27.7	80.0	19.2	80.0
	0.5	52.9	78.1	32.3	80.0	22.8	80.0
	1.0	58.8	80.0	34.5	80.0	23.9	80.0
1.0	−1.0	78.2	78.2	60.1	60.1	38.9	75.3
	−0.5	78.2	78.2	60.9	60.9	41.9	75.4
	0.0	78.6	78.6	64.0	64.0	47.1	76.2
	0.5	79.4	79.4	68.8	68.8	52.9	78.1
	1.0	80.0	80.0	80.0	80.0	58.8	80.0
1.5	−1.0	80.0	80.0	78.2	78.2	67.6	67.6
	−0.5	80.0	80.0	78.2	78.2	68.0	68.0
	0.0	80.0	80.0	78.6	78.6	70.1	70.1
	0.5	80.0	80.0	79.4	79.4	73.7	73.7
	1.0	80.0	80.0	80.0	80.0	80.0	80.0

5.4.3 Remarks

The traditional approach for the clinical evaluation of a test treatment under investigation is to power the study based on an efficacy endpoint. The test treatment is considered approvable if its safety and tolerability are acceptable provided that the efficacy has been established. In practice, in the interest of controlling the overall type I error rate at a prespecified level of significance, the type I error rate may be adjusted for multiple comparisons. It, however, should be noted that the overall type I error rate may be controlled at the risk of (1) decreasing the power and (2) increasing the sample size when switching from testing a single hypothesis (for efficacy) to testing a composite hypothesis (for both efficacy and safety).

In this chapter, for illustration purpose, we assume that the two study endpoints follow a bivariate normal distribution. In practice, both efficacy and safety endpoints could be either a continuous variable, a binary response, or time-to-event data. A similar idea can be applied to determine the impact on power and sample size calculation when switching from testing a single hypothesis to testing a composite hypothesis. It, however, should be noted that closed forms for the relationships of powers and formulas for sample size calculation between the single hypothesis and the composite hypothesis may not exist. In this case, clinical trial simulation may be useful.

5.5 Significant Digits

In practice statistical inference obtained based on data with different decimal places may lead to different conclusions. As an example, consider a parallel bioequivalence (BE) study. Suppose that there are 24 subjects in the group of test drug and 24 subjects in the group of reference drug. The data are given in Table 5.6. From the BE study results given in Table 5.7, it can be seen that keeping a different number of decimal digits can lead to different conclusions. Thus, the selection of the number of decimal places could be critical if the treatment effect is of marginal significance. Chow (2000) introduced the concept of signal-noise for determining the number of decimal places for results obtained from clinical trials. The idea is to select the

TABLE 5.6

Bioequivalence Example Data

X	X_0	X_1	X_2	Y	Y_0	Y_1	Y_2
1.169577	1	1.2	1.17	1.0722791	1	1.1	1.07
1.251990	1	1.3	1.25	1.0348811	1	1.0	1.03
1.449081	1	1.4	1.45	0.9020537	1	0.9	0.90
1.205818	1	1.2	1.21	1.1196368	1	1.1	1.12
1.355457	1	1.4	1.36	0.9736662	1	1.0	0.97
1.285863	1	1.3	1.29	1.1360977	1	1.1	1.14
1.519270	2	1.5	1.52	0.8531594	1	0.9	0.85
1.230438	1	1.2	1.23	1.1239591	1	1.1	1.12
1.374791	1	1.4	1.37	1.0642288	1	1.1	1.06
1.302860	1	1.3	1.30	0.9156539	1	0.9	0.92
1.396263	1	1.4	1.40	0.9044889	1	0.9	0.90
1.507581	2	1.5	1.51	0.9894644	1	1.0	0.99
1.337749	1	1.3	1.34	1.0281070	1	1.0	1.03
1.222744	1	1.2	1.22	0.8584933	1	0.9	0.86
1.235640	1	1.2	1.24	1.0074020	1	1.0	1.01
1.302359	1	1.3	1.30	0.9131539	1	0.9	0.91
1.379500	1	1.4	1.38	0.9563392	1	1.0	0.96
1.295147	1	1.3	1.30	1.2159481	1	1.2	1.22
1.376740	1	1.4	1.38	1.1442079	1	1.1	1.14
1.376414	1	1.4	1.38	1.0128952	1	1.0	1.01
1.321817	1	1.3	1.32	0.9561896	1	1.0	0.96
1.222626	1	1.2	1.22	0.8718494	1	0.9	0.87
1.140910	1	1.1	1.14	0.9620998	1	1.0	0.96
1.169492	1	1.2	1.17	0.9487145	1	0.9	0.95

Note: X, the original data from test drug; X_i, the original data with i decimal digit; Y, the data from reference drug; Y_i, the data with i decimal digit.

TABLE 5.7

Bioequivalence Study

Significant Digits	Confidence Interval	BE Limit	BE Result (Y/N)
0	(−0.013, 0.180)	(−0.2, 0.2)	Y
1	(0.261, 0.356)	(−0.2, 0.2)	N
2	(0.263, 0.362)	(−0.2, 0.2)	N

minimum number of decimal places in such a way that there is no statistically significant difference between the data set presented by using the minimum decimal places and any other data sets with more decimal places. In what follows, Chow's proposal is briefly described.

5.5.1 Chow's Proposal

The number of significant decimal digits of a given data set obtained from an analytical experiment is defined as the minimum number of decimal places of the data set which satisfies the following two conditions. First, the data set with the minimum number of decimal places will achieve the desired accuracy and precision. Second the data set with the minimum number of decimal places is not statistically distinguishable with those data sets with more decimal places than the minimum number of decimal places. In other words, the data set with significant decimal digits is not significantly different from those data sets where the number of decimal places exceeds the number of significant decimal digits.

Let X be a continuous random variable and X^* be its truncated value with d decimal digits. We would claim that X^* is not statistically different from X if we fail to reject the following null hypothesis at the α level of significance:

$$H_0 : \mu_X = \mu_{X^*} \quad \text{versus} \quad H_a : \mu_X \neq \mu_{X^*}, \tag{5.5}$$

where μ_X and μ_{X^*} are the population means for X and X^*, respectively. When X and X^* are not statistically distinguishable, the d decimal digits are considered significant decimal digits. Suppose X is a continuous random variable with standard deviation σ and X^* is its truncated value after rounding up to the dth decimal place. Then the maximum possible error due to the truncation would be less than 10^{1-d}. As an example, if $d = 3$, the smallest and largest values for a given number with three decimal places are $a.bc0$ and $a.bc9$, respectively. Hence, the maximum possible error is less than 0.01, which is 10^{-2}. Here −2 is obtained as $-2 = 1 - d = 1 - 3$ intuitively, if this worst-case error is small *enough*, the distortion of the distribution

TABLE 5.8

Significant Decimal Digits for Various
Selections of δ Given σ

	δ (%)				
α	1	5	10	15	20
0.01	4	4	3	3	3
0.10	3	3	2	2	2
0.50	3	2	2	2	1
1.00	2	2	1	1	1
2.00	2	1	1	1	1

due to the rounding error would be negligible. But the question is how small would be considered *enough*? An idea is to apply the concept of signal-noise in quality control and assurance to compare this error with X's standard deviation σ. The significant digits can then be chosen by taking the first d digits such that

$$\frac{10^{1-d}}{\sigma} < \delta' \quad \text{if and only if} \quad \frac{10^{-d}}{\sigma} < \delta'/10 = \delta,$$

where δ is a constant, which is to be chosen such that the truncated observation X^* is not statistically different from X at the α level of significance. In practice, a conventional choice of δ is δ = 10%. To provide a better understanding of the proposed procedure, the results for various choices of δ given σ are summarized in Table 5.8. As can be seen from Table 5.8, a smaller δ would require more decimal places to be used in order to achieve the desired accuracy and precision. Table 5.8 also indicates that more decimal places are needed for a smaller σ value.

5.5.2 Statistical Justification

Without loss of generality, we assume X follows a normal distribution with mean μ_X and variance σ^2, i.e., $X \sim N(\mu_X, \sigma^2)$. By proper truncation, X^* is still approximately normally distributed with mean μ_{X^*} and variance σ^2, where μ_{X^*} may be different from μ_X due to the rounding error. The following two-sample t can be used to test the null hypothesis given in (5.5)

$$T = \frac{\sqrt{n}(\bar{X} - \bar{X}^*)}{\sqrt{s_X^2 + s_{X^*}^2}},$$

where s_X^2 and $s_{X^*}^2$ are sample standard deviations of X and X^*, respectively. Under the null hypothesis that $H_0: \mu_X = \mu_{X^*}$, the two-sample T statistic follows a t distribution with $2(n-1)$ degrees of freedom. We reject the null hypothesis if $|T| > t_{\alpha/2,2(n-1)}$, where $t_{\alpha/2,2(n-1)}$ is the $(1 - \alpha/2)$th quantile for a t distribution with $2(n-1)$ degrees of freedom. Under the alternative hypothesis that $H_a: \mu_X \neq \mu_{X^*}$, the t statistic can be written as

$$T = \frac{\sqrt{n}(\bar{X} - \bar{X}^*)}{\sqrt{s_X^2 + s_{X^*}^2}} = \frac{\sqrt{n/2}[(\bar{X} - \mu_X)/\sigma - (\bar{X}^* - \mu_{X^*})/\sigma] + \sqrt{n/2}(\mu_X - \mu_{X^*})/\sigma}{\sqrt{s_X^2/2\sigma^2 + s_{X^*}^2/2\sigma^2}}$$

$$= \frac{N(0,1) + \delta}{\chi_{2(n-1)}^2/2(n-1)} \sim t_{2(n-1)}(\delta),$$

where $t_{2(n-1)}(\delta)$ denotes a t distribution with the noncentrality parameter of

$$\delta = \sqrt{\frac{n}{2}}\left(\frac{\mu_X - \mu_{X^*}}{\sigma}\right). \tag{5.6}$$

When $|\delta|$ is smaller, there is a lower probability that X^* will be different from X under t-test. On the other hand, since X^* is rounded at the dth decimal places, the maximum possible error due to truncation would be less than $10^{1-d} \geq |\mu_X - \mu_{X^*}|$. So a small value of $10^{-d}/\sigma$ would guarantee that X^* is not significantly different from X. The above argument can be applied similarly to a more general situation where a transformation is performed. Let $f(x)$ be the function of transformation of X. In this case, the hypotheses of interest become

$$H_0: f(\mu_X) = f(\mu_{X^*}) \quad \text{versus} \quad H_a: f(\mu_X) \neq f(\mu_{X^*}).$$

By Taylor's expansion, we have

$$\sqrt{n}(f(\bar{X}) - f(\bar{X}^*)) \approx f'(\mu_X)\sqrt{n}(\bar{X} - \bar{X}^*),$$

which approximately follows a normal distribution with mean $\sqrt{n}f'(\mu_X)(\mu_X - \mu_{X^*})$ and variance $2f'^2(\mu_X)\sigma^2$. As a result, the above null hypothesis can be tested by the following statistic:

$$T_f = \frac{\sqrt{n}f((\bar{X}) - f(\bar{X}^*))}{f'(\mu_X)\sqrt{s_X^2 + s_{X^*}^2}}.$$

Under the null hypothesis, T_f approximately follows a t distribution with $2(n - 1)$ degrees of freedom. Under the alternative hypothesis, T_f can be written as

$$T_f \approx \frac{N(\sqrt{n}\,f'(\mu_X)(\mu_X - \mu_{X^*}), 2f'^2(\mu_X)\sigma^2)}{f'(\mu_X)\sqrt{s_X^2 + s_{X^*}^2}}$$

$$= \frac{N\left(\dfrac{\sqrt{n}(\mu_X - \mu_{X^*})}{\sqrt{2}\sigma}, 1\right)}{\sqrt{\dfrac{s_X^2}{2\sigma^2} + \dfrac{s_{X^*}^2}{2\sigma^2}}} = \frac{N(0,1) + \delta}{\chi_{2(n-1)}^2 / 2(n-1)} \sim t_{2(n-1)}(\delta),$$

where δ is still the same noncentrality parameter as defined in (5.6). So if we choose significant digits properly, we can guarantee δ will be small and the probability that X^* is statistically different from X will be small as well. This shows that the proposed procedure works as well for data after transformation. To illustrate the use of the proposed procedure for transformed data, consider a log transformation, i.e., $f(x) = \log(x)$. Thus, the hypotheses become

$$H_0 : \log(\mu_X) = \log(\mu_{X^*}) \quad \text{versus} \quad H_a : \log(\mu_X) \neq \log(\mu_{X^*}).$$

Then $f'(\mu_X) = 1/\mu_X$ and the test statistic is given by

$$T_f = \frac{\sqrt{n}[\log(\bar{X}) - \log(\bar{X}^*)]}{\dfrac{1}{\mu_X}\sqrt{s_X^2 + s_{X^*}^2}} \sim t_{2(n-1)}(\delta).$$

A numerical study is conducted to demonstrate the use of the proposed procedure. Thirty analytical results were generated from $N(\pi, 0.01)$, which are given in Table 5.9. For convenience's sake, we keep six decimal digits as the original values. If we choose δ to be equal to 10%, we have

$$\frac{10^{-d}}{\sigma} = \frac{10^{-d}}{0.01} \leq 0.1.$$

It can be seen that the minimum number of d that satisfies the above expression is $d = 3$. Therefore, the number of significant decimal digits is chosen to be 3. Now consider four data sets $X_{ji}, j = 1, 2, 3, 4$, which are truncated at the jth decimal places, respectively. Then a two-sample t-test is performed to test

TABLE 5.9

Simulation Data Set for Two-Sample t-Test

i	X_i	X_{1i}	X_{2i}	X_{3i}	X_{4i}
1	3.145714	3.1	3.15	3.146	3.1457
2	3.140959	3.1	3.14	3.141	3.1410
3	3.141432	3.1	3.14	3.141	3.1414
4	3.127617	3.1	3.13	3.128	3.1276
5	3.142035	3.1	3.14	3.142	3.1420
6	3.146685	3.1	3.15	3.147	3.1467
7	3.146124	3.1	3.15	3.146	3.1461
8	3.138408	3.1	3.14	3.138	3.1384
9	3.125891	3.1	3.13	3.126	3.1259
10	3.136696	3.1	3.14	3.137	3.1367
11	3.133587	3.1	3.13	3.134	3.1336
12	3.158443	3.2	3.16	3.158	3.1584
13	3.140589	3.1	3.14	3.141	3.1406
14	3.128415	3.1	3.13	3.128	3.1284
15	3.149534	3.1	3.15	3.150	3.1495
16	3.153279	3.2	3.15	3.153	3.1532
17	3.147673	3.1	3.15	3.148	3.1477
18	3.140493	3.1	3.14	3.140	3.1405
19	3.150542	3.2	3.15	3.151	3.1505
20	3.123488	3.1	3.12	3.123	3.1235
21	3.161004	3.2	3.16	3.161	3.1610
22	3.140658	3.1	3.14	3.141	3.1407
23	3.151263	3.1	3.15	3.151	3.1512
24	3.124985	3.1	3.12	3.125	3.1250
25	3.140625	3.1	3.14	3.141	3.1406
26	3.168811	3.2	3.17	3.169	3.1688
27	3.159006	3.2	3.16	3.159	3.1590
28	3.143139	3.1	3.14	3.143	3.1431
29	3.123467	3.1	3.12	3.123	3.1235
30	3.146950	3.1	3.14	3.147	3.1470

if $X_{1i}, X_{2i}, X_{3i}, X_{4i}$ are significantly different from one another and are significantly different from the original X_i. The results are summarized in Table 5.10, from which we can see that X_{1i} are significantly different from the rest of the data sets. This shows that the rounding error can alter the distribution significantly. The results also indicate that X_{3i} is not significantly different from X_{4i}. It shows that the proposed procedure works well. It, however, should be noted that X_{2i} is also not significantly different from X_{3i} and X_{4i}. This indicates that the conventional choice of $\delta = 10\%$ may be conservative in this case.

TABLE 5.10

Pair-Wise Comparisons

Comparison	t-Statistic	p-Value
X_i versus X_{1i}	4.138	<0.001
X_i versus X_{2i}	0.116	0.908
X_i versus X_{3i}	0.008	0.994
X_i versus X_{4i}	0.003	0.997
X_{1i} versus X_{2i}	4.072	<0.001
X_{1i} versus X_{2i}	4.140	<0.001
X_{1i} versus X_{3i}	4.137	<0.001
X_{2i} versus X_{3i}	0.123	0.603
X_{2i} versus X_{4i}	0.112	0.911
X_{3i} versus X_{4i}	0.011	0.991

5.6 Concluding Remarks

Statistical justification of the proposed procedure for determining the number of significant decimal digits in observations obtained from studies conducted in analytical research was made under the assumption of normality. In practice, the observed analytical results may be described better by other distributions such as the Weibull distribution for dissolution results of the oral solid dosage form of a drug product. In this case, a similar concept can be carried out to provide a valid statistical justification. In many cases, log transformation is often considered for a better description or interpretation of the analytical results. For example, area under the plasma concentration–time curve (AUC) and time to achieve maximum concentration (C_{max}) in the studies of bioavailability and BE are known to be skewed to the right. As a result, a log transformation is recommended. In this case, the proposed procedure is useful for determining the number of significant decimal digits to maintain a certain degree of accuracy and precision for the assessment of BE. For the presentation of the analytical results, descriptive statistics such as mean, standard deviation, minimum, maximum, range, relative standard deviation (RSD) or coefficient of variation (CV) and statistical inferences such as confidence intervals and p-values are usually obtained. In practice, it is always a concern as to how many significant decimal digits should be used for descriptive statistics and statistical inferences to maintain the desired degree of accuracy and precision. In the interest of consistency, it is recommended that the same number of significant decimal digits be used for descriptive statistics and statistical inferences obtained from the analytical results.

In some cases, the analytical results may be expressed in a scientific form (e.g., 1.32×10^5 or 9.2×10^{-7}). The proposed procedure can be applied to its significant part (i.e., 1.32 for 1.32×10^5 or 9.2 for 9.2×10^{-7}) or its log (base 10) transformation. When analytical results involve different data sets, it is suggested that each data set keep its own significant decimal digits as determined by its standard deviation to maintain the same degree of accuracy and precision. A typical example is a dose proportionality study. The purpose of a dose proportionality study is usually to show that there is a linear relationship between dose and AUC within a given range. In other words, with a doubled dose, the AUC value is expected to be doubled. However, a high dose will generally produce a large variability in AUC values. As a result, low dose, median dose, and high dose are expected to have a different number of significant decimal digits to achieve the same degree of accuracy and precision. In the interest of keeping the same number of significant decimal digits, we may consider the AUC values adjusted for dose and then apply the proposed procedure to determining the number of significant decimal digits.

6

Instability of Sample Size Calculation

6.1 Introduction

In clinical trials, a pre-study power analysis for sample size calculation (estimation or determination) is often performed based on either (1) information obtained from small-scale pilot studies with limited number of subjects or (2) guess based on the best knowledge of the investigator (with or without scientific justification). The observed data and/or the investigator's best guess could be far from the truth. The deviation may bias the sample size calculation for reaching the desired power for achieving the study objectives at a prespecified level of significance. Sample size calculation is a key to the success of pharmaceutical/clinical research and development. Thus, how to select the minimum sample size for achieving the desired power at a prespecified significance level has become an important question for clinical scientists (Chow and Liu, 1998b; Chow et al., 2002b). A study without a sufficient number of subjects cannot guarantee the desired power (i.e., the probability of correctly detecting a clinically meaningful difference). On the other hand, an unnecessarily large sample size could be quite a waste to the limited resources.

In order to determine the minimum sample size required for achieving a desired power, one needs to have some information regarding study parameters such as variability associated with the observations and the difference (e.g., treatment effect) that the study is designed to detect. In practice, it is well recognized that sample size calculation depends upon the assumed variability associated with the observation, which is often unknown. Thus, the classical pre-study power analysis for sample size calculation based on information obtained from a small pilot study (with large variability) could vary widely and hence be unstable depending upon the sampling variability. As a result, one of the controversial issues regarding sample size calculation is the stability (sensitivity or robustness) of the obtained sample size. To overcome the instability of sample size calculation, alternatively, Lee et al. (2008) suggested that a bootstrap-median approach be considered to select a stable (required minimum) sample size. Such an improved stable sample size can be derived theoretically by the method of an Edgeworth-type expansion.

Lee et al. (2008) showed that the bootstrap-median approach performs quite well for providing a stable sample size in a clinical trial through an extensive simulation study.

It should be noted that procedures used for sample size calculation could be very different from one another according to different study objectives and hypotheses (e.g., testing for equality, testing for superiority, or testing for non-inferiority/equivalence) and different data types (e.g., continuous, binary, and time-to-event). For example, see Lachin and Foulkes (1986), Lakatos (1986), Wang and Chow (2002), Wang et al. (2002a), and Chow and Liu (2008). For a good introduction and summary, one can refer to Chow, Shao, and Wang (2008b). In this chapter, for simplicity, we will focus on the most commonly seen situation where the primary response is continuous and the hypotheses of interest are about the mean under the normality assumption. Most of our discussions thereafter focus on the one sample problem for the purpose of simplicity. However, the extension to the two-sample problem is straightforward.

The remainder of this chapter is organized as follows. In the next section, the classical sample size calculation is given. The instability of the classical sample size calculation and a proposed bootstrap-median approach are described in Section 6.3. Section 6.4 summarizes results from a simulation study. An example is discussed in Section 6.5. Section 6.6 provides some concluding remarks.

6.2 Sample Size Calculation

For simplicity and illustration purposes, consider the one-sample problem. Suppose there are a total of n independent and identically distributed responses from a clinical study. These responses are assumed to follow a normal distribution with mean μ and variance σ^2. Suppose that one of the study objectives is to detect a clinically meaningful difference, denoted by $\Delta = \mu - \mu_0$, where μ_0 is a prespecified reference point. Without loss of generality, we assume that μ_0 is zero, which implies that $\Delta = \mu$. Then, the one-sample t-test or the approximate z-test can be used to test the null hypothesis that $H_0: \mu = 0$. Under the alternative hypothesis that $H_a: \mu \neq 0$ and a significance level of α, the minimum sample size needed for achieving the desired power of $(1 - \beta)$ can be obtained as follows:

$$n_{\text{ideal}} = (z_{\alpha/2} + z_\beta)^2 \left[\frac{\sigma^2}{\mu^2} \right], \tag{6.1}$$

where z_α is the upper α-quantile of a standard normal distribution. For a detailed discussion about the above formula, one can refer to Chow et al. (2008b).

Ideally, if the value of μ and σ^2 are known, then the formula (6.1) can be used to determine the minimum sample size. In practice, however, the parameters μ and σ^2 are often unknown. Thus, a small pilot study is usually conducted to obtain information about the unknown parameters. Assume that a researcher conducts a small pilot study and obtains a small number of responses (say n_0) denoted by x_i, $i = 1, \ldots, n_0$. Based on the pilot study, sample mean $\hat{\mu} = \bar{x}$ and sample variance s^2 can be obtained and used to estimate the clinical efficacy μ and the associated variability σ^2. It is a common practice to replace the unknown parameters in (6.1) by its corresponding estimates to produce the following sample size estimator:

$$\hat{n} = (z_{\alpha/2} + z_{\beta})^2 \left[\frac{s^2}{\bar{x}^2} \right]. \tag{6.2}$$

In practice, formula (6.2) usually performs quite satisfactorily for a sufficiently large pilot sample size. However, if the size of the pilot study is relatively small, then the performance of (6.2) could be relatively instable and biased.

6.3 Instability and Bootstrap-Median Approach

In this section, the approximated sampling distribution and asymptotic bias of $s^2/\bar{x}^2$ under the normal population are derived to assess instability of classical sample size calculation. In addition, a bootstrap-median approach suggested by Lee et al. (2008) for a stable sample size determination is introduced.

6.3.1 Instability of Sample Size Calculation

By (6.2), sample size can be determined by the value of $s^2/\bar{x}^2$ at a prespecified level of significance. Thus, the stability of the traditional sample size formula depends upon the stability of $s^2/\bar{x}^2$. To provide a better understanding, the technique of an Edgeworth-type expansion is applied to approximate the sampling distribution of $s^2/\bar{x}^2$. From the sampling distribution of $s^2/\bar{x}^2$, the instability of the obtained sample size in terms of its bias can be studied when the size of the pilot study is relatively small.

Following the idea of Breunig (2001), we can approximate the finite sample distribution of $\hat{\theta} = s^2/\bar{x}^2$ by an Edgeworth-type expansion on the order of $n^{-1/2}$. Suppose $x_1, \ldots, x_n$ are independent samples from the normal distribution with a population mean ξ and variance σ^2. The parameter of interest is the population squared coefficient of variation ($CV^2 = \sigma^2/\xi^2$). The sample squared CV is defined as the plug-in estimator such that $\hat{\theta} = s^2/\bar{x}^2$, where $s^2 = \sum_i (x_i - \bar{x})^2/n$ is sample variance and $\bar{x}$ is sample mean. In practice, $\hat{\sigma}^2 = (n/n-1)s^2$ can be used to construct the sample squared CV. For convenience's sake, consider the superscript of the vector $x^{(i)}$ indicating the ith element of the real vector x. Note that the distribution of the standardized quantity $S_n = n^{1/2}(\hat{\theta} - \theta)$ can be written as

$$P\left[n^{1/2}(\hat{\theta} - \theta) \le x \right] = P\left[n^{1/2}\left(\frac{\sum_i x_i^2/n}{\bar{x}^2} - \frac{\sigma^2 + \xi^2}{\xi^2} \right) \le x \right].$$

Let $\theta = (\sigma^2 + \xi^2)/\xi^2$. Thus, the estimator of θ is given by $\hat{\theta} = (\sum_i x_i^2/n)/\bar{x}^2$. Define the real-valued function f on R^2 such that $f(w) = f(w^{(1)}, w^{(2)}) = w^{(2)}/(w^{(1)})^2$. Let $W = (X, X^2)$, $\bar{W} = (\sum_i x_i/n, \sum_i x_i^2/n)$, and $\mu = E(W) = (\xi, \sigma^2 + \xi^2)$. Then, $\theta = f(\mu)$ and $\hat{\theta} = f(W)$. Also put $Z = n^{1/2}(\bar{W} - \mu)$, and now by Taylor expansion we have the following expansion of $S_n = n^{1/2}(\hat{\theta} - \theta)$:

$$n^{1/2}(\hat{\theta} - \theta) = \sum_{i=1}^{2} f_i Z^{(i)} + n^{-1/2} \frac{1}{2} \sum_{i=1}^{2} \sum_{j=1}^{2} f_{ij} Z^{(i)} Z^{(j)} + O_p(n^{-1}),$$

where $f_{i_1, i_2, \ldots, i_p} = (\partial^p / \partial w^{(i_1)} w^{(i_2)} \ldots w^{(i_p)} f(w)) |_{w = \mu}$.

Put

$$\mu_{i_1, i_2, \ldots, i_p} = E\{(\bar{W} - \mu)^{(i_1)}, \ldots, (\bar{W} - \mu)^{(i_p)}\}$$

and define the pth central moment of X as $m_p = E(X - \xi)^p$ and the standardized pth central moment of X as $\gamma_p = E(X - \xi)^p/\sigma^p$. Then $\mu_i = 0$ for each i,

$$E(Z^{(i)} Z^{(j)}) = \mu_{ij},$$

$$E(Z^{(i)} Z^{(j)} Z^{(k)}) = n^{-1/2} \mu_{ijk},$$

$$E(Z^{(i)} Z^{(j)} Z^{(k)} Z^{(l)}) = \mu_{ij}\mu_{kl} + \mu_{ik}\mu_{jl} + \mu_{il}\mu_{jk} + O(n^{-1}).$$

Then,

$$E(S_n) = n^{-1/2} \frac{1}{2} \sum_{i=1}^{2} \sum_{j=1}^{2} f_{ij}\mu_{ij} + O(n^{-1}),$$

$$E(S_n^2) = n^{-1/2} \frac{1}{2} \sum_{i=1}^{2} \sum_{j=1}^{2} f_i f_j \mu_{ij} + O(n^{-1}),$$

$$E(S_n^3) = n^{-1/2} \frac{1}{2} \sum_{i=1}^{2} \sum_{j=1}^{2} \sum_{k=1}^{2} f_i f_j f_k \mu_{ijk}$$

$$+ \frac{3}{2} \sum_{i=1}^{2} \sum_{j=1}^{2} \sum_{k=1}^{2} \sum_{l=1}^{2} f_i f_j f_k f_l (\mu_{ij}\mu_{kl} + \mu_{ik}\mu_{jl} + \mu_{il}\mu_{jk})\} + O(n^{-1}).$$

The asymptotic expansions of the three cumulants of S_n are

$$\kappa_{1,n} = E(S_n) = n^{-1/2} A_1 + O(n^{-1}),$$

$$\kappa_{2,n} = E(S_n^2) - (E(S_n))^2 = \tau^2 + O(n^{-1}),$$

$$\kappa_{3,n} = E(S_n^3) - 3E(S_n^2)E(S_n) + 2(E(S_n))^3 = n^{-1/2} A_2 + O(n^{-1}),$$

where

$$\tau^2 = \sum_{i=1}^{2} \sum_{j=1}^{2} f_i f_j \mu_{ij},$$

$$A_1 = \frac{1}{2} \sum_{i=1}^{2} \sum_{j=1}^{2} f_{ij} \mu_{ij},$$

$$A_2 = \sum_{i=1}^{2} \sum_{j=1}^{2} \sum_{k=1}^{2} f_i f_j f_k \mu_{ijk} + 3 \sum_{i=1}^{2} \sum_{j=1}^{2} \sum_{k=1}^{2} \sum_{l=1}^{2} f_i f_j f_{kl} \mu_{ik}\mu_{jl}.$$

By using arguments in Hall (1992), we now have a one-term Edgeworth expansion of $S_n = n^{1/2}(\hat{\theta} - \theta)$ such that

$$P\left(\frac{n^{1/2}(\hat{\theta} - \theta)}{\tau} \leq x \right) = \Phi(x) + n^{-1/2} p_1(x)\phi(x) + O(n^{-1}),$$

where

$$p_1(x) = -\left\{ A_1 \tau^{-1} + \frac{1}{6} A_2 \tau^{-3}(x^2 - 1) \right\}.$$

In particular, if the population distribution is the normal distribution, we have

$$\gamma_3 = \gamma_5 = 0, \quad \gamma_4 = 3, \quad \text{and} \quad \gamma_6 = 15.$$

Therefore, under the normal distribution, after the tedious calculation involving cumulants we have

$$\tau^2 = 4\theta^3 + 2\theta^2, \quad A_1 = 3\theta^2 - \theta, \quad A_2 = 72\theta^5 + 48\theta^4 + 8\theta^3.$$

Thus, the asymptotic bias of $\hat{\theta}$ up to the first order can be obtained from the expansion of $E(S_n)$ above and is given by

$$E(\hat{\theta}) - \theta = n^{-1}(3\theta^2 - \theta) + O(n^{-2}).$$

Also, by Fisher–Cornish inversion, the α-quantile x_α of S_n/τ (i.e., $P(S_n/\tau \leq x_\alpha) = \alpha$) has the following expansion:

$$x_\alpha = z_\alpha - n^{-1/2}p_1(z_\alpha) + O(n^{-1}),$$

where z_α is the α-quantile of standard normal distribution.

Based on the above discussion, the asymptotic bias of $E(\hat{\theta})$ is given by

$$\text{Bias}(\hat{\theta}) = E(\hat{\theta}) - \theta = n^{-1}(3\theta^2 - \theta) = 3n^{-1}\theta^2\{1 + o(1)\}, \tag{6.3}$$

as min $\{n, \theta\} \to \infty$. Note that the primary term of the above bias is a quadratic function in θ. If both the pilot sample size n and the effective size μ/σ are relatively small, then the bias can be substantial. Table 6.1 summarizes the potential impact of bias of $\hat{\theta} = s^2/\Delta^2$ in the sample size calculation. As can be seen, the sample size calculation based on estimates from a small pilot study could be very significant—we may not reach the desired power for claiming that the treatment under investigation is efficacious. This becomes very critical especially when the treatment effect is considered marginally significant (positive).

TABLE 6.1

Instability of Sample Size

Δ	σ	$\theta = \sigma^2/\Delta^2$	Classic Sample Size N_0	Bias $3\theta^2/N_0$	Sample Size with Bias N
5	10	4	32	1.53	44
	20	16	126	6.12	174
	30	36	183	13.76	391
10	10	1	8	0.38	11
	20	4	32	1.53	44
	30	9	71	3.44	98

6.3.2 The Bootstrap-Median Approach

Since the bias of $E(\hat{\theta})$ is not negligible in many cases, alternatively, Lee et al. (2008) suggested considering the median of $s^2/\bar{x}^2$. Let $\eta_{0.5}$ be the median of sample CV squared such that $P(\hat{\theta} \le \eta_{0.5}) = 0.5$. Then, $\eta_{0.5}$ has a one-term expansion in terms of $n^{-1/2}$ as

$$\eta_{0.5} = \theta + n^{-1/2}\tau x_{0.5} = \theta + n^{-1}\left(A_1 - \frac{1}{6}A_2\tau^2\right) + O(n^{-2})$$

$$= \theta + n^{-1}\left\{(3\theta^2 - \theta) - \frac{72\theta^5 + 48\theta^4 + 8\theta^3}{24\theta^3 + 12\theta^2}\right\} + O(n^{-2}).$$

Thus,

$$\eta_{0.5} - \theta = n^{-1}\left\{(3\theta^2 - \theta) - \frac{72\theta^5 + 48\theta^4 + 8\theta^3}{24\theta^3 + 12\theta^2}\right\} + O(n^{-2})$$

$$= -n^{-1}\left\{\frac{36\theta^4 + 20\theta^3}{24\theta^3 + 12\theta^2}\right\} + O(n^{-2}) = -1.5n^{-1}\theta\{1 + o(1)\}, \tag{6.4}$$

whose leading term is linear in θ. It is a smaller order as compared with the bias of the mean (6.3).

As can be seen from (6.4), the bias incurred by the median could be substantially smaller than that of the mean for a small sample size and/or small effective size. In practice, however, we do not know the exact value for the median of $s^2/\bar{x}^2$. As a simple solution, Lee et al. (2008) proposed the use of bootstrap distribution to approximate the sampling distribution of $s^2/\bar{x}^2$, from where the median of $s^2/\bar{x}^2$ can be estimated by the bootstrap-median approach. Lee et al. (2008) referred to this approach as the bootstrap-median method.

6.4 Simulation Study

To evaluate the finite sample performances of the bootstrap-median approach for sample size determination, an extensive simulation study was conducted (Lee et al., 2008) based on 5000 simulation runs and 1000 bootstrap sample size.

6.4.1 One-Sample Problem

For the one-sample problem, a total of n_0 independent and identically distributed random variables are simulated within each simulation run to form the data from a pilot study. In this simulation study, Lee et al. (2008) considered

$$n_0 = 25, 50, 75, \text{ and } 100.$$

Also, eight different effective sizes are considered:

$$\frac{\mu}{\sigma} = 0.1, 0.15, 0.2, 0.25, 0.3, 0.4, 0.5, \text{ and } 0.75.$$

Both the traditional method and the bootstrap-median approach are used to estimate the minimum sample size (i.e., $\hat{n}$ and $\hat{n}_{BM}$, respectively) required for achieving a desired power $(1 - \beta)$ (say $\beta = 0.1$, i.e., 90% power) at a prespecified level of significance ($\alpha = 0.05$). At the same time, the ideal sample size n_{ideal} by using the true parameters is also computed.

In order to evaluate the performance of the two methods, we will investigate the quantiles of distribution of $\hat{n}/n_{ideal}$ and $\hat{n}_{BM}/n_{ideal}$. A comparison of various quantiles of $\hat{n}/n_{ideal}$ and $\hat{n}_{BM}/n_{ideal}$ can give us some insights into the difference in variability (stability) between the two methods. Hence, five quantiles (10%, 25%, 50%, 75%, 90%) of $\hat{n}/n_{ideal}$ and $\hat{n}_{BM}/n_{ideal}$ are obtained by simulations. For example, the 75% quantile of $\hat{n}/n_{ideal}$ and $\hat{n}_{BM}/n_{ideal}$, which are obtained from the equations $P(\hat{n}/n_{ideal} \leq k) = 0.75$ and $\hat{n}_{BM}/n_{ideal}$, respectively, are compared to investigate the stability of two methods.

Table 6.2 presents the simulated quantiles of $\hat{n}/n_{ideal}$ and $\hat{n}_{BM}/n_{ideal}$ under various combinations of effective size μ/σ and the sample size n_0 of the pilot study. If both the values of μ/σ and the pilot sample size n_0 are very small (i.e., $\mu/\sigma = 0.1$ and $n_0 = 25$), then the performances of both the traditional method and the bootstrap-median approach are poor because their medians of the ratios $\hat{n}/n_{ideal}$ and $\hat{n}_{BM}/n_{ideal}$ are far from 1.0. However, the variability of the bootstrap-median approach is substantially smaller than the traditional method. Considering 75% and 90% quantiles of two ratios, it can be seen that the distribution of the ratio $\hat{n}/n_{ideal}$ is extremely skewed to the right, which indicates that the traditional method could lead to extremely large sample size estimate with a higher probability than the bootstrap-median method. As the sample size increases to $n_0 = 100$, the performance of both methods become much better as the median value of the two ratios comes close to 1.0. Note that the inter-quantile range of the bootstrap-median method is only $1.69 - 0.34 = 1.35$, which is smaller than half the size of that of the traditional method ($3.82 - 0.34 = 3.48$).

As the value of μ/σ increases, the performance of both methods becomes similar. However, under situations that require a large ideal sample size ($n > 100$) such as $\mu/\sigma = 0.25$ or 0.30, the advantage of the bootstrap-median

TABLE 6.2

Simulated Quantiles of $\hat{n}/n_{\text{true}}$ and $\hat{n}_{\text{BM}}/n_{\text{true}}$ (One-Sample Problem with Normal Error)

μ/σ	n_{ideal}	n_0	Traditional Method: Quantiles of $\hat{n}/n_{\text{true}}$					Bootstrap-Median Methods: Quantiles of $\hat{n}_{\text{BM}}/n_{\text{true}}$				
			10%	25%	50%	75%	90%	10%	25%	50%	75%	90%
0.10	1051	25	0.07	0.14	0.42	1.90	10.74	0.06	0.13	0.30	0.46	0.52
		50	0.12	0.25	0.66	3.11	17.89	0.11	0.24	0.53	0.91	1.05
		75	0.16	0.30	0.83	3.31	21.48	0.16	0.30	0.72	1.31	1.57
		100	0.19	0.34	0.89	3.82	21.54	0.18	0.34	0.83	1.69	2.05
0.15	467	25	0.13	0.26	0.77	3.44	21.69	0.12	0.24	0.60	1.02	1.17
		50	0.19	0.35	0.94	3.66	23.24	0.19	0.34	0.86	1.80	2.30
		75	0.25	0.43	1.00	3.39	20.31	0.24	0.42	0.95	2.27	3.31
		100	0.28	0.47	0.99	3.03	14.97	0.28	0.46	0.97	2.48	4.24
0.20	263	25	0.17	0.33	0.90	3.89	25.63	0.16	0.32	0.79	1.65	2.05
		50	0.27	0.44	0.95	3.23	16.30	0.26	0.43	0.91	2.40	3.79
		75	0.32	0.52	1.00	2.67	10.98	0.32	0.51	0.98	2.43	4.99
		100	0.36	0.56	1.01	2.31	7.57	0.36	0.55	1.00	2.25	5.24
0.25	168	25	0.22	0.39	0.96	3.51	19.82	0.21	0.37	0.90	2.15	3.05
		50	0.32	0.51	0.98	2.38	9.67	0.31	0.50	0.94	2.24	4.89
		75	0.39	0.58	1.01	2.11	5.88	0.38	0.57	0.98	2.06	4.90
		100	0.42	0.61	1.00	1.90	4.22	0.42	0.60	0.99	1.87	4.13
0.30	117	25	0.26	0.45	0.97	3.05	15.63	0.25	0.43	0.92	2.37	4.15
		50	0.36	0.55	0.99	2.11	6.08	0.35	0.54	0.97	2.03	4.93
		75	0.43	0.62	0.99	1.78	3.83	0.42	0.61	0.98	1.74	3.64
		100	0.49	0.67	1.03	1.71	3.21	0.48	0.66	1.01	1.69	3.18

(*continued*)

TABLE 6.2 (continued)

Simulated Quantiles of $\hat{n}/n_{\text{true}}$ and $\hat{n}_{\text{BM}}/n_{\text{true}}$ (One-Sample Problem with Normal Error)

μ/σ	n_{ideal}	n_0	Traditional Method: Quantiles of $\hat{n}/n_{\text{true}}$					Bootstrap-Median Methods: Quantiles of $\hat{n}_{\text{BM}}/n_{\text{true}}$				
			10%	25%	50%	75%	90%	10%	25%	50%	75%	90%
0.40	66	25	0.32	0.51	0.96	2.12	6.59	0.31	0.49	0.92	2.02	4.70
		50	0.45	0.64	1.00	1.77	3.48	0.44	0.62	0.97	1.73	3.36
		75	0.51	0.70	1.00	1.55	2.55	0.50	0.68	0.98	1.52	2.48
		100	0.54	0.72	0.98	1.43	2.21	0.54	0.71	0.96	1.41	2.17
0.50	42	25	0.39	0.58	1.00	1.88	4.14	0.37	0.55	0.95	1.78	3.81
		50	0.50	0.67	0.98	1.52	2.44	0.49	0.65	0.95	1.47	2.37
		75	0.57	0.73	1.00	1.41	2.07	0.56	0.71	0.98	1.39	2.03
		100	0.60	0.76	0.99	1.34	1.86	0.59	0.74	0.97	1.32	1.83
0.75	19	25	0.48	0.66	0.96	1.49	2.33	0.46	0.62	0.92	1.41	2.24
		50	0.59	0.74	0.99	1.35	1.82	0.57	0.72	0.96	1.32	1.78
		75	0.65	0.79	1.00	1.26	1.62	0.64	0.77	0.98	1.24	1.59
		100	0.68	0.82	1.00	1.24	1.51	0.67	0.80	0.99	1.22	1.49

approach is clear under small pilot studies in terms of stability. When the effective size is large such as $\mu/\sigma = 0.5; 0.75$, sample sizes by both methods are almost identical, but the bootstrap-median approach prevents extremely large estimated sample size by the traditional method even though it is a small probability. Figure 6.1 shows the distributions of the absolute value of sample effective size $|\bar{x}/s|$ and two associated sample size estimates, one obtained by the traditional method and the other by the bootstrap-median approach under the sample size for the pilot study, $n_0 = 50$. The distribution of $|\bar{x}/s|$ is

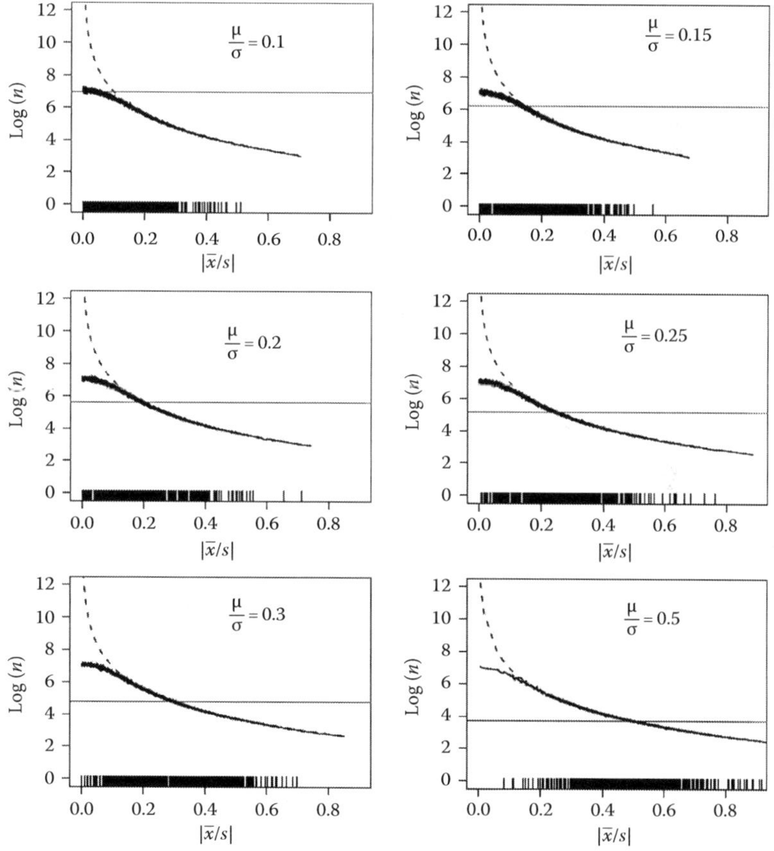

FIGURE 6.1

Distribution of absolute values of sample effective size in one-sample problem and associated sample sizes with log scale. (The solid line is sample size estimated by the bootstrap-median approach and the dashed line is sample size estimated by the traditional method. The red line indicates the ideal minimum required sample size. The sample size of the pilot study is $n_0 = 50$.)

illustrated by rug plot at the x-axis based on 500 simulation samples. Due to the finite bootstrap sample size ($B = 1000$), there is a little fluctuation in sample sizes obtained by the bootstrap-median approach, but it is small enough to ignore. It clearly shows that the bootstrap-median is much more stable than the traditional approach in which the bootstrap-median approach prevents the unreasonable large same size by the traditional method based on extremely skewed sampling distribution.

6.4.2 Two-Sample Problem

A similar simulation study was also conducted for the two-sample testing problem. Common variance σ^2 for two treatment groups is assumed and sample variance is obtained by pooling two sample variances $s^2 = [(n_{01} - 1)S_1^2 + (n_{02} - 1)S_2^2]/(n_{01} + n_{02} - 2)$. If μ_1 and μ_2 are population means for each group and σ^2 is the common variance under normal distribution, then the effective size is $\Delta/\sqrt{2}\sigma = (\mu_1 - \mu_2)/\sqrt{2}\sigma$ and the estimated sample size from a pilot study by the traditional method is given by $\hat{n} = (z_{\alpha/2} + z_\beta)^2 s^2/\hat{\Delta}^2$, where $\hat{\Delta}$ is the difference between two sample means. The bootstrap-median approach can be easily applied to this two-sample testing problem and the simulation result is very similar to the one-sample problem. Figure 6.2 presents the distribution of absolute values of sample effective size ($\bar{x}/\sqrt{2}s$) and associated sample size estimates by two methods with two different sample sizes for the pilot study ($n_0 = 25$ and $n_0 = 75$ per group). As seen in the simulation study for the one-sample problem, Figure 6.1 shows that the bootstrap-median is much more stable than the traditional approach.

6.5 An Example

For illustration purpose, consider the example given by Lee et al. (2008). A pharmaceutical company developed a new drug for lowering blood pressure in patients with essential hypertension. A pilot study was conducted to compare the efficacy of the newly developed drug (denoted by drug A) with a widely used existing drug (denoted by drug B). The primary efficacy variable is the change from baseline to Week 8 of systolic blood pressure. The data collected from the pilot study are summarized in Table 6.3. In this study, the estimated effective size is $\hat{\Delta}/\sqrt{2}s = -0.16$. If we applied the two methods for sample size calculation, the traditional method gives $n = 390$ per treatment group and the bootstrap-median approach with 3000 bootstrap simulations gives $n = 403$ per treatment group so that results from both methods are similar. If we assume that responses in this study are the true population, we may think that the required minimum sample size for

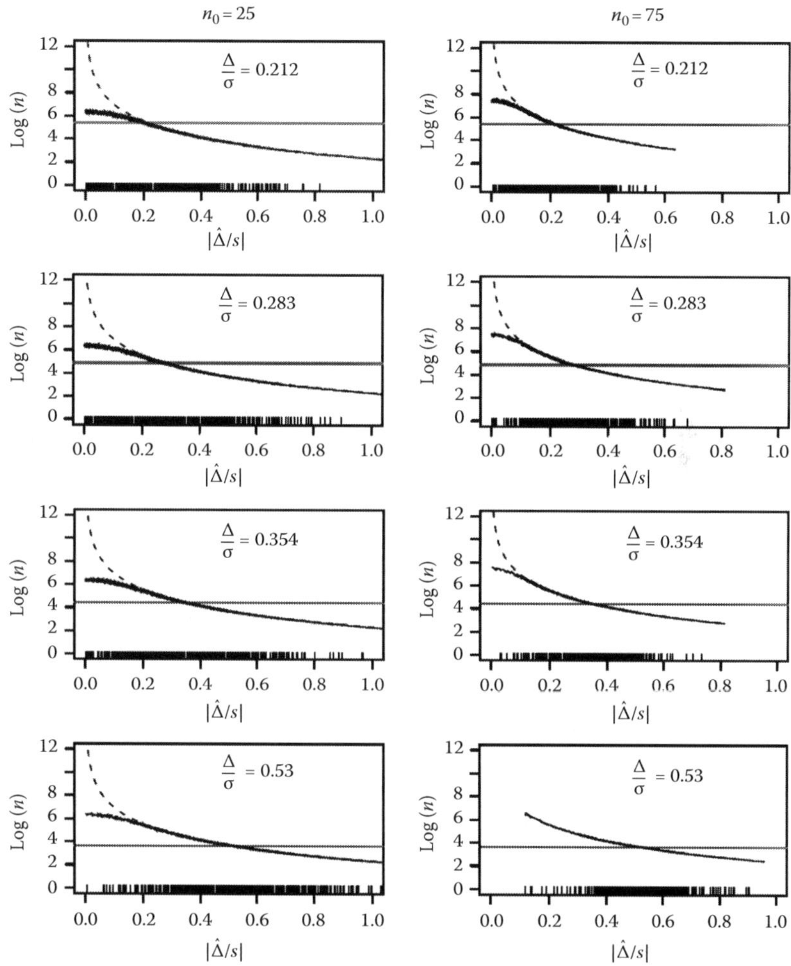

FIGURE 6.2

Distribution of absolute values of sample effective size in two-sample problem and associated sample sizes with log scale. (The solid line is sample size estimated by bootstrap-median approach and the dashed line is sample size estimated by the traditional method. The red line indicates the ideal minimum required sample size. The sample sizes of the pilot study are $n_0 = 25$ and $n_0 = 75$ per group.)

achieving a 90% power at the 5% level of significance is close to $n = 400$ per treatment.

Stability of the proposed bootstrap-median approach is investigated by considering study subjects in the given study above as the true responses of the population and then comparing sampling properties of two methods

TABLE 6.3

Information about Randomized Comparative Clinical Trial to Compare
Two Drugs for Lowering Blood Pressure of Hypertensive Patients

Treatment Group	Sample Size	Difference Baseline	Week 8	Change from Baseline	Difference of Change
Drug A	139	155.8 (12.4)	144.5 (16.2)	−11.4 (13.2)	−2.9 (12.3)
Drug B	131	155.4 (12.4)	146.9 (14.0)	−8.5 (11.2)	

Primary efficacy variable is the mean of sitting systolic blood pressure and the number in parentheses is standard deviation.

for sample size determination by simulation of repeated small pilot studies. Here, we will consider two scenarios: (1) equal sample size allocation for both treatment groups with $n_{0A} = 50$ and $n_{0B} = 50$ and (2) unequal sample size allocation with $n_{0A} = 75$ and $n_{0B} = 50$ to collect more information about the new drug. For each scenario, we consider 1000 independent small pilot studies whose study subjects are randomly selected from the given study by simple random sampling without replacement in each treatment group.

Furthermore, for each simulated pilot study, two sample sizes are determined by the traditional method and the bootstrap-median approach, respectively, with 1000 bootstrap sample size. Table 6.4 shows the summary statistics of the distribution of sample sizes from simulated small pilot studies under the two scenarios considered, and Figure 6.1 presents the sampling distribution of sample size by two methods. In the case of equal sample size allocation, as we can see in the simulation studies in the previous section, the bootstrap-median approach is much more stable than the traditional method. In the case of unequal sample size allocation, the sample size by the bootstrap-median approach is a little larger than by the traditional approach, but its stability is still superior (Figure 6.3).

TABLE 6.4

Summary Statistics for the Distribution of Sample Sizes by Two Methods Based on Simulated Small Pilot Studies under Two Scenarios

Scenario	Sample Size	Method	Min	First Quantile	Median	Mean	Third Quantile	Max
I	$n_{0A} = 50,$ $n_{0B} = 50$	Traditional	37	190	342	6,267	861	1,131,000
		BM	37	182	320	437	657	1,290
II	$n_{0A} = 50,$ $n_{0B} = 75$	Traditional	62	220	401	1,223,000	889	588,800,000
		BM	96	326	548	634	892	1,699

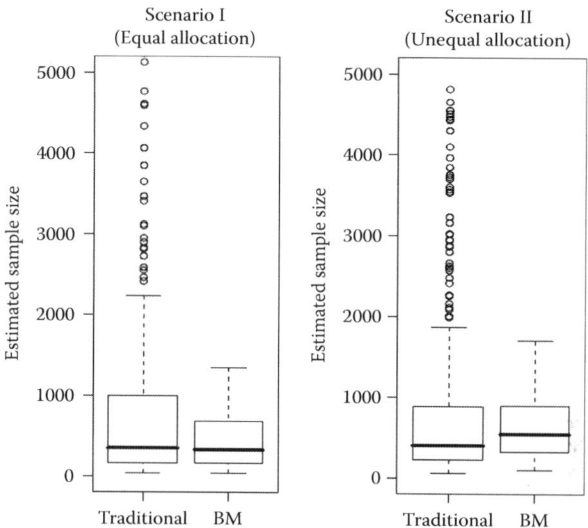

FIGURE 6.3

Distribution of estimated sample sizes from 1000 simulated small pilot studies based on the real comparative study. (Scenario I uses equal sample allocation $n_{0A} = n_{0B} = 50$ and Scenario II uses unequal sample allocation $n_{0A} = 50$ and $n_{0B} = 75$.)

6.6 Concluding Remarks

In this chapter, we have shown that sample size calculation based on data from a small pilot study by ignoring the sampling error using the traditional method could lead to unreasonable sample size estimate due to its instability, especially when the intended study is designed to detect a relatively small effect size. Alternatively, the proposed bootstrap-median approach could provide a relatively stable sample size estimates. The proposed bootstrap-median can be easily implemented to various study designs with different types of study endpoints. The proposed approach is based on the approximated sampling distribution of sample squared CV under the normal distribution, and the bootstrap approximation of median of the sample squared CV has its justification based on Edgeworth and Fisher–Cornish expansions. Based on extensive simulation studies and theoretical justification, it is suggested that the proposed bootstrap-median approach to estimate the minimum required sample size is much more stable than the traditional method; it is therefore recommended to use the proposed method whenever data from a small pilot study are available.

Although numerical experience and theory of Lee et al. (2008) are only limited to the standard one-sample and two-sample testing problems, the proposed procedure can be easily extended to other experimental designs (e.g., crossover design), other data types (e.g., binary endpoint, time-to-event data), and other hypotheses types (e.g., non-inferiority/equivalence test). Furthermore, it would be of great interest to see how the proposed method can be formulated into a Bayesian framework so that some valuable prior information can be used. All those problems are of great interest for both theory and practice. Further research along this line is definitely needed.

The above discussion justifies flexible sample size reestimation in standard group sequential design for clinical trials. For a group sequential design with some planned interim analyses, sample size adjustment (or reestimation) is usually performed at interim analyses to ensure that the study will achieve the desired power at a prespecified level of significance at the end of the study. Commonly considered sample size adjustment based on the ratio of the initial estimated effect size (E_0) to the observed effect size (E) is as follows:

$$N = \min\left\{N_{max}, \max\left(N_{min}, sign(E_0 E)\left|\frac{E_0}{E}\right|^a N_0\right)\right\},$$

where
    N is the sample size after adjustment
    N_{max} and N_{min} are the maximum (due to financial and/or other constraints) and minimum (the sample size for the interim analysis) sample sizes
    a is a constant (which is usually determined based on the review of the interim analysis results)
    $sign(x) = 1$ for $x > 0$; otherwise $sign(x) = -1$

Note that the above sample size adjustment can be applied to normal, binary, and survival study endpoints. Note that the above sample size adjustment reduces to the method proposed by the U.S. FDA statisticians for a normal study endpoint with $a = 2$ (see Cui et al., 1999).

Note that other controversial issues may be raised even after we have overcome the instability of initial sample size calculation (e.g., using a more robust bootstrap-median approach and applying a sample size reestimation method). First, the number used for sample size reestimation at interim is still an estimate. Thus, the original issue of instability persists. Second, how robust (stable) the obtained sample size is if there is a shift in target population during the conduct of the clinical trial is also a question. Further research is required in order to address these questions.

7

Integrity of Randomization/Blinding

7.1 Introduction

In double-blind clinical trials, randomization and blinding are often employed to prevent bias from clinical/statistical assessment of a test treatment under investigation. Randomization/blinding plays an important role in the conduct of clinical trials. Randomization/blinding not only generates comparable groups of patients who constitute representative samples from the intended (target) patient population but also enables valid statistical tests for clinical evaluation of the study drug. Randomization/blinding in clinical trials involves random recruitment of the patients from the targeted population and random assignment of patients to the treatments. For a valid statistical assessment of the efficacy and safety of a study drug, it is important that a representative sample of qualified patients be randomly selected from the target patient population. Randomization avoids subjective selection bias for the integrity and scientific and/or statistical validity of the intended clinical trials. Patients participating in the clinical trials are randomly assigned to one of the treatments under study, which avoids subjective assignment of treatments. On the other hand, blinding is the guard for preventing subjective evaluation bias and consequently ensures scientific and/or statistical validity of the intended clinical trials. When there is heterogeneity in demographics and/or patient characteristics, randomization with blocking and/or stratification is helpful in removing the potential bias that might occur due to the differences in demographics and/or patient characteristics. Under randomization and blinding, statistical inference can be drawn under some probability distribution assumption of the intended patient population. The probability distribution assumption depends on the method of randomization under a randomization (population) model. A study without randomization/blinding will result in the violation of the probability distribution assumption and consequently no accurate and reliable statistical inference on the study drug can be drawn.

In practice, however, there is no guarantee that subjective judgment in reporting, evaluation, data processing, and statistical analysis will be free of bias due to (1) possible mix-up of randomization and (2) the knowledge of the identity of the treatment codes. Since this subjective and judgmental bias is directly or indirectly related to treatment, it can seriously distort statistical inference on the treatment effect. However, if it is not impossible, it is often difficult to quantitatively assess such bias and its impact on the assessment of the treatment effect. For a given double-blind clinical trial, randomization schedule may be mix-up due to human error. In addition, it is human nature for both the patient and the investigator to guess what treatment the patient is receiving. To maintain the integrity of the randomization and to prevent treatment imbalance, a typical approach is to consider a larger blocking size in randomization. Thus, the following questions are commonly asked. First, what is the impact if we mix-up with the randomization schedule? Second, how do we test for the integrity of randomization and blinding in clinical trials? Third, what is the difference in the probability of guessing treatment code right for a blocking size of 2 as compared to that of the blocking size of 4 for a comparative clinical trial? In practice, even with the best intention for preserving blindness throughout a clinical trial, blindness can sometimes be breached for various reasons. One method to determine whether the blindness is seriously violated is to ask patients to guess their treatment codes during the study or at the conclusion of the trial prior to unblinding. In some cases, investigators may also be asked to guess patients' treatment codes. Once the guesses are recorded on the case report forms and entered into the database, the integrity of blinding can be tested (Chow and Shao, 2004).

In the next section, the effect of mix-up randomization is discussed. In Section 7.3, we study the probability of correctly guessing treatment assignments with various blocking sizes (e.g., 2 versus 4) for comparative clinical trials. Statistical tests for the integrity of blinding are described in Section 7.4. Section 7.5 discusses analysis under breached blindness. An example is given in the last section of this chapter.

7.2 The Effect of Mix-Up Randomization

A problem that is commonly encountered during the conduct of a clinical trial is that a proportion of treatment codes are mix-up in randomization schedules. Mixing up treatment codes can distort the statistical analysis based on the population or randomization model. In what follows we introduce a method proposed by Chow and Shao (2003) to quantitatively study the effect of mix-up treatment randomization codes. Consider a two-group parallel design for comparing a test drug and a control (placebo), where n_1 patients are randomly assigned to the treatment group and n_2 patients are randomly assigned to the control group. When randomization is properly applied, the population model holds and responses from patients are normally distributed. Consider first the simplest case where two patient populations (treatment and control) have the same variance σ^2, where σ^2 is known. Let μ_1 and μ_2 be the population means for the treatment and the control, respectively. The null hypothesis that $\mu_1 = \mu_2$ (i.e., there is no treatment effect) is rejected at the 5% level of significance if

$$\frac{|\bar{x}_1 - \bar{x}_2|}{\sigma\sqrt{1/n_1 + 1/n_2}} > z_{0.975}, \tag{7.1}$$

where

$\bar{x}_1$ is the sample mean of responses from patients in the treatment group
$\bar{x}_2$ is the sample mean of responses from patients in the control group
$z_{0.975}$ is the 97.5th percentile of the standard normal distribution

Intuitively, a mix-up of treatment codes does not affect the significance level of the test (7.1). The power of the test defined by (7.1), i.e., the probability of correctly detecting a treatment difference when $\mu_1 \neq \mu_2$, is

$$p(\theta) = P\left(\frac{|\bar{x}_1 - \bar{x}_2|}{\sigma\sqrt{1/n_1 + 1/n_2}} > z_{0.975}\right) = \Phi(\theta - z_{0.975}) + \Phi(-\theta - z_{0.975}),$$

where Φ is the standard normal distribution function and

$$\theta = \frac{\mu_1 - \mu_2}{\sigma\sqrt{1/n_1 + 1/n_2}}. \tag{7.2}$$

This follows from the fact that under the population model, $\bar{x}_1 - \bar{x}_2$ has a normal distribution with mean $\mu_1 - \mu_2$ and variance $\sigma^2(1/n_1 + 1/n_2)$.

Suppose that there are m patients whose treatment codes are randomly mix-up. A straightforward calculation shows that $\bar{x}_1 - \bar{x}_2$ is still normally distributed with variance $\sigma^2(1/n_1 + 1/n_2)$, but the mean of $\bar{x}_1 - \bar{x}_2$ is equal to

$$\left[1 - m\left(\frac{1}{n_1} + \frac{1}{n_2}\right)\right](\mu_1 - \mu_2).$$

It turns out that the power for the test defined by (7.1) is

$$p(\theta_m) = \Phi(\theta_m - z_{0.975}) + \Phi(-\theta_m - z_{0.975}),$$

where

$$\theta_m = \left[1 - m\left(\frac{1}{n_1} + \frac{1}{n_2}\right)\right]\frac{\mu_1 - \mu_2}{\sigma\sqrt{1/n_1 + 1/n_2}}. \tag{7.3}$$

Note that $\theta_m = \theta$ if $m = 0$, i.e., there is no mix-up. The effect of mix-up treatment codes can be measured by comparing $p(\theta)$ with $p(\theta_m)$. Suppose that $n_1 = n_2$. Then $p(\theta_m)$ depends on m/n_1, the proportion of mix-up treatment codes. For example, suppose that when there is no mix-up, $p(\theta) = 80\%$, which gives $|\theta| = 2.81$. When 5% of treatment codes are mix-up, i.e., $m/n_1 = 5\%$, $p(\theta_m) = 70.2\%$. When 10% of treatment codes are mix-up, $p(\theta_m) = 61.4\%$. Hence, a small proportion of mix-up treatment codes may seriously affect the probability of detecting treatment effect when such an effect exists. In this simple case, we may plan ahead to ensure a desired power when the maximum proportion of mix-up treatment codes is known. Assume that the maximum proportion of mix-up treatment codes is p and that the original sample size is $n_1 = n_2 = n_0$. Then

$$\theta_m = (1-2p)\theta = \frac{\mu_1 - \mu_2}{\sigma\sqrt{2}}\sqrt{(1-2p)^2 n_0}.$$

Thus, a new sample size $n_{\text{new}} = n_0/(1 - 2p)^2$ will maintain the desired power when the proportion of mix-up treatment codes is no larger than p. For example, if $p = 5\%$, then $n_{\text{new}} = 1.23n_0$, i.e., a 23% increase of the sample size will offset a 5% mix-up in randomization schedules.

The effect of mix-up treatment codes is higher when the study design becomes more complicated. Consider the two-group parallel design with an unknown σ^2. The test defined by (7.1) has to be modified by changing $z_{0.975}$ to $t_{0.975,n_1+n_2-2}$ and replacing σ^2 by its unbiased estimator

$$\hat{\sigma}^2 = \frac{(n_1-1)s_1^2 + (n_2-1)s_2^2}{n_1 + n_2 - 2},$$

where
  s_1^2 is the sample variance based on responses from patients in the treatment group
  s_2^2 is the sample variance based on responses from patients in the control group
  $t_{0.975,n_1+n_2-2}$ is the 97.5th percentile of the t-distribution with $n_1 + n_2 - 2$ degrees of freedom

The resulting test is known as the two-sample t-test. When randomization is properly applied without mix-up, the two-sample t-test has a 5% level of significance and the power is given by

$$1 - \Im_{n_1+n_2-2}(t_{0.975,n_1+n_2-2} \mid \theta) + \Im_{n_1+n_2-2}(-t_{0.975,n_1+n_2-2} \mid \theta),$$

where
  θ is defined by (7.2)
  $\Im_{n_1+n_2-2}(\bullet \mid \theta)$ is the noncentral t-distribution function with n_1+n_2-2 degrees of freedom and the noncentrality parameter θ

When there are m patients with mix-up treatment codes and $\mu_1 \neq \mu_2$, the effect on the distribution of $\bar{x}_1 - \bar{x}_2$ is the same as that in the case of known σ^2. In addition, the distribution of $\hat{\sigma}^2$ is also changed. A direct calculation shows that the expectation of $\hat{\sigma}^2$ is

$$E(\hat{\sigma}^2) = \sigma^2 + \frac{2(\mu_1 - \mu_2)^2 m}{n_1 + n_2 - 2}\left[2 - m\left(\frac{1}{n_1} + \frac{1}{n_2}\right)\right].$$

Hence, the actual power of the two-sample t-test is less than

$$1 - \Im_{n_1+n_2-2}\left(t_{0.975,n_1+n_2-2} \mid \theta_m\right) + \Im_{n_1+n_2-2}\left(-t_{0.975,n_1+n_2-2} \mid \theta_m\right),$$

where θ_m is given by (7.3).

Note that, in some situations, deliberate unequal allocation of patients between treatment groups may be desirable. For example, it may be of interest to allocate patients to the treatment and a control in a ratio of 2 to 1. Such situations include that (1) the patient population is small, (2) previous experience with the study drug is limited, (3) the response profile of the competitor is well known, and (4) there are missing values and the rates of missing depend on the treatment groups. Randomization is one of key elements for the success of clinical trials intended to address scientific and/or medical questions. It, however, should be noted that in many situations, randomization may not be feasible in clinical research. For example, nonrandomized observational or case-controlled studies are often conducted to study the relationship between smoking and cancer. However, if the randomization is not used due to some medical considerations, the FDA requires that statistical justification should be provided with respect to how systematic selection bias can be avoided. Clinical results may be directly or indirectly distorted when either the investigators or the patients know which treatment the patients are receiving, although randomization is applied to assign treatments. Blinding is commonly used to eliminate such a problem by blocking the identity of treatments.

7.3 Blocking Size in Randomization

In double-blind randomized clinical trials comparing two treatment groups, in the interest of treatment balance, a blocking size of 2 or 4 is usually employed in randomization. It is not uncommon that either the patients or the investigator may guess the treatment codes that patients are receiving. It is a concern that the use of blocking size of 2 may not prevent patients or the investigator from correctly guessing the treatment assignment. Correctly (or wrongly) guessing the treatment assignments will have an impact on the assessment of the effect of the treatment under investigation, especially for study endpoints that are evaluated subjectively. Thus, it is suggested to increase the blocking size to its maximum to decrease the probability of correctly guessing treatment assignments. However, increasing the blocking size may increase the chance of mixing up the randomization schedules. As a result, it is of interest to keep the blocking size within 4. Note that blocking size of 2 or 4 is commonly employed in double-blind randomized clinical trials for comparing two treatment groups.

In this section, we will study the probability of correctly guessing the treatment assignments with a blocking size of 2 as compared to that with a blocking size of 4 for a given sample size. In practice, since the patients normally do not have any idea what blocking size is used in the randomization, the probability of correctly guessing the treatment assignment for a given patient is equal to 1/2. However, the probability for the treating physician correctly guessing the treatment assignment is usually higher than 1/2 due to the knowledge of the blocking size and/or the observed clinical signs and symptoms of the patients. In what follows, we will calculate the probability of correctly guessing treatment assignments by the patients followed by the guess of the investigator.

To address the second question regarding the integrity of blinding, for a given sample size, the probabilities of guessing treatment codes right for blocking size 2 and blocking size 4 can be directly calculated and compared. For illustration purpose, probabilities of guessing treatment codes right for a small clinical trial are as follows.

Blocking Size	$N = 4$	$N = 8$	$N = 16$
2	0.2500	0.0625	0.0039
4	0.1667	0.0278	0.0008

In addition to the blocking size used, prior knowledge regarding the true blocking size may also be a factor which has an impact on the probability of correctly guessing. Hsieh et al. (2010) investigated six types of the possibilities of correctly guessing by considering the designs of the true blocking sizes of 4 and 2 as well as three types of prior knowledge on which the guesser bases his/her guesses. The three types of prior information include guess without prior knowledge, guess by thinking the true blocking size is 4, and guess by thinking the true blocking size is 2. The probability model for calculating the probabilities of correctly guessing is described in the next subsection followed by a numerical study to compare the above six types of probabilities for evaluating the impact of the blocking size and prior knowledge.

7.3.1 Probability of Correctly Guessing

Consider that a two-arm, balanced, randomized, and parallel design of the study is employed for comparing the test treatment with the reference treatment. For the purpose of comparing the probabilities of guessing the subjects' treatment right between the blocking randomization methods with blocking sizes of 4 and 2, the total sample size of N ($N/2$ subjects for each of two groups) is assumed to be a multiple of 4. Furthermore, the total numbers of the blocks corresponding to the blocking size of 4 and 2 are $N/4$ and $N/2$, respectively.

Let U_i be the event of guessing the i th subject's treatment right within the k th block for the design with the blocking size of 4, where $i = 1, 2, 3, 4$ and $k = 1, 2, \ldots, N/4$ are the possible events denoted by $X_0, X_1, \ldots, X_{15}$ for guessing m_k subjects' treatment right and the others wrong within the block where $m_k = 0, 1, 2, 3, 4$; these data are given in Table 7.1. On the other hand, if we hypothesize the rth two neighbor blocks for the design with the blocking size of 2 as a block which consists of four subjects (two from the first block and the other two from the second block of the two neighbor blocks), where $r = 1, 2, \ldots, N/4 = 1, 2, \ldots, N/4$, and treat the two subjects in the first block and the other two subjects in the second block of the two neighbor blocks as the first, second, third, and fourth subjects in this hypothesized block, respectively, U_i and the events given in Table 7.1 for the design with the blocking size of 4 can also be used to describe the behavior and events of correctly guessing for each of the two neighbor blocks for the design with the blocking size of 2.

TABLE 7.1

Possible Events of Guessing m_j Right within Each Block with $k = 1, \ldots, N/4$

m_k	X_i
0	$X_0 = U_1^C \cap U_2^C \cap U_3^C \cap U_4^C$
1	$X_1 = U_1 \cap U_2^C \cap U_3^C \cap U_4^C$
	$X_2 = U_1^C \cap U_2 \cap U_3^C \cap U_4^C$
	$X_3 = U_1^C \cap U_2^C \cap U_3 \cap U_4^C$
	$X_4 = U_1^C \cap U_2^C \cap U_3^C \cap U_4$
2	$X_5 = U_1 \cap U_2 \cap U_3^C \cap U_4^C$
	$X_6 = U_1 \cap U_2^C \cap U_3 \cap U_4^C$
	$X_7 = U_1 \cap U_2^C \cap U_3^C \cap U_4$
	$X_8 = U_1^C \cap U_2 \cap U_3 \cap U_4^C$
	$X_9 = U_1^C \cap U_2 \cap U_3^C \cap U_4$
	$X_{10} = U_1^C \cap U_2^C \cap U_3 \cap U_4$
3	$X_{11} = U_1 \cap U_2 \cap U_3 \cap U_4^C$
	$X_{12} = U_1 \cap U_2 \cap U_3^C \cap U_4$
	$X_{13} = U_1 \cap U_2^C \cap U_3 \cap U_4$
	$X_{14} = U_1^C \cap U_2 \cap U_3 \cap U_4$
4	$X_{15} = U_1 \cap U_2 \cap U_3 \cap U_4$

Let T_w and G_w be the true treatment received and the treatment guessed by the guesser of the wth subject within some block for the design with the blocking size of 4 (or the hypothesized block formed by the two neighbor blocks for the design with the blocking size of 2), respectively, where $w = 1$, 2, 3, 4. The event of guessing a subject's treatment right happens when the true treatment received is exactly what the guesser guessed, i.e., $T_w = G_w$. Thus, the probability of each event in Table 7.1 is equal to the probability of the union of some intersection of m_k's events of $T_w = G_w$ and $(4 - m_k)$'s events of $T_w \neq G_w$, where $m_k = 0, 1, 2, 3, 4$.

Now we consider the probability of guessing M subjects' treatment right among all N study subjects, where M is in fact equal to the sum of numbers of guessing right in each of total $N/4$ blocks for the design with the blocking size of 4 (or each of $N/4$ hypothesized blocks formed by each of the two neighborhood blocks for the design with the blocking size of 2), i.e., $M = \sum_{k=1}^{N/4} m_k$. In addition, the sum of the numbers of each possible event in Table 7.1 is equal to the total number of blocks, i.e., $N/4 = \sum_{i=0}^{15} y_i$, where y_i is the number of blocks with the event of X_i. The probability of guessing M subjects' treatment right among all N study subjects can then be given as

$$\binom{N/4}{y_0 \; y_1 \; \cdots \; y_{15}} p_{X_0}^{y_0} p_{X_1}^{y_1} \cdots p_{X_{15}}^{y_{15}} \tag{7.4}$$

with the restrictions given as

$$\frac{N}{4} = \sum_{i=0}^{15} y_i$$

$$M = \sum_{k=1}^{N/4} m_k = 0 y_0 + 1(y_1 + y_2 + y_3 + y_4)$$

$$+ 2(y_5 + y_6 + y_7 + y_8 + y_9 + y_{10}) + 3(y_{11} + y_{12} + y_{13} + y_{14}) + 4 y_{15},$$

where p_{x_i} is the probability corresponding to the event of X_i given in Table 7.1, where $i = 0, \ldots, 15$.

Different blocking sizes as well as prior knowledge about the blocking size the guesser had before guess will result in the different combinations of true treatment assignment, possible guesses by the guesser, and their corresponding probabilities within each block. For instance, if the true blocking size is 4, there are six possible combinations of treatment assignment for four subjects' treatment within each block including ABAB, ABBA, BAAB, BABA, AABB, and BBAA with the probability of 1/6 for each, where A and B denote the test treatment and the reference treatment, respectively, On the other hand, there are only four combinations of treatment assignments within the two neighbor blocks including ABAB, ABBA, BAAB, and BABA with the probability of 1/4 for each if the blocking size is 2. With respect to the impact of prior knowledge the guesser had before the guess, there will be six possible guesses including ABAB, ABBA, BAAB, BABA, AABB, and BBAA with the probability of 1/6 for each if the guesser thought the true blocking size is 4 before his/her guess. If the guesser had no prior knowledge about the true blocking size, the possible guesses are still these six combinations. However, the probability of his/her guess for each subject's treatment is 1/2, which results in the probability for each of the six possible guesses becoming 1/16 ($=1/2^4$). Table 7.2 summarizes the possible combinations of treatment assignments within each block for the design with the blocking size of 4 (or each hypothesized block formed by each two neighbor blocks if the blocking size is 2), the possible guesses by the guesser and their corresponding probabilities within each block under the different blocking sizes and prior information the guesser had before the guess.

7.3.2 Numerical Study

To evaluate the impact of the different blocking sizes on the probability of guessing the subject's treatment right by taking into consideration the prior information the guesser had, the following six kinds of probabilities denoted by $P_{4N}, P_{44,} P_{42}, P_{2N}, P_{24}$, and P_{22} are calculated by (7.4):

1. P_{4N}: P (Guess M subjects' treatment right with the true blocking size of 4|guesser has no prior knowledge about the true blocking size).
2. P_{44}: P (Guess M subjects' treatment right with the true blocking size of 4|guesser thinks that the true blocking size is 4).
3. P_{42}: P (Guess M subjects' treatment right with the true blocking size of 4|guesser thinks that the true blocking size is 2).
4. P_{2N}: P (Guess M subjects' treatment right with the true blocking size of 2|guesser has no prior knowledge about the true blocking size).
5. P_{24}: P (Guess M subjects' treatment right with the true blocking size of 2|guesser thinks that the true blocking size is 4).
6. P_{22}: P (Guess M subjects' treatment right with the true blocking size of 2|guesser thinks that the true blocking size is 2).

The values of each P_i in (7.4) correspond to the above six cases that are presented in Table 7.3. The detailed derivations for obtaining the value of each P_i can be found in the Appendix.

Table 7.4 presents the probabilities of guessing M subjects' treatment correctly for the total sample size of $N = 4, 8, 12, \ldots, 100$ with $M = 1, \ldots, N$. Denote the maximum value of M and $(N - M)$ by M_{Max}; the findings are summarized as the following:

TABLE 7.2

Possible Combinations of Treatment Assignment with the Corresponding Probabilities within Each Block by Considering the Different True Blocking Sizes and the Prior Knowledge the Guesser Had before the Guess

	True Blocking Size = 4			True Blocking Size = 2		
Prior Information	Category	Comb.	Prob.	Category	Comb.	Prob.
No	True	ABAB	1/6	True	ABAB	1/4
		ABBA			ABBA	
		BAAB			BAAB	
		BABA			BABA	
		AABB				
		BBAA				
	Guess	ABAB	1/16	Guess	ABAB	1/16
		ABBA			ABBA	
		BAAB			BAAB	
		BABA			BABA	
		AABB			AABB	
		BBAA			BBAA	
		ABBB				
		AABB				
		AAAB				
		BAAA				
		BBAA				
		BBBA				
		BABB				
		BBAB				
		ABAA				
		AABA				
Blocking size = 2	True	ABAB	1/6	True	ABAB	1/4
		ABBA			ABBA	
		BAAB			BAAB	
		BABA			BABA	
		AABB				
		BBAA				
	Guess	ABAB	1/4	Guess	ABAB	1/4
		ABBA			ABBA	
		BAAB			BAAB	
		BABA			BABA	
Blocking size = 4	True	ABAB	1/6	True	ABAB	1/4
		ABBA			ABBA	
		BAAB			BAAB	
		BABA			BABA	
		AABB				
		BBAA				

TABLE 7.2 (continued)

Possible Combinations of Treatment Assignment with the Corresponding Probabilities within Each Block by Considering the Different True Blocking Sizes and the Prior Knowledge the Guesser Had before the Guess

Prior Information	True Blocking Size = 4			True Blocking Size = 2		
	Category	Comb.	Prob.	Category	Comb.	Prob.
	Guess	ABAB	1/6	Guess	ABAB	1/6
		ABBA			ABBA	
		BAAB			BAAB	
		BABA			BABA	
		AABB			AABB	
		BBAA			BBAA	

True, true treatment assignment; Guess, treatment assignment the guesser guessed; Comb., combination of treatment assignment; Prob., probability.

TABLE 7.3

Value of Each P_{x_i} by Considering the Different True Blocking Sizes and the Prior Knowledge the Guesser Had before the Guess

P_{x_i}	True Blocking Size = 4			True Blocking Size = 2		
	No Prior Information	Prior Information of Blocking Size = 4	Information of Blocking Size = 2	No Prior Information	Prior Information of Blocking Size = 4	Prior Information of Blocking Size = 2
P_{x_0}	1/16	1/6	1/6	1/16	1/6	1/4
P_{x_1}	1/16	0	0	1/16	0	0
P_{x_2}	1/16	0	0	1/16	0	0
P_{x_3}	1/16	0	0	1/16	0	0
P_{x_4}	1/16	0	0	1/16	0	0
P_{x_5}	1/16	1/9	1/6	1/16	1/6	1/4
P_{x_6}	1/16	1/9	1/12	1/16	1/12	0
P_{x_7}	1/16	1/9	1/12	1/16	1/12	0
P_{x_8}	1/16	1/9	1/12	1/16	1/12	0
P_{x_9}	1/16	1/9	1/12	1/16	1/12	0
$P_{x_{10}}$	1/16	1/9	1/6	1/16	1/6	1/4
$P_{x_{11}}$	1/16	0	0	1/16	0	0
$P_{x_{12}}$	1/16	0	0	1/16	0	0
$P_{x_{13}}$	1/16	0	0	1/16	0	0
$P_{x_{14}}$	1/16	0	0	1/16	0	0
$P_{x_{15}}$	1/16	1/6	1/6	1/16	1/6	1/4

TABLE 7.4

Probabilities of Correctly Guessing for Different N and M by Considering the Different True Blocking Sizes and the Prior Knowledge the Guesser Had before the Guess

N	M_{Max}	M	$N - M$	True Blocking Size = 4			True Blocking Size = 2		
				P_{4N} (%)	P_{44} (%)	P_{42} (%)	P_{2N} (%)	P_{24} (%)	P_{22} (%)
4	2	2	2	37.50	66.67	66.67	37.50	66.67	50.00
	3	1	3	25.00	0.00	0.00	25.00	0.00	0.00
	4	4	0	6.25	16.67	16.67	6.25	16.67	25.00
8	4	4	4	27.34	50.00	50.00	27.34	50.00	37.50
	5	3	5	21.88	0.00	0.00	21.88	0.00	0.00
	6	2	6	10.94	22.22	22.22	10.94	22.22	25.00
	7	1	7	3.13	0.00	0.00	3.13	0.00	0.00
	8	8	0	0.39	2.78	2.78	0.39	2.78	6.25
12	6	6	6	22.56	40.74	40.74	22.56	40.74	31.25
	7	5	7	19.34	0.00	0.00	19.34	0.00	0.00
	8	4	8	12.09	23.61	23.61	12.09	23.61	23.44
	9	3	9	5.37	0.00	0.00	5.37	0.00	0.00
	10	2	10	1.61	5.56	5.56	1.61	5.56	9.38
	11	1	11	0.29	0.00	0.00	0.29	0.00	0.00
	12	12	0	0.02	0.46	0.46	0.02	0.46	1.56
16	8	8	8	19.64	35.03	35.03	19.64	35.03	27.34
	9	7	9	17.46	0.00	0.00	17.46	0.00	0.00
	10	6	10	12.22	23.46	23.46	12.22	23.46	21.88
	11	5	11	6.67	0.00	0.00	6.67	0.00	0.00
	12	4	12	2.78	7.72	7.72	2.78	7.72	10.94
	13	3	13	0.85	0.00	0.00	0.85	0.00	0.00
	14	2	14	0.18	1.24	1.24	0.18	1.24	3.13
	15	1	15	0.02	0.00	0.00	0.02	0.00	0.00
	16	16	0	0.00	0.08	0.08	0.00	0.08	0.39
20	10	10	10	17.62	31.17	31.17	17.62	31.17	24.61
	11	9	11	16.02	0.00	0.00	16.02	0.00	0.00
	12	8	12	12.01	22.76	22.76	12.01	22.76	20.51
	13	7	13	7.39	0.00	0.00	7.39	0.00	0.00
	14	6	14	3.70	9.26	9.26	3.70	9.26	11.72
	15	5	15	1.48	0.00	0.00	1.48	0.00	0.00
	16	4	16	0.46	2.12	2.12	0.46	2.12	4.40
	17	3	17	0.11	0.00	0.00	0.11	0.00	0.00
	18	2	18	0.02	0.26	0.26	0.02	0.26	0.98
	19	1	19	0.00	0.00	0.00	0.00	0.00	0.00
	20	20	0	0.00	0.01	0.01	0.00	0.01	0.10
28	15	15	13	13.95	0.00	0.00	13.95	0.00	0.00
	16	12	16	11.33	21.06	21.06	11.33	21.06	18.33
	18	18	10	4.89	11.03	11.03	4.89	11.03	12.22

TABLE 7.4 (continued)

Probabilities of Correctly Guessing for Different N and M by Considering the Different True Blocking Sizes and the Prior Knowledge the Guesser Had before the Guess

N	M_{Max}	M	$N-M$	True Blocking Size = 4			True Blocking Size = 2		
				P_{4N} (%)	P_{44} (%)	P_{42} (%)	P_{2N} (%)	P_{24} (%)	P_{22} (%)
	19	9	19	2.57	0.00	0.00	2.57	0.00	0.00
	20	8	20	1.16	3.81	3.81	1.16	3.81	6.11
	21	21	7	0.44	0.00	0.00	0.44	0.00	0.00
	22	6	22	0.14	0.86	0.86	0.14	0.86	2.22
	24	4	24	0.01	0.12	0.12	0.01	0.12	0.56
	25	3	25	0.00	0.00	0.00	0.00	0.00	0.00
	27	27	1	0.00	0.00	0.00	0.00	0.00	0.00
	28	28	0	0.00	0.00	0.00	0.00	0.00	0.01
36	18	18	18	13.21	23.08	23.08	13.21	23.08	18.55
	20	16	20	10.63	19.50	19.50	10.63	19.50	16.69
	21	15	21	8.10	0.00	0.00	8.10	0.00	0.00
	24	12	24	1.82	5.14	5.14	1.82	5.14	7.08
	27	9	27	0.14	0.00	0.00	0.14	0.00	0.00
	28	8	28	0.04	0.36	0.36	0.04	0.36	1.17
	30	6	30	0.00	0.06	0.06	0.00	0.06	0.31
	32	4	32	0.00	0.01	0.01	0.00	0.01	0.06
	33	3	33	0.00	0.00	0.00	0.00	0.00	0.00
	36	36	0	0.00	0.00	0.00	0.00	0.00	0.00
44	23	21	23	11.44	0.00	0.00	11.44	0.00	0.00
	24	20	24	10.01	18.18	18.18	10.01	18.18	15.42
	26	18	26	5.85	12.06	12.06	5.85	12.06	11.86
	27	27	17	3.90	0.00	0.00	3.90	0.00	0.00
	28	16	28	2.37	6.10	6.10	2.37	6.10	7.62
	29	15	29	1.31	0.00	0.00	1.31	0.00	0.00
	30	30	14	0.65	2.36	2.36	0.65	2.36	4.07
	32	12	32	0.12	0.69	0.69	0.12	0.69	1.78
	33	33	11	0.04	0.00	0.00	0.04	0.00	0.00
	35	9	35	0.00	0.00	0.00	0.00	0.00	0.00
	36	8	36	0.00	0.03	0.03	0.00	0.03	0.17
	38	6	38	0.00	0.00	0.00	0.00	0.00	0.04
	39	39	5	0.00	0.00	0.00	0.00	0.00	0.00
	40	4	40	0.00	0.00	0.00	0.00	0.00	0.01
	41	3	41	0.00	0.00	0.00	0.00	0.00	0.00
	42	42	2	0.00	0.00	0.00	0.00	0.00	0.00
	44	44	0	0.00	0.00	0.00	0.00	0.00	0.00
52	27	27	25	10.60	0.00	0.00	10.60	0.00	0.00
	28	24	28	9.47	17.08	17.08	9.47	17.08	14.39

(*continued*)

Controversial Statistical Issues in Clinical Trials

TABLE 7.4 (continued)

Probabilities of Correctly Guessing for Different N and M by Considering the Different True Blocking Sizes and the Prior Knowledge the Guesser Had before the Guess

N	M_{Max}	M	$N-M$	True Blocking Size = 4			True Blocking Size = 2		
				P_{4N} (%)	P_{44} (%)	P_{42} (%)	P_{2N} (%)	P_{24} (%)	P_{22} (%)
	30	30	22	6.01	12.07	12.07	6.01	12.07	11.51
	31	21	31	4.26	0.00	0.00	4.26	0.00	0.00
	32	20	32	2.80	6.78	6.78	2.80	6.78	7.92
	33	33	19	1.70	0.00	0.00	1.70	0.00	0.00
	34	18	34	0.95	3.03	3.03	0.95	3.03	4.66
	36	16	36	0.23	1.08	1.08	0.23	1.08	2.33
	37	15	37	0.10	0.00	0.00	0.10	0.00	0.00
	39	39	13	0.01	0.00	0.00	0.01	0.00	0.00
	40	12	40	0.01	0.07	0.07	0.01	0.07	0.34
	42	42	10	0.00	0.01	0.01	0.00	0.01	0.10
	43	9	43	0.00	0.00	0.00	0.00	0.00	0.00
	44	8	44	0.00	0.00	0.00	0.00	0.00	0.02
	45	45	7	0.00	0.00	0.00	0.00	0.00	0.00
	46	6	46	0.00	0.00	0.00	0.00	0.00	0.00
	48	4	48	0.00	0.00	0.00	0.00	0.00	0.00
	49	3	49	0.00	0.00	0.00	0.00	0.00	0.00
	51	51	1	0.00	0.00	0.00	0.00	0.00	0.00
	52	52	0	0.00	0.00	0.00	0.00	0.00	0.00
60	30	30	30	10.26	17.85	17.85	10.26	17.85	14.45
	32	28	32	9.00	16.15	16.15	9.00	16.15	13.54
	33	27	33	7.63	0.00	0.00	7.63	0.00	0.00
	36	24	36	3.13	7.25	7.25	3.13	7.25	8.06
	39	21	39	0.69	0.00	0.00	0.69	0.00	0.00
	40	20	40	0.36	1.47	1.47	0.36	1.47	2.80
	42	18	42	0.08	0.49	0.49	0.08	0.49	1.33
	44	16	44	0.01	0.13	0.13	0.01	0.13	0.55
60	45	15	45	0.01	0.00	0.00	0.01	0.00	0.00
	48	12	48	0.00	0.01	0.01	0.00	0.01	0.06
	51	9	51	0.00	0.00	0.00	0.00	0.00	0.00
	52	8	52	0.00	0.00	0.00	0.00	0.00	0.00
	54	6	54	0.00	0.00	0.00	0.00	0.00	0.00
	56	4	56	0.00	0.00	0.00	0.00	0.00	0.00
	57	3	57	0.00	0.00	0.00	0.00	0.00	0.00
	60	60	0	0.00	0.00	0.00	0.00	0.00	0.00
68	35	33	35	9.37	0.00	0.00	9.37	0.00	0.00
	36	32	36	8.58	15.35	15.35	8.58	15.35	12.83
	38	30	38	6.06	11.77	11.77	6.06	11.77	10.80
	39	39	29	4.66	0.00	0.00	4.66	0.00	0.00

TABLE 7.4 (continued)

Probabilities of Correctly Guessing for Different N and M by Considering the Different True Blocking Sizes and the Prior Knowledge the Guesser Had before the Guess

N	M_{Max}	M	$N - M$	True Blocking Size = 4			True Blocking Size = 2		
				P_{4N} (%)	P_{44} (%)	P_{42} (%)	P_{2N} (%)	P_{24} (%)	P_{22} (%)
	40	28	40	3.38	7.57	7.57	3.38	7.57	8.10
	41	27	41	2.31	0.00	0.00	2.31	0.00	0.00
	42	42	26	1.48	4.08	4.08	1.48	4.08	5.40
	44	24	44	0.51	1.85	1.85	0.51	1.85	3.19
	45	45	23	0.27	0.00	0.00	0.27	0.00	0.00
	47	21	47	0.06	0.00	0.00	0.06	0.00	0.00
	48	20	48	0.03	0.22	0.22	0.03	0.22	0.76
	50	18	50	0.00	0.06	0.06	0.00	0.06	0.31
	51	51	17	0.00	0.00	0.00	0.00	0.00	0.00
	52	16	52	0.00	0.01	0.01	0.00	0.01	0.11
	53	15	53	0.00	0.00	0.00	0.00	0.00	0.00
	54	54	14	0.00	0.00	0.00	0.00	0.00	0.03
	56	12	56	0.00	0.00	0.00	0.00	0.00	0.01
	57	57	11	0.00	0.00	0.00	0.00	0.00	0.00
	59	9	59	0.00	0.00	0.00	0.00	0.00	0.00
	60	8	60	0.00	0.00	0.00	0.00	0.00	0.00
	62	6	62	0.00	0.00	0.00	0.00	0.00	0.00
	63	63	5	0.00	0.00	0.00	0.00	0.00	0.00
	64	4	64	0.00	0.00	0.00	0.00	0.00	0.00
	65	3	65	0.00	0.00	0.00	0.00	0.00	0.00
	66	66	2	0.00	0.00	0.00	0.00	0.00	0.00
	68	68	0	0.00	0.00	0.00	0.00	0.00	0.00
76	40	36	40	8.22	14.65	14.65	8.22	14.65	12.22
	44	44	32	3.57	7.79	7.79	3.57	7.79	8.09
	48	28	48	0.66	2.20	2.20	0.66	2.20	3.52
	52	52	24	0.05	0.33	0.33	0.05	0.33	0.99
	56	20	56	0.00	0.03	0.03	0.00	0.03	0.17
	60	60	16	0.00	0.00	0.00	0.00	0.00	0.02
	64	12	64	0.00	0.00	0.00	0.00	0.00	0.00
	68	68	8	0.00	0.00	0.00	0.00	0.00	0.00
	72	4	72	0.00	0.00	0.00	0.00	0.00	0.00
	76	76	0	0.00	0.00	0.00	0.00	0.00	0.00
84	44	44	40	7.90	14.04	14.04	7.90	14.04	11.68
	48	36	48	3.71	7.93	7.93	3.71	7.93	8.04
	52	52	32	0.81	2.53	2.53	0.81	2.53	3.79
	56	28	56	0.08	0.46	0.46	0.08	0.46	1.20
	60	60	24	0.00	0.05	0.05	0.00	0.05	0.25

(*continued*)

TABLE 7.4 (continued)

Probabilities of Correctly Guessing for Different N and M by Considering the Different True Blocking Sizes and the Prior Knowledge the Guesser Had before the Guess

N	M_{Max}	M	$N - M$	True Blocking Size = 4			True Blocking Size = 2		
				P_{4N} (%)	P_{44} (%)	P_{42} (%)	P_{2N} (%)	P_{24} (%)	P_{22} (%)
	64	20	64	0.00	0.00	0.00	0.00	0.00	0.03
	68	68	16	0.00	0.00	0.00	0.00	0.00	0.00
	72	12	72	0.00	0.00	0.00	0.00	0.00	0.00
	76	76	8	0.00	0.00	0.00	0.00	0.00	0.00
	80	4	80	0.00	0.00	0.00	0.00	0.00	0.00
	84	84	0	0.00	0.00	0.00	0.00	0.00	0.00
92	48	44	48	7.61	13.50	13.50	7.61	13.50	11.21
	52	52	40	3.82	8.01	8.01	3.82	8.01	7.97
	56	36	56	0.95	2.82	2.82	0.95	2.82	4.01
	60	60	32	0.12	0.59	0.59	0.12	0.59	1.41
	64	28	64	0.01	0.07	0.07	0.01	0.07	0.34
	68	68	24	0.00	0.01	0.01	0.00	0.01	0.06
	72	20	72	0.00	0.00	0.00	0.00	0.00	0.01
	76	76	16	0.00	0.00	0.00	0.00	0.00	0.00
	80	12	80	0.00	0.00	0.00	0.00	0.00	0.00
	84	84	8	0.00	0.00	0.00	0.00	0.00	0.00
	88	4	88	0.00	0.00	0.00	0.00	0.00	0.00
	92	92	0	0.00	0.00	0.00	0.00	0.00	0.00
100	52	52	48	7.35	13.02	13.02	7.35	13.02	10.80
	56	44	56	3.90	8.05	8.05	3.90	8.05	7.88
	60	60	40	1.08	3.08	3.08	1.08	3.08	4.19
	64	36	64	0.16	0.73	0.73	0.16	0.73	1.60
	68	68	32	0.01	0.11	0.11	0.01	0.11	0.44
	72	28	72	0.00	0.01	0.01	0.00	0.01	0.08
	76	76	24	0.00	0.00	0.00	0.00	0.00	0.01
	80	20	80	0.00	0.00	0.00	0.00	0.00	0.00
	84	84	16	0.00	0.00	0.00	0.00	0.00	0.00
	88	12	88	0.00	0.00	0.00	0.00	0.00	0.00
	92	92	8	0.00	0.00	0.00	0.00	0.00	0.00
	96	4	96	0.00	0.00	0.00	0.00	0.00	0.00
	100	100	0	0.00	0.00	0.00	0.00	0.00	0.00

1. Common findings:
   a. The probabilities for guessing M subjects' treatment right are equal to those for guessing $(N - M)$ subjects' treatment right.
   b. P_{44}, P_{42}, P_{24}, and P_{22} are equal to 0 for the odd values of M.
2. Comparison of P_{4N}, P_{44}, and P_{42} for the design with the blocking size of 4:
   a. P_{44} is equal to P_{42}, which means there is no impact on the probability of correctly guessing whether the guesser thought the true blocking size is 2 or 4 before the guess.
   b. P_{44} and P_{42} are always greater than P_{4N} for all N and M_{Max} if M is even. In addition, the difference between P_{44} and P_{4N} becomes larger when N increases.
3. Comparison of P_{2N}, P_{22}, and P_{24} for the design with the blocking size of 2:
   a. P_{22} is always smaller than P_{2N} for all N and M_{Max}. In addition, the difference between P_{22} and P_{2N} becomes larger when N increases.
   b. P_{24} is greater than P_{22} for small M_{Max} for all N if M is even. However, it becomes smaller than P_{22} when M_{Max} is larger.
   c. P_{24} is always greater than P_{2N} for all N and M_{Max}. In addition, the difference between P_{24} and P_{2N} becomes larger when N increases.
4. Comparison of P_{4N}, P_{44}, and P_{42} for the design with the blocking size of 4 with P_{2N}, P_{22}, and P_{24} for the design with the blocking size of 2:
   a. P_{4N} is equal to P_{2N}, which means there is no difference between correctly guessing the designs with the blocking sizes of 4 and 2 if the guesser guessed without any prior knowledge.
   b. Since $P_{4N} = P_{2N}$, the comparison between P_{4N} and P_{22} and that between P_{4N} and P_{24} are the same as the comparison between P_{2N} and P_{22} and that between P_{2N} and P_{24}, respectively.
   c. P_{44} is greater than P_{22} for a small M_{Max}, while P_{22} becomes larger than P_{44} when M_{Max} becomes larger.
   d. P_{44} and P_{42} are equal to P_{24}, which means the probabilities of correctly guessing are the same when the design with the blocking size of 4 is chosen or the blocking size the guesser thought of before the guess was 4.
5. The comparison between P_{42} and P_{2N} and that between P_{42} and P_{22} are the same as the comparison between P_{24} and P_{2N} and that between P_{24} and P_{22} since $P_{42} = P_{24}$.

7.3.3 Remarks

The design with the blocking size of 4 is usually considered to have less selection bias than the design with the blocking size of 2 because the true treatment is harder to guess. However, prior knowledge about the true blocking size the guesser had before the guess is also a factor that has an impact on the probability of correctly guessing. As the results show in the numerical study, the probabilities of correctly guessing the designs with the true blocking sizes of 4 and 2 are smallest and equal if there is no prior knowledge about the true blocking size before the guess. The results are obvious since the probability of guessing right for the true treatment of each individual subject without any prior knowledge is $1/2$, which is just like tossing a fair coin. However, the probability of correctly guessing between two types of probabilities with the true blocking size is the same as what the guesser thought of before the guess, i.e., P_{44} versus P_{22}, P_{44} is greater than P_{22} for small M_{Max} while the results are opposite when M_{Max} becomes larger. On the other hand, P_{44} is even equal to P_{42} and P_{24}, i.e., the probability of correctly guessing if the true blocking size is 4 with the prior knowledge the guesser had is the blocking size of 2 and the true blocking size is 2 with the prior knowledge the guesser had is the blocking size of 4, respectively. The results seem different from what we think usually, i.e., the probabilities of correctly guessing for the design with the blocking size of 4 is always lower than that for the design with the blocking size of 2. However, not only does it show what the true blocking size is, but the prior knowledge the guesser had before the guess also has great impact on the probabilities of guessing correctly.

The choice of blocking size for randomized trials depends not only on the number of treatments but also on the sample size for the clinical trials. In practice, the probabilities of correctly guessing will be reduced if the blocking size becomes larger but it may result in the imbalance of treatment assignment, especially if patient characteristics change with time. On the other hand, with respect to the two-arm trial with a small sample size, the probabilities of correctly guessing are not small for both the designs with the blocking sizes of 4 and 2, in particular when M_{Max} is small. Therefore, the design with the blocking size of at least 6 or the design with mixed blocking sizes rather than only the blocking size of 4 may need to be suggested for a two-arm trial.

7.4 Test for Integrity of Blinding

Consider the following example given in Karlowski et al. (1975). A double-blind placebo-controlled study was conducted by the National Institutes of Health to evaluate the difference between the prophylactic and therapeutic effects of ascorbic acid for the common cold. At the completion of the study, a questionnaire was distributed to every subject enrolled in the study so that they could guess which treatment they received. The results from the 190 subjects (101 subjects are in the actual treatment group and 89 subjects are in the placebo group) who completed the study are summarized as follows. Among the 101 subjects in the actual treatment group, 40 subjects guessed right, 12 subjects guessed wrong, and 49 subjects indicated "Do not know." For the 89 subjects in the placebo group, 39 subjects guessed right, 11 subjects guessed wrong, and 39 subjects indicated "Do not know."

To test the integrity of blinding we need to define a null hypothesis H_0. If patients guess their treatment codes randomly, then blindness is considered to be preserved. Thus, we consider

$$H_0 : \text{patients guess their treatment codes randomly.}$$

Let A_i be the event that a patient guesses he/she is in the ith group and B_j be the event that a patient is assigned to the jth group. If a patient guesses his/her treatment code randomly, then the events A_i and B_j are independent for any i and j, and $P(A_i) = 1/2$. Assume that patients who answered "Do not know" did not guess their treatment codes throughout the study. Let m_j be the number of patients in group j who guessed their treatment codes, $j = 1, 2$. Then, under the null hypothesis H_0, we have

$$P \text{ (a patient in group } j \text{ guesses that he/she is in group } i)$$

$$= P(A_i \cap B_j) = P(A_i)P(B_j) = \frac{m_j}{2(m_1 + m_2)}, \quad j = 1, 2. \tag{7.5}$$

Let a_{ij} be the observed number (frequency) of the patients who are in the jth group and guessed that they are in the ith group. Then the integrity of blinding can be tested by analyzing a contingency table (Table 7.5), where the numbers in parentheses are the expected frequencies under H_0 computed according to (7.5).

For example, with the data given in Table 7.5, we obtain a contingency table (Table 7.6).

Based on Table 7.6, we can use either Fisher's exact test or Person's chi-square test to test for the integrity of blinding. This test for the integrity of blinding can be generalized to the case where there are a treatment groups, which leads to an $a \times a$ contingency table. Analyses on investigators' guesses of patients' treatment codes can be performed similarly.

Consider a single-site parallel design comparing $a \geq 2$ treatments. Let A_{ij} be the event that a patient in the jth treatment group guesses that he/she is in

TABLE 7.5

Contingency Table for the Integrity of Blinding

Patient's Guess	Actual Assignment		
	Group 1	Group 2	Total
Group 1	$o_{11}\left(\dfrac{m_1}{2}\right)$	$o_{12}\left(\dfrac{m_1}{2}\right)$	$o_{11}+o_{12}\left(\dfrac{m_1+m_2}{2}\right)$
Group 2	$o_{21}\left(\dfrac{m_1}{2}\right)$	$o_{22}\left(\dfrac{m_1}{2}\right)$	$o_{21}+o_{22}\left(\dfrac{m_1+m_2}{2}\right)$
Total	m_1	m_2	

TABLE 7.6

Contingency Table for Patients' Guess

Patient's Guess	Actual Assignment		
	Active Treatment	Placebo	Total
Active treatment	40 (26)	11 (25)	51 (25.5)
Placebo	12 (26)	39 (25)	51 (25.5)
Total	52	50	

the ith group; $i = 1, \ldots, a, a + 1$, where $i = a + 1$ defines the event that a patient does not guess (or answers "do not know"). If the hypothesis

$$H_0 : P(A_{ij}) = P(A_{1j}) \quad \text{for any } i \text{ and } j$$

holds, then the blindness is considered to be preserved. We can test H_0 using the well-known Pearson chi-square test (with $a(a - 1)$ degrees of freedom) under the contingency tables constructed based on the observed counts. A straightforward calculation using data results in the observed Pearson's chi-square statistic of 31.3, which results in a p-value smaller than 0.001. Thus, we conclude with a very high significance level that the blindness is not preserved. Hence, the integrity of blinding is in doubt.

7.5 Analysis under Breached Blindness

When the test of the integrity of blinding produces a significant result, analyzing the data by ignoring this result may lead to a biased result (i.e., the integrity of blinding is doubtful). In what follows we introduce a method of testing treatment effects by incorporating the data of patients' guesses of their treatment codes (Chow and Shao, 2003, 2004). The idea is to include a patient's guess as a factor in the analysis of variance (ANOVA) for the treatment effects.

Suppose that the study design is a single-site parallel design comparing $a \geq 2$ treatments. If the blindness is preserved, then the treatment effects can be tested using the one-way ANOVA table. If we add patients' or investigators' guessing treatment codes as a factor, then we can test treatment effects by using a two-way ANOVA table. If we add both patients' and investigators' guessing treatment codes as factors, then we can test treatment effects by using a three-way ANOVA table. If the study is a multicenter trial, then including guessing factors leads to a three-way or two-way ANOVA. For illustration purpose, consider adding one guessing factor, γ with b levels, into a single study site (i.e., one-way ANOVA is used if the guessing factor is ignored). There are different ways for constructing the variable γ. One way is to use the guessing treatment i, $i = 1, ..., a$, as the first a levels of γ and not guessing (do not know) as the last level. Hence $b = a + 1$. Another way is to use guessing correctly, guessing incorrectly, and not guessing as three levels for γ and thus, $b = 3$. Even if the original design is balanced, i.e., each treatment (and center) has the same number of patients, the two-way ANOVA or three-way ANOVA after including factor γ is not balanced. Hence methods for unbalanced ANOVA are necessarily considered.

Let x_{ijk} be the response from the kth patient under the ith treatment with the jth guessing status, where $i = 1, ..., a$, $j = 1, ..., b$, $k = 1, ..., n_{ij}$, and n_{ij} is the number of patients in the (i, j)th cell. Let $\bar{x}_{ij.}$ be the sample mean of the patients in the (i, j)th cell, $\bar{x}_{i..}$ be the sample mean of the patients under treatment i, $\bar{x}_{.j.}$ be the sample mean of patients with guessing status j, $\bar{x}$ be the sample mean of all patients, $n_{i.}$ be the number of patients under treatment i, $n_{.j}$ be the number of patients with guessing status j, and n be the total number of patients. Define $R(\mu) = n\bar{x}^2$, $R(\mu, \tau) = \sum_{i=1}^{a} n_{i.}\bar{x}_{i..}^2$ (where τ denotes treatment effect and μ denotes the overall mean), $R(\mu, \gamma) = \sum_{j=1}^{b} n_{.j}\bar{x}_{.j.}^2$, $R(\mu, \tau, \tau \times \gamma) = \sum_{i=1}^{a} \sum_{j=1}^{b} n_{ij}\bar{x}_{ij.}^2$ (where $\tau \times \gamma$ denotes the interaction between τ and γ), and $R(\mu, \tau, \gamma) = \sum_{i=1}^{a} n_{i.}\bar{x}_{i..}^2 + Z'C^{-1}Z$, where Z is a $(b - 1)$-vector whose jth component is $n_{.j}\bar{x}_{.j.} - \sum_{i=1}^{a} n_{ij}\bar{x}_{i..}$, $j = 1, ..., b-1$ and C is a $(b - 1) \times (b - 1)$ matrix whose jth diagonal element is $n_{.j} - \sum_{i=1}^{a} n_{ij}^2/n_{i.}$ and (j, l)th off-diagonal element is $-\sum_{i=1}^{a} n_{ij}n_{il}/n_{i.}$. Now let

$$R(\tau \times \gamma \mid \mu, \tau, \gamma) = R(\mu, \tau, \gamma, \tau \times \gamma) - R(\mu, \tau, \gamma),$$

$$R(\tau \mid \mu) = R(\mu, \tau) - R(\mu),$$

$$R(\gamma \mid \mu) = R(\mu, \gamma) - R(\mu),$$

$$R(\tau \mid \mu, \gamma) = R(\mu, \tau, \gamma) - R(\mu, \gamma),$$

$$R(\gamma \mid \mu, \tau) = R(\mu, \tau, \gamma) - R(\mu, \tau),$$

$$\text{SSE} = \sum_{i=1}^{a} \sum_{j=1}^{b} \sum_{k=1}^{n_{ij}} x_{ijk}^2 - R(\mu, \tau, \gamma, \tau \times \gamma)$$

and s be the number of nonzero n_{ij}'s. An ANOVA table according to Searle (1971) can be constructed.

An F-ratio (in the last column of Table 7.7) is said to be significant at level α if it is larger than the $(1 - \alpha)$th quantile of the F-distribution with denominator degrees of freedom $n - s$ and the numerator degrees of freedom given by the number in the third column of the same row. Note that $F(\tau \mid \mu)$ is the F-ratio for testing τ-effect (treatment effect) adjusted for μ and ignoring γ, whereas $F(\tau \mid \mu, \gamma)$ is the F-ratio for testing τ-effect adjusted for both μ and γ. These two F-ratios are the same in a balanced model but are different in an unbalanced model. A similar discussion can be made for $F(\gamma \mid \mu)$ and $F(\gamma \mid \mu, \tau)$.

Because of the imbalance, the interpretation of the results given by F-ratios in the ANOVA table is not straightforward. Table 7.8 lists a total of 14 possible cases according to the significance of F-ratios $F(\tau \mid \mu)$, $F(\tau \mid \mu, \gamma)$, $F(\gamma \mid \mu)$,

TABLE 7.7

ANOVA for Treatment Effects under Breached Blindness

Source	Sum of Squares	df	F-Ratio
τ after μ	$R(\tau \mid \mu)$	$a - 1$	$F(\tau \mid \mu) = \dfrac{R(\tau \mid \mu)/(a-1)}{\text{SSE}/(n-s)}$
γ after μ and τ	$R(\gamma \mid \mu, \tau)$	$b - 1$	$F(\gamma \mid \mu, \tau) = \dfrac{R(\gamma \mid \mu, \tau)/(b-1)}{\text{SSE}/(n-s)}$
γ after μ	$R(\gamma \mid \mu)$	$a - 1$	$F(\gamma \mid \mu) = \dfrac{R(\gamma \mid \mu)/(b-1)}{\text{SSE}/(n-s)}$
τ after μ and γ	$R(\tau \mid \mu, \gamma)$	$b - 1$	$F(\tau \mid \mu, \gamma) = \dfrac{R(\tau \mid \mu, \gamma)/(a-1)}{\text{SSE}/(n-s)}$
Interaction	$R(\tau \times \gamma \mid \mu, \tau, \gamma)$	$s - a - b + 1$	$F(\tau \times \gamma \mid \mu, \tau, \gamma) = \dfrac{R(\tau \times \gamma \mid \mu, \tau, \gamma)/(s-a-b+1)}{\text{SSE}/(n-s)}$
Error	SSE	$n - s$	
Total	SS(TO)	$n - 1$	

s, number of nonzero n_{ij}'s.

TABLE 7.8

Conclusions on the Significance of the Treatment Effect When $F(\tau \times \gamma \mid \mu, \tau, \gamma)$ Is Insignificant

Significance of F-Ratio					
Fitting τ and Then γ After τ		Fitting γ and Then τ After γ		Effects to Be Included in the Model According to Chow and Shao (2004)	Conclusion: Significance of the Treatment Effect
$F(\tau \mid \mu)$	$F(\gamma \mid \mu, \tau)$	$F(\gamma \mid \mu)$	$F(\tau \mid \mu, \gamma)$		
Yes	Yes	Yes	Yes	τ, γ	Yes
Yes	Yes	No	Yes	τ, γ	Yes
Yes	No	Yes	Yes	τ	Yes
Yes	No	No	Yes	τ	Yes
No	Yes	Yes	Yes	τ, γ	Yes
No	Yes	No	Yes	τ, γ	Yes
No	No	No	Yes	τ, γ	Yes
No	Yes	Yes	No	γ	No
No	No	Yes	No	γ	No
Yes	Yes	Yes	No	γ	No
Yes	No	Yes	No	τ, γ	No
Yes	No	No	No	τ	Yes
No	Yes	No	No	τ, γ	No
No	No	No	No	None	No

Source: Chow, S.C. and Shao, J., *Statistics in Medicine*, 23, 1185, 2004.

and $F(\gamma \mid \mu, \tau)$. The suggestion from Searle (1971, Chapter 7) regarding which effects should be included in the model is given in the second last column of Table 7.5. However, our purpose is slightly different, i.e., we are interested in whether the treatment effect τ is significant regardless of the presence of the effect γ. Our recommendations in these 14 cases are given in the last column of Table 7.8, which is interpreted as follows. When both $F(\tau \mid \mu)$ and $F(\tau \mid \mu, \gamma)$ are significant (rows 1 through 4 of Table 7.8), regardless of whether the γ effect is significant or not, the conclusion is easy to make, i.e., the treatment effect is significant. In the next three cases (rows 5 through 7 of Table 7.8), $F(\tau \mid \mu)$ is not significant but $F(\tau \mid \mu, \gamma)$ is significant, indicating that the treatment effect cannot be clearly detected by ignoring γ but once γ is included in the model as a blocking variable, the treatment effect is significant. In these three cases, we conclude that the treatment effect is significant. When $F(\tau \mid \mu, \gamma)$ is not significant but $F(\gamma \mid \mu)$ is significant, it indicates that once γ is fitted into the model, the treatment effect is not significant, i.e., the treatment effect is distorted by the γ effect. In such cases (rows 8 through 11 of Table 7.8), we cannot conclude that the treatment effect is significant. In the last three cases (rows 12 through 14 of Table 7.6), both $F(\gamma \mid \mu)$ and $F(\tau \mid \mu, \gamma)$ are not significant. If $F(\tau \mid \mu)$ is significant but $F(\gamma|\mu, \tau)$ is not (row 12 of Table 7.8), it indicates that γ has no effect and the treatment effect is significant. On the other hand, if $F(\gamma|\mu, \tau)$ is significant but $F(\tau|\mu)$ is not (row 13 of Table 7.8)—a case that *should happen somewhat infrequently* according to Searle (1971)—we cannot conclude that the treatment effect is significant. Finally, when neither $F(\tau|\mu)$ nor $F(\gamma|\mu, \tau)$ is significant, we cannot conclude that the treatment effect is significant (row 14 of Table 7.8).

The analysis is difficult when the interaction $F(\tau \times \gamma|\mu, \tau, \gamma)$ is significant. In general, we cannot conclude that the treatment effect is significant when $F(\tau \times \gamma|\mu, \tau, \gamma)$ is significant. An analysis conditional on the value of γ may be carried out to draw some partial conclusions.

Note that we only focus on the analysis of a single response variable for treatment effects. Although our main idea of adding the guessing treatment code factors into the analysis can be applied to more complex cases (e.g., when there are other response variables or covariates that may be influenced by guessing treatment codes), further research is needed.

7.6 An Example

Consider a double-blind placebo-controlled trial with a two-group parallel design for the evaluation of the effectiveness of an appetite suppressant in weight loss in obese women (see Brownell and Stunkard, 1982). Table 7.9 lists the data on patients' guesses of the treatment codes. Observed mean weight loss (kg) is summarized in Table 7.10.

In this example, the blindness is not preserved with a high significance. If patients' guessing is ignored, then a simple two-sample t-test (which is the same as the one-way ANOVA) results in the observed t-statistic of 2.45 and p-value of 0.009. Hence the treatment effect is significant when patients' guessing is ignored.

Suppose that one would like to know whether the significant result is a biased result due to breached blindness. The method described in the previous section can be applied to reanalyze the data in Table 7.10. First, consider

TABLE 7.9

Results of Patients' Guesses

	Actual Treatment Assignment	
Patient's Guess	**Active Drug**	**Placebo**
Active drug	19	3
Placebo	3	16
Do not know	2	6
Total	24	25

Source: Brownell, K.D. and Stunkard, A.J., *Am. J. Psychiatr.*, 139, 1487, 1982. With permission.

TABLE 7.10

Sample Mean Weight Loss (kg)

Patient's Guess	Actual Treatment Assignment	
	Active Drug	**Placebo**
Active drug	9.6	2.6
Placebo	3.9	6.1
Do not know	12.2	5.8
Total	9.1	5.6

Source: Brownell, K.D. and Stunkard, A.J., *Am. J. Psychiatry*, 139, 1487, 1982. With permission.

the analysis with γ = guessing correctly, guessing incorrectly, and not guessing. The sample means (with estimated standard deviation in parentheses) and sample sizes are given by

$$\bar{x}_{11.} = 9.6(1.14),\ n_{11} = 19,\quad \bar{x}_{21.} = 6.1(1.25),\ n_{21} = 16,$$

$$\bar{x}_{12.} = 3.9(2.88),\ n_{12} = 3,\quad \bar{x}_{22.} = 2.6(2.88),\ n_{22} = 3,$$

$$\bar{x}_{13.} = 12.2(3.53),\ n_{13} = 2,\quad \bar{x}_{23.} = 5.8(2.04),\ n_{23} = 6,$$

$$\bar{x}_{1..} = 9.1(1.02),\ n_{1.} = 24,\quad \bar{x}_{.1.} = 8.0(0.84),\ n_{.1} = 35,$$

$$\bar{x}_{2..} = 5.6(1.00),\ n_{2.} = 25,\quad \bar{x}_{.2.} = 3.3(2.04),\ n_{.2} = 6,$$

$$\bar{x} = 7.3(0.71),\ n = 49,\quad \bar{x}_{.3.} = 7.4(1.76),\ n_{.3} = 8.$$

The resulting ANOVA table is summarized here.

Source	R	df	R/df	F-Ratio	p-Value
τ after μ	$R(\tau\,|\,\mu)$	1	160.2	6.43	0.015
γ after μ and τ	$R(\gamma\,|\,\mu,\tau)$	2	57.6	2.31	0.111
γ after μ	$R(\gamma\,|\,\mu)$	2	66.1	2.65	0.082
τ after μ and γ	$R(\tau\,|\,\mu,\gamma)$	1	143.2	5.75	0.021
Interaction	$R(\tau\times\gamma\,|\,\mu,\tau,\gamma)$	2	12.6	0.51	0.604
Error	SSE	43	24.9		

Source: Chow, S.C. and Shao, J., *Stat. Med.*, 23, 1185, 2004. © 2004 by John Wiley & Sons Ltd. With permission.

It seems that the interaction $F(\tau \times \gamma\,|\,\mu, \tau, \gamma)$ is not significant and both treatment effect F-ratios $F(\tau\,|\,\mu)$ and $F(\tau\,|\,\mu, \gamma)$ are significant. Thus, according to the previous section (see Table 7.8), we can conclude that the treatment effect is significant, regardless of whether the effect of γ is significant or not. However, the conclusion may be different if we consider the levels of γ, the sample means (with estimated standard deviation in parentheses), and the sample sizes given by

$$\bar{x}_{11.} = 9.6(1.14),\ n_{11} = 19,\quad \bar{x}_{21.} = 2.6(2.88),\ n_{21} = 3,$$

$$\bar{x}_{12.} = 3.9(2.88),\ n_{12} = 3,\quad \bar{x}_{22.} = 6.1(1.25),\ n_{22} = 16,$$

$$\bar{x}_{13.} = 12.2(3.53),\ n_{13} = 2,\quad \bar{x}_{23.} = 5.8(2.04),\ n_{23} = 6,$$

$$\bar{x}_{1..} = 9.1(1.02),\ n_{1.} = 24,\quad \bar{x}_{.1.} = 8.7(1.06),\ n_{.1} = 22,$$

$$\bar{x}_{2..} = 5.6(1.00),\ n_{2.} = 25,\quad \bar{x}_{.2.} = 5.7(1.14),\ n_{.2} = 19,$$

$$\bar{x} = 7.3(0.71),\ n = 49,\quad \bar{x}_{.3.} = 7.4\,(1.76),\ n_{.3} = 8.$$

Note that $\bar{x}_{1i.}$ are unchanged but $\bar{x}_{2j.}$ have changed with this new choice of levels of γ. The corresponding ANOVA table is given below. As can be seen from the ANOVA table, although $F(\tau|\mu)$ remains the same, the value of $F(\tau \times \gamma|\mu, \tau, \gamma)$ is much larger than that in the previous case. The p-value corresponding to $F(\tau \times \gamma|\mu, \tau, \gamma)$ is 0.097, which indicates that the interaction between the treatment and γ is marginally significant. If this interaction is ignored, then we may conclude that the treatment effect is significant, since the results are the same as those in the previous case except that $F(\tau|\mu, \gamma)$ is less significant. But no conclusion can be made if the interaction effect cannot be ignored.

Source	R	df	R/df	F-Ratio	p-Value
τ after μ	$R(\tau\|\mu)$	1	160.2	6.43	0.015
γ after μ and τ	$R(\gamma\|\mu,\tau)$	2	8.9	0.36	0.700
γ after μ	$R(\gamma\|\mu)$	2	54.7	2.20	0.123
τ after μ and γ	$R(\tau\|\mu,\gamma)$	1	68.6	2.76	0.104
Interaction	$R(\tau\times\gamma\|\mu,\tau,\gamma)$	2	61.3	2.46	0.097
Error	SSE	43	24.9		

Source: Chow, S.C. and Shao, J., *Stat. Med.*, 23, 1185, 2004. © 2004 by John Wiley & Sons Ltd. With permission.

It can be seen from this example that the choice of levels of γ is important. Different ways of constructing the levels of γ may lead to different conclusions. In this example, it seems that the first method of constructing the level of γ (guessing correctly, guessing incorrectly, and not guessing) is better, since the guessing factor has less interaction with the treatment effect.

In the presence of interaction, however, a subgroup analysis (according to the levels of γ) may be useful. Subgroup sample mean comparisons can

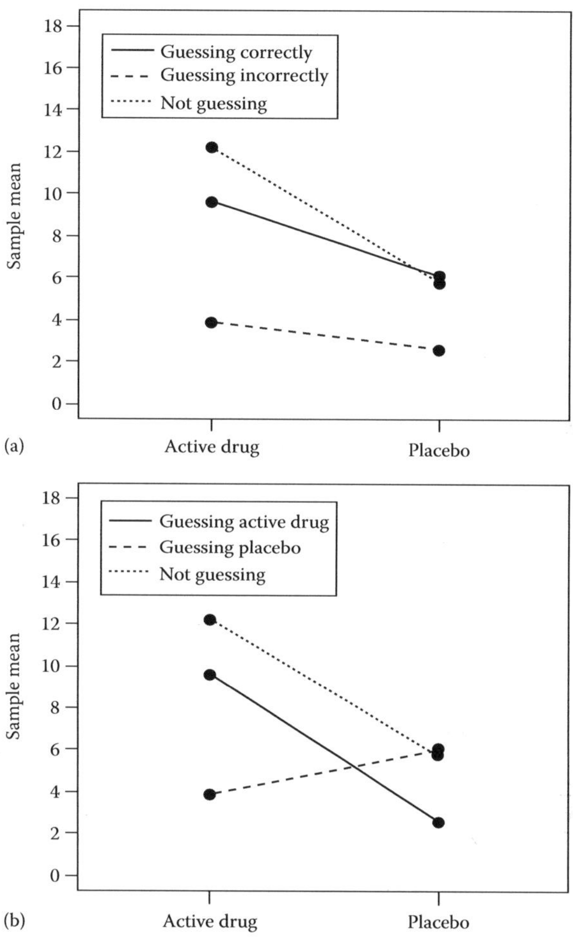

FIGURE 7.1
Subgroup sample mean weight loss (kg).

be made as indicated in Figure 7.1. Figure 7.1 displays six subgroup sample means $\bar{x}_{ij}$, $i = 1, 2, j = 1, 2, 3$. The first part of Figure 7.1 considers the situation where γ = guessing active drug, guessing placebo, and not guessing. The two sample means (dots) corresponding to the same γ level are connected by a straight line segment. In the first part of the figure, although the three line segments have different slopes, the slopes have the same sign. Furthermore, every pair of two line segments either does not cross or crosses slightly. This indicates that in the situation considered in the first part of the figure, there is no significant interaction and the treatment effect is evident. On the other hand, the slopes of the line segments in the second part of the figure have different signs and two line segments cross considerably, which indicates that interaction is significant and we cannot draw an overall conclusion on the treatment effect in the situation considered in the second part of the figure. A partial conclusion that can be drawn from the second part of the figure is that the treatment effect is significant when we focus on patients not guessing their treatment codes.

7.7 Concluding Remarks

When the integrity of blinding is doubtful, adjustments to statistical analysis should be made. One of the controversial issues regarding the blinding is whether a formal statistical test for the integrity of the blinding should be performed at the end of the clinical trial (especially when significantly positive results are observed). In addition, what action should be taken if a positive clinical trial fails to pass the test for the integrity of the blinding? Should the positive clinical trial be questioned and/or challenged? On the other hand, what action should be taken if a negative clinical trial fails to pass the test for the integrity of the blinding? In this case, should the data (or subgroup) be reanalyzed for a more accurate and reliable assessment of the treatment effect?

Regarding the impact of different blocking sizes in the randomization of a clinical trial, it should be noted that the knowledge of the blocking size may increase the probability of guessing the treatment codes right for the investigator. Although the increase of the blocking size may decrease the probability of guessing the treatment codes right, it will also increase the probability of mixing up the randomization schedule and the possibility of treatment imbalance at the end of the trial. Note that the discussions given in the previous sections are based on an unbiased coin design. Analysis based on a biased coin design can be performed similarly.

8

Clinical Strategy for Endpoint Selection

8.1 Introduction

In clinical trials, it is important to determine the primary response variables for addressing the scientific and/or medical questions of interest. The response variables, which are also known as the clinical endpoints, are usually chosen to fulfill the study objectives. Once the response variables are chosen, the possible outcomes of treatment are defined and the corresponding information would be used to assess the efficacy and safety of the study drug under investigation. Typically, to assess the efficacy and safety of a study drug, the study drug is first shown to be statistically significantly different from a placebo control. If there is a statistically significant difference, the trial is demonstrated to have a high probability of correctly detecting a clinically meaningful difference, which is known as the (statistical) power of the trial. Therefore, in practice, a pre-study power analysis for sample size estimation is usually performed to ensure that the trail with the intended sample size has a desired power, say 80%, for addressing the scientific/medical question of interest. The purpose is to find an appropriate sample size based on the information (the desired power, variability, and clinically meaningful differences, etc.) provided by clinical scientists.

In many clinical studies, it is not uncommon that the sample size of a study is determined based on expected absolute change from the baseline of a primary study endpoint but the collected data are analyzed based on relative change from the baseline (e.g., percent change from baseline) of the primary study endpoint, or based on the percentage of patients who show some improvement (i.e., responder analysis). The definition of a responder could be based on either absolute change from the baseline or relative change from the baseline of the primary study endpoint. It is very controversial in terms of the interpretation of the analysis results, especially when a significant result is observed based on a particular study endpoint (e.g., absolute change from baseline, relative change from baseline, or responder analysis) but not on other study endpoints (e.g., absolute change from baseline, relative change from baseline, or responder analysis). In practice, it is then of interest to explore how an observed significant difference of a study endpoint (e.g., absolute change from baseline, relative change from baseline, or responder

TABLE 8.1

Weight Data from 10 Female Subjects

Pretreatment	Posttreatment	Absolute Change	Relative Change
110	106	4	3.6
90	80	10	11.1
105	100	5	4.8
95	93	2	2.2
170	163	7	4.1
90	84	8	8.9
150	145	5	3.3
135	131	4	3.0
160	159	1	0.6
100	91	9	9.0
120.5 (30.5)	115.2 (31.53)	5.3	5.1

Note: Numbers in the parentheses are the corresponding standard deviation.

analysis) can be translated to that of another study endpoint (e.g., absolute change from baseline, relative change from baseline, or responder analysis). An immediate impact on the assessment of treatment effect based on different study endpoints is the power analysis for sample size calculation. For example, sample size required for achieving a desired power based on the absolute change could be very different from that obtained based on the percent change, or the percentage of patients who show an improvement based on the absolute change or relative change at α level of significance. As an example, consider a clinical trial for the evaluation of possible weight reduction of a test treatment in female patients. Weight data from 10 subjects are given in Table 8.1.

As can be seen from Table 8.1, mean absolute change and mean percent change from pretreatment are 5.3 lb and 5.1%, respectively. If a subject is considered a responder when there is weight reduction of more than 5 lb (absolute change) or by more than 5% (relative change), the response rates based on absolute change and relative change are given by 60% and 30%, respectively. It should be noted that sample sizes required for achieving a desired power for detecting a clinically meaningful difference, say, by an absolute change of 5.0 lb and a relative change of 5.0%, for the two study endpoints would not be the same. Similarly, the required sample sizes are also different using the response rates based on absolute change and relative change. Table 8.2 summarizes sample size calculation based on absolute change, relative change, and responders (defined based on either absolute change or relative change).

In clinical trials, one of the most controversial issues regarding clinical endpoint selection is which clinical endpoint is telling the truth. The other controversial issue is how to translate clinical results among the study endpoints. In practice, the sponsors always choose the clinical endpoints in their best interest.

TABLE 8.2

Sample Size Calculation

Study Endpoint	Clinical Meaningful Difference	Sample Size Required
Absolute change	5 lb	262
Relative change	5%	146
Responder (based on absolute change)[a]	>5 lb	12
Responder (based on relative change)[b]	>5%	19

[a] Response rate based on absolute change greater than 5 lb is 60%.
[b] Response rate based on relative change greater than 5% is 30%.

The regulatory agencies, however, require the primary clinical endpoint to be specified in the study protocol. Positive results from other clinical endpoints will not be considered as the primary analysis results for regulatory approval. This, however, does not have any scientific or statistical justification for the assessment of the treatment effect of the test drug under investigation.

In this chapter, we attempt to provide some insight to the above issues. In particular, the focus is to evaluate the effect on the power of the test when the sample size of the clinical study is determined by an alternative clinical strategy based on different study endpoints and non-inferiority margins. In the next section, models and assumptions for studying the relationship among these study endpoints are described. Under the model, translations among different study endpoints are studied. Section 8.4 provides a comparison of different clinical strategies for endpoint selections in terms of sample size and the corresponding power. A numerical study is given in Section 8.5 to provide some insight regarding the effect to the different clinical strategies for endpoint selection. A brief concluding remark is presented in the last section.

8.2 Clinical Strategy for Endpoint Selection

In clinical trials, for a given primary response variable, commonly considered study endpoints include (1) measure based on absolute change (e.g., endpoint change from baseline), (2) measure based on relative change, (3) proportion of responders based on absolute change, and (4) proportion of responders based on relative change. We will refer to these study endpoints as the *derived study endpoints* because they are derived from the original data collected from the same patient population. In practice, it will be more complicated if the intended trial is to establish non-inferiority of a test treatment to an active control (reference) treatment. In this case, sample size calculation will also depend on the size of the non-inferiority margin, which may be based on either absolute change or relative change of the derived study endpoint. For example, based

TABLE 8.3

Clinical Strategy for Endpoint Selection in Non-Inferiority Trials

Study Endpoint	Non-Inferiority Absolute Difference (δ_1)	Margin Relative Difference (δ_2)
Absolute change (E_1)	$I = E_1\delta_1$	$II = E_1\delta_2$
Absolute change (E_2)	$III = E_2\delta_1$	$IV = E_2\delta_2$
Responder based on absolute change (E_3)	$V = E_3\delta_1$	$VI = E_3\delta_2$
Responder based on relative change (E_4)	$VII = E_4\delta_1$	$VII = E_4\delta_2$

on the responder analysis, we may want to detect a 30% difference in response rate or to detect a 50% relative improvement in response rate. Thus, in addition to the four types of derived study endpoints, there are also two different ways to define a non-inferiority margin. Thus, there are many possible clinical strategies with different combinations of the derived study endpoint and the selection of non-inferiority margin for the assessment of the treatment effect. These clinical strategies are summarized in Table 8.3.

To ensure the success of an intended clinical trial, the sponsor will usually carefully evaluate all possible clinical strategies for selecting the type of study endpoint, clinically meaningful difference, and non-inferiority margin during the stage of protocol development. In practice, some strategies may lead to the success of the intended clinical trial (i.e., achieve the study objectives with the desired power), while others may not. A common practice for the sponsor is to choose a strategy in their best interest. However, regulatory agencies such as the FDA may challenge the sponsor regarding inconsistent results. This has raised the following questions. First, which study endpoint is telling the truth regarding the efficacy and safety of the test treatment under study? Second, how to translate the clinical information among different derived study endpoints since they are obtained based on the same data collected from the same patient population? Tse and Chow (2011) made an attempt to address these questions in the following sections. These questions, however, remain unanswered.

8.3 Translations among Clinical Endpoints

Suppose that there are two test treatments, namely, a test treatment (T) and a reference treatment (R). Denote the corresponding measurements of the ith subject in the jth treatment group before and after the treatment by W_{1ij} and W_{2ij}, respectively, where $j = T$ or R corresponds to the test and the reference treatment, respectively. Assume that the measurement W_{1ij} is lognormal distributed with parameters μ_j and σ_{1j}^2, i.e.,

$$W_{1ij} \sim \text{lognormal}\,(\mu_j, \sigma_{1j}^2).$$

Let $W_{2ij} = W_{1ij}(1 + \Delta_{ij})$, where Δ_{ij} denotes the percentage change after receiving the treatment. In addition, assume that Δ_{ij} is lognormally distributed with parameters μ_{Δ_j} and $\sigma^2_{\Delta_j}$, i.e.,

$$\Delta_{ij} \sim \text{lognormal} \left(\mu_{\Delta_j}, \sigma^2_{\Delta_j} \right).$$

Thus, the difference and the relative difference between the measurements before and after the treatment are given by $W_{2ij} - W_{1ij}$ and $(W_{2ij} - W_{1ij})/W_{1ij}$, respectively. In particular,

$$W_{2ij} - W_{1ij} = W_{1ij}\Delta_{ij} \sim \text{lognormal} \left(\mu_j + \mu_{\Delta_j}, \sigma^2_j + \sigma^2_{\Delta_j} \right),$$

and

$$\frac{W_{2ij} - W_{1ij}}{W_{1ij}} \sim \text{lognormal} \left(\mu_{\Delta_j}, \sigma^2_{\Delta_j} \right).$$

To simplify the notations, define X_{ij} and Y_{ij} as $X_{ij} = \log(W_{2ij} - W_{1ij})$, $Y_{ij} = \log((W_{2ij} - W_{1ij})/W_{1ij})$. Then, both X_{ij} and Y_{ij} are normally distributed with means $\mu_j + \mu_{\Delta_j}$ and μ_{Δ_j}, $i = 1, 2, \ldots, n_j$, $j = T, R$, respectively.

Thus, possible derived study endpoints based on the responses observed before and after the treatment as described earlier include X_{ij}, the absolute difference between "before treatment" and "after treatment" responses of the subjects, Y_{ij}, the relative difference between "before treatment" and "after treatment" responses of the subjects, $r_{A_j} = \#\{x_{ij} > c_1, i = 1, \ldots, n_j\}/n_j$, the proportion of responders, which is defined as a subject whose absolute difference between "before treatment" and "after treatment" responses is larger than a prespecified value c_1, $r_{R_j} = \#\{y_{ij} > c_2, i = 1, \ldots, n_j\}/n_j$, the proportion of responders, which is defined as a subject whose relative difference between "before treatment" and "after treatment" responses is larger than a prespecified value c_2.

To define the notation, for $j = T, R$, let $p_{A_j} = E(r_{A_j})$ and $p_{R_j} = E(r_{R_j})$. Given the above possible types of derived study endpoints, we may consider the following hypotheses for testing non-inferiority with non-inferiority margins determined based on either absolute difference or relative difference:

1. The absolute difference of the responses

$$H_0 : (\mu_R - \mu_{\Delta_R}) - (\mu_T - \mu_{\Delta_T}) \geq \delta_1 \quad \text{versus} \quad H_a : (\mu_R - \mu_{\Delta_R}) - (\mu_T - \mu_{\Delta_T}) < \delta_1 \tag{8.1}$$

2. The relative difference of the responses

$$H_0 : (\mu_{\Delta_R} - \mu_{\Delta_T}) \geq \delta_2 \quad \text{versus} \quad H_a : (\mu_{\Delta_R} - \mu_{\Delta_T}) < \delta_2 \tag{8.2}$$

3. The difference of responders' rates based on the absolute difference of the responses

$$H_0 : p_{A_R} - p_{A_T} \geq \delta_3 \quad \text{versus} \quad H_a : p_{A_R} - p_{A_T} < \delta_3 \tag{8.3}$$

4. The relative difference of responders' rates based on the absolute difference of the responses

$$H_0 : \frac{p_{A_R} - p_{A_T}}{p_{A_R}} \geq \delta_4 \quad \text{versus} \quad H_a : \frac{p_{A_R} - p_{A_T}}{p_{A_R}} < \delta_4 \tag{8.4}$$

5. The absolute difference of responders' rates based on the relative difference of the responses

$$H_0 : p_{R_R} - p_{R_T} \geq \delta_5 \quad \text{versus} \quad H_a : p_{R_R} - p_{R_T} < \delta_5 \tag{8.5}$$

6. The relative difference of responders' rates based on the relative difference of the responses

$$H_0 : \frac{p_{R_R} - p_{R_T}}{p_{R_R}} \geq \delta_6 \quad \text{versus} \quad H_a : \frac{p_{R_R} - p_{R_T}}{p_{R_R}} < \delta_6 \tag{8.6}$$

For a given clinical study, the above are the possible clinical strategies for the assessment of the treatment effect. Practitioners or sponsors of the study often choose the strategy in their best interest. It should be noted that the current regulatory position is to require the sponsor to prespecify which study endpoint will be used for the assessment of the treatment effect in the study protocol without any scientific justification.

In practice, however, it is of particular interest to study the effect of power analysis for sample size calculation based on different clinical strategies. As pointed out earlier, the required sample size for achieving a desired power based on the absolute difference of a given primary study endpoint may be quite different from that obtained based on the relative difference of the given primary study endpoint. Thus, it is of interest to the clinician or clinical scientist to investigate this issue under various scenarios. In particular, the following settings are often considered in practice.

	Settings					
Strategy Used for	1	2	3	4	5	6
Sample size determination	2.1	2.2	2.3	2.4	2.5	2.6
Testing treatment effect	2.2	2.1	2.4	2.3	2.6	2.5

There are certainly other possible settings besides those considered above. For example, hypotheses (8.1) may be used for sample size determination but hypotheses (8.3) are used for testing treatment effect. However, the comparison of these two clinical strategies would be affected by the value of c_1, which is used to determine the proportion of responders. However, in the interest of a simple and easier comparison, the number of parameters is kept as minimal as possible. Details of the comparison of the above six settings are given in the next section.

8.4 Comparison of Different Clinical Strategies

8.4.1 Test Statistics, Power, and Sample Size Determination

Note that X_{ij} denotes the absolute difference between "before treatment" and "after treatment" responses of the ith subjects under the jth treatment, and Y_{ij} denotes the relative difference between "before treatment" and "after treatment" responses of the ith subjects under the jth treatment. Let $\bar{x}_{.j} = 1/n_j = \sum_{i=1}^{n_j} x_{ij}$ and $\bar{y}_{.j} = 1/n_j = \sum_{i=1}^{n_j} y_{ij}$ be the sample means of X_{ij} and Y_{ij} for the jth treatment group, $j = T, R$, respectively.

Based on normal distribution, the null hypothesis in (8.1) is rejected at a level α of significance if

$$\frac{\bar{x}_{.R} - \bar{x}_{.T} + \delta_1}{\sqrt{(1/n_T + 1/n_R)[(\sigma_T^2 + \sigma_{\Delta_T}^2) + (\sigma_R^2 + \sigma_{\Delta_R}^2)]}} > z_\alpha. \tag{8.7}$$

Thus, the power of the corresponding test is given as

$$\Phi\left(\frac{(\mu_T + \mu_{\Delta_T}) - (\mu_R + \mu_{\Delta_R}) + \delta_1}{\sqrt{(n_T^{-1} + n_R^{-1})[(\sigma_T^2 + \sigma_{\Delta_T}^2) + (\sigma_R^2 + \sigma_{\Delta_R}^2)]}} - z_\alpha\right), \tag{8.8}$$

where $\Phi(\cdot)$ is the cumulative distribution function of the standard normal distribution. Suppose that the sample sizes allocated to the reference and test treatments are in the ratio of r, where r is a known constant. Using these results, the required total sample size for the test hypotheses (8.1) with a power level of $(1 - \beta)$ is $N = n_T + n_R$, with

$$n_T = \frac{(z_\alpha + z_\beta)^2 (\sigma_1^2 + \sigma_2^2)(1 + 1/\rho)}{\left[(\mu_R + \mu_{\Delta_R}) - (\mu_T + \mu_{\Delta_T}) - \delta_1\right]^2}, \tag{8.9}$$

$n_R = \rho n_T$ and z_u is $1 - u$ quantile of the standard normal distribution.

Note that y_{ij}'s are normally distributed. The test statistic based on $\bar{y}_{.j}$ would be similar to the above case. In particular, the null hypothesis in (8.2) is rejected at a significance level α if

$$\frac{\bar{y}_{T.} - \bar{y}_{R.} + \delta_2}{\sqrt{(1/n_T) + (1/n_R)(\sigma_{\Delta T}^2 + \sigma_{\Delta R}^2)}} > z_\alpha. \tag{8.10}$$

The power of the corresponding test is given as

$$\Phi\left(\frac{\mu_{\Delta T} - \mu_{\Delta R} + \delta_2}{\sqrt{(n_T^{-1} + n_R^{-1})(\sigma_{\Delta T}^2 + \sigma_{\Delta R}^2)}} - z_\alpha\right). \tag{8.11}$$

Suppose that $n_R = \rho n_T$, where R is a known constant. Then the required total sample size to test hypotheses (8.2) with a power of $(1 - \beta)$ is $(1 + \rho)n_T$, where

$$n_T = \frac{(z_\alpha + z_\beta)^2(\sigma_{\Delta T}^2 + \sigma_{\Delta R}^2)(1 + 1/\rho)}{\left[(\mu_R + \mu_{\Delta R}) - (\mu_T + \mu_{\Delta T}) - \delta_2\right]^2}. \tag{8.12}$$

For a sufficiently large sample size n_j, r_{A_j} is asymptotically normal with mean p_{A_j} and variance $p_{A_j}(1 - p_{A_j})/n_j$, $j = T, R$. Thus, based on the Slutsky theorem, the null hypothesis in (8.3) is rejected at an approximate α level of significance if

$$\frac{r_{A_T} - r_{A_R} + \delta_3}{\sqrt{(1/n_T)r_{A_T}(1 - r_{A_T}) + (1/n_R)r_{A_R}(1 - r_{A_R})}} > z_\alpha. \tag{8.13}$$

The power of the above test can be approximated by

$$\Phi\left(\frac{p_{A_T} - p_{A_R} + \delta_3}{\sqrt{n_T^{-1}p_{A_T}(1 - p_{A_T}) + n_R^{-1}r_{A_R}(1 - p_{A_R})}} - z_\alpha\right). \tag{8.14}$$

if $n_R = \rho n_T$, where r is a known constant. Then, the required sample size to test hypotheses (8.3) with a power level of $(1 - \beta)$ is $(1 + \rho)n_T$, where

$$n_T = \frac{(z_\alpha + z_\beta)^2\left[p_{A_T}(1 - p_{A_T}) + p_{A_R}(1 - p_{A_R})/\rho\right]}{(p_{A_R} - p_{A_T} - \delta_3)^2}. \tag{8.15}$$

Note that, by definition, $p_{A_j} = 1 - \Phi\left((c_1 - (\mu_j + \mu_{\Delta_j}))/\sqrt{\sigma_j^2 + \sigma_{\Delta_j}^2}\right)$, where $j = T, R$. Therefore, following similar arguments, the above results also apply to test hypotheses (8.5) with p_{A_j} replaced by $p_{R_j} = 1 - \Phi((c_2 - \mu_{\Delta_j})/\sigma_{\Delta_j})$ and δ_3 replaced by δ_5.

The hypotheses in (8.4) are equivalent to

$$H_0 : (1 - \delta_4)p_{A_R} - p_{A_T} \geq 0 \quad \text{versus} \quad H_a : (1 - \delta_4)p_{A_R} - p_{A_T} < 0. \tag{8.16}$$

Therefore, the null hypothesis in (8.4) is rejected at an approximate level of significance if

$$\frac{r_{A_T} - (1 - \delta_4)r_{A_R}}{\sqrt{(1/n_T)r_{A_T}(1 - r_{A_T}) + [(1 - \delta_4)^2/n_R]r_{A_R}(1 - r_{A_R})}} > z_\alpha. \tag{8.17}$$

Using normal approximation to the test statistic when both n_T and n_R are sufficiently large, the power of the above test can be approximated by

$$\Phi\left(\frac{p_{A_T} - (1 - \delta_4)p_{A_R}}{\sqrt{n_T^{-1}p_{A_T}(1 - p_{A_T}) + n_R^{-1}(1 - \delta_4)^2 p_{A_R}(1 - p_{A_R})}} - z_\alpha\right) \tag{8.18}$$

Suppose that $n_R = \rho n_T$, where r is a known constant. Then the required total sample size to test hypotheses (8.10), or equivalently (8.16), with a power level of $(1 - \beta)$ is $(1 + \rho)n_T$, where

$$n_T = \frac{(z_\alpha + z_\beta)^2 \left[p_{A_T}(1 - p_{A_T}) + (1 - \delta_4)^2 p_{A_R}(1 - p_{A_R})/\rho\right]}{\left[p_{A_T} - (1 - \delta_4)p_{A_R}\right]^2}. \tag{8.19}$$

Similarly, the results derived in (8.17) through (8.19) for the hypotheses (8.4) also apply to the hypotheses in (8.6) with p_{A_j} replaced by $p_{R_j} = 1 - \Phi((c_2 - \mu_{\Delta_j})/\sigma_{\Delta_j})$ and δ_4 replaced by δ_6.

8.4.2 Determination of the Non-Inferiority Margin

Based on the results derived in the previous section, the non-inferiority margins corresponding to the tests based on the absolute difference and the relative difference can be chosen in such a way that the two tests would have the same power. In particular, hypotheses (8.1) and (8.2) would give the power level if the power function given in (8.8) is the same as that given in (8.11). Consequently, the non-inferiority margins δ_1 and δ_2 would satisfy the following equation:

$$\frac{(\sigma_T^2 + \sigma_{\Delta_T}^2) + (\sigma_R^2 + \sigma_{\Delta_R}^2)}{\left[(\mu_T + \mu_{\Delta_T}) - (\mu_R + \mu_{\Delta_R}) + \delta_1\right]^2} = \frac{(\sigma_{\Delta_T}^2 + \sigma_{\Delta_R}^2)}{\left[(\mu_{\Delta_T} - \mu_{\Delta_R}) + \delta_2\right]^2}. \tag{8.20}$$

Similarly for hypotheses (8.3) and (8.4), the non-inferiority margins δ_3 and δ_4 would satisfy the following relationship:

$$\frac{p_{A_T}(1-p_{A_T})+p_{A_R}(1-p_{A_R})/\rho}{(p_{A_R}-p_{A_T}-\delta_3)^2} = \frac{p_{A_T}(1-p_{A_T})+(1-\delta_4)^2 p_{A_R}(1-p_{A_R})/\rho}{\left[p_{A_R}-(1-\delta_4)p_{A_T}\right]^2}. \tag{8.21}$$

For hypotheses (8.5) and (8.6), the non-inferiority margins δ_5 and δ_6 satisfy

$$\frac{p_{R_T}(1-p_{R_T})+p_{R_R}(1-p_{R_R})/\rho}{(p_{R_R}-p_{R_T}-\delta_5)^2} = \frac{p_{R_T}(1-p_{R_T})+(1-\delta_6)^2 p_{R_R}(1-p_{R_R})/\rho}{\left[p_{R_R}-(1-\delta_6)p_{R_T}\right]^2}. \tag{8.22}$$

The results given in (8.20), (8.21), and (8.22) provide a way of translating the non-inferiority margins between the endpoints based on the difference and the relative difference. In the next section, we present a numerical study to provide some insight into how the power level of these tests would be affected by the choices of different study endpoints for various combinations of parameter values.

8.5 A Numerical Study

In this section, a numerical study was conducted to provide some insight about the effect on different clinical strategies.

8.5.1 Absolute Difference versus Relative Difference

In Table 8.4, the required sample sizes for the test of non-inferiority are based on the absolute difference (X_{ij}) and relative difference (Y_{ij}). In particular, the nominal power level ($1 - \beta$) is chosen to be 0.80 and α is 0.05. The corresponding sample sizes are calculated using the formulae in (8.9) and (8.12). It is difficult to conduct any comparison because the corresponding non-inferiority margins are based on different measurement scales. However, to provide some idea to assess the impact of switching from a clinical endpoint based on absolute difference to that based on relative difference, a numerical study on the power of the test was conducted. In particular, Table 8.5 presents the power of the test for non-inferiority based on the relative difference (Y) with the sample sizes determined by the power based on the absolute difference (X). The power was calculated using the result given in (8.11). The results demonstrate that the effect is, in general, very significant. In many cases, the power is much smaller than the nominal level 0.8.

TABLE 8.4

Sample Sizes for Non-Inferiority Testing Based on Absolute Difference and Relative Difference ($\alpha = 0.05$, $\beta = 0.20$, $\rho = 1$)

	$(\mu_R + \mu_{\Delta_R}) - (\mu_T + \mu_{\Delta_T}) = 0.20$									$(\mu_R + \mu_{\Delta_R}) - (\mu_T + \mu_{\Delta_T}) = 0.30$								
$s_T^2 + s_R^2$	1.0			2.0			3.0			1.0			2.0			3.0		
$s_{D_T}^2 + s_{D_R}^2$	1.0	1.5	2.0	1.0	1.5	2.0	1.0	1.5	2.0	1.0	1.5	2.0	1.0	1.5	2.0	1.0	1.5	2.0
Absolute difference																		
$\delta_1 = 0.50$	275	344	413	413	481	550	550	619	687	619	773	928	928	1082	1237	1237	1392	1546
$\delta_1 = 0.55$	202	253	303	303	354	404	404	455	505	396	495	594	594	693	792	792	891	990
$\delta_1 = 0.60$	155	194	232	232	271	310	310	348	387	275	344	413	413	481	550	550	619	687
$\delta_1 = 0.65$	123	153	184	184	214	245	245	275	306	202	253	303	303	354	404	404	455	505
$\delta_1 = 0.70$	99	124	149	149	174	198	198	223	248	155	194	232	232	271	310	310	348	387
Relative difference																		
$\delta_2 = 0.40$	310	464	619	310	464	619	310	464	619	1237	1855	2474	1237	1855	2474	1237	1855	2474
$\delta_2 = 0.50$	138	207	275	138	207	275	138	207	275	310	464	619	310	464	619	310	464	619
$\delta_2 = 0.60$	78	116	155	78	116	155	78	116	155	138	207	275	138	207	275	138	207	275

TABLE 8.5

Power of the Test of Non-Inferiority Based on Relative Difference

		$(\mu_R + \mu_{\Delta_R}) - (\mu_T + \mu_{\Delta_T}) = 0.20$									$(\mu_R + \mu_{\Delta_R}) - (\mu_T + \mu_{\Delta_T}) = 0.30$								
	$s_T^2 + s_R^2$	1.0			2.0			3.0			1.0			2.0			3.0		
	$s_{D_T}^2 + s_{D_R}^2$	1.0	1.5	2.0	1.0	1.5	2.0	1.0	1.5	2.0	1.0	1.5	2.0	1.0	1.5	2.0	1.0	1.5	2.0
$\delta_1 = 0.50$	$\delta_2 = 0.4$	75.8	69.0	65.1	89.0	81.3	75.8	95.3	89.0	83.6	54.6	48.4	45.2	69.5	60.0	54.5	80.0	69.5	62.6
	$\delta_2 = 0.5$	96.9	94.2	92.0	99.6	98.4	96.9	100.0	99.6	98.9	97.0	94.1	91.9	99.6	98.4	96.9	100.0	99.6	98.9
	$\delta_2 = 0.6$	99.9	99.6	99.2	100.0	100.0	99.9	100.0	100.0	100.0	100.0	99.9	99.8	100.0	100.0	100.0	100.0	100.0	100.0
$\delta_1 = 0.55$	$\delta_2 = 0.4$	64.2	57.6	53.8	79.3	70.1	64.2	88.4	79.3	72.7	40.6	35.9	33.5	53.1	45.0	40.6	63.5	53.1	47.1
	$\delta_2 = 0.5$	91.5	86.7	83.3	98.0	94.7	91.5	99.6	98.0	95.8	87.9	82.2	78.6	96.4	91.8	87.9	99.0	96.4	93.3
	$\delta_2 = 0.6$	99.1	97.9	96.7	99.9	99.7	99.1	100.0	99.9	99.8	99.5	98.6	97.8	100.0	99.8	99.5	100.0	100.0	99.9
$\delta_1 = 0.60$	$\delta_2 = 0.4$	54.6	48.5	45.2	69.5	60.1	54.6	80.1	69.5	62.6	31.8	28.3	26.5	41.8	35.2	31.8	50.5	41.7	36.9
	$\delta_2 = 0.5$	84.0	77.9	73.9	94.4	88.6	84.0	98.2	94.4	90.4	75.8	69.0	65.1	89.0	81.3	75.8	95.3	89.0	83.6
	$\delta_2 = 0.6$	97.0	94.2	91.9	99.6	98.4	97.0	100.0	99.6	98.9	96.9	94.2	92.0	99.6	98.4	96.9	100.0	99.6	98.9
$\delta_1 = 0.65$	$\delta_2 = 0.4$	47.0	41.4	38.7	60.8	51.8	46.8	71.5	60.6	54.2	26.1	23.4	21.9	33.9	28.8	26.1	41.2	34.0	30.1
	$\delta_2 = 0.5$	76.0	69.1	65.2	89.1	81.3	75.9	95.3	89.0	83.6	64.2	57.6	53.8	79.3	70.1	64.2	88.4	79.3	72.7
	$\delta_2 = 0.6$	93.2	88.7	85.7	98.6	95.8	93.1	99.7	98.6	96.8	91.5	86.7	83.3	98.0	94.7	91.5	99.6	98.0	95.8
$\delta_1 = 0.70$	$\delta_2 = 0.4$	40.6	36.0	33.6	53.2	45.2	40.6	63.5	53.2	47.2	22.2	20.0	18.9	28.5	24.4	22.2	34.5	28.5	25.4
	$\delta_2 = 0.5$	67.9	61.2	57.4	82.8	73.9	67.9	91.0	82.7	76.3	54.6	48.5	45.2	69.5	60.1	54.6	80.1	69.5	62.6
	$\delta_2 = 0.6$	87.9	82.3	78.7	96.5	91.9	87.9	99.0	96.4	93.4	84.0	77.9	73.9	94.4	88.6	84.0	98.2	94.4	90.4

8.5.2 Responders' Rate Based on Absolute Difference

Similar computation was conducted for the case when the hypotheses are defined in terms of the responders' rate based on the absolute difference, i.e., hypotheses defined in (8.3) and (8.4). Table 8.6 gives the required sample sizes, with the derived results given in (8.15) and (8.19), for the corresponding hypotheses with non-inferiority margins given both in terms of absolute difference and relative difference of the responders' rates. Similarly, Table 8.7 presents the power of the test for non-inferiority based on the relative difference of the responders' rate with the sample sizes determined by the power based on the absolute difference of the responders' rate. The power was calculated using the result given in (8.14). Again, the results demonstrate that the effect is, in general, very significant. In many cases, the power is much smaller than the nominal level 0.8.

8.5.3 Responders' Rate Based on Relative Difference

Similar to the issues considered in the above paragraph with the exception that the responders' rate is defined based on the relative difference, the required sample sizes for the corresponding hypotheses with non-inferiority margins given both in terms of absolute difference and relative difference of the responders' rates are defined based on the relative difference, i.e., the hypotheses defined in (8.5) and (8.6). The results are shown in Table 8.8. Following similar steps, Table 8.9 presents the power of the test for non-inferiority based on the relative difference of the responders' rate with the sample sizes determined by the power based on the absolute difference of the responders' rate. A similar pattern emerges and the results demonstrate that the power is usually much smaller than the nominal level 0.8.

8.6 Concluding Remarks

In clinical trials, it is not uncommon that a study is powered based on expected absolute change from the baseline of a primary study endpoint but the collected data are analyzed based on relative change from the baseline (e.g., percent change from baseline) of the primary study endpoint, or the collected data are analyzed based on the percentage of patients who show some improvement (i.e., responder analysis). The definition of a responder could be based on either absolute change from baseline or relative change from baseline of the primary study endpoint. It is very controversial in terms of the interpretation of the analysis results, especially when a significant result

TABLE 8.6

Sample Sizes for Non-Inferiority Testing Based on Absolute Difference and Relative Difference of Response Rates Defined by the Absolute Difference (X_{ij}) $(\alpha = 0.05, \beta = 0.20, \rho = 1, c_1 - (\mu_T + \mu_{\Delta_T}) = 0)$

$s_T^2 + s_R^2$		$c_1 - (m_R + m_{D_R}) = -0.60$									$c_1 - (m_R + m_{D_R}) = -0.80$							
	1.0			2.0			3.0			1.0			2.0			3.0		
$s_{D_T}^2 + s_{D_R}^2$	1.0	1.5	2.0	1.0	1.5	2.0	1.0	1.5	2.0	1.0	1.5	2.0	1.0	1.5	2.0	1.0	1.5	2.0
Absolute difference																		
$\delta_3 = 0.25$	399	284	228	228	195	173	173	157	146	2191	898	558	558	410	329	329	279	245
$\delta_3 = 0.30$	159	128	111	111	99	91	91	85	81	382	253	195	195	162	141	141	127	116
$\delta_3 = 0.35$	85	73	65	65	60	56	56	53	51	153	117	98	98	86	78	78	72	68
$\delta_3 = 0.40$	53	47	43	43	40	38	38	37	35	82	68	59	59	54	50	50	47	44
$\delta_3 = 0.45$	36	33	31	31	29	28	28	27	26	51	44	40	40	37	34	34	33	31
Relative difference																		
$\delta_4 = 0.35$	458	344	285	285	249	224	224	206	193	1625	869	601	601	469	391	391	340	304
$\delta_4 = 0.40$	199	166	147	147	134	124	124	117	112	392	288	234	234	202	180	180	165	153
$\delta_4 = 0.45$	109	96	88	88	82	78	78	75	72	168	139	121	121	110	102	102	95	91

TABLE 8.7

Power of the Test of Non-Inferiority Based on Relative Difference of Response Rates ($\alpha = 0.05$, $\beta = 0.20$, $\rho = 1$, $c_1 - (\mu_T + \mu_{\Delta_T}) = 0$)

| | | $c_1 - (\mu_R + \mu_{\Delta_R}) = -0.60$ | | | | | | | | | $c_1 - (\mu_R + \mu_{\Delta_R}) = -0.80$ | | | | | | | | |
| | $s_R^2 + s_T^2$ | 1.0 | | | 2.0 | | | 3.0 | | | 1.0 | | | 2.0 | | | 3.0 | | |
	$s_{D_R}^2 + s_{D_T}^2$	1.0	1.5	2.0	1.0	1.5	2.0	1.0	1.5	2.0	1.0	1.5	2.0	1.0	1.5	2.0	1.0	1.5	2.0
$\delta_3 = 0.25$	$\delta_4 = 0.35$	75.1	73.1	71.9	71.9	71.2	70.6	70.6	70.1	69.9	89.3	81.2	77.4	77.4	75.2	73.8	73.8	72.9	72.3
	$\delta_4 = 0.40$	97.0	94.6	92.8	92.8	91.4	90.2	90.2	89.2	88.5	100.0	99.7	98.6	98.6	97.1	95.7	95.7	94.5	93.4
	$\delta_4 = 0.45$	99.9	99.6	99.1	99.1	98.6	98.1	98.1	97.6	97.2	100.0	100.0	100.0	100.0	99.9	99.8	99.8	99.6	99.3
$\delta_3 = 0.30$	$\delta_4 = 0.35$	42.9	44.9	46.3	46.3	47.0	47.7	47.7	48.1	48.8	33.0	38.1	41.0	41.0	42.8	44.0	44.0	45.1	45.7
	$\delta_4 = 0.40$	71.9	70.5	69.9	69.9	69.1	68.6	68.6	68.3	68.3	79.1	75.5	73.5	73.5	72.1	71.1	71.1	70.6	70.0
	$\delta_4 = 0.45$	91.4	89.1	87.6	87.6	86.3	85.3	85.3	84.5	84.1	98.2	95.7	93.5	93.5	91.6	90.2	90.2	89.1	88.0
$\delta_3 = 0.35$	$\delta_4 = 0.35$	28.3	30.9	32.4	32.4	33.6	34.4	34.4	35.1	35.8	18.9	23.2	26.1	26.1	28.1	29.7	29.7	30.9	32.0
	$\delta_4 = 0.40$	49.3	50.2	50.5	50.5	50.9	51.0	51.0	51.2	51.5	46.4	47.7	48.6	48.6	49.2	49.7	49.7	50.1	50.6
	$\delta_4 = 0.45$	71.2	70.2	69.1	69.1	68.7	68.0	68.0	67.6	67.5	76.7	74.0	72.4	72.4	71.2	70.5	70.5	69.9	69.7
$\delta_3 = 0.40$	$\delta_4 = 0.35$	21.2	23.4	24.9	24.9	25.9	26.8	26.8	27.7	28.0	13.9	17.1	19.3	19.3	21.2	22.5	22.5	23.6	24.3
	$\delta_4 = 0.40$	35.9	37.4	38.3	38.3	38.9	39.4	39.4	40.3	40.1	30.6	33.2	34.6	34.6	36.0	36.9	36.9	37.6	37.8
	$\delta_4 = 0.45$	53.8	54.0	54.0	54.0	53.8	53.8	53.8	54.4	53.7	53.7	53.9	53.7	53.7	54.1	54.1	54.1	54.2	53.7
$\delta_3 = 0.45$	$\delta_4 = 0.35$	17.2	19.1	20.5	20.5	21.3	22.2	22.2	22.8	23.3	11.4	13.9	15.8	15.8	17.2	18.1	18.1	19.2	19.8
	$\delta_4 = 0.40$	27.9	29.6	30.8	30.8	31.4	32.2	32.2	32.6	32.9	22.7	25.1	26.9	26.9	28.1	28.7	28.7	29.8	30.0
	$\delta_4 = 0.45$	41.6	42.7	43.5	43.5	43.5	44.0	44.0	44.2	44.2	39.2	40.4	41.5	41.5	42.1	42.0	42.0	42.9	42.6

TABLE 8.8

Sample Sizes for Non-Inferiority Testing Based on Absolute Difference and Relative Difference of Response Rates Defined by the Relative Difference (Y_{ij}) ($\alpha = 0.05$, $\beta = 0.20$, $\rho = 1$, $c_2 - \mu_{\Delta_T} = 0$)

$s_{D_R}^2 + s_{D_T}^2$	$c_2 - \mu_{\Delta_R} = -0.30$				$c_2 - \mu_{\Delta_R} = -0.40$				$c_2 - \mu_{\Delta_R} = -0.50$				$c_2 - \mu_{\Delta_R} = -0.60$			
	1.0	1.5	2.0	2.5	1.0	1.5	2.0	2.5	1.0	1.5	2.0	2.5	1.0	1.5	2.0	2.5
Absolute difference																
$\delta_5 = 0.25$	173	130	111	101	329	201	157	135	836	351	238	189	4720	745	399	284
$\delta_5 = 0.30$	91	74	66	61	141	102	85	76	244	147	114	97	504	229	159	128
$\delta_5 = 0.35$	56	48	44	41	78	61	53	49	114	81	67	59	180	110	85	73
$\delta_5 = 0.40$	38	33	31	29	50	41	37	34	66	51	44	40	92	64	53	47
$\delta_5 = 0.45$	28	25	23	22	34	29	27	25	43	35	31	29	56	42	36	33
Relative difference																
$\delta_6 = 0.35$	224	173	151	138	391	256	206	180	823	412	297	243	2586	754	458	344
$\delta_6 = 0.40$	124	104	94	88	180	137	117	106	279	186	151	132	478	266	199	166
$\delta_6 = 0.45$	78	68	63	60	102	83	75	69	136	104	90	81	189	132	109	96

TABLE 8.9

Power of the Test of Non-Inferiority Based on Relative Difference of Response Rates ($\alpha = 0.05$, $\beta = 0.20$, $\rho = 1$, $c_2 - \mu_{\Delta_T} = 0$)

		$c_2 - \mu_{\Delta_R} = -0.30$				$c_2 - \mu_{\Delta_R} = -0.40$				$c_2 - \mu_{\Delta_R} = -0.50$				$c_2 - \mu_{\Delta_R} = -0.60$			
	$s_{D_R}^2 + s_{D_T}^2$	1.0	1.5	2.0	2.5	1.0	1.5	2.0	2.5	1.0	1.5	2.0	2.5	1.0	1.5	2.0	2.5
$\delta_5 = 0.25$	$\delta_6 = 0.35$	70.6	69.5	68.8	68.7	73.8	71.2	70.1	69.6	80.5	74.3	72.1	70.9	95.7	79.6	75.1	73.1
	$\delta_6 = 0.40$	90.2	87.4	85.7	84.9	95.7	91.6	89.2	87.7	99.6	96.2	93.1	91.0	100.0	99.4	97.0	94.6
	$\delta_6 = 0.45$	98.1	96.4	95.2	94.5	99.8	98.7	97.6	96.7	100.0	99.8	99.2	98.5	100.0	100.0	99.9	99.6
$\delta_5 = 0.30$	$\delta_6 = 0.35$	47.7	49.3	50.1	50.5	44.0	47.0	48.1	49.0	38.6	43.7	45.9	47.1	29.2	39.2	42.9	44.9
	$\delta_6 = 0.40$	68.6	67.7	67.3	66.8	71.1	69.4	68.3	67.7	75.2	71.4	69.9	68.9	81.9	74.6	71.9	70.5
	$\delta_6 = 0.45$	85.3	83.1	81.8	80.9	90.2	86.7	84.5	83.3	95.4	90.6	87.9	86.0	99.2	94.9	91.4	89.1
$\delta_5 = 0.35$	$\delta_6 = 0.35$	34.4	36.9	38.2	38.7	29.7	33.3	35.1	36.5	23.6	29.4	32.2	33.8	16.1	24.4	28.3	30.9
	$\delta_6 = 0.40$	51.0	52.0	52.5	52.4	49.7	50.8	51.2	51.8	47.8	49.9	50.6	50.9	45.3	48.2	49.3	50.2
	$\delta_6 = 0.45$	68.0	67.4	67.0	66.3	70.5	68.7	67.6	67.4	73.7	71.1	69.6	68.5	78.4	73.5	71.2	70.2
$\delta_5 = 0.40$	$\delta_6 = 0.35$	26.8	28.8	30.3	30.8	22.5	25.8	27.7	28.7	17.3	22.1	24.6	26.3	12.0	17.9	21.2	23.4
	$\delta_6 = 0.40$	39.4	40.5	41.6	41.7	36.9	39.0	40.3	40.7	33.2	36.6	38.2	39.3	29.0	33.6	35.9	37.4
	$\delta_6 = 0.45$	53.8	53.7	54.2	53.7	54.1	54.1	54.4	54.0	53.5	54.0	54.1	54.2	53.7	53.6	53.8	54.0
$\delta_5 = 0.45$	$\delta_6 = 0.35$	22.2	24.2	25.1	25.8	18.1	21.0	22.8	23.7	14.1	17.9	20.0	21.6	10.0	14.5	17.2	19.1
	$\delta_6 = 0.40$	32.2	33.7	34.1	34.6	28.7	31.0	32.6	33.1	25.2	28.6	30.3	31.7	21.4	25.6	27.9	29.6
	$\delta_6 = 0.45$	44.0	44.7	44.5	44.8	42.0	43.1	44.2	44.1	40.3	42.1	42.9	43.8	38.6	40.5	41.6	42.7

is observed based on a study endpoint (e.g., absolute change from baseline, relative change from baseline, or responder analysis) but not on another study endpoint (e.g., absolute change from baseline, relative change from baseline, or responder analysis). Based on the numerical results of this study, it is evident that the power of the test can be decreased drastically when the study endpoint is changed. However, when switching from a study endpoint based on absolute difference to the one based on relative difference, one possible way to maintain the power level is to modify the corresponding non-inferiority margin, as suggested by the results given in Section 8.2.

9

Protocol Amendments

9.1 Introduction

In clinical trials, it is not uncommon to issue protocol amendments during the conduct of a clinical trial due to various reasons such as slow enrollment and/or safety concerns. For slow enrollment, the investigator may modify the entry (inclusion/exclusion) criteria in order to expedite patient enrollment in a timely fashion. On the other hand, during the conduct of a clinical trial, it is possible that additional safety information may become available. This additional safety information may come either from similar clinical trials conducted simultaneously or from publications newly published in leading medical journals. With this additional safety information, protocol amendment is necessarily issued for patient protection. For good clinical practice (GCP), before protocol amendments can be issued, description, rationales, and clinical/statistical justification regarding the changes made should be provided to ensure the validity and integrity of the clinical trial. As a result of the changes or modifications, the original target patient population under study could have become a similar but different patient population. If the changes or modifications are made frequently during the conduct of the trial, the target patient population is in fact a moving target patient population. This raises the controversial issue regarding the validity of the statistical inference drawn based on data collected before and after protocol amendment.

In practice, there is a risk that major (or significant) modifications made to the trial and/or statistical procedures could lead to a totally different trial, which cannot address the scientific/medical questions that the clinical trial is intended to answer. In clinical trials, most investigators consider protocol amendment a God-sent gift which allows the investigator certain degree of flexibility to make any changes/modifications to the ongoing clinical trials. It, however, should be noted that protocol amendments have potential risks for introducing additional bias/variation to the ongoing clinical trial. Thus, it is important to identify, control, and hopefully eliminate/minimize the sources of bias/variation. Thus, it is of interest to measure the impact of changes or modifications that are made to the trial procedures and/or statistical methods after the protocol amendment. This raises another controversial issue regarding (1) the impact of changes made and (2) the degree of changes that are allowed in a protocol amendment.

In current practice, standard statistical methods are applied to the data collected from the actual patient population regardless of the frequency of changes (protocol amendments) that have been made during the conduct of the trial assuming that the overall type I error is controlled at the pre-specified level of significance. This, however, has raised a serious regulatory/statistical concern as to whether the resultant statistical inference (e.g., independent estimates, confidence intervals, and p values) drawn on the originally planned target patient population based on the clinical data from the actual patient population (as a result of the modifications made via protocol amendments) is accurate and reliable? After some modifications are made to the trial and/or statistical methods, not only may the target patient population have become a similar but different patient population, but also the sample size may not achieve the desired power for detection of a clinically important effect size of the test treatment at the end of the study. In practice, we expect to lose power when the modifications have led to a shift in mean response and/or inflation of variability of the response of the primary study endpoint. As a result, the originally planned sample size may have to be adjusted. Thus, it is suggested that the relative efficiency at each protocol amendment be taken into consideration for derivation of an adjusted factor for sample size in order to achieve the desired power.

In the next section, the concept of moving the target patient population as the result of protocol amendments is introduced. Also included in the section is the derivation of a sensitivity index for measuring the degree of population shift. Section 9.3 discusses the method with covariate adjustment proposed by Chow and Shao (2005). Inference based on mixture distribution is described in Section 9.4. In Section 9.5, sample size adjustment after protocol amendment is discussed. A brief concluding remark is given in the last section.

9.2 Moving Target Patient Population

In practice, for a given clinical trial, it is not uncommon to have three to five protocol amendments after the initiation of the clinical trial. One of the major impacts of many protocol amendments is that the target patient population may have been shifted during the process, which may have resulted in a totally different target patient population at the end of the trial. A typical example is the case when significant adaptation (modification) is applied to inclusion/exclusion criteria of the study. Denote by (μ, σ) the *target* patient population. After a given protocol amendment, the resultant (actual) patient population may have been shifted to (μ_1, σ_1), where $\mu_1 = \mu + \varepsilon$ is the population mean of the primary study endpoint and $\sigma_1 = C\sigma (C > 0)$ is the population standard deviation of the primary study endpoint. The shift in target patient population can be characterized by

$$E_1 = \left|\frac{\mu_1}{\sigma_1}\right| = \left|\frac{\mu+\varepsilon}{C\sigma}\right| = |\Delta|\left|\frac{\mu}{\sigma}\right| = |\Delta|E,$$

where
$\Delta = (1 + \varepsilon/\mu)/C,$
E and E_1 are the effect size before and after population shift, respectively.

Chow et al. (2002a) and Chow and Chang (2006) refer to Δ as a sensitivity index measuring the change in effect size between the actual patient population and the original target patient population.

Similarly, denote by (μ_i, σ_i) the actual patient population after the ith modification of trial procedure, where $\mu_i = \mu + \varepsilon_i$ and $\sigma_i = C_i\sigma$, $i = 0, 1, \ldots, K$. Note that $i = 0$ reduces to the original target patient population (μ, σ). That is, when $i = 0$, $\varepsilon_0 = 0$ and $C_0 = 1$. After K protocol amendments, the resultant actual patient population becomes (μ_K, σ_K), where

$$\mu_K = \mu + \sum_{i=1}^{K} \varepsilon_i \quad \text{and} \quad \sigma_K = \prod_{i=1}^{K} C_i\sigma.$$

It should be noted that (ε_i, C_i), $i = 1, \ldots, K$ are in fact random variables. As a result, the resultant actual patient population is a *moving* target patient population rather than a *fixed* target patient population. In addition, sample sizes before and after protocol amendments and the number of protocol amendments issued for a given clinical trial are also random variables. Thus, one of the controversial issues that commonly encountered in clinical trials with several protocol amendments during the conduct of the trials is *How to assess the treatment effect while the target patient population is a moving target?*

Table 9.1 provides a summary of the impacts of various scenarios of location shift (i.e., change in ε) and scale shift (change in C, either inflation or deflation of variability). As can be seen from Table 9.1, there is a masking effect between location shift and scale shift. In other words, shift in location could be offset by the inflation or deflation of variability. As a result, the sensitivity index remains unchanged while the target patient population has been shifted. One of the controversial issues in this regard is whether the conclusion drawn (by ignoring the population shift) at the end of the trial is accurate and reliable.

As indicated by Chow and Chang (2006), the impact of protocol amendments on statistical inference due to shift in target patient population

TABLE 9.1

Changes in Sensitivity Index

	Inflation of Variability		Deflation of Variability	
$\varepsilon/\mu(\%)$	$C(\%)$	Δ	$C(\%)$	Δ
−20	120	0.667	80	1.000
−10	120	0.750	80	1.125
−5	120	0.792	80	1.188
0	120	0.833	80	1.250
5	120	0.875	80	1.313
10	120	0.917	80	1.375
20	120	1.000	80	1.500

(moving target patient population) can be studied through a model that links the moving population means with some covariates (Chow and Shao, 2005). However, in many cases, such covariates may not exist or exist but are not observed. In this case, it is suggested that inference on Δ be considered to measure the degree of shift in location and scale of patient population based on a mixture distribution by assuming that the location or scale parameter is random (Chow et al., 2005). These methods will be described in the subsequent sections.

9.3 Analysis with Covariate Adjustment

As indicated earlier, statistical methods for analyzing clinical data should be modified when there are protocol amendments during the trial, since any protocol deviations and/or violations may introduce bias to the trial. As a result, conclusion drawn based on the analysis of data ignoring there are possible shift in target patient population could be biased and hence misleading. To overcome this problem, Chow and Shao (2005) proposed to model the population deviations due to protocol amendments using some relevant covariates and developed a valid statistical inference which is described in the following sections.

9.3.1 Continuous Study Endpoint

Suppose that there are a total of K possible protocol amendments. Let μ_k be the mean of the study endpoint after the kth protocol amendment, $k = 1, \ldots, K$. Suppose that, for each k, clinical data are observed from n_k patients so that the sample mean $\bar{y}_k$ is an unbiased estimator of μ_k, $k = 0, 1, \ldots, K$. Now, let x be a (possibly multivariate) covariate whose values are distinct from different protocol amendments. To derive statistical inference for μ_0 (the population mean for the original target patient population), Chow and Shao (2005) assumed the following:

$$\mu_k = \beta_0 + \beta' x_k, \quad k = 0, 1, \ldots, K, \tag{9.1}$$

where
  β_0 is an unknown parameter,
  β is an unknown parameter vector whose dimension is the same as x,
  β' denotes the transpose of β,
  x_k is the value of x under the kth amendment (or the original protocol
    when $k = 0$).

If values of x are different within a fixed population (say P_k, patient population after the kth protocol amendment), then x_k is a characteristic of x such as the average of all values of x within P_k.

Under model (9.1), parameters β_0 and β can be unbiasedly estimated by

$$\begin{pmatrix} \hat{\beta}_0 \\ \hat{\beta} \end{pmatrix} = (X'WX)^{-1} X'W\bar{y}, \tag{9.2}$$

where
  $\bar{y} = (\bar{y}_0, \bar{y}_1, \ldots, \bar{y}_K)'$,
  X is a matrix whose kth row is $(1, x'_k)$, $k = 0, 1, \ldots, K$,
  W is a diagonal matrix whose diagonal elements are $n_0, n_1, \ldots, n_K$.

It is assumed that the dimension of x is less or equal to K so that $(X'WX)^{-1}$ is well defined. To estimate μ_0, we consider the following unbiased estimator $\hat{\mu}_0 = \hat{\beta}_0 + \hat{\beta}' x_0$. Chow and Shao (2005) indicated that $\hat{\mu}_0$ is distributed as $N(\mu_0, \sigma^2 c_0)$ with $c_0 = (1, x_0)(X'WX)^{-1}(1, x_0)'$. Let s_k^2 be the sample variance based on the data from population P_k, $k = 0, 1, \ldots, K$. Then, $(n_k - 1)s_k^2/\sigma^2$ has the chi-square distribution with $n_k - 1$ degrees of freedom and consequently, $(N - K)s^2/\sigma^2$ has the chi-square distribution with $N - K$ degrees of freedom, where

$$s^2 = \sum_{k=0}^{K} \frac{(n_k - 1)s_k^2}{(N - K)}$$

and $N = \sum_k n_k$. Confidence intervals for μ_0 and testing hypotheses related to μ_0 can be carried out using the t-statistic $t = (\hat{\mu}_0 - \mu_0)/\sqrt{c_0 s^2}$.

Note that when P_k's have different standard deviations and/or data from P_k are not normally distributed, we may consider an approximation by assuming

                    Controversial Statistical Issues in Clinical Trials

that all n_k's are large. Thus, by the central limit theorem, it can be shown that $\hat{\mu}_0$ is approximately normally distributed with mean μ_0 and variance

$$\tau^2 = (1, x_0)(X'WX)^{-1} X'W\Sigma X(X'WX)^{-1}(1, x_0)', \qquad (9.3)$$

where Σ is the diagonal matrix whose kth diagonal element is the population variance of P_k, $k = 0, 1, \ldots, K$. Large sample statistical inference can be made by using the z-statistic $z = (\hat{\mu}_0 - \mu_0)/\hat{\tau}$ (which is approximately distributed as the standard normal), where $\hat{\tau}$ is the same as τ with the kth diagonal element of σ estimated by s_k^2, $k = 0, 1, \ldots, K$.

Note that the above statistical inference for μ_0 is a conditional inference. In their paper, Chow and Shao (2005) also derived unconditional inference for μ_0 under certain assumptions. In addition, Chow, Chang, and Pong (2009) considered alternative approach with random coefficients under model (9.1) and proposed a Bayesian approach for obtaining inference on μ_0.

9.3.2 Binary Response

As indicated, the statistical inference for μ_0 described above is for a continuous endpoint. Following a similar idea, Yang et al. (2011) derived statistical inference for μ_0 assuming that the study endpoint is a binary response. Their method is briefly summarized as follows:

Let Y_{ij} be the binary response from the jth subject after the ith amendment; $Y_{ij} = 1$ if subject j after amendment i exhibits the response of interest, and 0 otherwise, for $i = 0, 1, \ldots, k$ and $j = 1, \ldots, n_i$. Note that the subscript 0 for i indicates that the values are related to the original patient population. Let p_i denote the response rate of the patient population after the ith amendment. Ignoring the possible population deviations results in a pooled estimator

$$\bar{p} = \frac{\sum_{i=0}^{k} \sum_{j=1}^{n_i} Y_{ij}}{\sum_{i=0}^{k} n_i},$$

which may be biased for the original defined response rate p_0. In many clinical trials, the protocol amendments are made with respect to one or a few relevant covariates. Modifying entry criteria, for example, may involve patient demographics such as age or body weight and patient characteristics such as disease status or medical history. This section develops a statistical inference procedure for the original response rate p_0 based on a covariate-adjusted model.

9.3.2.1 Estimation of the Single Response Rate

Let X_{ij} be the corresponding covariate for the jth subject after the ith amendment (or the original protocol when $i = 0$). Throughout this section we assume that the response rates for different patient populations can be related by the following model:

$$p_i = \frac{\exp(\beta_0 + \beta_1 v_i)}{1 + \exp(\beta_0 + \beta_1 v_i)}, \quad i = 0, 1, \ldots, k,$$

where
β_0 and β_1 are unknown parameters,
v_i is the true mean of the random covariate under the ith amendment.

Under the above model, the maximum likelihood estimates for the parameters β_0 and β_1, however, cannot be obtained directly because the v_i's are unknown. One approach to estimate β_0 and β_1 is to replace v_i by $\bar{X}_i$, the sample mean under the ith amendment (see Chow and Shao, 2005). Consequently, we specify a logistic model for estimating $\beta = (\beta_0, \beta_1)^T$ as

$$P(Y_{ij} = 1 \mid \bar{X}_i = \bar{x}_i) = \frac{\exp(\beta_0 + \beta_1 \bar{x}_i)}{1 + \exp(\beta_0 + \beta_1 \bar{x}_i)}. \tag{9.4}$$

Suppose that $X_{ij}, j = 1, 2, \ldots, n_i, i = 0, 1, \ldots, k$, are independent random variables with means v_i. Thus, the sample means $\bar{X}_i, i = 0, 1, \ldots, k$ are independent random variables with means v_i. Let $f_{\bar{X}_i}(\bar{x}_i)$ denote the probability density function of $\bar{X}_i$. In the development that follows, the $f_{\bar{X}_i}(\bar{x}_i)$ are assumed independent of β_0 or β_1.

Since the conditional distribution of Y_{ij} given $\bar{x}_i$ is a Bernoulli distribution with the parameter defined in (9.4) and $f_{\bar{X}_i}(\bar{x}_i)$ is the probability density function of $\bar{X}_i$, the likelihood function of observing $y_{ij}(j = 1, 2, \ldots, n_i)$ and $\bar{x}_i$ under the ith amendment is given by

$$\ell_i = \prod_{j=1}^{n_i} \left[\left(\frac{\exp(\beta_0 + \beta_1 \bar{x}_i)}{1 + \exp(\beta_0 + \beta_1 \bar{x}_i)} \right)^{y_{ij}} \left(\frac{1}{1 + \exp(\beta_0 + \beta_1 \bar{x}_i)} \right)^{1-y_{ij}} \right] f_{\bar{X}_i}(\bar{x}_i).$$

Therefore, the joint likelihood function is $\ell = \prod_{i=0}^{k} \ell_i$ and the log-likelihood function is given by

$$l(\beta) = l_1(\beta) + \sum_{i=0}^{k} \ln f_{\bar{X}_i}(\bar{x}_i), \tag{9.5}$$

where

$$l_1(\beta) = \sum_{i=0}^{k} \sum_{j=1}^{n_i} \left[y_{ij} \ln \left(\frac{\exp(\beta_0 + \beta_1 \bar{x}_i)}{1 + \exp(\beta_0 + \beta_1 \bar{x}_i)} \right) + (1 - y_{ij}) \ln \left(\frac{1}{1 + \exp(\beta_0 + \beta_1 \bar{x}_i)} \right) \right].$$

Because $f_{\bar{X}_i}(\bar{x}_i)$ does not depend on β_0 or β_1, the maximum likelihood estimate $\beta = (\beta_0, \beta_1)^T$, which maximizes $l_1(\beta)$ also maximizes $l(\beta)$. Thus, the data can be analyzed using a fixed-covariate model. By considering the covariate as a random variable, a simple closed-form estimate of the asymptotic covariance matrix of maximum likelihood estimate of the parameters can be obtained to calculate the sample size required to test the hypotheses about the parameters (see Demidenko, 2007). On the basis of the estimate $\hat{\beta}$, we propose to estimate p_0 by

$$\hat{p}_0 = \frac{\exp(\hat{\beta}_0 + \hat{\beta}_1 \bar{X}_0)}{1 + \exp(\hat{\beta}_0 + \hat{\beta}_1 \bar{X}_0)}.$$

For inference on p_0, we need to derive the asymptotic distribution of $\hat{p}_0$. In this case, the limiting results regarding the maximum likelihood estimators are obtained as the number of protocol amendments is finite and the numbers of observations from the distinct amendments become large. Assuming that $n_i/N \to r_i$ as $n_i \to \infty$, where $N = \sum_{i=0}^{k} n_i$, and k is a finite constant, it can be shown that

$$\sqrt{N}(\hat{\beta} - \beta) \xrightarrow{d} N(0, \mathbf{I}^{-1}), \tag{9.6}$$

where

$$\mathbf{I} = \begin{bmatrix} \displaystyle\sum_{i=0}^{k} r_i \frac{\exp(\beta_0 + \beta_1 v_i)}{(1 + \exp(\beta_0 + \beta_1 v_i))^2} & \displaystyle\sum_{i=0}^{k} r_i \frac{v_i \exp(\beta_0 + \beta_1 v_i)}{(1 + \exp(\beta_0 + \beta_1 v_i))^2} \\[2em] \displaystyle\sum_{i=0}^{k} r_i \frac{v_i \exp(\beta_0 + \beta_1 v_i)}{(1 + \exp(\beta_0 + \beta_1 v_i))^2} & \displaystyle\sum_{i=0}^{k} r_i \frac{v_i^2 \exp(\beta_0 + \beta_1 v_i)}{(1 + \exp(\beta_0 + \beta_1 v_i))^2} \end{bmatrix}.$$

Moreover, by the delta method and Slutsky's theorem, it follows that $\sqrt{N}(\hat{p}_0 - p_0)$ is asymptotically normally distributed with mean 0 and variance

$$V = \left[\frac{\exp(\beta_0 + \beta_1 v_0)}{(1 + \exp(\beta_0 + \beta_1 v_0))^2} \right]^2 (1, v_0) \mathbf{I}^{-1} (1, v_0)^T.$$

Let $\hat{V}$ be the maximum likelihood estimator of V with β_0, β_1, v_i, and r_i replaced by $\hat{\beta}_0$, $\hat{\beta}_1$, $\bar{X}_i$, and n_i/N, respectively. It is known that $\bar{X}_i \xrightarrow{p} v_i$ and $\hat{\beta} \xrightarrow{p} \beta$ by the Weak Law of Large Number and the consistency of a maximum likelihood estimator. Thus, we have $\hat{V} \xrightarrow{p} V$. Then, it can be shown that $\sqrt{N}(\hat{p}_0 - p_0)/\sqrt{\hat{V}}$ is asymptotically distributed as a standard normal distribution by Slutsky's theorem. Based on this result, an approximate $100(1 - \alpha)\%$ confidence interval of p_0 is given by $\left(\hat{p}_0 - z_{\alpha/2}\sqrt{\hat{V}/N},\ \hat{p}_0 + z_{\alpha/2}\sqrt{\hat{V}/N}\right)$, where $z_{\alpha/2}$ is the $100(1 - \alpha/2)$th percentile of a standard normal distribution.

9.3.2.2 Comparison for Two Treatments

In clinical trials, it is often of interest to compare two treatments, that is, a test treatment versus an active control or placebo. Let Y_{tij} and X_{tij} be the response and the corresponding relevant covariate for the jth subject after the ith amendment under the tth treatment ($t = 1, 2, i = 0, 1, \ldots, k, j = 1, 2, \ldots, n_{tj}$). For each amendment, patients selected by the same criteria are randomly allocated to either the test treatment $D_1 = 1$ or control treatment $D_2 = 0$ groups. In this particular case, the true mean values of the covariate for the two treatment groups are the same under each amendment. Therefore, the relationships between the binary response and the covariate for both treatment groups can be described by a single model,

$$p_{ti} = \frac{\exp(\beta_1 + \beta_2 D_t + \beta_3 v_i + \beta_4 D_t v_i)}{1 + \exp(\beta_1 + \beta_2 D_t + \beta_3 v_i + \beta_4 D_t v_i)}, \quad t = 1, 2, \quad i = 0, 1, \ldots, k.$$

Hence, the response rates for the test treatment and the control treatment are

$$p_{1i} = \frac{\exp(\beta_1 + \beta_2 + (\beta_3 + \beta_4)v_i)}{1 + \exp(\beta_1 + \beta_2 + (\beta_3 + \beta_4)v_i)} \quad \text{and} \quad p_{2i} = \frac{\exp(\beta_1 + \beta_3 v_i)}{1 + \exp(\beta_1 + \beta_3 v_i)},$$

respectively.

Similar to the single treatment study described previously, the joint likelihood function of $\beta = (\beta_1, \ldots, \beta_4)^T$ is given by

$$\prod_{t=1}^{2} \prod_{i=0}^{k} \prod_{j=1}^{n_{ti}} \left[\left(\frac{\exp(\beta^T z^{(ti)})}{1 + \exp(\beta^T z^{(ti)})} \right)^{y_{tij}} \left(\frac{1}{1 + \exp(\beta^T z^{(ti)})} \right)^{1 - y_{tij}} f_{\bar{X}_{.i}}(\bar{x}_{.i}) \right],$$

where $f_{\bar{X}_{.i}}(\bar{x}_{.i})$ is the probability density function of $\bar{X}_{.i} = \sum_{t=1}^{2} \sum_{j=1}^{n_{ti}} X_{tij}$ and $z^{(ti)} = (1, D_t, \bar{x}_{.i}, D_t \bar{x}_{.i})^T$. The log-likelihood function is then given by

$$l(\beta) = \sum_{t=1}^{2} \sum_{i=0}^{k} \sum_{j=1}^{n_{ti}} \left[y_{tij} \ln \left(\frac{\exp(\beta^T z^{(ti)})}{1 + \exp(\beta^T z^{(ti)})} \right) \right.$$

$$\left. + (1 - y_{tij}) \ln \left(\frac{1}{1 + \exp(\beta^T z^{(ti)})} \right) + \ln f_{\bar{X}_{.i}}(\bar{x}_{.i}) \right]. \tag{9.7}$$

Given the resulting maximum likelihood estimate $\hat{\beta} = (\hat{\beta}_1, \ldots, \hat{\beta}_4)^T$, we obtain the estimate of p_{10} and p_{20} as follows:

$$\hat{p}_{10} = \frac{\exp(\hat{\beta}_1 + \hat{\beta}_2 + (\hat{\beta}_3 + \hat{\beta}_4)\bar{X}_{\cdot 0})}{1 + \exp(\hat{\beta}_1 + \hat{\beta}_2 + (\hat{\beta}_3 + \hat{\beta}_4)\bar{X}_{\cdot 0})}, \quad \hat{p}_{20} = \frac{\exp(\hat{\beta}_1 + \hat{\beta}_3\bar{X}_{\cdot 0})}{1 + \exp(\hat{\beta}_1 + \hat{\beta}_3\bar{X}_{\cdot 0})}.$$

Let $n_{t\cdot} = \sum_{i=0}^{k} n_{ti}$ be the sample size for the tth treatment group, and let $N = n_1 + n_2$ be the total sample size. When $n_{ti}/n_{t\cdot} \to r_{ti}$ and $n_{i\cdot}/N \to c$ as all n_{ti} tend to infinity, it is shown by a similar derivation for a single response rate as shown above that

$$\frac{\sqrt{N}((\hat{p}_{10} - \hat{p}_{20}) - (p_{10} - p_{20}))}{\sqrt{\hat{V}_d}} \xrightarrow{d} N(0, 1),$$

where $\hat{V}_d = \varphi^T \left(\sum_{t=1}^{2} \sum_{i=0}^{k} n_{ti}\hat{\mathbf{I}}^{(ti)}/N \right)^{-1} \varphi,$

$$\varphi^T = \begin{pmatrix} \hat{p}_{10}(1-\hat{p}_{10}) - \hat{p}_{20}(1-\hat{p}_{20}) \\ \hat{p}_{10}(1-\hat{p}_{10}) \\ \bar{X}_{\cdot 0}(\hat{p}_{10}(1-\hat{p}_{10}) - \hat{p}_{20}(1-\hat{p}_{20})) \\ \bar{X}_{\cdot 0}(\hat{p}_{10}(1-\hat{p}_{10})) \end{pmatrix},$$

and

$$\hat{\mathbf{I}}^{(ti)} = \hat{p}_{ti}(1-\hat{p}_{ti}) \begin{bmatrix} 1 & D_t & \bar{X}_{\cdot i} & D_t\bar{X}_{\cdot i} \\ D_t & D_t^2 & D_t\bar{X}_{\cdot i} & D_t^2\bar{X}_{\cdot i} \\ \bar{X}_{\cdot i} & D_t\bar{X}_{\cdot i} & \bar{X}_{\cdot i}^2 & D_t\bar{X}_{\cdot i}^2 \\ D_t\bar{X}_{\cdot i} & D_t^2\bar{X}_{\cdot i} & D_t\bar{X}_{\cdot i}^2 & D_t^2\bar{X}_{\cdot i}^2 \end{bmatrix}.$$

As indicated by Chow, Shao, and Wang (2008), the problem of testing superiority and non-inferiority can be unified by the following hypotheses:

$$H_0 : p_{10} - p_{20} \leq \delta \quad \text{versus} \quad H_a : p_{10} - p_{20} > \delta, \tag{9.8}$$

where δ is the (clinical) superiority or non-inferiority margin. When $\delta > 0$, the rejection of the null hypothesis indicates the superiority of the test treatment over the control. When $\delta < 0$, the rejection of the null hypothesis indicates the non-inferiority of the test treatment against the control. Under the null hypothesis, the test statistic

$$T = \frac{\sqrt{N}(\hat{p}_{10} - \hat{p}_{20} - \delta)}{\sqrt{\hat{V}_d}} \qquad (9.9)$$

approximately follows a standard normal distribution when all n_{ti} are sufficiently large. Thus, we reject the null hypothesis at the α level of significance if $T > z_\alpha$. For testing equivalence, the following hypotheses are considered:

$$H_0 : |p_{10} - p_{20}| \geq \delta \quad \text{versus} \quad H_a : |p_{10} - p_{20}| < \delta, \qquad (9.10)$$

where δ is the equivalence limit. Thus, the null hypothesis is rejected at a significance level α and the test treatment is concluded to be equivalent to the control if

$$\frac{\sqrt{N}(\hat{p}_{10} - \hat{p}_{20} - \delta)}{\sqrt{\hat{V}_d}} < -z_\alpha \quad \text{and} \quad \frac{\sqrt{N}(\hat{p}_{10} - \hat{p}_{20} + \delta)}{\sqrt{\hat{V}_d}} > z_\alpha.$$

9.4 Assessment of Sensitivity Index

The primary assumption of the above approaches is that there is a relationship between μ_{ik}'s and a covariate vector x. As indicated earlier, such covariates may not exist or may not be observed in practice. In this case, Chow and Shao (2005) suggested assessing the sensitivity index and consequently deriving an unconditional inference for the original target patient population assuming that the shift parameter (i.e., ε) and/or the scale parameter (i.e., C) is random. Thus, the shift and scale parameters (i.e., ε and C) of the target population after a protocol amendment is made can be estimated by

$$\hat{\varepsilon} = \hat{\mu}_{\text{actual}} - \hat{\mu} \quad \text{and} \quad \hat{C} = \frac{\hat{\sigma}_{\text{actual}}}{\hat{\sigma}},$$

respectively, where $(\hat{\mu}, \hat{\sigma})$ and $(\hat{\mu}_{\text{actual}}, \hat{\sigma}_{\text{actual}})$ are some estimates of (μ, σ) and $(\mu_{\text{actual}}, \sigma_{\text{actual}})$, respectively. As a result, the sensitivity index can be estimated by

$$\hat{\Delta} = \frac{1 + \hat{\varepsilon}/\hat{\mu}}{\hat{C}}.$$

9.4.1 The Case Where ε Is Random and C Is Fixed

Estimates for μ and σ can be obtained based on data collected prior to any protocol amendments issued. Assume that the response variable x is distributed as $N(\mu, \sigma^2)$. Let x_{ji}, $i = 1, \ldots, n_j$; $j = 0, \ldots, m$ be the response of the ith patient after the jth protocol amendment. As a result, the total number of patients is given by $n = \sum_{j=0}^{m} n_j$. Note that n_0 is the number of patients in the study prior to any protocol amendments. Based on x_{0i}, $i = 1, \ldots, n_0$, the maximum likelihood estimates of μ and σ^2 can be obtained as follows:

$$\hat{\mu} = \frac{1}{n_0} \sum_{i=1}^{n_0} x_{0i} \quad \text{and} \quad \hat{\sigma}^2 = \frac{1}{n_0} \sum_{i=1}^{n_0} (x_{0i} - \hat{\mu})^2.$$

To obtain estimates for μ_{actual} and σ_{actual}, Chow and Shao (2005) considered the case where μ_{actual} is random and σ_{actual} is fixed. For convenience's sake, we set $\mu_{\text{actual}} = \mu$ and $\sigma_{\text{actual}} = \sigma$ for the derivation of ε and C. Assume that x is conditional on μ, i.e., $x|_{\mu=\mu_{\text{actual}}}$ follows a normal distribution $N(\mu, \sigma^2)$. That is,

$$x|_{\mu=\mu_{\text{actual}}} \sim N(\mu, \sigma^2),$$

where
  μ is distributed as $N(\mu_\mu, \sigma_\mu^2)$,
  σ, μ_μ, and σ_μ are some unknown constants.

Thus, the unconditional distribution of x is a mixed normal distribution given as

$$\int N(x; \mu, \sigma^2) N(\mu; \mu_\mu, \sigma_\mu^2) d\mu = \frac{1}{\sqrt{2\pi\sigma^2}} \frac{1}{\sqrt{2\pi\sigma_\mu^2}} \int_{-\infty}^{\infty} e^{-\frac{(x-\mu)^2}{2\sigma^2} - \frac{(\mu-\mu_\mu)^2}{2\sigma_\mu^2}} d\mu,$$

where $x \in (-\infty, \infty)$. It can be verified that the above mixed normal distribution is a normal distribution with mean μ_μ and variance $\sigma^2 + \sigma_\mu^2$. In other words, x is distributed as $N(\mu_\mu, \sigma^2 + \sigma_\mu^2)$. See Theorem 9.1.

Theorem 9.1

Suppose that $X|_\mu \sim N(\mu, \sigma^2)$ and $\mu \sim N(\mu_\mu, \sigma_\mu^2)$, then we have

$$X \sim N(\mu_\mu, \sigma^2 + \sigma_\mu^2). \tag{9.11}$$

Proof

Consider the following characteristic function of a normal distribution $N(t;\mu,\sigma^2)$:

$$\phi_0(w) = \frac{1}{\sqrt{2\pi\sigma^2}} \int_{-\infty}^{\infty} e^{iwt - \frac{1}{2\sigma^2}(t-\mu)^2}\, dt = e^{iw\mu - \frac{1}{2}\sigma^2 w^2}.$$

For distributions $X\mid_\mu \sim N(\mu,\sigma^2)$ and $\mu \sim N(\mu_\mu,\sigma_\mu^2)$, the characteristic function after exchanging the order of the two integrations is given by

$$\phi(w) = \int_{-\infty}^{\infty} e^{iw\mu - 1/2\sigma^2 w^2} N(\mu;\mu_\mu,\sigma_\mu^2)\, d\mu$$

$$= \int_{-\infty}^{\infty} e^{iw\mu - (\mu - \mu_\mu/2\sigma_\mu^2) - 1/2\sigma^2 w^2}\, d\mu.$$

Note that

$$\int_{-\infty}^{\infty} e^{iw\mu - \frac{\mu - \mu_\mu}{2\sigma_\mu^2}}\, d\mu = e^{iw\mu - \frac{1}{2}\sigma^2 w^2}$$

is the characteristic function of the normal distribution. It follows that

$$\phi(w) = e^{iw\mu - 1/2\sigma^2 w^2},$$

which is the characteristic function of $N(\mu_\mu, \sigma^2 + \sigma_\mu^2)$. This completes the proof.

Based on the above theorem, the maximum likelihood estimates (MLEs) of σ^2, μ_μ, and σ_μ^2 can be obtained as follows:

$$\tilde\mu_\mu = \frac{1}{m+1}\sum_{j=0}^{m}\tilde\mu_j, \quad \tilde\sigma_\mu^2 = \frac{1}{m+1}\sum_{j=0}^{m}(\tilde\mu_j - \tilde\mu_\mu)^2, \tag{9.12}$$

and

$$\tilde\sigma^2 = \frac{1}{n}\sum_{j=0}^{m}\sum_{i=1}^{n_j}(x_{ji} - \tilde\mu_j)^2,$$

where

$$\tilde{\mu}_j = \frac{1}{n_j} \sum_{i=1}^{n_j} x_{ji}.$$

Based on these maximum likelihood estimates, estimates of the shift parameter (i.e., ε) and the scale parameter (i.e., C) can be obtained as follows: $\tilde{\varepsilon} = \tilde{\mu} - \hat{\mu}$ and $\tilde{C} = \tilde{\sigma}/\hat{\sigma}$, respectively. Consequently, the sensitivity index can be estimated by simply replacing ε, μ, and C with their corresponding estimates $\tilde{\varepsilon}$, $\tilde{\mu}$, and $\tilde{C}$.

9.4.2 The Case Where ε Is Fixed and C Is Random

Similarly, let $\mu_{\text{actual}} = \mu$ and $\sigma_{\text{actual}} = \sigma$ and assume that $x|_{\sigma=\sigma_{\text{actual}}}$ follows a normal distribution $N(\mu, \sigma^2)$, that is,

$$x|_{\sigma=\sigma_{\text{actual}}} \sim N(\mu, \sigma^2)$$

where σ^2 is distributed as an inverse gamma distribution denoted by $IG(\alpha, \lambda)$, where μ, α, and λ are unknown parameters.

Theorem 9.2

Suppose that $x|_{\sigma=\sigma_{\text{actual}}} \sim N(\mu, \sigma^2)$ and $\sigma^2 \sim IG(\alpha, \lambda)$, then

$$x \sim f(x) = \frac{\Gamma(\alpha + 1/2)}{\Gamma(\alpha)} \frac{1}{\sqrt{2\pi\lambda}} \left[1 + \frac{(x-\mu)^2}{2\lambda} \right]^{-(\alpha+1/2)}. \tag{9.13}$$

That is, x is a noncentral t-distribution, where $\mu \in R$ is the location parameter, λ/α is the scale parameter, and 2α is the degree of freedom.

Proof

$$f(x, \sigma^2) = f(x \mid \sigma^2) f(\sigma^2)$$

$$= \frac{1}{\sqrt{2\pi}\sigma} \frac{\lambda^\alpha}{\Gamma(a)} \left(\frac{1}{\sigma^2} \right)^{\alpha+1} \exp\left\{ -\frac{(x-\mu)^2 + 2\lambda}{2\sigma^2} \right\}$$

$$f(x) = \int_0^{+\infty} f(x,\sigma^2)\,d\sigma^2$$

$$= \frac{1}{\sqrt{2\pi}}\,\frac{\lambda^\alpha}{\Gamma(a)}\int_0^{+\infty}\left(\frac{1}{\sigma^2}\right)^{\alpha+3/2}\exp\left\{-\frac{(x-\mu)^2+2\lambda}{2\sigma^2}\right\}d\sigma^2$$

$$= \frac{1}{\sqrt{2\pi}}\,\frac{\lambda^\alpha}{\Gamma(a)}\int_0^{+\infty} t^{\alpha-1/2}\exp\left\{-\frac{(x-\mu)^2+2\lambda}{2}t\right\}dt$$

$$= \frac{\Gamma(\alpha+1/2)}{\Gamma(\alpha)}\,\frac{1}{\sqrt{2\pi\lambda}}\left[1+\frac{(x-\mu)^2}{2\lambda}\right]^{-(\alpha+1/2)}.$$

Thus, X follows a noncentral t-distribution. Hence, we have $E(x) = \mu$ and $\mathrm{var}(x) = \lambda/(\alpha-1)$. This completes the proof.

Based on the above theorem, the maximum likelihood estimation of the parameters μ, α, and λ can be obtained as follows. Suppose that the observations satisfy the following conditions:

1. $(x_{ji}\,|\,\mu,\sigma_i^2) \sim N(\mu,\sigma_i^2)$, $\ i = 0,\ldots,m,\, j = 1,\ldots,n_i$, and given $\sigma_i^2, x_{1i},\ldots,x_{n_i i}$, are independent and identically distributed (i.i.d.)
2. $\{x_{ji},\, j = 1,\ldots,n_i\},\, i = 0,\ldots,m$ are independent
3. $\sigma_i^2 \sim IG(\alpha,\lambda)$

The likelihood function is given by

$$f(x_{01},\ldots,x_{mn_m}) = \prod_{i=0}^{m}\int_0^{\infty}\prod_{j=1}^{n_i} f(x_{ij}\,|\,\sigma_i^2)f(\sigma_i^2)\,d\sigma_i^2$$

$$= \prod_{i=0}^{m}\int_0^{\infty}\prod_{j=1}^{n_i}\frac{1}{\sqrt{2\pi}\sigma_i}\exp\left\{-\frac{(x_{ij}-\mu)^2}{2\sigma_i^2}\right\}\frac{\lambda^\alpha}{\Gamma(\alpha)}\exp\left\{-\frac{\lambda}{\sigma_i^2}\right\}d\sigma_i^2$$

$$= \prod_{i=0}^{m}\prod_{j=1}^{n_i}\frac{\Gamma(\alpha+1/2)}{\Gamma(\alpha)}\,\frac{1}{\sqrt{2\pi\lambda}}\left[1+\frac{(x_{ij}-\mu)^2}{2\lambda}\right]^{-(\alpha+1/2)}. \tag{9.14}$$

Thus, the log-likelihood function is

$$
L = \ln f(x_{01}, \ldots, x_{mn_m})
$$

$$
= n \ln \Gamma\left(\alpha + \frac{1}{2}\right) - n \ln \Gamma(\alpha) - \frac{n}{2} \ln 2\pi\lambda - \left(\alpha + \frac{1}{2}\right) \sum_{i=0}^{m} \sum_{j=1}^{n_i} \ln\left[1 + \frac{(x_{ij} - \mu)^2}{2\lambda}\right].
$$

$$(9.15)$$

Based on (9.15), we can obtain the derivatives of the unknown parameters μ, α, and λ, as follows:

$$
\frac{\partial L}{\partial \mu} = \sum_{i=0}^{m} \sum_{j=1}^{n_i} \frac{(x_{ij} - \mu)}{1 + (x_{ij} - \mu)^2/2\lambda} = 0
$$

$$
\frac{\partial L}{\partial \alpha} = n\psi\left(\alpha + \frac{1}{2}\right) - n\psi(\alpha) - \sum_{i=0}^{m} \sum_{j=1}^{n_i} \ln\left[1 + \frac{(x_{ij} - \mu)^2}{2\lambda}\right] = 0
$$

$$
\frac{\partial L}{\partial \lambda} = -n + \frac{(\alpha + 1/2)}{\lambda} \sum_{i=0}^{m} \sum_{j=1}^{n_i} \frac{(x_{ij} - \mu)^2}{1 + (x_{ij} - \mu)^2/2\lambda}
$$

where $\Psi(\alpha) = \Gamma'(\alpha)/\Gamma(\alpha)$ is a digamma function.
Define

$$
w_{ij} = \left[\frac{1 + (x_{ij} - \mu)^2}{2\lambda}\right]^{-1}.
$$

$$(9.16)$$

Then the maximum likelihood estimation of the parameters μ, α, and λ can be decided by

$$
\hat{\mu} = \frac{\sum_{i=0}^{m} \sum_{j=1}^{n_i} w_{ij} x_{ij}}{\sum_{i=0}^{m} \sum_{j=1}^{n_i} w_{ij}},
$$

$$(9.17)$$

$$
\hat{\lambda} = \left(\hat{\alpha} + \frac{1}{2}\right) \frac{1}{n} \sum_{i=0}^{m} \sum_{j=1}^{n_i} w_{ij}(x_{ij} - \hat{\mu})^2.
$$

$$(9.18)$$

The digamma function may be approximated as in Johnson and Kotz (1972) as $\psi(\alpha) = \ln(\alpha - 0.5)$, and employing a Taylor expansion we have

$$\hat{\alpha} = 0.5 + \frac{n}{2}\sum_{i=0}^{m}\sum_{j=1}^{n_i} \ln w_{ij}^{-1}. \tag{9.19}$$

The maximum likelihood estimates of μ, α, and λ can be obtained by (9.17) through (9.19). In fact, it is difficult to solve the equation from (9.17) through (9.19) directly, but there are some published results giving the maximum likelihood estimation of the location parameter and freedom degree in a central t-distribution, and according to (9.17) through (9.19), the estimation of the scale parameter in a noncentral t-distribution could be obtained.

Lu et al. (2010) used the moment estimation to obtain the estimation of the parameters μ, α, and λ. The observations

$$(x_{ij} \mid \mu, \sigma_i^2) \sim N(\mu, \sigma_i^2), \quad i = 0, \ldots, m, \quad j = 1, \ldots, n_i,$$

and x_{ij} independent, according to Theorem 9.1, x is a noncentral t-distribution, mean $= E(x) = \mu$ and variance $= \text{var}(x) = \lambda/(\alpha - 1)$, if $\alpha > 1$; even the central moment

$$\mu_k(x) = \mu_{k-2}(x)\left[\frac{2\lambda(k-1)}{(2\alpha - k)}\right] \quad \text{if } \alpha > \frac{k}{2},$$

since the fourth moment does not exist for $\alpha \le 2$, moreover the variance of the estimator of α is infinite if $\alpha \le 4$. Under the background of medical research, we assume that if $\alpha > 4$ is held, and the obvious choices are sample mean, variance, and the fourth moment employed, the moment estimation of the parameters could be obtained:

$$\hat{\mu} = \frac{1}{n}\sum_{i=1}^{n} x_i,$$

$$\hat{\alpha} = \frac{[3(S_n^2)^2 - 2S_n^4]}{[3(S_n^2)^2 - S_n^4]},$$

$$\hat{\lambda} = \frac{-S_n^2 S_n^4}{[3(S_n^2)^2 - S_n^4]}.$$

We now examine the large-sample behavior of maximum likelihood estimates. Further differentiability assumption is required, and under the conditions of normal distribution and IG-distribution, that requirement can be satisfied. Cox and Snell (1968) have derived a general formula for the second-order bias of the maximum likelihood estimator of the vector

$$b(\hat{\beta}_s) = \sum_{r,t,u} k^{s,r} k^{t,u} \left\{ \frac{1}{2} k_{rtu} + k_{rt,u} \right\}, \tag{9.20}$$

where the set parameter vector $\theta = (\beta_r, \beta_s, \beta_t) = (\mu, \alpha, \lambda)^T$ and r, s, t, u index the parameter space (μ, α, λ), and we use the standard notation for the moments of the derivatives of the log-likelihood function: $k_{rs} = E[U_{rs}]$, $k_{rst} = E[U_{rst}]$, $k_{rs,t} = E[U_{rs}U_t]$, where $U_r = \partial l/\partial \beta_r$, $U_{rs} = \partial^2 l/\partial \beta_r \partial \beta_s$, $U_{rst} = \partial^3 l/\partial \beta_r \partial \beta_s \partial \beta_t$. Also, $k^{r,s}$ denotes the general (r, s) element of the inverse of the information matrix, the information matrix itself having its general (r, s) element given by $k_{rs} = -E[U_{rs}]$. Let the fisher information matrix be

$$I(\theta) = n \begin{pmatrix} \dfrac{\alpha(2\alpha+1)}{\lambda(2\alpha+3)} & 0 & 0 \\[2ex] 0 & \Psi'(\alpha) - \Psi'\left(\alpha + \dfrac{1}{2}\right) & \dfrac{\alpha(\lambda-1)-1}{\lambda(\alpha+1)} \\[2ex] 0 & \dfrac{\alpha(\lambda-1)-1}{\lambda(\alpha+1)} & \dfrac{\alpha}{\lambda^2(2\alpha+3)} \end{pmatrix}, \tag{9.21}$$

so that $k_{\lambda\lambda\alpha} = k_{\lambda\alpha\lambda} = k_{\alpha\lambda\lambda} = -(4\alpha + 3)/(2\alpha + 1)(2\alpha + 3)\lambda^2$, $k_{\alpha\alpha\alpha} = \Psi''(\alpha + 1/2) - \Psi''(\alpha)$, $k_{\mu\mu\lambda} = k_{\mu\lambda\mu} = k_{\lambda\mu\mu} = 4\alpha(\alpha + 1)^2/\lambda^2(2\alpha + 3)(2\alpha + 5)$, $k_{\mu\mu\alpha} = k_{\mu\alpha\mu} = k_{\alpha\mu\mu} = -2\alpha/\lambda$ $(2\alpha + 3)$ when r, s, t take other values in the parameter space except those enumerated above such as $k_{rst} = 0$ and $k_{rs,t} = 0$, where r, s, t index the parameter space. The bias of the maximum likelihood estimate of the parameter α is

$$b(\hat{\alpha}) = \frac{A_1\{B_1 C_1 - D_1 + E_1 F_1\}}{nM^2} \tag{9.22}$$

where $M = \{[\psi'(\alpha) - \psi'(\alpha + 1/2)][\alpha/\lambda^2(2\alpha + 3) - 1/\lambda^2(2\alpha + 1)^2]\}(\alpha/\lambda)(2\alpha + 1)/(2\alpha + 3)$ is the determinant of the inverse information matrix $I^{-1}(\theta)$, $A_1 = \alpha^2/2\lambda^6(2\alpha + 3)^3$, $B_1 = \alpha(2\alpha + 1)(12\alpha + 21)/(2\alpha + 5)$, $C_1 = (\psi'(\alpha) - \psi'(\alpha + 1/2))$, $D_1 = 2(4\alpha + 3)/(2\alpha + 1)$, $E_1 = \alpha^2(2\alpha + 1)^2/(2\alpha + 3)$, $F_1 = \psi''(\alpha + 1/2) - \psi''(\alpha)$.

At the same time we have

$$b(\hat{\lambda}) = \frac{(A_2 C_1^{\,2} + B_2 F_1 - E_2 C_1)}{nM^2},$$ (9.23)

where

$A_2 = 2\alpha^3 (2\alpha + 1)^2 (5\alpha + 8)/\lambda^5 (2\alpha + 3)^3 (2\alpha + 5),$

$B_2 = \alpha^3 (2\alpha + 1)/\lambda^5 (2\alpha + 3)^3,$

$C_2 = \alpha^2 (14\alpha + 9)/\lambda^5 (2\alpha + 3)^3.$

The maximum likelihood estimator of α has an n^{-1} order bias, which is the same for the estimator λ, and we also obtain the bias of parameter μ as $b(\hat{\mu}) = 0$, which is obviously the unbiased estimate of the parameter μ.

In the case where μ_{actual} is fixed and σ_{actual} is random we will focus on the statistical inference on ε, C, and Δ to illustrate the impact on the statistical inference of the actual patient population after m protocol amendment.

9.5 Sample Size Adjustment

In clinical trials, for a given target patient population, sample size calculation is usually performed based on a test statistic (which is derived under the null hypothesis) evaluated under an alternative hypothesis. After protocol amendments, the target patient population may have been shifted to an *actual* patient population. In this case, the original sample size may have to be adjusted in order to achieve the desired power for the assessment of the treatment effect for the *original* patient population. For the clinical evaluation of efficacy and safety, statistical inference such as hypotheses testing is usually considered. In practice, the commonly considered testing hypotheses include (1) testing for equality, (2) testing for non-inferiority, (3) testing for superiority, and (4) testing for equivalence. The hypotheses are summarized as follows:

$$\text{Equality:} \quad H_0 : \mu_1 = \mu_2 \quad \text{versus} \quad H_a : \mu_1 - \mu_2 = \delta \neq 0,$$ (9.24)

$$\text{Non-inferiority:} \quad H_0 : \mu_1 - \mu_2 \leq \delta \quad \text{versus} \quad H_a : \mu_1 - \mu_2 > \delta,$$

$$\text{Superiority:} \quad H_0 : \mu_1 - \mu_2 \leq \delta \quad \text{versus} \quad H_a : \mu_1 - \mu_2 > \delta,$$

$$\text{Equivalence:} \quad H_0 : |\mu_1 - \mu_2| > \delta \quad \text{versus} \quad H_a : |\mu_1 - \mu_2| \leq \delta,$$

where δ is a clinically meaningful difference (for testing equality), a non-inferiority margin (for testing non-inferiority), a superiority margin (for testing superiority), and an equivalence limit (for testing equivalence), respectively.

Let $n_{classic}$ and n_{actual} be the sample sizes based on the original patient population and the actual patient population, respectively, as the result of protocol amendments. Also, let $n_{actual} = Rn_{classic}$, where R is the adjustment factor. Following the procedures described by Chow, Shao, and Wang (2008), sample sizes for both $n_{classic}$ and n_{actual} can be obtained. For example, Table 9.2 provides formulas for sample size adjustment based on covariate-adjusted model for binary response endpoint, while Tables 9.3 and 9.4 give sample size adjustments based on random location shift and random scale shift, respectively.

9.6 Concluding Remarks

As indicated, the investigator has the flexibility to modify or change the study protocol during the conduct of the clinical trial by issuing protocol amendments. This flexibility gives the investigator (1) the opportunity to correct (minor changes) the assumptions early and (2) the chance to redesign (major changes) the study. It is well recognized that the abuse of this flexibility may result in a moving target patient population, which makes it almost impossible for the intended trial to address the medical or scientific questions that the study intends to answer. Thus far, regulatory agencies do not have regulations regarding the issue of protocol amendments after the initiation of a clinical trial. It is suggested that regulatory guideline/guidance regarding (1) levels of changes and (2) number of protocol amendments that are allowed be developed in order to maintain the validity and integrity of the intended study. In addition, it is also suggested that a sensitivity analysis should be conducted for evaluating the possible impact due to protocol amendments.

As pointed out by Chow and Chang (2006), the impact on statistical inference due to protocol amendments could be substantial, especially when there are major modifications which have resulted in a significant shift in mean response and/or inflation of the variability of response of the study parameters. It is suggested that a sensitivity analysis with respect to changes in study parameters be performed to provide a better understanding of the impact of changes (protocol amendments) in study parameters on statistical inference. Thus, regulatory guidance on what range of changes in study parameters is considered acceptable is necessarily developed. As indicated earlier, adaptive design methods are very attractive to the clinical researchers and/or sponsors due to their flexibility, especially in clinical trials of early clinical development. It, however, should be noted that there is a high risk that a clinical trial using adaptive design methods may fail in terms of its scientific validity and/or its limitation of providing useful information with a desired power, especially when the sizes of the trials are relatively small and there are a number of protocol amendments.

TABLE 9.2

Sample Size Adjustment Based on Covariate-Adjusted Model

Test	Hypothesis	Non-Adjustment	Adjustment
Superiority	$H_0 : p_{10} - p_{20} \leq \delta$ $H_1 : p_{10} - p_{20} > \delta$	$N_{\text{classic}} = \dfrac{(z_\alpha + z_\gamma)^2}{(p_{10} - p_{20} - \delta)^2}\left[\dfrac{p_{10}(1-p_{10})}{w} + \dfrac{p_{20}(1-p_{20})}{1-w}\right]$	$N_{\text{actual}} = \dfrac{(z_\alpha + z_\gamma)^2 \tilde{V}_d}{(p_{10} - p_{20} - \delta)^2}$
Non-inferiority	$H_0 : p_{10} - p_{20} \leq -\delta$ $H_1 : p_{10} - p_{20} > -\delta$	$N_{\text{classic}} = \dfrac{(z_\alpha + z_\gamma)^2}{(p_{10} - p_{20} + \delta)^2}\left[\dfrac{p_{10}(1-p_{10})}{w} + \dfrac{p_{20}(1-p_{20})}{1-w}\right]$	$N_{\text{actual}} = \dfrac{(z_\alpha + z_\gamma)^2 \tilde{V}_d}{(p_{10} - p_{20} + \delta)^2}$
Equivalence	$H_0 : |p_{10} - p_{20}| \geq \delta$ $H_1 : |p_{10} - p_{20}| < \delta$	$N_{\text{classic}} = \dfrac{(z_\alpha + z_\gamma)^2}{(\delta - |p_{10} - p_{20}|)^2}\left[\dfrac{p_{10}(1-p_{10})}{w} + \dfrac{p_{20}(1-p_{20})}{1-w}\right]$	$N_{\text{actual}} = \dfrac{(z_\alpha + z_{\gamma/2})^2 \tilde{V}_d}{\left(\delta - |p_{10} - p_{20}|\right)^2}$

w is the proportion of patients for the first treatment

$$\tilde{V}_d = [g'(\mathbf{b})]^T \left(w \sum_{i=0}^{k} \rho_{1i}\mathbf{I}^{(1i)} + (1-w)\sum_{i=0}^{k} \rho_{2i}\mathbf{I}^{(2i)} \right)^{-1} g'(\mathbf{b})$$

where $w = n_1/N$, $\rho_{ti} = n_{ti}/n_{t.}$ and

$$g'(\mathbf{b}) = \begin{pmatrix} p_{10}(1-p_{10}) - p_{20}(1-p_{20}) \\ p_{10}(1-p_{10}) \\ v_0\left(p_{10}(1-p_{10}) - p_{20}(1-p_{20})\right) \\ v_0\left(p_{10}(1-p_{10})\right) \end{pmatrix}.$$

TABLE 9.3

Sample Size Adjustment Based on Random Location Shift

Test	Hypothesis	Non-Adjustment	Adjustment
Equality	$H_0 : \mu_1 - \mu_2 = 0$ $H_a : \mu_1 - \mu_2 \neq 0$	$N_{\text{classic}} = \dfrac{2(z_{\alpha/2} + z_\beta)^2 \tilde{\sigma}^2}{(\mu_1 - \mu_2)^2}$	$N_{\text{actual}} = \dfrac{2(m+1)(z_{\alpha/2} + z_\beta)^2 \tilde{\sigma}^2}{(m+1)(\mu_1 - \mu_2)^2 - 2(z_{\alpha/2} + z_\beta)^2 \tilde{\sigma}_\mu^2}$
Non-inferiority/ superiority	$H_0 : \mu_1 - \mu_2 \leq \delta$ $H_a : \mu_1 - \mu_2 > \delta$	$N_{\text{classic}} = \dfrac{2(z_\alpha + z_\beta)^2 \tilde{\sigma}^2}{(\mu_1 - \mu_2 - \delta)^2}$	$N_{\text{actual}} = \dfrac{2(m+1)(z_\alpha + z_\beta)^2 \tilde{\sigma}^2}{(m+1)(\mu_1 - \mu_2 - \delta)^2 - (z_\alpha + z_\beta)^2 \tilde{\sigma}_\mu^2}$
Equivalence	$H_0 : |\mu_1 - \mu_2| \geq \delta$ $H_a : |\mu_1 - \mu_2| < \delta$	$N_{\text{classic}} = \dfrac{2(z_\alpha + z_{\beta/2})^2 \tilde{\sigma}^2}{(|\mu_1 - \mu_2| - \delta)^2}$	$N_{\text{actual}} = \dfrac{2(m+1)(z_{\alpha/2} + z_\beta)^2 \tilde{\sigma}^2}{(m+1)(|\mu_1 - \mu_2| - \delta)^2 - (z_{\alpha/2} + z_\beta)^2 \tilde{\sigma}_\mu^2}$

TABLE 9.4

Sample Size Adjustment Based on Random Scale Shift

Test	Hypothesis	Non-Adjustment	Adjustment
Equality	$H_0 : \mu_1 - \mu_2 = 0$ $H_a : \mu_1 - \mu_2 \neq 0$	$N_{\text{classic}} = \dfrac{2(z_{\alpha/2} + z_\beta)^2 \tilde{\sigma}^2}{(\mu_1 - \mu_2)^2}$	$N_{\text{actual}} = \dfrac{2(z_{1-\alpha/2} + z_{1-\beta})^2 (m+1)\tilde{v}\tilde{\sigma}^2 \sum_{j=0}^{m} (V_{1j}^{(t)})^2}{(\mu_1 - \mu_2)^2 (\tilde{v} - 2)\left(\sum_{j=0}^{m} V_{1j}^{(t)}\right)^2}$
Non-inferiority/ superiority	$H_0 : \mu_1 - \mu_2 \leq \delta$ $H_a : \mu_1 - \mu_2 > \delta$	$N_{\text{classic}} = \dfrac{2(z_\alpha + z_\beta)^2 \tilde{\sigma}^2}{(\mu_1 - \mu_2 - \delta)^2}$	$N_{\text{actual}} = \dfrac{2(z_\alpha + z_\beta)^2 (m+1)\tilde{v}\tilde{\sigma}^2 \sum_{j=0}^{m} (V_{1j}^{(t)})^2}{(\mu_1 - \mu_2 - \delta)^2 (\tilde{v} - 2)\left(\sum_{j=0}^{m} V_{1j}^{(t)}\right)^2}$
Equivalence	$H_0 : \lvert \mu_1 - \mu_2 \rvert \geq \delta$ $H_a : \lvert \mu_1 - \mu_2 \rvert < \delta$	$N_{\text{classic}} = \dfrac{2(z_\alpha + z_{\beta/2})^2 \tilde{\sigma}^2}{(\lvert \mu_1 - \mu_2 \rvert - \delta)^2}$	$N_{\text{actual}} = \dfrac{2(z_\alpha + z_{\beta/2})^2 (m+1)\tilde{v}\tilde{\sigma}^2 \sum_{j=0}^{m} (V_{1j}^{(t)})^2}{(\lvert \mu_1 - \mu_2 \rvert - \delta)^2 (\tilde{v} - 2)\left(\sum_{j=0}^{m} V_{1j}^{(t)}\right)^2}$

$V_{1j}^{(t)} = \dfrac{v^{(t)}(\sigma^{(t)})^2 + n_j(\sigma^{(t)})^2}{v^{(t)}(\sigma^{(t)})^2 + \sum_{i=1}^{n_j}(x_{ji} - \mu^{(t)})^2}$ where $\{\mu^{(t)}, \sigma^{(t)}, v^{(t)}\}$ is the tth step estimate in the EM algorithm.

As indicated in the previous sections, analysis with covariate adjustment and the assessment of sensitivity index are the two commonly considered approaches when there is population shift due to protocol amendment. For the method of analysis with covariate adjustment, an alternative approach considering random coefficients in model (9.1) and/or a Bayesian approach may be useful for obtaining an accurate and reliable estimate of the treatment effect of the compound under study. For the assessment of sensitivity index, in addition to the cases where (1) ε is random and C is fixed, and (2) ε is fixed and C is random, there are other cases such as (1) both ε and C are random, (2) sample sizes before and after protocol amendments are random variables, and (3) the number of protocol amendments is also a random variable remain unchanged.

In addition, statistically, it is a challenge to clinical researchers when there are missing values. These could be due to the causes that are related to or unrelated to the changes or modifications made in the protocol amendments. In this case, missing values must be handled carefully to provide an unbiased assessment and interpretation of the treatment effect. When there is a population shift either in location parameter or scale parameter, the standard methods for the assessment of treatment effect are necessarily modified. For example, the standard methods such as the O'Brien–Fleming method in typical group sequential design for controlling the overall type I error rate are not appropriate when there is a population shift due to protocol amendments.

10

Seamless Adaptive Trial Designs

10.1 Introduction

In recent years, the use of adaptive design methods in clinical research and development based on accrued data and/or external information has become very popular due to its flexibility and efficiency (Liu and Chi, 2001; Chow and Chang, 2005, 2006; Krams et al. 2006; EMEA, 2007; FDA, 2010b). An adaptive design is defined as a clinical trial design that allows adaptations (modifications or changes) to trial and/or statistical procedure of the trial after its initiation without undermining the validity and integrity of the trial. In their recent publication, with the emphasis of the feature of design adaptations only (rather than ad hoc adaptations), the Pharmaceutical Research Manufacturer Association (PhRMA) Working Group on Adaptive Design defines an adaptive design as a study design that uses accumulating data to decide on how to modify aspects of the study as it continues, without undermining the validity and integrity of the trial. On the other hand, the FDA defines an adaptive design as a study that includes a prospectively planned opportunity for modification of one or more specified aspects of the study design and hypotheses based on analysis of data (usually interim data) from subjects in the study (FDA, 2010b). Based on the adaptations applied, adaptive designs can be classified into three categories: prospective, concurrent, and retrospective adaptive designs. Chow and Chang (2006) indicate that commonly considered adaptive designs in these categories include, but are not limited to, (1) an adaptive randomization design, (2) a group sequential design (Jennison and Turnball, 2000; Kelly, 2005a, 2005b), (3) a flexible sample size reestimation design, (4) a drop-the-loser (or pick-the-winner) design (Sampson and Sill, 2005), (5) an adaptive dose-finding design (Chang and Chow, 2005), (6) a biomarker-adaptive design (Chang, 2005a, 2005b), (7) an adaptive treatment-switching design (Branson and Whitehead, 2002; Shao et al., 2005), (8) a hypothesis-adaptive design, (9) a seamless adaptive trial design (Maca et al., 2006), and (10) a multiple adaptive design, which is any combinations of the above-mentioned adaptive designs. Among these, group sequential design, adaptive dose-finding design, and (two-stage) seamless adaptive design are probably the most commonly employed adaptive designs in clinical trials. In this chapter, however, we will only focus on the two-stage seamless adaptive trial design.

A seamless trial design is referred to as a program that addresses study objectives within a single trial that are normally achieved through separate trials in clinical development (Bauer and Kieser, 1999; Maca et al., 2006). An adaptive seamless design is a seamless trial design that would use data from patients enrolled before and after the adaptation in the final analysis. Thus, a two-stage seamless adaptive design consists of two phases (stages), namely a learning (or exploratory) phase (Stage 1) and a confirmatory phase (Stage 2). The learning phase provides the opportunity for adaptations such as stopping the trial early due to safety and/or futility/efficacy based on accrued data at the end of the learning phase. A two-stage seamless adaptive trial design reduces lead time between the learning (i.e., the first study for the traditional approach) and confirmatory (i.e., the second study for the traditional approach) phases. Most importantly, data collected at the learning phase are combined with those obtained at the confirmatory phase for the final analysis.

In the next section, controversial issues regarding the flexibility, efficiency, validity, and integrity of clinical trials utilizing adaptive trial designs are discussed. Also included in the section are regulatory perspectives of the use of adaptive design methods in clinical trials. Types of two-stage seamless adaptive trial designs depending upon whether the study objectives and/or the study endpoints at different stages are the same are described. Section 10.4 summarizes statistical methods for the analysis of the type of two-stage seamless designs with different study endpoints. Statistical methods for the analysis of the type of two-stage seamless designs with different study objectives/endpoints are developed in Section 10.5. Some concluding remarks are provided in the last section of this chapter.

10.2 Controversial Issues

The use of adaptive design methods for modifying the trial and/or statistical procedures of ongoing clinical trials based on accrued data has been practiced for years in clinical research. Adaptive design methods in clinical research are very attractive to clinical scientists due to the following reasons. First, it reflects medical practice in the real world. Second, it is ethical with respect to both efficacy and safety (toxicity) of the test treatment under investigation. Third, it is not only flexible, but also efficient in the early phase of clinical development. However, some concerns regarding the validity and integrity of the clinical trials utilizing adaptive trial designs have been raised and discussed tremendously within the pharmaceutical industry and the regulatory agencies. In what follows, controversial issues regarding the flexibility, efficiency, validity, and integrity of a clinical trial utilizing adaptive trial design are briefly described.

10.2.1 Flexibility and Efficiency

A two-stage adaptive seamless design is considered a more efficient and flexible study design as compared to the traditional approach of having separate studies in terms of controlling type I error rate and power. For controlling the overall type I error rate, as an example, consider a two-stage adaptive seamless phase II/III design. Let α_{II} and α_{III} be the type I error rate for phase II and phase III studies, respectively. Then, the overall α for the traditional approach of having two separate studies is given by $\alpha = \alpha_{II}\alpha_{III}$. In the two-stage adaptive seamless phase II/III design, on the other hand, the actual α is given by $\alpha = \alpha_{III}$. Thus, the α for a two-stage adaptive seamless phase II/III design is actually $1/\alpha_{II}$ times larger than the traditional approach for having two separate phase II and phase III studies. Similarly, for the evaluation of power, let $Power_{II}$ and $Power_{III}$ be the power for phase II and phase III studies, respectively. Then, the overall α for the traditional approach of having two separate studies is given by $Power = Power_{II} \times Power_{III}$. In the two-stage adaptive seamless phase II/III design, the actual power is given by $Power = Power_{III}$. Thus, the power for a two-stage adaptive seamless phase II/III design is actually $1/Power_{II}$ times larger than the traditional approach for having two separate phase II and phase III studies.

In addition, a two-stage seamless adaptive trial design that combines two separate (independent) studies can help in reducing lead time between studies. In practice, the lead time between studies is estimated to be about 6 months to 1 year. As a common clinical practice, the phase III study will not be initiated until the final report of the phase II trial is reviewed and issued. After the completion of a phase II study, on average, it will usually take about 4 months to lock database (including data entry/verification and data query/validation), programming and data analysis, and final integrated statistical/clinical report. During the preparation of the phase III trial, the development of a study protocol and Institutional Review Board (IRB) review/approval will also take some time. As a result, the application of a two-stage phase II/III seamless adaptive trial design will not only reduce the lead time between studies, but also allow the sponsor (investigator) to make a go/no-go decision at the end of the first stage (phase II study) early. In some case, a two-stage phase II/III seamless adaptive trial design may require a smaller sample size as compared to the traditional approach of two separate studies for phase II and phase III since data collected from both stages would be combined for a final assessment of the test treatment effect under investigation.

10.2.2 Validity and Integrity

In practice, before an adaptive design can be implemented, some practical issues such as feasibility, validity, and robustness are necessarily addressed. For feasibility, several questions arise. For example, does the adaptive design require extra efforts in implementation? Do the level of difficulty and the associated cost justify the gain from implementing the adaptive design?

Does the implementation of the adaptive design delay patient recruitment and prolong study duration? How often are the unblinded analyses practical and to whom should the data should be unblinded? How should the impact of the data monitoring committee's (DMC) decision regarding the trial (e.g., recommending an early stopping or other adaptations due to safety concerns) be considered at the design stage?

For the issue of validity, it is reasonable to ask the following questions. Does the unblinding cause potential bias in treatment assessment? Does the implementation of an adaptive design destroy the randomness? For example, response-adaptive randomization is used to assign more patients to the superior treatment groups by changing the randomization schedule. However, for ethical reasons, the patients should be informed that the later they come into the study, the greater is their chance of being assigned to the superior groups. For this reason, patients may prefer to wait for a late entry into the study. This could cause bias because sicker patients might enroll earlier just because they cannot wait. When this happens, the treatment effect is confounded by the patient's disease background. The bias could occur for a drop-losers design and other adaptive designs.

Regarding the issue of robustness, without virtually any exception, a trial cannot be conducted exactly as specified in the protocol. Would protocol deviations invalidate the adaptive method? For example, if an actual interim analysis were performed at a different (information) time than the scheduled one, how does it impact the type I error of the adaptive design? How does an unexpected DMC action affect the power and validity of the design? Would a protocol amendment such as endpoint change or inclusion/exclusion change invalidate the design and analysis? Would delayed responses diminish the advantage of implementing an adaptive design such as continual reassessment method (CRM) in an adaptive dose-escalation design and trials with a survival endpoint?

Adaptive designs usually involve multiple comparisons and often invoke a dependent sampling procedure or an adaptive combination of subsamples from different stages. Therefore, studies with adaptive designs are much more complicated than those with classic designs. The theoretical challenges arise from a typical adaptive design include (1) α adjustment to control overall type I error rate for multiple comparisons, (2) the p-value adjustment due to the dependent sampling procedure, (3) finding a robust unbiased point estimate, and (4) finding a reliable confidence interval. In practice, it is not always easy to derive an analytical form for correct adjusted alpha and p-value due to the flexibility of adaptations. However, they can be addressed through computer simulations regardless of the complexity of the adaptive designs. To do this, it is necessary to define an appropriate test statistic that can be applied before and after adaptations. A simulation can then be conducted under the null hypothesis for obtaining the sampling distribution of the test statistic. Based on the simulated distribution, the rejection region, adjusted alpha, and adjusted p-values can be obtained. The simulations can be done during protocol design to provide justification for choosing an appropriate design.

10.2.3 Regulatory Concerns

As it is recognized by the regulatory agencies, there are some possible benefits when utilizing adaptive design methods in clinical trials. For example, the use of adaptive design methods in clinical trials allows the investigator to correct wrong assumptions and select the most promising option early. In addition, adaptive designs make use of cumulative information of the ongoing trial and emerging external information to the trial, which allow the investigator to react earlier to *surprises* regardless of positive or negative results. As a result, the use of adaptive design methods may speed up the development process.

Although the investigator may have a second chance to redesign the trial after seeing the data from the trial itself at interim (or externally), it is flexible but more problematic operationally due to potential bias that may have been introduced to the conduct of the trial. For example, it is a major concern that unblinding during an interim analysis may have introduced potential bias by a change in clinical practice resulting from feedback from the analysis. As a result, we may have compromised scientific integrity of trial conduct due to operational bias. As indicated by the United States Food and Drug Administration (FDA), operational biases commonly occur when adaptations in trial and/or statistical procedures are applied. Trial procedures are referred to as eligibility criteria, dose/dose regimen and duration, assessment of study endpoints, and/or diagnostic/laboratory testing procedures that are employed during the conduct of the trial. Statistical procedures include (1) selection and/or modification of study design; (2) formulation and/or modification of statistical hypotheses (according to study objectives); (3) selection and/or modification of study endpoints; (4) sample size calculation, reestimation, and/or adjustment; (5) generation of randomization schedules; and (6) development of statistical analysis plan (SAP). As a result, commonly seen operational biases due to adaptations include (1) sample size reestimation at interim analysis; (2) sample size allocation to treatments (e.g., change from 1:1 ratio to an unequal ratio); (3) delete, add, or change treatment arms after the review of interim analysis results; (4) shift in patient population after the application of adaptations (e.g., change in inclusion/exclusion criteria and/or subgroups); (5) change in statistical test strategy (e.g., change log-rank to other tests); (6) change study endpoints (e.g., change survival to time-to-disease progression and/or response rate in cancer trials); and (7) change study objectives (e.g., switch a superiority hypothesis to a non-inferiority hypothesis).

In summary, regulatory agencies do not object to the use of the adaptive design methods in clinical trials due to its flexibility, efficiency, and potential benefits as described above. However, the validity and integrity of the clinical trials after the implementation of various adaptations have raised critical concerns about the drug evaluation and approval process. These concerns include, but are not limited to, the following: (1) that we may not be able to control (preserve) the overall type I error rate at a prespecified level of significance, (2) that the obtained p-values may not be correct, (3) that the obtained confidence interval may not be reliable, and (4) that major (significant) adaptations may have resulted in a totally different trial that is unable to address the scientific/medical questions the original study intended to answer.

10.3 Types of Two-Stage Seamless Adaptive Designs

In practice, two-stage seamless adaptive trial designs can be classified into the following four categories depending upon study objectives and study endpoints at different stages (Chow and Tu, 2009). See also Table 10.1.

In other words, we have (1) Category I (SS)—same study objectives and same study endpoints, (2) Category II (SD)—same study objectives but different study endpoints, (3) Category III (DS)—different study objectives but same study endpoints, and (4) Category IV (DD)—different study objectives and different study endpoints. Note that different study objectives are usually referred to dose finding (selection) at the first stage and efficacy confirmation at the second stage, while different study endpoints are directed to biomarker versus clinical endpoint or the same clinical endpoint with different treatment durations. Category I trial design is often viewed as a similar design to a group sequential design with one interim analysis despite the fact that there are differences between a group sequential design and a two-stage seamless design. In this chapter, our emphasis will be placed on Category II designs. The results obtained can be similarly applied to Category III and Category IV designs with some modification for controlling the overall type I error rate at a prespecified level. In practice, typical examples for a two-stage adaptive seamless design include a two-stage adaptive seamless phase I/II design and a two-stage adaptive seamless phase II/III design. For the two-stage adaptive seamless phase I/II design, the objective in the first stage is biomarker development and the study objective in the second stage is to establish early efficacy. For a two-stage adaptive seamless phase II/III design, the study objective is for treatment selection (or dose finding) while the study objective at the second stage is efficacy confirmation.

TABLE 10.1

Types of Two-Stage Seamless Adaptive
Designs

	Study Endpoint	
Study Objectives	**Same (S)**	**Different (D)**
Same (S)	I = SS	II = SD
Different (D)	III = DS	IV = DD

Statistical consideration for the first kind of two-stage seamless designs is similar to that of a group sequential design with one interim analysis. Sample size calculation and statistical analysis for this kind of study designs can be found in Chow and Chang (2006). For other kinds of two-stage seamless trial designs, standard statistical methods for group sequential design are not appropriate and hence should not be applied directly. In this chapter, statistical methods for a two-stage adaptive seamless design with different study endpoints (e.g., biomarker versus clinical endpoint or the same clinical endpoint with different treatment durations) but same study endpoint will be developed. Modification to the derived results is necessary if the study endpoints and study objectives are different at different stages.

One of the questions that are commonly asked when applying a two-stage adaptive seamless design in clinical trials is sample size calculation/allocation. For the first kind of two-stage seamless designs, the methods based on individual p-values as described by Chow and Chang (2006) can be applied. However, these methods are not appropriate for Category IV (DD) trial designs with different study objectives and endpoints at different stages. For Category IV (DD) trial designs, the following issues are challenging to the investigator and the biostatistician. First, how do we control the overall type I error rate at a pre-specified level of significance? Second, is the typical O'Brien–Fleming type of boundaries feasible? Third, how to perform a valid final analysis that combines data collected from different stages? Cheng and Chow (2010) attempt to address these questions by proposing a new multiple-stage transitional seamless adaptive design accompanied with valid statistical tests to incorporate different study endpoints for achieving different study objectives at different stages.

10.4 Analysis for Seamless Design with Same Study Objectives/Endpoints

In practice, since a two-stage seamless design with the same study objectives and same study endpoints at different stages is similar to a typical group sequential design with one planned interim analysis, standard statistical methods for group sequential design are often employed. With various adaptations that are applied, many interesting methods have been developed in the literature. For example, the following is a list of methods that are commonly employed: (1) Fisher's criterion for combination of independent p-values from subsamples collected between two consecutive adaptations (Bauer and Kohne, 1994; Bauer and Rohmel, 1995; Posch and Bauer, 2000), (2) weighting the samples differently before and after each adaptation (Cui et al., 1999), (3) the conditional error function approach (Proschan and Hunsberger, 1995; Liu and Chi, 2001), and (4) conditional power approaches (Li et al., 2005). The method using Fisher's combination of p-values provides great flexibility in the selection of statistical methods for individual hypothesis testing based on subsamples. However, as pointed out by Muller and Schafer (2001), the method lacks flexibility in the choice of boundaries. Among other interesting studies, Proschan and Wittes (2000) constructed an unbiased estimate that uses all of the data from the trial. Adaptive designs featuring response-adaptive randomization were studied by Rosenberger and Lachin (2003). The impact of study population changes due to protocol amendments was studied by Chow et al. (2005). An adaptive design with a survival endpoint was studied by Li et al. (2005). Hommel et al. (2005) studied a two-stage adaptive design with correlated data. An adaptive approach for a bivariate-endpoint was studied by Todd (2003). Tsiatis and Mehta (2003) showed that for any adaptive design with sample size adjustment, there exists a more powerful group sequential design.

In what follows, for illustration purpose, we will introduce the method based on the sum of p-values (MSP) by Chow and Chang (2006) and Chang (2007). The MSP follows the idea of considering a linear combination of the p-values calculated using subsamples from the current and previous stages. Because of the simplicity of this method, it has been widely used in clinical trials. The theoretical framework of the MSP is described in the following section.

10.4.1 Theoretical Framework

Consider a clinical trial with K interim analyses. The final analysis is treated as the Kth interim analysis. Suppose that at each interim analysis, a hypothesis test is performed followed by some actions that are dependent on the analysis results. Such actions could result in an early stopping due to futility/efficacy or safety, sample size reestimation, modification of randomization, or other adaptations. In this setting, the objective of the trial can be formulated using a global hypothesis test, which is an intersection of the individual hypothesis tests from the interim analyses

$$H_0 : H_{0i} \cap \cdots \cap H_{0K},$$

where H_{0i}, $i = 1, \ldots, K$ is the null hypothesis to be tested at the ith interim analysis. Note that there are some restrictions on H_{0i}, that is, rejection of any H_{0i}, $i = 1, \ldots, K$ will lead to the same clinical implication (e.g., drug is efficacious); hence all H_{0i}, $i = 1, \ldots, K$ are constructed for testing the *same* endpoint within a trial. Otherwise the global hypothesis cannot be interpreted.

In practice, H_{0i} is tested based on a subsample from each stage, and without loss of generality, assume H_{0i} is a test for the efficacy of a test treatment under investigation, which can be written as

$$H_{0i} : \eta_{i1} \geq \eta_{i2} \quad \text{versus} \quad H_{ai} : \eta_{i1} < \eta_{i2},$$

where η_{i1} and η_{i2} are the responses of the two treatment groups at the ith stage. It is often the case that when $\eta_{i1} = \eta_{i2}$, the p-value p_i for the subsample at the

ith stage is uniformly distributed on [0, 1] under H_0 (Bauer and Kohne, 1994). This desirable property can be used to construct a test statistic for multiple-stage seamless adaptive designs. As an example, Bauer and Kohne (1994) used Fisher's combination of the p-values. Similarly, Chang (2007) considered a linear combination of the p-values as follows:

$$T_k = \sum_{i=1}^{K} w_{ki} p_i, \quad i = 1, \ldots, K, \tag{10.1}$$

where
$w_{ki} > 0$
K is the number of analyses planned in the trial

For simplicity, consider the case where $w_{ki} = 1$. This leads to

$$T_k = \sum_{i=1}^{K} p_i, \quad i = 1, \ldots, K. \tag{10.2}$$

The test statistic T_k can be viewed as cumulative evidence against H_0. The smaller the T_k is, the stronger the evidence is. Equivalently, we can define the test statistic as $T_k = \sum_{i=1}^{K} p_i / K$, which can be viewed as an average of the evidence against H_0. The stopping rules are given by

$$\begin{cases} \text{Stop for efficacy} & \text{if } T_k \leq \alpha_k, \\ \text{Stop for futility} & \text{if } T_k \geq \beta_k, \\ \text{Continue} & \text{otherwise,} \end{cases} \tag{10.3}$$

where T_k, α_k, and β_k are monotonic increasing functions of k, $\alpha_k < \beta_k$, $k = 1, \ldots, K - 1$, and $\alpha_K = \beta_K$. Note that α_k and β_k are referred to as the efficacy and futility boundaries, respectively. To reach the kth stage, a trial has to pass 1 to $(k-1)$th stages. Therefore, a so-called proceeding probability can be defined as the following unconditional probability:

$$\psi_k(t) = P\left(T_k < t, \alpha_1 < T_1 < \beta_1, \ldots, \alpha_{k-1} < T_{k-1} < \beta_{k-1}\right)$$

$$= \int_{\alpha_1}^{\beta_1} \cdots \int_{\alpha_{k-1}}^{\beta_{k-1}} \int_{-\infty}^{t} f_{T_1,\ldots,T_k}(t_1, \ldots, t_k) dt_k dt_{k-1}, \ldots, dt_1, \tag{10.4}$$

where
$t \geq 0$, t_i, $i = 1, \ldots, k$ is the test statistic at the ith stage
$f_{T_1,\ldots,T_k}$ is the joint probability density function

The error rate at the kth stage is given by

$$\pi_k = \psi_k(\alpha_k). \tag{10.5}$$

When efficacy is claimed at a certain stage, the trial is stopped. Therefore, the type I error rates at different stages are mutually exclusive. Hence, the experiment-wise type I error rate can be written as follows:

$$\alpha = \sum_{k=1}^{K} \pi_k. \tag{10.6}$$

Note that (10.4) through (10.6) are the keys to determine the stopping boundaries, which will be illustrated in the next subsection with two-stage seamless adaptive designs. The adjusted p-value calculation is the same as the one in a classic group sequential design (see, e.g., Jennison and Turnbull, 2000). The key idea is that when the test statistic at the kth stage $T_k = t = \alpha_k$ (i.e., just on the efficacy stopping boundary), the p-value is equal to α spent $\sum_{i=1}^{k} \pi_i$. This is true regardless of which error spending function is used and consistent with the p-value definition of the classic design. The adjusted p-value corresponding to an observed test statistic $T_k = t$ at the kth stage can be defined as

$$p(t;k) = \sum_{i=1}^{k-1} \pi_i + \psi_k(t), \quad k = 1, \dots, K. \tag{10.7}$$

This adjusted p-value indicates weak evidence against H_0, if the H_0 is rejected at a late stage because one has spent some α at previous stages. On the other hand, if the H_0 was rejected at an early stage, it indicates strong evidence against H_0 because there is a large portion of overall alpha that has not been spent yet. Note that p_i in (10.1) is the stage-wise naive (unadjusted) p-value from a subsample at the ith stage, while $p(t;k)$ are adjusted p-values calculated from the test statistic, which are based on the cumulative sample up to the kth stage where the trial stops; Equations 10.6 and 10.7 are valid regardless of how p_i is calculated.

10.4.2 Two-Stage Adaptive Design

In this subsection, we will apply the general framework to the two-stage designs. Chang (2007) derived the stopping boundaries and p-value formula for three different types of adaptive designs that allow (1) early efficacy stopping, (2) early stopping for both efficacy and futility, and (3) early futility stopping. The formulation can be applied to both superiority and non-inferiority trials with or without sample size adjustment.

10.4.2.1 Early Efficacy Stopping

For a two-stage design ($K = 2$) allowing for early efficacy stopping ($\beta_1 = 1$), the type I error rates to spend at Stage 1 and Stage 2 are

$$\pi_1 = \psi_1(\alpha_1) = \int_0^{\alpha_1} dt_1 = \alpha_1, \tag{10.8}$$

and

$$\pi_2 = \psi_2(\alpha_2) = \int_{\alpha_1}^{\alpha_2} \int_t^{\alpha_1} dt_2 dt_1 = \frac{1}{2}(\alpha_2 - \alpha_1)^2, \tag{10.9}$$

respectively. Using (10.8) and (10.9), (10.6) becomes

$$\alpha = \alpha_1 + \frac{1}{2}(\alpha_2 - \alpha_1)^2. \tag{10.10}$$

Solving for α_2, we obtain

$$\alpha_2 = \sqrt{2(\alpha - \alpha_1)} + \alpha_1. \tag{10.11}$$

Note that when the test statistic $t_1 = p_1 > \alpha_2$, it is certain that $t_2 = p_1 + p_2 > \alpha_2$. Therefore, the trial should stop when $p_1 > \alpha_2$ for futility. The clarity of the method in this respect is unique, and the futility stopping boundary is often hidden in other methods. Furthermore, α_1 is the stopping probability (error spent) at the first stage under the null hypothesis condition and $\alpha - \alpha_1$ is the error spent at the second stage. Table 10.2 provides some examples of the stopping boundaries from (10.11).

TABLE 10.2

Stopping Boundaries for Two-Stage Efficacy Designs

One-sided α		0.005	0.010	0.015	0.020	0.025	0.030
0.025	α_1						
0.025	α_2	0.2050	0.1832	0.1564	0.1200	0.0250	—
0.05	α_2	0.3050	0.2928	0.2796	0.2649	0.2486	0.2300

Source: Chang, M., *Stat. Med.*, 26, 2772, 2007. With permission.

The adjusted p-value is given by

$$p(t; k) = \begin{cases} t & \text{if } k = 1, \\ \alpha_1 + \dfrac{1}{2}(t - \alpha_1)^2 & \text{if } k = 2, \end{cases} \tag{10.12}$$

where
$t = p_1$ if the trial stops at Stage 1
$t = p_1 + p_2$ if the trial stops at Stage 2

10.4.2.2 Early Efficacy or Futility Stopping

It is obvious that if $\beta_1 \geq \alpha_2$, the stopping boundary is the same as it is for the design with early efficacy stopping. However, futility boundary β_1 when $\beta_1 \geq \alpha_2$ is expected to affect the power of the hypothesis testing. Therefore,

$$\pi_1 = \int_0^{\alpha_1} dt_1 = \alpha_1, \tag{10.13}$$

and

$$\pi_2 = \begin{cases} \displaystyle\int_{\alpha_1}^{\beta_1} \int_{t_1}^{\alpha_2} dt_2 dt_1 & \text{for } \beta_1 \leq \alpha_2, \\ \displaystyle\int_{\alpha_1}^{\alpha_2} \int_{t_1}^{\alpha_2} dt_2 dt_1 & \text{for } \beta_1 > \alpha_2. \end{cases} \tag{10.14}$$

Carrying out the integrations in (10.13) and substituting the results into (10.6), we have

$$\alpha = \begin{cases} \alpha_1 + \alpha_2(\beta_1 - \alpha_1) - \dfrac{1}{2}(\beta_1^2 - \alpha_1^2) & \text{for } \beta_1 < \alpha_2, \\ \alpha_1 + \dfrac{1}{2}(\alpha_2 - \alpha_1)^2 & \text{for } \beta_1 \geq \alpha_2. \end{cases} \tag{10.15}$$

Various stopping boundaries can be chosen from (10.15). See Table 10.3 for examples of the stopping boundaries. The adjusted p-value is given by

TABLE 10.3

Stopping Boundaries for Two-Stage Efficacy and Futility Designs

One-Sided α		$\beta_1 = 0.15$				
0.025	α_1	0.005	0.010	0.015	0.020	0.025
	α_2	0.2154	0.1871	0.1566	0.1200	0.0250
		$\beta_1 = 0.2$				
0.05	α_1	0.005	0.010	0.015	0.020	0.025
	α_2	0.3333	0.3155	0.2967	0.2767	0.2554

Source: Chang, M., *Stat. Med.*, 26, 2772, 2007. With permission.

$$
p(t;\,k) = \begin{cases}
t & \text{if } k = 1, \\
\alpha_1 + t(\beta_1 - \alpha_1) - \dfrac{1}{2}(\beta_1^2 - \alpha_1^2) & \text{if } k = 2 \text{ and } \beta_1 < \alpha_2, \\
\alpha_1 + \dfrac{1}{2}(t - \alpha_1)^2 & \text{if } k = 2 \ \beta_1 \geq \alpha_2.
\end{cases} \tag{10.16}
$$

where

$t = p_1$ if the trial stops at Stage 1

$t = p_1 + p_2$ if the trial stops at Stage 2

10.4.2.3 Early Futility Stopping

A trial featuring early futility stopping is a special case of the previous design, where $\alpha_1 = 0$ in (10.15). Hence, we have

$$
\alpha = \begin{cases}
\alpha_2 \beta_1 - \dfrac{1}{2}\beta_1^2 & \text{for } \beta_1 < \alpha_2, \\
\dfrac{1}{2}\alpha_2^2 & \text{for } \beta_1 \geq \alpha_2.
\end{cases} \tag{10.17}
$$

Solving for α_2, it can be obtained that

$$
\alpha_2 = \begin{cases}
\dfrac{\alpha}{\beta_1} + \dfrac{1}{2}\beta_1 & \text{for } \beta_1 < \sqrt{2\alpha}, \\
\sqrt{2\alpha} & \text{for } \beta_1 \geq \alpha_2.
\end{cases} \tag{10.18}
$$

TABLE 10.4

Stopping Boundaries for Two-Stage Futility Design

One-Sided α	β_1	0.1	0.2	0.3	≥ 0.4
0.025	α_2	0.3000	0.2250	0.2236	0.2236
0.05	α_2	0.5500	0.3500	0.3167	0.3162

Source: Chang, M., *Stat. Med.*, 26, 2772, 2007. With permission.

Examples of the stopping boundaries generated using (10.18) are presented in Table 10.4. The adjusted p-value can be obtained from (10.16), where $\alpha_1 = 0$, that is,

$$p(t; k) = \begin{cases} t & \text{if } k = 1, \\ \alpha_1 + t\beta_1 - \dfrac{1}{2}\beta_1^2 & \text{if } k = 2 \text{ and } \beta_1 < \alpha_2, \\ \alpha_1 + \dfrac{1}{2}t^2 & \text{if } k = 2 \ \beta_1 \geq \alpha_2. \end{cases} \qquad (10.19)$$

10.4.3 Conditional Power

Conditional power is a very useful operating characteristic of adaptive designs. It can be used for interim decision-making and drawing comparisons among different designs and different statistical methods for adaptive designs. Because the stopping boundaries for the most existing methods are either based on z-scale or p-scale, for the purpose of comparison, we will use the transformation $p_k = 1 - \Phi(z_k)$ and, inversely, $z_k = \Phi^{-1}(1 - p_k)$, where z_k and p_k are the normal z-score and the naive p-value from the subsample at the kth stage, respectively. Note that z_2 has asymptotically normal distribution with $N(\delta/se(\hat{\delta}_2), 1)$ under the alternative hypothesis, where $\hat{\delta}_2$ is the estimation of treatment difference in the second stage and

$$se(\hat{\delta}_2) = \sqrt{\frac{2\hat{\sigma}^2}{n_2}} \approx \sqrt{\frac{2\sigma^2}{n_2}}.$$

To derive the conditional power, we express the criterion for rejecting H_0 as

$$z_2 \geq B(\alpha_2, p_1). \qquad (10.20)$$

From (10.20), we can immediately obtain the conditional probability given the first stage naive p-value, p_1, in the second stage as

$$P_C(p_1, \delta) = 1 - \Phi\left(B(\alpha_2, p_1) - \frac{\delta}{\sigma}\sqrt{\frac{n_2}{2}}\right), \quad \alpha_1 < p_1 \leq \beta_1. \qquad (10.21)$$

For the method based on the product of stage-wise p-values (MPP), the rejection criterion for the second stage is $p_1 p_2 \leq \alpha_2$, i.e., $z_2 \geq \Phi^{-1}(1 - \alpha_2/p_1)$. Therefore, $B(\alpha_2, p_1) = \Phi^{-1}(1 - \alpha_2/p_1)$. Similarly, for the MSP, the rejection criterion for the second stage is $p_1 + p_2 \leq \alpha_2$, i.e., $z_2 = B(\alpha_2, p_1) = \Phi^{-1}(1 - \max(0, \alpha_2 - p_1))$. For the inverse-normal method (Lehmacher and Wassmer, 1999), the rejection criterion for the second stage is $w_1 z_1 + w_2 z_2 \geq \Phi^{-1}(1 - \alpha_2)$, i.e., $z_2 \geq (\Phi^{-1}(1 - \alpha_2) - w_1 \Phi^{-1}(1 - p_1))/w_2$, where w_1 and w_2 are prefixed weights satisfying the condition of $w_1^2 + w_2^2 = 1$. Note that the group sequential design and the Cui–Hung–Wang (CHW) method (Cui et al., 1999) are special cases of the inverse-normal method. For simplicity, we will compare only MPP and MSP analytically because the third method also depends on two additional parameters, w_1 and w_2. To compare the conditional power, the same α_1 should be used for both methods; otherwise the comparison will be much less informative. From (10.21), we can see that the comparison of the conditional power is equivalent to the comparison of function $B(\alpha_2, p_1)$. Equating the two $B(\alpha_2, p_1)$, we have

$$\frac{\hat{\alpha}_2}{p_1} = \tilde{\alpha}_2 - p_1, \tag{10.22}$$

where $\hat{\alpha}_2$ and $\tilde{\alpha}_2$ are the final rejection boundaries for MPP and MSP, respectively. Solving (10.22) for p_1, we obtain the critical point for p_1

$$\eta = \frac{\tilde{\alpha}_2 \mp \sqrt{\tilde{\alpha}_2^2 - 4\tilde{\alpha}_2}}{2} \tag{10.23}$$

such that when $p_1 < \eta_1$ or $p_2 > \eta_2$ MPP has a higher conditional power than MSP. When $\eta_1 < p_1 < \eta_2$, MSP has a higher conditional power than MPP. For example, for overall one-sided $\alpha = 0.025$, if we choose $\alpha_1 = 0.01$ and $\beta_1 = 0.3$, then $\hat{\alpha}_2 = 0.0044$ and $\tilde{\alpha}_2 = 0.2236$, and finally $\eta_1 = 0.0218$ and $\eta_2 = 0.2018$ from (10.23). The unconditional power P_w is the expectation of conditional power, i.e.,

$$P_w = E_\delta[P_C(p_1, \delta)]. \tag{10.24}$$

Therefore, the difference in unconditional power between MSP and MPP is dependent on the distribution of p_1 and, consequently, dependent on the true difference δ and the stopping boundaries at the first stage (α_1, β_1).

Note that in Bauer and Kohne's (1994) method using Fisher's combination, which leads to the equation $\alpha_1 + \ln(\beta_1/\alpha_1)e^{-(1/2)\chi_{4,1-\alpha}^2} = \alpha$, it is obvious that the determination of β_1 leads to a unique α_1 and, consequently, α_2. This is a non-flexible approach. However, it can be verified that the method can be generalized to $\alpha_1 + \alpha_2 \ln \beta_1/\alpha_1 = \alpha$, where α_2 does not have to be $e^{-(1/2)\chi_{4,1-\alpha}^2}$.

Note that Tsiatis and Mehta (2003) indicated that there is an optimal (uniformly more powerful) design for any class of sequential design with a specified error spending function. In other words, for any adaptive design, one can always construct a classic group sequential test statistic that, for any parameter value in the space of alternatives, will reject the null hypothesis earlier with equal or higher probability, and, for any parameter value not in the space of alternatives, will accept the null hypothesis earlier with equal or higher probability. However, the efficacy gain by the classic group sequential design comes with a cost—for example, an increased number of interim analyses increases (e.g., from 3 to 10), which definitely has an associated cost practically. Also, the optimal design is under the condition of a prespecified error-spending function, but adaptive designs do not require in general a fixed error-spending function.

10.5 Analysis for Seamless Design with Different Endpoints

For illustration purpose, consider a two-stage phase II/III seamless adaptive trial design with different (continuous) study endpoints. Let x_i be the observation of one study endpoint (e.g., a biomarker) from the ith subject in phase II, $i = 1, \ldots, n$ and y_j be the observation of another study endpoint (the primary clinical endpoint) from the jth subject in phase III, $j = 1, \ldots, m$. Assume that x_i's are independently and identically distributed with $E(x_i) = \nu$ and $\mathrm{Var}(x_i) = \tau^2$, and y_j's are independently and identically distributed with $E(y_j) = \mu$ and $\mathrm{Var}(y_j) = \sigma^2$. Chow et al. (2007) proposed using the established functional relationship to obtain predicted values of the clinical endpoint based on data collected from the biomarker (or surrogate endpoint). Thus, these predicted values can be combined with the data collected at the confirmatory phase to develop a valid statistical inference for the treatment effect under study. Suppose that x and y can be related in a straight-line relationship

$$y = \beta_0 + \beta_1 x + \varepsilon, \tag{10.25}$$

where ε is an error term with zero mean and variance ς^2. Furthermore, ε is independent of x. In practice, we assume that this relationship is well-explored and the parameters β_0 and β_1 are known. Based on (10.25), the observations x_i observed in the learning phase would be translated to $\beta_0 + \beta_1 x_i$ (denoted by $\hat{y}_i$) and are combined with those observations y_i collected in the confirmatory phase. Therefore, $\hat{y}_i$'s and y_i's are combined for the estimation of the treatment mean μ. Consider the following weighted-mean estimator:

$$\hat{\mu} = \omega \bar{\hat{y}} + (1 - \omega)\bar{y}, \tag{10.26}$$

where

$$\bar{\hat{y}} = (1/n) \sum_{i=1}^{n} \hat{y}$$

$$\bar{y} = (1/m) \sum_{j=1}^{m} y_j$$

$$0 \le \omega \le 1$$

It should be noted that $\hat{\mu}$ is the minimum variance unbiased estimator among all weighted-mean estimators when the weight is given by

$$\omega = \frac{n/(\beta_1^2 \tau^2)}{n/(\beta_1^2 \tau^2) + m/\sigma^2} \tag{10.27}$$

if β_1, τ^2, and σ^2 are known. In practice, τ^2 and σ^2 are usually unknown and ω is commonly estimated by

$$\hat{\omega} = \frac{n/s_1^2}{n/s_1^2 + m/s_2^2}, \tag{10.28}$$

where s_1^2 and s_2^2 are the sample variances of $\hat{y}_i$'s and y_j's, respectively. The corresponding estimator of μ, which is denoted by

$$\hat{\mu}_{GD} = \hat{\omega}\bar{\hat{y}} + (1-\hat{\omega})\bar{y}, \tag{10.29}$$

is called the Graybill–Deal (GD) estimator of μ. The GD estimator is often called the weighted mean in metrology. Khatri and Shah (1974) gave an exact expression of the variance of this estimator in the form of an infinite series. An approximate unbiased estimator of the variance of the GD estimator, which has bias of order $O(n^{-2} + m^{-2})$, was proposed by Meier (1953). In particular, it is given as

$$\widehat{\mathrm{Var}}(\hat{\mu}_{GD}) = \frac{1}{n/S_1^2 + m/S_2^2} \left[1 + 4\hat{\omega}(1-\hat{\omega}) \left(\frac{1}{n-1} + \frac{1}{m-1} \right) \right].$$

For the comparison of the two treatments, the following hypotheses are considered:

$$H_0: \mu_1 = \mu_2 \quad \text{versus} \quad H_1: \mu_1 \ne \mu_2. \tag{10.30}$$

Let $\hat{y}_{ij}$ be the predicted value $\beta_0 + \beta_1 x_{ij}$, which is used as the prediction of y for the jth subject under the ith treatment in phase II. From (10.29), the GD estimator of μ_i is given as

$$\hat{\mu}_{GDi} = \hat{\omega}_i \bar{\hat{y}}_i + (1 - \hat{\omega}_i)\bar{y}_i, \tag{10.31}$$

where

$$\bar{\hat{y}}_i = (1/n_i)\sum_{j=1}^{n_i} \hat{y}_{ij}$$

$$\bar{y}_i = (1/m_i)\sum_{j=1}^{m_i} y_{ij}$$

$\hat{\omega}_i = n_i/S_{1i}^2/(n_i/S_{1i}^2 + m_i/S_{2i}^2)$ with S_{1i}^2 and S_{2i}^2 being the sample variances of $(\hat{y}_{i1}, \ldots, \hat{y}_{in_i})$ and $(y_{i1}, \ldots, y_{im_i})$, respectively

For hypotheses (10.30), consider the following test statistic:

$$\tilde{T}_1 = \frac{\hat{\mu}_{GD1} - \hat{\mu}_{GD2}}{\sqrt{\widehat{Var}(\hat{\mu}_{GD1}) + \widehat{Var}(\hat{\mu}_{GD2})}}, \tag{10.32}$$

where

$$\widehat{Var}(\hat{\mu}_{GDi}) = \frac{1}{n_i/S_{1i}^2 + m_i/S_{2i}^2}\left[1 + 4\hat{\omega}_i(1 - \hat{\omega}_i)\left(\frac{1}{n_i - 1} + \frac{1}{m_i - 1}\right)\right]$$

is an estimator of $Var(\hat{\mu}_{GDi})$, $i = 1, 2$. Using arguments similar to those in Section 2.1, it can be verified that $\tilde{T}_1$ has a limiting standard normal distribution under the null hypothesis H_0 if $Var(S_{1i}^2)$ and $Var(S_{2i}^2) \to 0$ as n_i and $m_i \to \infty$. Consequently, an approximate $100(1 - \alpha)\%$ confidence interval of $\mu_1 - \mu_2$ is given as

$$\left(\hat{\mu}_{GD1} - \hat{\mu}_{GD2} - z_{\alpha/2}\sqrt{V_T}, \quad \hat{\mu}_{GD1} - \hat{\mu}_{GD2} + z_{\alpha/2}\sqrt{V_T}\right), \tag{10.33}$$

where $V_T = \widehat{Var}(\hat{\mu}_{GD1}) + \widehat{Var}(\hat{\mu}_{GD2})$. Therefore, hypothesis H_0 is rejected if the confidence interval (9) does not contain 0. Thus, under the local alternative hypothesis that $H_1: \mu_1 - \mu_2 = \delta \neq 0$, the required sample size to achieve a $1 - \beta$ power satisfies

$$\frac{-z_{\alpha/2} + |\delta|}{\sqrt{Var(\hat{\mu}_{GD1}) + Var(\hat{\mu}_{GD2})}} = z_\beta.$$

Let $m_i = \rho n_i$ and $n_2 = \gamma n_1$. Then, denoted by N_T the total sample size for two treatment groups is $(1 + \rho)(1 + \gamma)n_1$ with n_1 given as

$$n_1 = \frac{1}{2} AB\left(1 + \sqrt{1 + 8(1+\rho)A^{-1}C}\right), \tag{10.34}$$

where
$A = (z_{\alpha/2} + z_\beta)^2/\delta^2$
$B = \sigma_1^2/(\rho + r_1^{-1}) + \sigma_2^2/\gamma(\rho + r_2^{-1})$
$C = B^{-2}[\sigma_1^2/r_1(\rho + r_1^{-1})^3 + \sigma_2^2/\gamma^2 r_2(\rho + r_2^{-1})^3]$ with $r_i = \beta_1^2 \tau_i^2/\sigma_i^2$, $\quad i = 1, 2$

For the case of testing for superiority, consider the following local alternative hypothesis:

$$H_1 : \mu_1 - \mu_2 = \delta_1 > \delta.$$

The required sample size to achieve $1 - \beta$ power satisfies

$$-z_\alpha + (\delta_1 - \delta)\Big/\sqrt{\mathrm{Var}(\hat{\mu}_{\mathrm{GD1}}) + \mathrm{Var}(\hat{\mu}_{\mathrm{GD2}})} = z_\beta.$$

Using the notations in the above paragraph, the total sample size for two treatment groups is $(1 + \rho)(1 + \gamma)n_1$ with n_1 given as

$$n_1 = \frac{1}{2} DB\left(1 + \sqrt{1 + 8(1+\rho)D^{-1}C}\right), \tag{10.35}$$

where $D = (z_\alpha + z_\beta)^2/(\delta_1 - \delta)^2$. For the case of testing for equivalence with a significance level α, consider the local alternative hypothesis $H_1 : \mu_1 - \mu_2 = \delta_1$ with $|\delta_1| < \delta$. The required sample size to achieve $1 - \beta$ power satisfies

$$-z_\alpha + (\delta - \delta_1)\Big/\sqrt{\mathrm{Var}(\hat{\mu}_{\mathrm{GD1}}) + \mathrm{Var}(\hat{\mu}_{\mathrm{GD2}})} = z_\beta.$$

Thus, the total sample size for two treatment groups is $(1 + \rho)(1 + \gamma)n_1$ with n_1 given

$$n_1 = \frac{1}{2} EB\left(1 + \sqrt{1 + 8(1+\rho)E^{-1}C}\right), \tag{10.36}$$

where $E = (z_\alpha + z_{\beta/2})^2/(\delta - |\delta_1|)^2$.

Note that following a similar idea as described above, statistical tests and formulas for sample size calculation for testing hypotheses of equality, non-inferiority, superiority, and equivalence for binary response and time-to-event endpoints can be obtained.

10.6 Analysis for Seamless Design with Different Objectives/Endpoints

In this section, we will focus on statistical inference for the scenario where the study objectives at different stages are different (e.g., dose selection versus efficacy confirmation) and study endpoints at different stages are different (e.g., biomarker or surrogate endpoint versus regular clinical study endpoint).

As indicated earlier, one of the major concerns when applying adaptive design methods in clinical trials is probably how to control the overall type I error rate at a prespecified level of significance. It is also a concern as to how the data collected from both stages should be combined for the final analysis. Besides, it is of interest to know how the sample size calculation/allocation should be done for achieving individual study objectives originally set for the two stages (separate studies). In this chapter, a multiple-stage transitional seamless trial design with different study objectives and different study endpoints and with and without adaptations is proposed. The impact of the adaptive design methods on the control of the overall type I error rate under the proposed trial design is examined. Valid statistical tests and the corresponding formulas for sample size calculation/allocation are derived under the proposed trial design.

As indicated earlier, a two-stage seamless trial design that combines two independent studies (e.g., a phase II study and a phase III study) is often considered in clinical research and development. Under such a trial design, the investigator may be interested in having one planned interim analysis at each stage. In this case, the two-stage seamless trial design becomes a four-stage trial design if we consider the time point at which the planned interim analysis will be conducted as the end of the specific stage. In this chapter, we will refer to such a trial design as a multiple-stage transitional seamless design to emphasize the importance of smooth transition from stage to stage. In what follows, we will focus on the proposed multiple-stage transitional seamless design with (adaptive version) and without (nonadaptive version) adaptations.

10.6.1 Nonadaptive Version

Consider a clinical trial comparing k treatments groups, $E_1, \ldots, E_k$, with a control group C. One early surrogate endpoint and one subsequent primary endpoint are potentially available for assessing the treatment effect. Let θ_i and ψ_i, $i = 1, \ldots, k$ be the treatment effect comparing E_i with C measured by the surrogate endpoint and the primary endpoint, respectively. The ultimate hypothesis of interest is

$$H_{0,2}: \psi_1 = \cdots = \psi_k, \tag{10.37}$$

which is formulated in terms of the primary endpoint. However, along the way, the hypothesis

$$H_{0,1} : \theta_1 = \cdots = \theta_k, \tag{10.38}$$

in terms of the short-term surrogate endpoint will also be assessed. Cheng and Chow (2010) assumed that ψ_i is a monotone increasing function of the corresponding θ_i. The trial is conducted as a group sequential trial with the accrued data analyzed at three stages (i.e., Stage 1, Stage 2a, Stage 2b, and Stage 3) with four interim analyses, which are briefly described in the following. For simplicity, consider the case where the variances of the surrogate endpoint and the primary outcomes, denoted as σ^2 and τ^2, are known.

At *Stage 1* of the study, $(k + 1)n_1$ subjects will be randomized equally to receive either one of the k treatments or the control. As a result, there are n_1 subjects in each group. At the first interim analysis, the most promising treatment will be selected and used in the subsequent stages based on the surrogate endpoint. Let $\hat{\theta}_{i,1}, i = 1, ..., k$ be the pair-wise test statistics, and $S = \arg\max_{1 \le i \le k} \hat{\theta}_{i,1}$ then if $\hat{\theta}_{S,1} \le c_1$ for some c_1, the trial is stopped and $H_{0,1}$ is accepted. Otherwise, if $\hat{\theta}_{S,1} > c_{1,1}$, then the treatment E_S is recommended as the most promising treatment and will be used in all the subsequent stages. Note that only the subjects receiving either the promising treatment or the control will be followed formally for the primary endpoint. The treatment assessment on all other subjects will be terminated and the subjects will receive standard care and undergo necessary safety monitoring.

At *Stage 2a*, $2n_2$ additional subjects will be equally randomized to receive either the treatment E_S or the control C. The second interim analysis is scheduled when the short-term surrogate measures from these $2n_2$ Stage 2 subjects and the primary endpoint measures from those $2n_1$ Stage 1 subjects who receive either the treatment E_S or the control C become available. Let $T_{1,1} = \hat{\theta}_{S,1}$ and $T_{1,2} = \hat{\psi}_{S,1}$ be the pair-wise test statistics from Stage 1 based on the surrogate endpoint and the primary endpoint, respectively, and $\hat{\theta}_{S,2}$ be the statistic from Stage 2 based on the surrogate. If

$$T_{2,1} = \sqrt{\frac{n_1}{n_1 + n_2}} \hat{\theta}_{S,1} + \sqrt{\frac{n_2}{n_1 + n_2}} \hat{\theta}_{S,2} \le c_{2,1},$$

stop the trial and accept $H_{0,1}$. If $T_{2,1} > c_{2,1}$ and $T_{1,2} > c_{1,2}$, stop the trial and reject both $H_{0,1}$ and $H_{0,2}$. Otherwise, if $T_{2,1} > c_{2,1}$ but $T_{1,2} \le c_{1,2}$, we will move on to Stage 2b.

At *Stage 2b*, no additional subjects will be recruited. The third interim analysis will be performed when the subjects in Stage 2a complete their primary endpoints. Let

$$T_{2,2} = \sqrt{\frac{n_1}{n_1 + n_2}}\, \hat{\psi}_{S,1} + \sqrt{\frac{n_2}{n_1 + n_2}}\, \hat{\psi}_{S,2},$$

where $\hat{\psi}_{S,2}$ is the pair-wise test statistic from Stage 2b. If $T_{2,2} > c_{2,2}$, stop the trial and reject $H_{0,2}$. Otherwise, move on to Stage 3.

At *Stage 3*, the final stage, $2n_3$ additional subjects will be recruited and followed till their primary endpoints. For the fourth interim analysis, define

$$T_3 = \sqrt{\frac{n_1}{n_1 + n_2 + n_3}}\, \hat{\psi}_{S,1} + \sqrt{\frac{n_2}{n_1 + n_2 + n_3}}\, \hat{\psi}_{S,2} + \sqrt{\frac{n_1}{n_1 + n_2 + n_3}}\, \hat{\psi}_{S,3},$$

where $\hat{\psi}_{S,3}$ is the pair-wise test statistic from Stage 3. If $T_3 > c_3$, stop the trial and reject $H_{0,2}$; otherwise, accept $H_{0,2}$. The parameters in the above designs, n_1, n_2, n_3, $c_{1,1}$, $c_{1,2}$, $c_{2,1}$, $c_{2,2}$, and c_3, are determined such that the procedure will have a controlled type I error rate of α and a target power of $1 - \beta$. The determination of these parameters will be given in the next section.

In the above design, the surrogate data in the first stage are used to estimate the most promising treatment rather than assessing $H_{0,1}$. This means that upon completion of Stage 1, a dose does not need to be significant in order to be recommended for the subsequent stages. This feature is important since it does not suffer from any lack of power due to limited sample sizes.

There are two sets of hypotheses to be tested, namely $H_{0,1}$ and $H_{0,2}$. To claim efficacy, $H_{0,2}$ has to be rejected, and hence is the hypothesis of primary interest. However, to ensure appropriate control of the type I error rate associated with the sequential design with change of endpoints, $H_{0,1}$ has to be assessed along the way according to the closed testing principle. The proposed two-stage seamless design is attractive due to its efficiency (e.g., reduces the lead time between a phase II trial and a phase III study) and flexibility (e.g., allows to make decision early and take appropriate actions such as stopping the trial early or deleting/adding dose groups). At the first stage, with a limited number of subjects, the goal is to detect any signals for safety and/or evidence for early efficacy. With a limited number of subjects, there will not be any power for detecting a small clinically meaningful difference. This justifies the use of precision analysis for achieving statistical significance as a criterion for dose selection.

10.6.2 Adaptive Version

The proposed design approach in the previous section is a group sequential procedure with treatment selection. There is no adaptation involved in the above procedure. Tsiatis and Mehta (2003) and Jennison and Turnbull (2006) argue that adaptive designs typically suffer from loss of efficiency and hence are typically not recommended in regular practice. However, as pointed out by Proschan et al. (2006), in some scenarios, particularly when there is no enough primary outcome information available, it is appealing to use an adaptive procedure as long as it is statistically justified. For the trials we are considering, since the primary outcome takes much longer time to observe compared to its surrogate, we feel that an adaptive procedure is useful in our setting. And the transitional feature of our proposed design make it possible to modify the design adaptively upon completion of the second interim analysis (i.e., Stage 2a). One possible adaptation is the correlation between the surrogate endpoint and the primary outcome. As a nuisance parameter, it plays an important role in the power calculation of the procedure. This nuisance parameter can be estimated using the first stage patients who are followed for their primary outcomes.

Another possible modification is to recalibrate the treatment effect of the primary out come by exploring the relationship between the surrogate endpoint and the primary outcome. Specifically, assuming there is a local linear relationship between ψ and θ, a reasonable assumption when focusing only on their values at a neighborhood of the most promising treatment E_S, then at the end of Stage 2a, the treatment effect in term of the primary endpoint can be reestimated as

$$\hat{\delta}_S = \frac{\hat{\psi}_{S,1}}{\hat{\theta}_{S,1}} T_{2,1}.$$

Then we could reestimate the Stage 3 sample size based on a modified treatment effect of the primary outcome $\delta = \max\{\delta_S, \delta_0\}$, where δ_0 is a minimally clinically relevant treatment effect agreed upon prior to the trial. The reason we choose the modified treatment this way is to ensure the clinical relevance of the test procedure. Let m be the reestimated Stage 3 sample size based on δ. If $m \leq n_3$, then there is no modification for the procedure. If $m > n_3$, then m (instead of the originally planned n_3) patients per arm will be recruited at Stage 3. The justification of the above adaptation can be found in Cheng and Chow (2010).

10.6.3 An Example

A pharmaceutical company is interested in conducting a clinical trial utilizing a two-stage seamless adaptive design for evaluation of safety (tolerability) and efficacy of a test treatment for patients with hepatitis C infection. The trial will combine two independent studies (one for dose selection and the other one for efficacy confirmation) into a single study. The study will consist of two stages at which the first stage is for dose selection and the second stage is for establishment of non-inferiority of the selected dose from the first stage as compared to the standard of care therapy (control). The primary objectives of the study then contain study objectives at both stages. For the first stage, the primary objective is to select the optimal dose as compared to the standard of care therapy, while the primary objective of the second stage is to establish non-inferiority of the selected dose as compared to the standard of care therapy. The treatment duration is 48 weeks of treatment followed by a 24 weeks follow-up. The primary study endpoint is the sustained virologic response (SVR) at Week 72, which is defined as an undetectable HCV RNA level (<10 IU/mL) at Week 72. The proposed two-stage seamless adaptive design is briefly outlined, as follows: Stage 1—this stage is a five-arm randomized evaluation of four active dose levels of the test treatment. Qualified subjects will be randomly assigned to one of the five treatment groups at a 1:1:1:1:1 ratio. After all Stage 1 subjects have completed Week 12 of the study, an interim analysis was performed. Based upon the safety results of this analysis as well as virologic response at Weeks 12 and 24, Stage 1 subjects who have not yet completed the study protocol will continue with their assigned therapies for the remainder of the planned 48 weeks, with final follow-up at Week 72. An optimal dose will be selected based on the interim analysis results of the 12 week early virologic response (EVR), which is defined as 2-log10 reduction in HCV RNA level at Week 12, assuming that the 12 week EVR is predictive of 72 week SVR. The 12 week EVR is considered as a surrogate endpoint for the primary endpoint of 72 week SVR. Under this assumption, an optimal dose will be selected using precision analysis under some pre specified selection criteria. In other words, the dose group with highest confidence level for achieving statistical significance (i.e., the observed difference is not by chance alone) will be selected. The selected dose will then proceed to testing for non-inferiority compared to standard of care in Stage 2. Stage 2—this stage will be a non-inferiority comparison of the selected dose from Stage 1. A separate cohort of subjects will be randomized to receive either the selected dose from Stage 1 or the standard of care treatment as given in Stage 1 in a 1:1 ratio. A second interim analysis will be performed when all Stage 2 subjects have completed Week 12 and 50% of the subjects (Stage 1 and Stage 2 combined) have completed 48 weeks treatment and follow-up of 24 weeks. Depending on the results of this analysis, including the virologic response at Weeks 12 and 24, sample size reestimation will be performed to whether additional subjects are needed in order for achieving the desired power for establishment of non-inferiority for the selected dose.

In both stages, subjects who do not meet the study criteria for virologic response at Weeks 12 and 24, and those who do meet these criteria but then relapse at any later time through study Week 72, will discontinue study treatment and will be offered treatment, off protocol, with standard of care. For the two planned interim analyses, the incidence of EVR as well as safety data, will be reviewed by an independent data safety monitoring board (DSMB). The commonly used O'Brien–Fleming boundaries will be applied for controlling the overall type I error rate at 5% (O'Brien and Fleming, 1979). Adaptations such as stopping the trial early, discontinuing selected treatment arms, and reestimating the sample size may be applied as recommended by the DSMB. Stopping rules for the study will be designated by the DSMB, based on their ongoing analyses of the data and as per their charter.

10.7 Concluding Remarks

As indicated earlier, in practice, statistical methods for a standard group sequential trial design with one planned interim analysis is often applied to the two-stage seamless adaptive design regardless whether the study objectives and/or the study endpoints at different stages are the same. It is then a concern whether the obtained p-value and confidence interval for assessment of the treatment effect are correct or reliable. Sample size needed for achieving a desired power that obtained under a standard group sequential design may not be sufficient for achieving the study objectives under the two-stage seamless adaptive trial design especially when the study objectives and/or study endpoints at different stages are different. More discussions regarding adaptive design methods in clinical trials can be found in Chapter 26.

In its recent draft guidance on adaptive clinical trial design, the U.S. FDA classifies adaptive designs as either *well understood designs* or *less well understood designs* depending upon the nature of adaptations either blinded or unblinded (FDA, 2010b). In practice, however, most of the adaptive designs (including seamless adaptive designs described in this chapter) are considered less well understood designs. As a result, one of the major challenges is not only the development of a set of criteria for choosing a good design among these less well understood designs, but also the development of appropriate statistical methods under the selected less well understood designs for valid statistical inference of the test treatment under investigation.

11

Multiplicity in Clinical Trials

11.1 General Concept

In clinical trials, one of the ultimate goals is to demonstrate that the observed difference of a given study endpoint (e.g., the primary efficacy endpoint) is not only of clinical importance (or a clinically meaningful difference) with statistical meaning (or of statistically significance). A study endpoint is said to have statistical meaning when the observed difference is not by chance alone and is reproducible if we are to conduct a similar study under similar experimental conditions. In practice, the observed clinically meaningful difference that has achieved statistical significance is also known as *statistical difference*. Thus, a statistical difference means that the difference is not by chance alone and it is reproducible. In drug research and evaluation, it is of interest to control the chances of false negative (or making type I error) and to minimize the chances of false positive (or making type II error) at a prespecified level of significance. As a result, based on a given study endpoint, controlling the overall type I error rate at a prespecified level of significance for achieving a designed power (i.e., the probability of correctly detecting a clinically meaningful difference if such a difference truly exists) has been a common practice for sample size determination.

In practice, the investigator may consider more than one endpoint (say two study endpoints) as the primary study endpoints. In this case, our goal is to demonstrate that the observed differences of the two study endpoints are clinically meaningful differences with statistical meaning. In other words, the observed differences are not by chance alone and they are reproducible. In this case, the level of significance is necessarily adjusted for controlling the *overall* type I error rate at a prespecified level of significance for multiple endpoints. This has raised the critical issue of multiplicity in clinical research and development. In clinical trials, *multiplicity* is usually referred to as multiple inferences that are made in simultaneous context (Westfall and Bretz, 2010). As a result, α adjustment for multiple comparisons is to make sure that the *simultaneously* observed differences are not by chance alone. In clinical trials, commonly seen multiplicity includes comparison of (1) multiple treatments (dose groups), (2) multiple endpoints, (3) multiple time points,

(4) interim analyses, (5) multiple tests of the sample hypothesis, (6) variable/ model selection, and (7) subgroup analyses.

In general, if there are k treatments, there are $k(k - 1)/2$ possible pair-wise comparisons. In practice, two types of error rates are commonly considered (Lakshminarayanan, 2010). The first type is a comparison-wise error rate (CWE), which is a type I error rate for each comparison. That is, it is the probability of erroneously rejecting the null hypothesis between treatments involved in the comparison. The other type of error rate is an experiment-wise error rate (EWE) or family-wise error rate (FWER), which is the error rate associated with one or more type I errors for all comparisons included in the experiment. Thus, for k comparisons, CWE $= \alpha$ and FWER $= 1 - (1 - \alpha)^k$. As a result, the FWER could be much larger than the significance level associated with each test if multiple statistical tests are performed using the same data set. In practice, thus, it is of interest to control the FWER. In the past several decades, several procedures for controlling FWER have been suggested in the literature. These procedures can be classified into either single-step procedures or stepwise (e.g., step-up and step-down) procedures. Note that an alternative approach to multiplicity control is to consider the false discovery rate (FDR) (see Benjamini and Hochberg, 1995).

In the next section, regulatory perspectives regarding multiplicity adjustment are discussed. Also included are some commonly seen controversial issues of multiplicity in clinical trials. Section 11.3 provides a summary of commonly considered statistical methods for multiplicity adjustment for controlling the overall type I error rate. An example concerning a dose-finding study is given in Section 11.4. A brief concluding remark is given in the last section of this chapter.

11.2 Regulatory Perspective and Controversial Issues

11.2.1 Regulatory Perspectives

Regulatory position regarding adjustment for multiplicity is not clear. In 1998, the International Conference on Harmonization (ICH) E9 published guidelines regarding Statistical Principles in Clinical Trials. These guidelines have several comments reflecting concern over the multiplicity problem. The ICH E9 guidelines recommend that the analysis of clinical trial data may necessitate an adjustment to the type I error. In addition, the ICH E9 suggests details of any adjustment procedure or an explanation of why adjustment is not thought necessary to be set out in the analysis plan. The European Agency for the Evaluation of Medicinal Products (EMEA), on the other hand, in its Committee for Proprietary Medicinal Products (CPMP) draft guidance "Points to Consider on Multiplicity Issues in Clinical Trials" indicates that multiplicity can have a substantial influence on the rate of false positive conclusions whenever there is an opportunity to choose the most favorable results from two or more analyses. The EMEA guidance also echoes the ICH recommendation for stating details of the multiple comparisons procedure in the analysis plan.

11.2.2 Controversial Issues

When conducting clinical trials involving multiple comparisons, the following questions are always raised (see also Hung and Wang, 2009):

1. Why do we need to adjust for multiplicity?
2. When do we need to adjust for multiplicity?
3. How do we adjust for multiplicity?
4. Is the FWER well controlled?

To address the first question, it is suggested that the null/alternative hypotheses be clarified since the type I error rate and the corresponding power are evaluated under the null hypothesis and the alternative hypothesis, respectively.

Regarding the second question, it should be noted that adjustment for multiplicity is to ensure that the simultaneously observed differences are not by chance alone. For example, for the evaluation of a test treatment under investigation, if regulatory approval is based on single endpoint, then no α adjustment is necessary. However, if regulatory approval is based on multiple endpoints, then α adjustment is a must in order to make sure that the simultaneously observed differences are not by chance alone and they are reproducible. Conceptually, it is not correct that α needs to be adjusted if more than one statistical test (e.g., primary hypothesis and secondary hypothesis) is to be performed. Whether α should be adjusted depends upon the null hypothesis (e.g., a single hypothesis with one primary endpoint or a composite hypothesis with multiple endpoints) to be tested. The interpretations of the test results for single null hypothesis and composite null hypothesis are different.

For questions (3) and (4), several useful methods for multiplicity adjustment are available in the literature (see Hsu, 1996; Chow and Liu, 1998b; Westfall et al., 1999). These methods are either single-step methods (e.g., Bonferroni's method), step-down methods (e.g., Holm's method), or step-up methods (e.g., Hochberg's method). In the next section, some commonly employed methods for multiplicity adjustment are briefly described.

As pointed out by Westfall and Bretz (2010), the controversial issues of multiplicity in clinical trials that are commonly encountered include (1) penalizing for doing more or good job (i.e., performing additional test), (2) adjusting α for all possible tests conducted in the trial, and (3) the family of hypotheses to be tested. Penalizing for doing good job is referred to as adjustment for multiplicity for dose-finding trials that include more dose groups. For adjusting α for all possible tests conducted in the trial, although the α is controlled at the prespecified level, it is over-killed because it is not in the investigator's best interest to show that all of the observed differences simultaneously are not by chance alone. In practice, it is very controversial to select an appropriate family of hypotheses (e.g., primary endpoints and secondary endpoints for efficacy or safety or both) for multiplicity adjustment for clinical evaluation of the test treatment under investigation.

It should be noted that the most worrisome impact of multiplicity on the inference for clinical trials is not only the control of FWER though that can be problematic but also the power for correctly detecting a clinically meaningful treatment effect. One of the most controversial issues in multiplicity is having adequate control of FWER but failing to achieve the desired power due to multiplicity.

11.3 Statistical Method for Adjustment of Multiplicity

As indicated earlier, commonly considered procedures or methods for controlling the FWER at some prespecified level of significance can be classified into two categories: (1) single-step methods (e.g., Bonferroni's correction) and (2) stepwise procedures, which include step-down methods (e.g., Holm's method) and step-up methods (e.g., Hochberg's method). In practice, commonly used procedures for controlling the FWER in clinical trials are classic multiple comparison procedures (MCPs), which include Bonferroni, Tukey, and Dunnett procedures. These procedures and a few others are briefly described in the following.

11.3.1 Bonferroni's Method

Among the above mentioned procedures, the method of Bonferroni is probably the most commonly considered procedure for addressing multiplicity in clinical trials though it is somewhat conservative.

Suppose there are k treatments and we are interested in testing the following hypothesis:

$$H_0 : \mu_1 = \mu_2 = \cdots = \mu_k ,$$

where μ_i, $i = 1, \ldots, k$ is the mean for the ith treatment. Let y_{ij}, $j = 1, \ldots, n_i$, $i = 1, \ldots, k$ be the jth observation obtained in the ith treatment. Also, let $\bar{y}_i$ and

$$s^2 = \frac{\sum_{i=1}^{k} \sum_{j=1}^{n_i} (y_{ij} - \bar{y}_i)^2}{\sum_{i=1}^{k} (n_i - 1)}$$

be the least square mean for the ith treatment and an estimate of the variance obtained from an analysis of variance (ANOVA), respectively. n_i is the sample size of the ith treatment. We then reject the null hypothesis in favor of the alternative hypothesis that the treatment means μ_i and μ_j are different for every $i \neq j$ if

$$\left| \bar{y}_i - \bar{y}_j \right| > t_{\alpha/2}(v) \left[s^2 (n_i^{-1} + n_j^{-1}) \right]^{1/2}, \tag{11.1}$$

where $t_{\alpha/2}(v)$ denotes a critical value for the t distribution with $v = \Sigma(n_i - 1)$ degrees of freedom and an upper tail probability of $\alpha/2$. Bonferroni's method simply requires that if there are k inferences in a family, then all inferences should be performed at the α/k significance level rather than at the α level.

Note that the application of Bonferroni's correction to ensure that the probability of declaring one or more false positives is no more than α. However, this method is not recommended when there are a large number of pair-wise comparisons. In this case, the following multiple range test procedures are useful.

11.3.2 Tukey's Multiple Range Testing Procedure

Similar to (11.1), we can declare that the treatment means μ_i and μ_j are different for every $i \neq j$ if

$$\left| \bar{y}_i - \bar{y}_j \right| > q(\alpha, k, v) \left[s^2 \frac{(n_i^{-1} + n_j^{-1})}{2} \right]^{1/2}, \tag{11.2}$$

where $q(\alpha, k, v)$ is the studentized range statistic. This method is known as Tukey's multiple range test procedure. It should be noted that simultaneous confidence intervals on all pairs of mean differences $\mu_i - \mu_j$ can be obtained based on the following:

$$P\left\{ \mu_i - \mu_j \in \bar{y}_i - \bar{y}_j \pm |q| \left[s^2 \frac{(n_i^{-1} + n_j^{-1})}{2} \right]^{1/2} \text{ for all } i \neq j \right\} = 1 - \alpha. \tag{11.3}$$

Note that tables of critical values for the studentized range statistic are widely available. As an alternative to Tukey's multiple range testing procedure, Duncan's multiple range testing procedure is often considered. Duncan's multiple testing procedure is to conclude that the largest and smallest of the treatment means are significantly different if

$$\left| \bar{y}_i - \bar{y}_j \right| > q(\alpha_p, p, v) \left[\frac{\text{MSE}}{n} \right]^{1/2}, \tag{11.4}$$

where

    p is the number of averages,

    $q(\alpha_p, p, v)$ is the critical value from the studentized range statistic with an FWER of α_p.

11.3.3 Dunnett's Test

When comparing several treatments with a control, Dunnett's test is probably the most popular method. Suppose there are $k - 1$ treatments and one control. Denote by μ_i, $i = 1, \ldots, k - 1$ and μ_k the mean of the ith treatment and the control, respectively. Further, supposes that the treatment groups can be described by the following balanced one-way ANOVA model:

$$y_{ij} = \mu_i + \varepsilon_{ij}, \quad i = 1, \ldots, k; \quad j = 1, \ldots, n.$$

It is assumed that ε_{ij} are normally distributed with mean 0 and unknown variance σ^2. Under this assumption, μ_i and σ^2 can be estimated. Consequently, one-sided and two-sided simultaneous confidence intervals for $\mu_i - \mu_k$ can be obtained.

For the one-sided simultaneous confidence interval of $\mu_i - \mu_k$, $i = 1, \ldots, k - 1$, the lower bound is given by

$$\hat{\mu}_i - \hat{\mu}_k - T\hat{\sigma}\sqrt{\frac{2}{n}} \quad \text{for } i = 1, \ldots, k-1, \tag{11.5}$$

where $T = T_{k-1, v}\{\rho_{ij}\}(\alpha)$ satisfies

$$\int_0^\infty \int_{-\infty}^\infty \left[\Phi\left(z - \sqrt{2}Tu\right)\right]^{k-1} d\Phi(z)\gamma(u)du = 1 - \alpha,$$

where Φ is the distribution function of the standard normal. It should be noted that $T = T_{k-1, v}\{\rho_{ij}\}(\alpha)$ are the critical values of the distribution of max T_i, where $T_1, T_2, \ldots, T_k$ multivariate t distributed with v degrees of freedom and correlation matrix $\{\rho_{ij}\}$.

For the two-sided simultaneous confidence interval $\mu_i - \mu_k$, $i = 1, \ldots, k - 1$, the lower bound is given by

$$\hat{\mu}_i - \hat{\mu}_k \pm |h|\hat{\sigma}\sqrt{\frac{2}{n}} \quad \text{for } i = 1, \ldots, k-1, \tag{11.6}$$

where $|h|$ satisfies

$$\int_0^\infty \int_{-\infty}^\infty \left[\Phi\left(z + \sqrt{2} \mid h \mid t\right) - \Phi\left(z - \sqrt{2} \mid h \mid t\right) \right]^{k-1} d\Phi(z)\gamma(t)dt = 1 - \alpha.$$

Similarly, $|h|$ are the critical values of the distribution of max T_i, where $T_1, T_2, \ldots, T_k$ multivariate t distributed with v degrees of freedom and correlation matrix $\{\rho_{ij}\}$.

11.3.4 Closed Testing Procedure

In clinical trials involving multiple comparisons, as an alternative, the use of the closed testing procedure has become very popular since it was introduced by Marcus et al. (1976). The closed testing procedure can be described as follows. First, form all intersections of elementary hypothesis H_i, then test all intersections using non-multiplicity adjusted tests. An elementary hypothesis H_i is then declared significant if all intersections which include the elementary hypothesis as a component of the intersection are significant. More specifically, suppose there is a family of hypotheses denoted by $\{H_i, 1 \le i \le k\}$. Let $H_P = \cap_{j \in P} H_j$ where $P = \{1, 2, \ldots, k\}$. H_P is rejected if and only if every H_Q is rejected for all $Q \subset P$ assuming that an α-level test for each hypothesis H_P is available. Marcus et al. (1976) showed that this testing procedure controls the FWER.

In practice, the closed testing procedure is commonly employed in a dose-finding study with several doses of a test treatment under investigation. As an example, consider the following family of hypotheses:

$$H_i : \mu_i - \mu_k \le 0, \quad 1 \le i \le k-1$$

against one-sided alternatives, where the kth treatment group is the placebo group. Assume that the sample sizes in the treatment groups are equal (say n) and the sample size for the placebo group is n_k. Let

$$\rho = \frac{n}{n + n_k}.$$

Then, the closed testing procedure can be carried out by the following steps:

Step 1: Calculate T_i, the t-statistics for $1 \le i \le k - 1$. Let the ordered t-statistics be $T_{(1)} \le T_{(2)} \le \cdots \le T_{(k-1)}$ with their corresponding hypotheses denoted by $H_{(1)}, H_{(2)}, \ldots, H_{(k-1)}$.

Step 2: Reject $H_{(j)}$ if $T_{(i)} > T_{i,v,\rho}(\alpha)$ for $i = k - 1, k - 2, \ldots, j$. If we fail to reject $H_{(j)}$, then conclude that $H_{(j-1)}, \ldots, H_{(1)}$ are also to be retained.

The closed testing procedures have been shown to be more powerful than the classic multiple comparisons procedures, such as the classic Bonferroni, Tukey, and Dunnett procedures. Note that the above step-down testing procedure is more powerful than that of Dunnett's testing procedure given in (11.5). There is considerable flexibility in the choice of tests for the intersection hypotheses, leading to the wide variety of procedures that fall within the closed testing umbrella. In practice, a closed testing procedure generally starts with the global null hypothesis and proceeds sequentially toward intersection hypotheses involving fewer endpoints. However, it can begin with the individual hypotheses and move toward the global null hypothesis.

11.3.5 Other Tests

In addition to the testing procedures described above, there are several tests (p-value based stepwise test procedures) that are also commonly considered in clinical trials involving multiple comparisons. These methods include, but are limited to, Simes' method (see Hochberg and Tamhane, 1987; Hsu, 1996; Sarkar and Chang, 1997), Holm's method (Holm, 1979), Hochberg's method (Hochberg, 1988; Hochberg and Benjamini, 1990), Hommel's method (Hommel, 1988), and Rom's method (Rom, 1990), which are briefly summarized in the following.

Simes' method is designed to reject the global null hypothesis if $p_{(i)} \leq i\alpha/m$ for at least one $i = 1, \ldots, m$. The adjusted p-value for the global hypothesis is given by

$$p = m \min\{p_{(1)}/1, \ldots, p_{(m)}/m\}.$$

Note that Simes' method improves Bonferroni's method in controlling the global type I error rate under independence (Sarkar and Chang, 1997). One of the limitations of Simes' method is that it cannot be used to draw inferences on individual hypotheses since it only tests the global hypothesis.

Holm's method is a sequentially rejective procedure, which sequentially contrasts ordered unadjusted p-values with a set of critical values and rejects a null hypothesis if the p-value and each of the smaller p-values are less than their corresponding critical values. Holm's method not only improves the sensitivity of Bonferroni's correction method to detect real differences but also increases in power and provides a strong control of the FWER.

Hochberg's method applies exactly the same set of critical values as Holm's method but performs the test procedure in a step-up fashion. Hochberg's method enables to identify more significant endpoints and hence is more powerful than Holm's method. In practice, Hochberg's method is somewhat conservative when individual p-values are independent. In the case where the endpoints are negatively correlated, the FWER control is not guaranteed for all types of dependence among p-values (i.e., the size could potentially exceed α).

Following the principle of closed testing procedure and Simes' test, Hommel's method is a powerful sequentially rejective method that allows for inferences on individual endpoints. It is shown to be marginally more powerful than Hochberg's method. However, the Hommel procedure also suffers from the disadvantage of not preserving the FWER. It does protect the FWER when the individual tests are independent or positively dependent (Sarkar and Chang, 1997).

Rom's method is a step-up procedure which is slightly more powerful as compared to Hochberg's method. Rom's procedure controls the FWER at the α level under the independence of p-values. More details can be found in Rom (1990).

11.4 Gatekeeping Procedures

11.4.1 Multiple Endpoints

Consider a dose–response study comparing m doses of a test drug to a placebo or an active control agent. Suppose that the efficacy of the test drug will be assessed using a primary endpoint and $s - 1$ ordered secondary endpoints. Suppose that the sponsor is interested in testing null hypotheses of no treatment effect with respect to each endpoint against one-sided alternatives. Thus, there are a total of ms null hypotheses, which can be grouped into s families to reflect the ordering of the endpoints. Now, let y_{ijk} denote the measurement of the ith endpoint collected in the jth dose group from the kth patient, where $k = 1, \ldots, n$, $i = 1, \ldots, s$, and $j = 0$ (control), $1, \ldots, m$. The mean of y_{ijk} is denoted by μ_{ij}. Also, let t_{ij} be the t-statistic for comparing the jth dose group to the control with respect to the ith endpoint. It is assumed that the t-statistics follow a multivariate t distribution. Furthermore, y_{ijk}'s are normally distributed. Denote by $\Im_i$ the family of null hypotheses for the ith endpoint, $i = 1, \ldots, s$, i.e., $\Im_i = \{H_{i1} : \mu_{i0} = \mu_{i1}, \ldots, H_{im} : \mu_{i0} = \mu_{im}\}$. The s families of null hypotheses are tested in a sequential manner.

Family $\Im_1$ (the primary endpoint) is examined first and testing continues to family $\Im_2$ (most important secondary endpoint) if at least one null hypothesis has been rejected in the first family. This approach is consistent with a regulatory view that findings with respect to secondary outcome variables are meaningful only when the primary analysis is significant. The same principle can be applied to the analysis of ordered secondary endpoints. Dmitrienko et al. (2006) suggest focusing on testing procedures that meet the following condition:

Condition A: Null hypotheses in $\Im_{i+1}$ can be tested only after at least one null hypothesis was rejected in $\Im_i$, $i = 1, \ldots, s - 1$. Secondly, it is important to ensure that the outcome of the multiple tests early in the sequence does not depend on the subsequent analyses.

Condition B: Rejection or acceptance of null hypotheses in $\Im_i$ does not depend on the test statistics associated with $\Im_{i+1}, \ldots, \Im_s$, $i = 1, \ldots, s - 1$. Finally, one ought to account for the hierarchical structure of this multiple testing problem and examine secondary dose–control contrasts only if the corresponding primary dose–control contrast was found significant.

Condition C: The null hypothesis H_{ij}, $i \geq 2$ can be rejected only if H_{1j} is rejected, $j = 1, \ldots, m$. It is important to point out that the logical restrictions for secondary analyses in condition C are caused only by the primary endpoint. This requirement helps clinical researchers streamline drug labeling and improves the power of secondary tests at the doses for which the primary endpoint was significant.

Within each of the s families, multiple comparisons can be carried out using Dunnett's test as follows. Reject H_{ij} if the corresponding t-statistic (t_{ij}) is greater than a critical value c for which the null probability of $\max(t_{i1}, \ldots, t_{im}) > c$ is α. Note that Dunnett's test protects the type I error rate only within each family. Dmitrienko et al. (2006) extended Dunnett's test for controlling the FWER for all ms null hypotheses.

11.4.2 Gatekeeping Testing Procedures

Dmitrienko et al. (2006) considered the following example to illustrate the process of constructing a gatekeeping testing procedure for dose–response studies. For simplicity, Dmitrienko et al. (2006) focused on the case where $m = 2$ and $s = 2$. In this example, it is assumed that the treatment groups are balanced with n patients per group. The four (i.e., $ms = 4$) null hypotheses are grouped into two ($s = 2$) families, i.e., $\mathfrak{I}_1 = \{H_{11}, H_{12}\}$ and $\mathfrak{I}_2 = \{H_{21}, H_{22}\}$. Note that $\mathfrak{I}_1$ consists of hypotheses for comparing low and high doses to placebo with respect to the primary endpoint, while $\mathfrak{I}_2$ contains hypotheses for comparing low and high doses to placebo with respect to the secondary endpoint.

Now let t_{11}, t_{12}, t_{21}, and t_{22} denote the t-statistics for testing H_{11}, H_{12}, H_{21}, and H_{22}. We can then apply the principle of the closed testing for constructing gatekeeping procedures. According to this principle, one first considers all possible nonempty intersections of the four null hypotheses (this family of 15 intersection hypotheses is known as the closed family) and then sets up tests for each intersection hypothesis. Each of these tests controls the type I error rate at the individual hypothesis level and the tests are chosen to meet conditions A, B, and C described above. To define tests for each of the 15 intersection hypotheses in the closed family, let H denote an arbitrary intersection hypothesis and consider the following rules:

1. If H includes both primary hypotheses, the decision rule for H should not include t_{21} or t_{22}. This is done to ensure that a secondary hypothesis cannot be rejected unless at least one primary hypothesis is rejected (condition A).

2. The same critical value should be used for testing the two primary hypotheses. This way, the rejection of primary hypotheses is not affected by the secondary test statistics (condition B).

3. If H includes a primary hypothesis and a matching secondary hypothesis (e.g., $H = H_{11} \cap H_{21}$), the decision rule for H should not depend on the test statistic for the secondary hypothesis. This guarantees that H_{21} cannot be rejected unless H_{11} is rejected (condition C).

Note that similar rules used in gatekeeping procedures based on the Bonferroni's test can be found in Dmitrienko et al. (2003) and Chen et al. (2005). To implement these rules, it is convenient to utilize the decision matrix approach (Dmitrienko et al., 2003). For the sake of compact notation, we will adopt the following binary representation of the intersection hypotheses. If an intersection hypothesis equals H_{11}, it will be denoted by H^*_{1000}. Similarly, $H^*_{1100} = H_{11} \cap H_{12}, H^*_{1010} = H_{11} \cap H_{21}$, etc.

Table 11.1 (reproduced from Table I of Dmitrienko et al., 2006) displays the resulting decision matrix that specifies a rejection rule for each intersection hypothesis in the closed family. The three constants (c_1, c_2, and c_3)

TABLE 11.1

Decision Matrix for a Clinical Trial with
Two Dose–Placebo Comparisons and
Two Endpoints ($m = 2$, $s = 2$)

Intersection Hypothesis	Rejection Rule
H^*_{1111}	$t_{11} > c_1$ or $t_{12} > c_1$
H^*_{1110}	$t_{11} > c_1$ or $t_{12} > c_1$
H^*_{1101}	$t_{11} > c_1$ or $t_{12} > c_1$
H^*_{1100}	$t_{11} > c_1$ or $t_{12} > c_1$
H^*_{1011}	$t_{11} > c_1$ or $t_{22} > c_2$
H^*_{1010}	$t_{11} > c_1$
H^*_{1001}	$t_{11} > c_1$ or $t_{22} > c_2$
H^*_{1000}	$t_{11} > c_1$
H^*_{0111}	$t_{12} > c_1$ or $t_{21} > c_2$
H^*_{0110}	$t_{12} > c_1$ or $t_{21} > c_2$
H^*_{0101}	$t_{12} > c_1$
H^*_{0100}	$t_{12} > c_1$
H^*_{0011}	$t_{21} > c_1$ or $t_{22} > c_1$
H^*_{0010}	$t_{21} > c_3$
H^*_{0001}	$t_{22} > c_3$

The test associated with this matrix rejects a null hypothesis if all intersection hypotheses containing it are rejected. For example, the test rejects H_{11} if H^*_{1111}, H^*_{1110}, H^*_{1101}, H^*_{1100}, H^*_{1011}, H^*_{1010}, H^*_{1001} and H^*_{1000} are rejected.

TABLE 11.2

Critical Values for Individual Intersection
Hypotheses in a Clinical Trial with Two
Dose–Placebo Comparisons and Two
Endpoints ($m = 2, s = 2$)

Correlation between the Endpoints (ρ)	c_1	c_2	c_3
0.01	2.249	2.309	1.988
0.1	2.249	2.307	1.988
0.5	2.249	2.291	1.988
0.9	2.249	2.260	1.988
0.99	2.249	2.250	1.988

Source: Dmitrienko, A. et al., *Pharm. Stat.*, 5, 19, 2006.
The correlation between the two endpoints (ρ) ranges
between 0.01 and 0.99, overall one-sided type I error probability is 0.025 and sample size per treatment group is 30
patients. With permission from John Wiley & Sons, Ltd.

in Table 11.2 (reproduced from Table II of Dmitrienko et al., 2006) represent
critical values for the intersection hypothesis tests. The values are chosen in
such a way that, under the global null hypothesis of no treatment effect, the
probability of rejecting each individual intersection hypothesis is α. Note
that the constants are computed in a sequential manner (c_1 is computed first,
followed by c_2, etc.) and thus c_1 is the one-sided $100(1 - \alpha)$th percentile of
Dunnett's distribution with 2 and $3(n - 1)$ degrees of freedom. Secondly,
the other two critical values (c_2 and c_3) depend on the correlation between
the primary and secondary endpoints, which is estimated from the data.
Calculation of these critical values is illustrated later.

The decision matrix in Table 11.1 defines a multiple testing procedure
that rejects a null hypothesis if all intersection hypotheses containing the
selected null hypothesis are rejected. For example, H_{12} will be rejected if
$H^*_{1111}, H^*_{1110}, H^*_{1101}, H^*_{1111}, H^*_{0111}, H^*_{0110}, H^*_{0101}$, and H^*_{0100} are all rejected. By the
closed testing principle, the resulting procedure protects the FWER in the
strong sense at the α level. It is easy to verify that the proposed procedure
possesses the following properties and thus meets the criteria that define a
gatekeeping strategy based on Dunnett's test:

1. The secondary hypotheses, H_{21} and H_{22}, cannot be rejected when the
   primary test statistics, t_{11} and t_{12}, are nonsignificant (condition A).

2. The outcome of the primary analyses (based on H_{11} and H_{12}) does
   not depend on the significance of the secondary dose–placebo comparisons (condition B). In fact, the procedure rejects H_{11} if and only
   if $t_{11} > c_1$. Likewise, H_{12} is rejected if and only if $t_{12} > c_1$. Since c_1 is a
   critical value of Dunnett's test, the primary dose–placebo compari-
   sons are carried out using the regular Dunnett test.

3. The null hypothesis H_{21} cannot be rejected unless H_{11} is rejected and thus the procedure compares the low dose to placebo for the secondary endpoint only if the corresponding primary comparison is significant. The same is true for the other secondary dose–placebo comparison (condition C).

Under the global null hypothesis, the four statistics follow a central multivariate t distribution. The three critical values in Table 11.1 can be found using the algorithm for computing multivariate t probabilities proposed by Genz and Bretz (2002). Table 11.2 shows the values of c_1, c_2, and c_3 selected values of ρ (correlation between the two endpoints). It is assumed in Table 11.2 that the overall one-sided type I error rate is 0.025 and the sample size per group is 30 patients.

The information presented in Tables 11.1 and 11.2 helps evaluate the effect of the described gatekeeping approach on the secondary tests. Suppose, for example, that the two dose–placebo comparisons for the primary endpoint are significant after Dunnett's adjustment for multiplicity ($t_{11} > 2.249$ and $t_{12} > 2.249$). A close examination of the decision matrix in Table 11.1 reveals that the null hypotheses in the second family will be rejected if their t-statistics are greater than 2.249. In other words, the resulting multiplicity adjustment ignores the multiple tests in the primary family.

However, if the low dose does not separate from the placebo for the primary endpoint ($t_{11} \leq 2.249$ and $t_{12} > 2.249$), it will be more difficult to find significant outcomes in the secondary analyses. First of all, the low dose–placebo comparison is automatically declared nonsignificant. Secondly, the high dose will be significantly different from the placebo for the secondary endpoint if $t_{22} > c_2$. Note that c_2, which lies between 2.250 and 2.309 when $0.01 \leq \rho \leq 0.99$, is greater than Dunnett's critical value $c_1 = 2.249$ (in general, $c_2 > c_1 > c_3$). The larger critical value is the price of sequential testing. Note, however, that the penalty becomes smaller with increasing correlation.

11.5 Concluding Remarks

When conducting a clinical trial involving one or more doses (e.g., dose-finding study) or one or more study endpoints (e.g., efficacy versus safety endpoint), the first dilemma at the planning stage of the clinical trial is the establishment of a *family* of hypotheses a priori in the study protocol for achieving the study objective of the intended clinical trial. Based on the study design and various underlying hypotheses, clinical strategies are usually explored for testing various hypotheses for achieving the study objectives. One such set of hypotheses (e.g., drug versus placebo, positive control agent versus placebo, primary endpoint versus secondary primary endpoint) would help to conclude whether both the drug and positive control agent are superior to placebo or the drug is efficacious in terms of the primary endpoint, secondary primary endpoint, or both. Under the family of hypotheses, valid MCPs for controlling the overall type I error rate should be proposed in the study protocol.

The other dilemma at the planning stage of the clinical trial is sample size calculation. A typical procedure is to obtain required sample size under either an ANOVA method or an analysis of covariance (ANCOVA) model based on an overall F test. This approach may not be appropriate if the primary objective involves multiple comparisons. In practice, when multiple comparisons are involved, the method of Bonferroni is usually performed to adjust the type I error rate. Again, the Bonferroni's method is conservative and may require more patients than are actually needed. Alternatively, Hsu (1996) suggested a confidence interval approach as follows. Given a confidence interval approach with level of $1 - \alpha$, perform sample size calculations so that with a prespecified power $1 - \beta(<1 - \alpha)$, the confidence intervals will cover the true parameter value and be sufficiently narrow (Hsu, 1996).

As indicated, multiple comparisons are commonly encountered in clinical trials. Multiple comparisons may involve comparisons of multiple treatments (dose groups), multiple endpoints, multiple time points, interim analyses, multiple tests of the sample hypothesis, variable/model selection, and subgroup analyses in a study. In this case, statistical methods for controlling error rates such as CWE, FWER, or FDR are necessary for multiple comparisons. The closed testing procedure is useful for addressing multiplicity issues in dose-finding studies. In the case where there are a large number of tests involved such as tests for safety data, it is suggested that the method using FDR for controlling the overall type I error rate be considered.

12

Independence of Data Monitoring Committee

12.1 Introduction

In clinical trials, an independent data monitoring committee (DMC) is often established to serve as a guard for validity and integrity of an intended clinical trial (see NIH 1998, 2000; Ellenberg et al., 2002). The DMC is independent of any activities related to the clinical operation of the study, which is comprised of experienced physicians and statisticians. Depending on the study objectives and needs of the sponsor, the primary responsibility of the independent DMC include, but are not limited to, (1) ensuring the validity and integrity of the intended clinical trial, (2) performing ongoing safety monitoring, and (3) performing interim analysis for efficacy. An established independent DMC will perform its function and activity according to a written charter, which is usually developed and approved by the sponsor, the investigator, and the DMC. In practice, there is separate staff supporting the functions and activities of DMC. This separate staff is usually referred to as DMC support staff. The DMC support staff is responsible for performing unblinded interim analysis and presenting the results to the DMC.

The use of DMC in clinical trials can be traced back to the early 1960s (FDA, 2006b). However, the DMC did not appear in pharmaceutical trials until early the 1990s (Herson, 2009). As more and more clinical trials sponsored by the pharmaceutical/device industry utilizing DMC for study monitoring, in 2001, the United States Food and Drug Administration (FDA) published a draft guidance on DMC to assist the sponsor in (1) determining the need for DMC, (2) establishing a DMC, and (3) setting up standard operation procedures for DMC's function and activity. The FDA draft guidance, however, was not finalized until 2006. Although the intention of the independent DMC is good, some controversial issues inevitably occur. These controversial issues include, but are not limited to, (1) the challenge of the independence of an independent DMC, (2) the issue regarding the direct communication between DMC and FDA, and (3) the use of DMC for clinical trials utilizing adaptive design methods. In this chapter, some insight regarding these controversial issues will be provided.

The remainder of this chapter is organized as follows. In the next section, regulatory requirements regarding the establishment, role/responsibility, function/activity of an independent DMC are described. Section 12.3 discusses the composition of a DMC and the development of the charter. DMC functional activities are outlined in Section 12.4. Some observations regarding the independence of an independent DMC are summarized in Section 12.5. Brief concluding remarks are given in the last section of this chapter.

12.2 Regulatory Requirements

The FDA published a draft guidance on DMC to assist clinical trial sponsors in determining *when* a DMC may be useful for clinical trial monitoring and *how* such a committee should operate in 2001. This draft guidance, which covers studies required for evaluating new drugs, biologics, and devices, was not finalized until 2006 (FDA, 2006b; Dixon et al., 2006). As indicated in the FDA guidance on DMC, a clinical trial DMC is defined as a group of individuals with pertinent expertise that reviews on a regular basis accumulating data from one or more ongoing clinical trials. The DMC advises the sponsor regarding the continuing safety of trial subjects and those yet to be recruited for the trial, as well as the continuing validity and scientific merit of the trial.

As indicated in the FDA guidance on DMC, the use of DMC can be traced back to the early 1960s. In 1967, a National Institutes of Health (NIH) external advisory group first introduced the concept of a formal committee charged with reviewing the accumulating data as the trial progressed to monitor safety, effectiveness, and trial conduct issues in a set of recommendations to the then NHI. However, few trials sponsored by the pharmaceutical/medical device industry incorporated DMC oversight until relatively recently. Although government agencies such as the NIH that sponsor clinical research have required the use of DMCs in certain trials, current FDA regulations, however, impose no requirements for the use of DMCs in trials except under 21 CFR 50.24(a)(7)(iv) for research studies in emergency settings in which the informed consent requirement is excepted. The DMC was also mentioned in several ICH guidelines such as (1) the ICH E3 guideline on clinical study reports (ICH, 1995), (2) the ICH E6 guideline on good clinical practices (ICH, 1996a), and (3) the ICH E9 guideline on statistical principles (ICH, 1998) since then. In the past decade, the role/responsibility and function/activity of an independent DMC have been discussed in the literature. For example, Hemmings and Day (2004) provided a good discussion of regulatory issues related to DMCs. The books by Ellenberg et al. (2002) and DeMets et al. (2006) gave a comprehensive overview of composition, role/responsibility, function/activity, and impact of an independent DMC in clinical trials from both regulatory and academic perspectives. Most recently, as indicated by Herson (2009), some leading medical journals have adopted a policy not to publish results of industry-sponsored trials unless an independent DMC was involved (see also Fontanarosa et al., 2005).

12.2.1 Determining Need for a DMC

As indicated in the FDA guidance, DMCs are generally established for large, randomized multisite studies that evaluate treatments intended to prolong life or reduce risk of a major adverse health outcome such as a cardiovascular event or recurrence of cancer. DMCs are recommended for any controlled trial of any size that will compare rates of mortality or major morbidity, but a DMC is not required or recommended for most clinical studies. DMCs are generally not needed, for example, for trials at early stages of product development. They are also generally not needed for trials addressing lesser outcomes, such as relief of symptoms, unless the trial population is at elevated risk of more severe outcomes. Although the value of a DMC is well recognized, the FDA suggested the following factors be assessed when determining whether to establish a DMC for a particular trial. These factors include the following: (1) What is the risk to trial participants? (2) Is a DMC review practical? (3) Will a DMC help assure the scientific validity of the trial? These factors are related primarily to safety, practicality, and scientific validity.

12.2.2 Confidentiality of Interim Data and Analysis

In clinical trials, knowledge of treatment codes will introduce bias to the clinical data collected from the study. As described in 21CFR314.126(b)(5) (for drug products) and 21CFR860.7(f)(1) (for devices), sponsors should make every attempt to minimize bias. Thus, it is suggested that unblinded interim data and the results of comparative interim analyses should not be assessed by anyone other than DMC members or the statistician performing the analysis and presenting the results to the DMC. As a result, the FDA guidance strongly recommends that procedures be established to safeguard confidential interim data from the project team, investigators, sponsor representatives, or anyone else outside the DMC and the statistician performing the interim analyses.

The FDA also recommends that any part of the interim report to the DMC that includes comparative effectiveness and safety data presented by the study group, whether coded or completely unblinded, be available only to DMC members during the course of the trial, including any follow-up period—that is, until the trial is completed and the blind is broken for the sponsor and investigators. The FDA emphasizes that if interim reports are shared with the sponsor, it may become impossible for the sponsor to make potentially warranted changes in the trial design or analysis plan in an unbiased manner.

12.2.3 Desirability of an Independent DMC

The FDA guidance on DMC emphasizes the importance of the independence of the DMC. Independence could be defined as DMC members (1) have no involvement in the design and conduct of the trial except through their role on the DMC and (2) have no financial or other important connections with the sponsor or other trial organizers that could influence their objectivity in evaluating trial data. The FDA also pointed out that the independence of an independent DMC has the following advantages. First, independence from the sponsor helps ensure that sponsor interests do not unduly influence the DMC, promoting objectivity that benefits the subjects and the trial. Second, through enhancement of objectivity and reduction of the possibilities for bias, independence of the DMC increases the credibility of the trial's conclusion. Third, independence of the DMC and complete blinding of the sponsor to interim outcome data preserve the ability of the sponsor to make certain modifications to a trial in response to new external information without introducing bias. Finally, in a commercially sponsored trial, independence of the DMC may shield the sponsor from security issues by maintaining the sponsor in a fully blinded situation.

It, however, should be noted that as pointed out by the FDA, DMCs are rarely, if ever, entirely independent of the sponsor, as the sponsor generally selects the DMC members and pays the committee members for their expenses and services.

12.3 DMC Composition and Charter

As indicated earlier, when conducting a clinical trial, an independent DMC is established not only to ensure the validity and integrity of the clinical trial but also to monitor ongoing safety data and perform interim analysis for efficacy. Thus, the selection of DMC members is extremely important as DMC responsibilities relate to the safety of trial participants. DMCs typically operate under a written charter that includes well-defined standard operating procedures. The DMC composition and charter are briefly described in the following sections.

12.3.1 DMC Composition and Support Staff

As indicated in the FDA guidance on DMC, most DMCs are composed of clinicians with expertise in relevant clinical specialties and at least one bio-statistician who is knowledgeable about statistical methods for clinical trials and sequential analysis of trial data. For trials with unusually high risks or with broad public health implications, the DMC may include a medical ethicist knowledgeable about the design, conduct, and interpretation of clinical trials.

Since DMC members are usually from different organizations with no administrative/statistical support, the sponsor may provide administrative/statistical support to the DMC. This administrative/statistical support is usually referred to as DMC support staff. The DMC support staff may consist of a statistician/programmer and/or a data manager. In practice, the responsibilities of the DMC support staff may include, but are not limited to, (1) assisting in DMC charter development, (2) coordinating with DMC statistician for development of DMC statistical analysis plan (SAP), (3) performing unblinded safety data reviews and interim analyses, (4) organizing DMC meetings, (5) preparing open and closed reports for DMC meetings, (6) preparing minutes of DMC meeting, (7) alerting DMC for any safety issues and/or unusual patterns of the data, (8) performing additional analyses as requested by the DMC, and (9) documenting all DMC activities, correspondences, and recommendations.

12.3.2 DMC Charter

When an independent DMC is established, a charter that describes the role, responsibility, function, and activity of the DMC is necessarily developed. The DMC support staff is responsible for providing any assistance that may be needed for the development of the charter. However, the DMC may dedicate the responsibility to the DMC support staff. In practice, a DMC charter is often developed according to the following principles.

During the development of the charter, the DMC support staff will interact with DMC members for any requirements, inputs, and comments that they may have. The DMC charter is usually developed according to the operating guideline or manual provided by the sponsor (if available). The draft charter will also be reviewed by the sponsor before it is submitted to the DMC for review and final approval. In practice, the draft DMC charter will be reviewed at the DMC organizational meeting for approval. Table 12.1 lists a table of contents for a typical DMC charter.

TABLE 12.1

Table of Contents for a DMC

1. Introduction
2. Role of the committee
3. Organizational flow
4. Committee membership
   4.1 Members
   4.2 Financial disclosure
   4.3 Duration of DMC membership
5. Committee meetings
   5.1 Organizational meeting
   5.2 Scheduled interim analysis meeting
   5.3 Unscheduled meetings
6. Communication
   6.1 Open reports
   6.2 Closed reports
   6.3 Committee minutes
   6.4 Committee recommendations
   6.5 Sponsor decision
   6.6 DMC additional data request
7. Timetable

12.4 DMC's Functions and Activities

The DMC's functions and activities include administrative look, safety data monitoring, and interim analysis for efficacy. These functions and activities are necessary to ensure quality, validity, and integrity of the clinical trial. Some of these functions and activities are briefly described in the following sections.

12.4.1 Randomization

The study statistician from the sponsor is responsible for providing specifications for randomization schedule generation according to the final approved protocol. The DMC statistician or support statistician/programmer is responsible for overseeing the generation of patient and study drug randomization schedules based on the specifications. The DMC support statistician is responsible for the review of the generated randomization schedules. The DMC support statistician/programmer is responsible for the transfer and implementation of the generated randomization codes to the designated area for drug packaging, shipment, and distribution.

TABLE 12.2

Data for DMC Review

CRF data
Adverse event and SAE data
Central laboratory data
Any data from other sources
Protocol deviation/violation
Endpoints adjudication

12.4.2 Critical Data Flow

In order to perform safety data reviews and interim analyses in a blinded or unblinded fashion, the following data are usually provided to the DMC for review (see Table 12.2). These safety data are critical for DMC safety review.

Serious adverse event (SAE) data are instant data and considered as real-time data, which is provided by the SAE reporting system within 24 h. Others are considered case report form (CRF) data. There are lag times between CRF and CRF in a database. The safety data review will be based on real-time data while interim analyses for efficacy will be based on CRF data. Before performing safety data reviews and interim analyses, the DMC support statistician will examine critical data flows according to the following principles.

The DMC support staff are responsible for conducting test data transfers for critical data described above. The DMC support statistician and programmer are responsible for the validation of the data transfer procedure prior to the conduct of safety data reviews and interim analyses.

12.4.3 DMC Report and Analysis Plan

The DMC SAP is usually prepared by the study statistician (sponsor) or a statistician from contract research organization (CRO) if a CRO is used for the trial. The DMC support statistician is responsible for the development of DMC SAP. The DMC SAP will go through an internal review process. The DMC SAP will be reviewed by the study statistician before it is submitted to the DMC (statistician) for review and comments. The DMC SAP will be reviewed at the DMC organizational meeting for approval.

In practice, DMC SAP is usually developed based on the report and analysis plan (RAP) by focusing on SAP with mock-up tables, listings, and graphs for critical safety data and efficacy data. The DMC support staff will conduct safety data reviews and perform interim analyses according to the following principles: (1) review and/or analyses based on pooled data in which the treatment groups are combined; (2) review and/or analyses based on unblinded data in which the treatment groups are separated and identified as treatment A, treatment B, etc.; (3) review and/or analyses based on partially unblinded data for which the reviewers are aware of the treatment codes for A, B, etc.; and (4) safety data reviews and interim analyses will be performed as specified in the protocol and/or DMC charter or as requested by the DMC.

In case of interim analysis for efficacy, rules and/or boundaries for stopping early due to safety and efficacy/futility should be specified in the DMC SAP. In addition, procedures for sample size reestimation based on either (1) variability, (2) conditional power, or (3) reproducibility probability in a blinded fashion should be described in detail in the DMC SAP.

12.4.4 Sensitivity Analysis

In addition to safety monitoring, the DMC may be asked to review interim analysis results for efficacy. The DMC may recommend stopping the trial early due to safety and/or futility/efficacy based on the review of interim analysis results. In practice, a two-stage optimal design is often used in cancer trials to fulfill this purpose (Simon, 1989). The concept of a two-stage optimal design is to stop a trial early if the test treatment is not effective and not to stop the trial early if the test treatment is promising. A typical two-stage design, which is often expressed as $(r_1/n_1, r_2/n)$, is to test

$$H_0 : p < p_0 \quad \text{versus} \quad H_a : p \geq p_1,$$

where p_0 and p_1 are undesirable and desirable response rates, respectively. Thus, at the first stage, n_1 subjects are tested. If there are less than r_1 responses, the trial stops; otherwise, proceed to the second stage and additional $n_2 = n - n_1$ subjects are recruited. At the end of the second stage, data collected from both stages are combined for a final analysis. We claim the test treatment has reached the desired response rate if there are more than r_2 responses.

In practice, at the end of the first stage, if less than r_1 responses are observed, before a recommendation to stop the trial is made, a sensitivity analysis is often requested by the DMC. In other words, it is of interest to evaluate the probability of observing $k(<r_1)$ given a sample size of n_1 assuming that the true response rate $p(\geq 0)$. If the true response rate p is indeed greater than the undesirable response rate p_0, then the probability of observing $k(<r_1)$ responses is expected to be small. At the same time, the DMC will also evaluate the probability of achieving the desirable response rate p_1 as the sample size approaches to n. This sensitivity analysis will provide the DMC with a better understanding/picture of the possible true response.

12.4.5 Executive Summary/Report

The DMC support staff is responsible for preparing an executive summary and reports usually at least 1 week prior to the DMC meeting. The executive summary and reports will be prepared according to the following principles.

An executive summary/report on safety and efficacy is necessarily submitted to the DMC for review at each DMC meeting except for the DMC organizational meeting. Key findings with some tables and listings will be included in the executive summary/report. The executive summary/report will go through an internal review process before it is submitted to the DMC for review. During the preparation of the safety reports, the DMC support statistician will alert the DMC regarding any safety issues and/or unusual patterns that may occur in order to maintain the integrity of the clinical trial of safety profiles.

Note that the dissemination of the executive summary and reports is subjective to the approval of the chairman of the DMC.

12.4.6 DMC Meetings

Normally the DMC will have an organizational meeting and subsequent meetings depending upon the nature of the intended clinical trial. The DMC organizational meeting is not only to review the final protocol but also to approve the DMC charter, which describes the role, responsibility, function, and activity of the DMC in the intended clinical trial.

Subsequent DMC meetings are to review results of safety data reviews and interim analyses in an unblinded fashion. The DMC may make recommendations to the sponsor regarding process improvement of the clinical trial or an early stop of the clinical trial. The DMC chairman is responsible for initiating DMC meetings. The DMC support staff are responsible for providing assistance for organizing the meeting and preparing the meeting agenda and material which should be sent to all attendees usually 2 weeks prior to the DMC meetings. Note that the DMC support staff will attend DMC meetings as nonvoting members. Representatives from the sponsor may be invited to DMC open sessions as specified in the DMC charter.

The minutes of the DMC meetings summarize the discussions that took place at DMC meetings. Hence, these meeting minutes are considered official documents for DMC activities of the intended clinical trial. The DMC minutes will be handled according to the following principles.

The chairman of DMC is responsible for assigning an individual to take minutes of the meeting at the beginning of the meeting. The DMC support statistician is usually delegated for this responsibility by the chairman of the DMC. If the DMC support staff are delegated with this responsibility, the support physician and statistician will summarize any medical and statistical issues discussed at the meeting. The draft meeting minutes will be submitted to the DMC for review and approval. The DMC support staff will distribute the approved meeting minutes to DMC members and other attendees as deemed appropriate by the DMC chairman.

12.4.7 DMC Documents and Information Dissemination

The DMC support staff will maintain a master file containing all documents related to DMC activities according to the following principles. The DMC master file includes study protocol, CRF, report analysis plan (RAP), DMC charter, any correspondences including fax, e-mail, meeting minutes, executive summaries and interim reports, as well as DMC decisions and recommendations. The DMC support staff are responsible for updating the master file. At the end of the study, a copy of the complete DMC master file will be submitted to the chairman of the DMC and the sponsor.

All information related to DMC activities may be disseminated according to the following principles.

The DMC support staff are prohibited to disseminate any information regarding DMC activities without approval from the chairman of the DMC. Open session reports may be distributed as deemed appropriate by the DMC.

12.4.8 DMC Recommendations

To maintain the integrity of the clinical trial, the DMC may make recommendations during the conduct of the trial based on safety data reviews and/or interim analyses according to the following principles. The DMC may alert the sponsor on any safety related issues and make recommendations regarding the improvement of the clinical trial. The DMC may make a recommendation to stop the trial early based on stopping rules as specified in the DMC charter. The DMC chairman is responsible for communicating with the sponsor regarding the DMC recommendations. All recommendations must be documented.

Note that the DMC may request the DMC support staff to perform additional analyses which are not preplanned in the DMC charter or SAP according to the following principles. Requests for additional analyses should be made by the chairman of the DMC or his/her designates. Upon the receipt of the requests, the DMC support statistician will prepare mock-up tables for the DMC statistician's review and approval before proceeding with the analyses. The results of additional analyses will go through the same review process as those of safety data reviews and interim analyses.

12.4.9 DMC Organizational Flow

The relationship between the sponsor (and CRO) and DMC (and DMC support staff) is summarized in the DMC organizational flowchart in Figure 12.1.

In practice, it is not uncommon that the sponsor uses a CRO to assist in the conduct of a clinical trial. In this case, there may be four statisticians involved: (1) study or lead statistician from the sponsor, (2) project statistician from the CRO, (3) DMC support statistician either from the CRO or affiliated with the DMC statistician, and (4) DMC statistician. If the DMC support staff is from the CRO, the CRO should build up a firewall to make sure that the

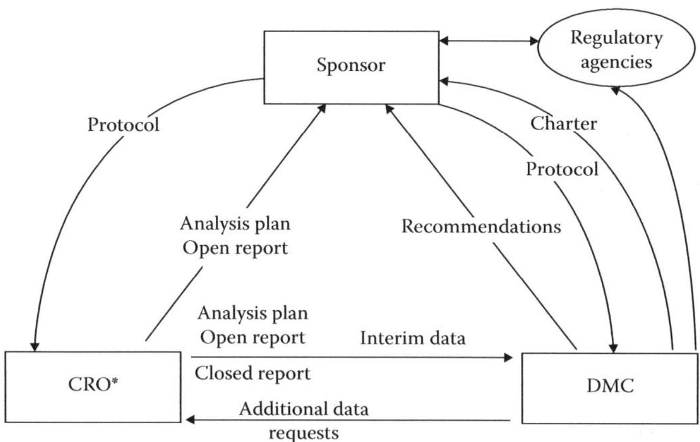

FIGURE 12.1
Organizational flowchart for DMC. *Not otherwise involved with study.

DMC support staff is truly independent of any personnel who are involved in the conduct and project management of the clinical trial. The DMC support staff will only be answerable to the DMC and should not communicate with the project teams either from the sponsor or the CRO.

12.5 Independence of DMC

One of the major concerns in clinical trials when utilizing a DMC is probably the *independence* of the DMC. As indicated earlier, the DMC's primary responsibilities include data safety monitoring and possibly interim analysis for efficacy. The DMC has the authority to stop the trial early due to safety, efficacy/futility, or both after the review of the accumulated data at interim. Most DMCs prefer a blinded review regardless of safety and efficacy data with an option to unblind the treatment codes if significant findings regarding safety and/or efficacy are observed at the closed session of the DMC meeting. The DMC will make recommendations to the sponsor although the sponsor may or may not accept the DMC's recommendation. The good intention of the DMC will ensure the quality, validity, and integrity of the clinical trial. In practice, however, some sponsors will make every attempt to direct (or influence) the function and activity of the DMC. In some cases, they are successful and in many cases they have failed. The following is a summary of the issues that are commonly seen in clinical trials utilizing DMCs across therapeutic areas.

12.5.1 Some Observations

DMC member selection: DMC members are appointed by the sponsor. The sponsor will usually select opinion leaders in the subject area who are in favor of their products. In many cases, DMC members are closely related to the principal investigator or sponsor. For convenience's sake, the sponsor may ask the identified DMC member (e.g., medical expert) to identify a statistician from his/her organization. In this case, it is most likely that the identified DMC statistician may administratively report to the DMC member. This has raised the controversial issue of independence of the DMC. Thus, it is suggested that the selection procedure and the qualification of the selected DMC members be documented.

Replacement of DMC members: In the case where the DMC members have strong opinions regarding the design and analysis of the study protocol and/or charter, the sponsor should communicate with these DMC members rather than replace them. For creditability and to avoid selection bias it is suggested that the reasons for replacing DMC members be documented and submitted to the regulatory agency for review. However, none of the sponsors is in compliance to this suggestion, especially when the DMC members are replaced before the DMC is officially established.

Development of DMC charter: Once the DMC is established, it is a common practice for the sponsor to take the lead to assist the DMC to develop a charter, which will outline the role/responsibility and function/activity of the DMC without consulting with the DMC members. DMC members usually will not have the chance to review it until a few days before or at the first DMC organizational meeting. In many cases, the initial DMC is a teleconference call rather than a face-to-face meeting in order to save cost. Since DMC members are usually key opinion leaders in the subject area, they may not have the chance to thoroughly review the charter prior to the meeting. As a result, the charter is usually approved in a hurry. Consequently, the DMC charter developed by the sponsor may have influenced the procedures and/or direction of safety monitoring and interim analysis for efficacy.

Communication with DMC: In practice, it is not common that the sponsor will seek advice from individual DMC members without the knowledge of the chairman of the DMC. This has caused an issue among the DMC members within the DMC. Thus, it is suggested that the chairman of the DMC be the primary contact person between the sponsor and the DMC. In some cases, when communicating with the DMC, the sponsor may argue with the DMC based on *informal* communication with medical/statistical reviewers from regulatory agencies. This has a negative impact on the function and activity of the DMC. Thus, it is suggested that a written communication with the regulatory agencies be provided to the DMC when a debatable issue is encountered.

Operations without the knowledge of DMC: In some cases, the sponsor may have begun to enroll patients prior to the initial DMC meeting. In this case, the

DMC is asked to endorse the study protocol without reviewing the study protocol. This is definitely not a good clinical practice. However, it does happen. This has seriously affected the function and activity of the DMC. Thus, it is suggested that the clinical trial be suspended until the initial DMC meeting has taken place for maintaining the integrity of the clinical trial.

Protocol amendments: In many cases, the sponsor may have issued protocol amendments or modified randomization schedules without consulting with DMC members. This has made it very difficult for the DMC to perform their job responsibilities. Thus, it is suggested that detailed information regarding the description, rationales, and impact of the changes made to the study protocol be provided to the DMC prior to the issue of protocol amendments.

Integrity of blinding: In some cases, it was found that the project statistician and the unblinding statistician (i.e., DMC support statistician) are the same person. In this case, the double-blind study is considered unblinded. This is a serious violation of good clinical practice. The integrity of blinding is seriously in doubt. Thus, it is suggested that the data collected after the unblinding should not be used for clinical evaluation of the test treatment under investigation.

Operational bias: For a clinical trial with planned interim analyses, the DMC will usually make recommendations after the review of the interim analysis results. The chairman of the DMC will communicate with the sponsor regarding the recommendations. In the case where the sponsor disagrees with the DMC's recommendation, the sponsor may request a second opinion from an independent medical/statistical expert. This is fine if it is agreed by the DMC. However, in some cases, the sponsor may seek a second opinion (with a different data set by including data after the interim analysis) to overrule the DMC's recommendation without consulting with the DMC. The second opinion may not be aware of the interim analysis results conducted by the DMC (which was conducted based on a different data set). It should be noted that data collected after the interim analysis has been contaminated by the operational bias.

12.5.2 Controversial Issues

As discussed in the previous section, one of most controversial issues regarding the DMC is probably: "Is an independent DMC really independent?" To ensure the integrity/success of the clinical trial, the DMC plays an important role. The DMC means to be independent of the project team for providing a fair and unbiased safety data monitoring and/or interim analysis for efficacy. In practice, the sponsor will make every attempt to influence the DMC's functions and activities. In clinical trials, the independence of the established DMC has been challenged by clinical researchers. The loss of independence could have a negative impact on the quality, validity, and integrity of the clinical trial.

One of the other controversial issues is whether it is appropriate to allow the DMC to communicate with the regulatory agency directly. From the sponsor's point of view, it is not desirable to reveal the information (which might be against regulatory review/approval) to the regulatory agency, especially if the observation could be limited and/or has not yet been verified. From the DMC's point of view, it is important to bring it to the regulatory agency's attention if a critical concern regarding the safety and/or integrity of the trial has occurred. In addition, the regulatory agency may not have the resources to take care of the issues reported by individual DMCs.

Another controversial issue that has been discussed tremendously is: "What if the sponsor decides not to accept the DMC's recommendation?" As an example, the DMC may recommend stopping the trial early due to futility after the review of the interim analysis results. However, the sponsor may argue that the DMC's recommendation is drawn based on limited information observed at interim. The sponsor may take the following actions to argue against the DMC's recommendation. First, the sponsor may perform a sensitivity analysis with respect to various study parameter specifications. In addition, the sponsor may request a second opinion from an independent medical or statistical expert to justify the DMC's recommendation.

In recent years, the use of adaptive design methods in clinical trials has become very popular due to its flexibility and efficiency for identifying clinical benefits in a timely fashion. However, one of the major concerns is that the use of adaptive design methods may introduce so-called operational bias and/or variation. To ensure the success of the adaptive design methods, it is suggested that the established DMC should take more responsibilities (beyond that described in the DMC charter) for preventing operational bias when implementing the adaptive design methods in clinical trials. It is very controversial whether we should put additional burden on the already over-loaded DMC.

12.6 Concluding Remarks

As discussed above, controversial issues regarding the IDMC have been raised. These controversial issues, which are briefly summarized in the following, have an impact on the quality, integrity, and success of clinical trials conducted at various phases of clinical development.

First, is an IDMC really independent? As pointed out by the FDA, DMCs are rarely entirely independent of the sponsor due to the fact that (1) the sponsor selects the DMC members, (2) the sponsor gives the DMC its charge, and, most importantly, (3) the sponsor pays for the DMC's expenses and services. The true independence may result in eliminating from consideration the most knowledgeable clinical researchers/scientists who are likely to have had some past interaction with others sponsoring or performing research in their area of expertise.

Second, should the IDMC be encouraged to communicate with regulatory agencies for any wrongdoing in the conduct of the intended clinical trial? As indicated in the FDA guidance, for trials that may be terminated due to safety concerns, timely communication with the FDA is required. In this case, the FDA strongly recommends that sponsors initiate discussion with the FDA prior to early termination of any trial implemented specifically to investigate a potential safety concern. As the FDA pointed out, in rare cases, the FDA wishes to interact with a DMC of an ongoing trial to ensure that specific issues of urgent concern to the FDA are fully considered by the DMC or to address questions to the DMC regarding the consistency of the safety data in the ongoing trial to that in the earlier trials, to optimize regulatory decision making.

Finally, should we put additional burden on the DMC if adaptive design methods are used? As more and more clinical trials are utilizing adaptive design methods, there is a discussion regarding whether we should put additional burden on the existing DMC or establish a separate DMC in order to monitor scientific validity and integrity of the clinical trials utilizing adaptive design methods. There is no universal agreement on this issue.

13

Two-Way ANOVA versus One-Way ANOVA with Repeated Measures

13.1 Introduction

In clinical research, a parallel-group design with multiple assessments at a number of prespecified time points post treatment is usually employed to compare treatment difference between a test compound and a control (e.g., a placebo control, a standard therapy, or an active control). Under such a study design, a one-way (treatment) analysis of variance (ANOVA) with repeated measures is a valid statistical method for assessment of treatment difference. In practice, however, it is not uncommon that a two-way ANOVA is *wrongly* used to assess treatment difference by treating the prespecified time points as a class variable rather than a covariate from the same subject. It, however, should be noted that one of the primary assumptions for a two-way (treatment by time) ANOVA is independence among observations observed at different time points. As a result, the use of a two-way ANOVA is inappropriate since observations observed at different time points from the same subject are correlated. Thus, it is of interest to evaluate the validity of statistical inference obtained from the two-way ANOVA under the one-way ANOVA model with repeated measures.

The remainder of this chapter is organized as follows. In Section 13.2, the one-way ANOVA with repeated measures will be briefly outlined. Also included in this section is the correct statistical procedure for assessment of treatment difference. In Section 13.3, the standard two-way ANOVA model will be introduced. Also included in this section is the standard statistical procedure under such a model. In Section 13.4, the statistical property of statistical inference obtained from the two-way ANOVA will be evaluated under the correct one-way ANOVA with repeated measures model in terms of type I error. In Section 13.5, a simulation study is performed to confirm the results obtained in Section 13.4. In Section 13.6, a real example concerning a clinical study is given for illustration purpose. Finally, the chapter is concluded with a discussion in Section 13.7.

13.2 One-Way ANOVA with Repeated Measures

For the purpose of simplicity, we will only consider two treatment groups. The following statistical model is usually considered for data from a one-way ANOVA with repeated measures:

$$y_{ijk} = \mu + \alpha_i + S_{ij} + b_{ij}t_k + e_{ijk},\tag{13.1}$$

where

y_{ijk} is the kth observation from the jth subject in the ith treatment group,

α_i is the fixed effect for the ith treatment $\left(\sum_{i=1}^{2}\alpha_i = 0\right)$,

S_{ij} is the random effect due to the jth subject in the ith treatment group,

b_{ij} is the coefficient of the jth subject in the ith treatment group,

e_{ijk} is the random error in observing y_{ijk}.

In the above model, it is assumed that (1) S_{ij}'s are independently distributed as $N(0,\sigma_S^2)$, (2) b_{ij}'s are independently distributed as $N(b_i,\sigma_b^2)$, and (3) e_{ijk}'s are independently distributed as $N(0, \sigma^2)$. For any given subject j within treatment i, $\{y_{ijk}\}$ are correlated and can be described by a regression line with slope b_{ij} conditioned on S_{ij}, i.e.,

$$y_{ijk} = \mu_{ij} + b_{ij}t_k + e_{ijk}, \quad k = 1, \ldots, m,$$

where $\mu_{ij} = \mu + \alpha_i + S_{ij}$. When conditioned on S_{ij}, unbiased estimators of the coefficient can be obtained by the method of ordinary least squares (OLS), which are given by

$$\hat{\mu}_{ij} = \frac{\sum_{k=1}^{m} y_k \sum_{k=1}^{m} t_k^2 - \sum_{k=1}^{m} t_k \sum_{k=1}^{m} y_k t_k}{m \sum_{k=1}^{m} t_k^2 - \left(\sum_{k=1}^{m} t_k\right)^2},$$

$$\hat{b}_{ij} = \frac{m \sum_{k=1}^{m} y_k t_k - \sum_{k=1}^{m} y_k \sum_{k=1}^{m} t_k}{m \sum_{k=1}^{m} t_k^2 - \left(\sum_{k=1}^{m} t_k\right)^2}.$$

Conditioning on S_{ij} and b_{ij}, we have

$$\hat{\mu}_{ij} \sim N\left(\mu_{ij}, \frac{\sigma^2 \sum_{k=1}^{m} t_k^2}{m \sum_{k=1}^{m} t_k^2 - \left(\sum_{k=1}^{m} t_k\right)^2}\right),$$

$$\hat{b}_{ij} \sim N\left(b_{ij}, \frac{\sigma^2 m}{m\sum_{k=1}^{m} t_k^2 - \left(\sum_{k=1}^{m} t_k\right)^2}\right).$$

Thus, unconditionally,

$$\hat{\mu}_{ij} \sim N\left(\mu + \alpha_i, \sigma_S^2 + \sigma^2 \frac{\sum_{k=1}^{m} t_k^2}{m\sum_{k=1}^{m} t_k^2 - \left(\sum_{k=1}^{m} t_k\right)^2}\right),$$

$$\hat{b}_{ij} \sim N\left(b_i, \sigma_b^2 + \sigma^2 \frac{m}{m\sum_{k=1}^{m} t_k^2 - \left(\sum_{k=1}^{m} t_k\right)^2}\right).$$

Then, intuitive estimators for μ_i and b_i can be obtained as

$$\hat{\mu}_{i.} = \frac{1}{n}\sum_{j=1}^{n} \hat{\mu}_{ij}, \quad \hat{b}_{i.} = \frac{1}{n}\sum_{j=1}^{n} \hat{b}_{ij}.$$

Since the objective is to compare the treatment effect, it is of interest to test the following hypotheses:

$$H_0 : \alpha_1 = \alpha_2 \quad \text{versus} \quad H_a : \alpha_1 \neq \alpha_2.$$

A significant difference between α_1 and α_2 usually indicates a significant baseline difference between treatment groups. The above hypotheses can be tested by using the statistic

$$T_1 = \frac{\sqrt{n}(\hat{\mu}_{1.} - \hat{\mu}_{2.})}{\sqrt{\sum_{i=1}^{2}\sum_{j=1}^{n} (\hat{\mu}_{ij} - \hat{\mu}_i)^2/(n-1)}}.$$

Under the null hypotheses of no treatment difference between two treatment groups, T_1 follows a t distribution with $2n - 2$ degrees of freedom. Hence, we reject the null hypothesis at the α level of significance if

$$|T_1| > t_{\alpha/2, 2n-2}.$$

Furthermore, it is also of interest to test the following null hypothesis of equal slopes (i.e., rate of change in study endpoint over the time period):

$$H_0 : b_1 = b_2 \quad \text{versus} \quad H_a : b_1 \neq b_2.$$

The above hypotheses can be tested using the following statistic:

$$T_2 = \frac{\sqrt{n}\left(\hat{b}_{1.} - \hat{b}_{2.}\right)}{\sqrt{\sum_{i=1}^{2}\sum_{j=1}^{n}(\hat{b}_{ij} - \hat{b}_i)^2 / (n-1)}}.$$

Under the null hypotheses, T_2 follows a t distribution with $2n - 2$ degrees of freedom. Hence, we would reject the null hypotheses of no difference in rate of change in the study endpoint between treatment groups at the α level of significance if

$$|T_2| > t_{\alpha/2, 2n-2}.$$

13.3 Two-Way ANOVA

In practice, time is often wrongly treated as another factor by ignoring the correlation structure of the observations from the same subject. As a result, the following two-way ANOVA model is used to describe the data:

$$y_{ijk} = \mu + \alpha_i + \gamma_k + \eta_{ij} + e_{ijk}, \tag{13.2}$$

where y_{ijk} is the observation from the jth subject in the ith treatment in the kth visit. It is also assumed that $\sum_{i=1}^{2} \alpha_i = 0$ and $\sum_{k=1}^{m} \gamma_k = 0$, and e_{ijk} is independent and identically distributed as $N(0, \sigma^2)$. In order to test for the treatment effect, the following quantities are defined:

$$SSE = \sum_{i=1}^{2}\sum_{k=1}^{m}\sum_{j=1}^{n}\left(y_{ijk} - \bar{y}_{i.k}\right)^2, \tag{13.3}$$

$$SSA = nm\sum_{i=1}^{2}\left(\bar{y}_{i..} - \bar{y}_{...}\right)^2, \tag{13.4}$$

where

$$\bar{y}_{i.k} = \frac{1}{n}\sum_{j=1}^{n} y_{ijk}, \quad \bar{y}_{i..} = \frac{1}{nm}\sum_{j=1}^{n}\sum_{k=1}^{m} y_{ijk}, \quad \bar{y}_{...} = \frac{1}{2nm}\sum_{i=1}^{2}\sum_{j=1}^{n}\sum_{k=1}^{m} y_{ijk}.$$

The test statistic is given by

$$T = \frac{\text{SSA}}{\text{SSE}/(2m(n-1))}. \tag{13.5}$$

Under the null hypothesis that $\alpha_1 = \alpha_2$, T is distributed as an F-random variable with 1 and $2m(n-1)$ degrees of freedom. Hence, the null hypothesis of no treatment effect is rejected at the α level of significance if $T > F_{1-\alpha,2m(n-1)}$, where $F_{1-\alpha,2m(n-1)}$ is the αth percentile of a standard F distribution with 1 and $2m(n-1)$ degrees of freedom.

13.4 Statistical Evaluation

Compare model (13.1) and model (13.2), it can be seen that one important difference between these two models is that model (13.2) ignores the correlation structure of the observations from the same subject. Thus, it is of interest to evaluate statistical properties of the statistical inferences obtained from model (13.2) under model (13.1).

Under model (13.2), SSE and SSA are independent. Under model (13.1), it is of interest to determine whether they are still independent. It can be noted that SSA is actually a function of $\{\bar{y}_{i..}\}$ and SSE is a function of $\{y_{ijk} - \bar{y}_{i.k}\}$. Hence, if we can establish independence between $\{\bar{y}_{i0..}\}$ and $\{y_{ijk} - \bar{y}_{i.k}\}$ for all i_0, i, j, k, then we can conclude that SSE and SSA are independent for each other. To see this, it should be noted that, under model (13.1),

$$\bar{y}_{i0..} = \mu + \alpha_{i0} + \bar{S}_{i0.} + \bar{b}_{i0.}\bar{t} + \bar{e}_{i0..}, \quad y_{ijk} - \bar{y}_{i.k} = (S_{ij} - \bar{S}_{i.}) + (b_{ij} - \bar{b}_{i.})t_k + (e_{ijk} - \bar{e}_{i.k}),$$

where

$$\bar{S}_{i.} = \frac{1}{n}\sum_{j=1}^{n} S_{ij}, \quad \bar{b}_{i.} = \frac{1}{n}\sum_{j=1}^{n} b_{ij}, \quad \bar{e}_{i.k} = \frac{1}{n}\sum_{j=1}^{n} e_{ijk}.$$

Note that if $i_0 \neq i$, then $\{\bar{y}_{i0..}\}$ and $\{y_{ijk} - \bar{y}_{i.k}\}$ are independent of each other because of the fact that they are statistics based on observations from different treatment groups. On the other hand, if $i_0 = i$, it can be noted that $\{S_{ij}\}$,

$j = 1, \ldots, n$ are independent and identically distributed as a normal random variable. In this case, $S_{ij} - \bar{S}_{i\cdot}$ and $\bar{S}_{i\cdot}$ are independent of each other. On the other hand, according to model (13.1) $\bar{S}_{i\cdot}$ is independent of b_{ij} and e_{ijk}. Hence, $\bar{S}_{i\cdot}$ is independent of $y_{ijk} - \bar{y}_{i\cdot k}$. A similar argument can also be applied to $\bar{b}_{i\cdot}$ and $\bar{e}_{i\cdot k}$. Hence, $\bar{y}_{i\cdot\cdot}$ and $y_{ijk} - \bar{y}_{i\cdot k}$ are independent of each other. This leads to the conclusion that SSE and SSA are independent of each other.

The next question of interest is to find out the distributions of SSE and SSA under model (13.1). Under model (13.1), for a fixed i and k, y_{ijk} are independent and identically distributed as a normal random variable with mean $\mu + \alpha_i + b_i t_k$ and variance $\sigma_k^2 = \sigma_S^2 + \sigma_b^2 t_k^2 + \sigma^2$. As a result,

$$\sum_{i=1}^{2} \sum_{j=1}^{n} (y_{ijk} - \bar{y}_{i.k})^2$$

is distributed as $\sigma_k^2 \chi^2(2n-2)$, where $\chi^2(2n-2)$ denotes a chi-square random variable with $(2n-2)$ degrees of freedom. However, it should be noted for different k, the quantity

$$\sum_{i=1}^{2} \sum_{j=1}^{n} (y_{ijk} - \bar{y}_{i.k})^2$$

usually are dependent of each other because of the fact that they have the observations from the same subject. As indicated by Chow et al. (2002b) and Lee et al. (2002a), SSE is distributed as a weighted chi-square random variable. More specifically,

$$\text{SSE} \sim \sum_{k=1}^{m} \lambda_k \chi^2(2n-2),$$

where the exact formula for λ_k can be derived using the methodology developed in Lee et al. (2002a). Although the exact formula for λ_k is not provided here, by matching the first-order moment, we know that the following condition should be satisfied:

$$\sum_{k=1}^{m} \lambda_k = \sum_{k=1}^{m} \sigma_k^2. \tag{13.6}$$

On the other hand, under model (13.1), it can be obtained that

$$\bar{y}_{i\cdot\cdot} = \frac{1}{nm} \sum_{j=1}^{n} \sum_{k=1}^{m} y_{ijk} = \frac{1}{nm} \sum_{j=1}^{n} \sum_{k=1}^{m} (\mu + \alpha_i + S_{ij} + b_{ij} t_k + e_{ijk}) = \mu + \alpha_i + \bar{S}_{i\cdot} + \bar{b}_{i\cdot} \bar{t} + \bar{e}_{i\cdot\cdot},$$

where

$$\bar{t} = \frac{1}{m}\sum_{k=1}^{m} t_k, \quad \bar{e}_{i..} = \frac{1}{nm}\sum_{j=1}^{n}\sum_{k=1}^{m} e_{ijk}.$$

Under the null hypothesis that $\alpha_1 = \alpha_2 = 0$, $\{\bar{y}_{i..}\}$ are independent and identically distributed as normal random variables with mean $\mu + b\bar{t}$ and variance $\sigma_S^2/n + \sigma_b^2 \bar{t}^{\,2}/n + \sigma^2/nm$. As a result, SSA defined in (13.4) is distributed as a scaled chi-square random variable. More specifically,

$$\text{SSA} \sim (m\sigma_S^2 + m\bar{t}^{\,2}\sigma_b^2 + \sigma^2)\chi^2(1).$$

As a result, T is not distributed as a standard F distribution. Instead,

$$T \sim \frac{(m\sigma_S^2 + m\bar{t}^{\,2}\sigma_b^2 + \sigma^2)\chi^2(1)}{\sum_{k=1}^{m}\lambda_k\chi^2(2n-2)}.$$

As it can be seen, due to the fact that SSE is distributed as a weighted chi-square random variable, T is not distributed as any standard distribution commonly encountered in practice. Its statistical property can be studied by exploring its exact distribution by either simulation or numerical methods. However, those methods have the disadvantage not only of being complicated but also of having lack of insight. Here we provide an alternative. The idea is to find a scaled chi-square distribution, which is "similar" to the exact distribution of SSE, and then approximate SSE's true distribution by this approximate distribution. More specifically, compare SSE with $\sigma_*^2\chi^2(2mn-2m)$, where

$$\sigma_*^2 = \frac{1}{m}\sum_{k=1}^{m}\sigma_k^2 = \frac{1}{m}\sum_{k=1}^{m}(\sigma_S^2 + \sigma_b^2 t_k^2 + \sigma^2) = \sigma_S^2 + \frac{\sigma_b^2\left(\sum t_k^2\right)}{m + \sigma^2}.$$

Note that these two random variables share the following two common characteristics: (1) both of their distributions belong to the family of weighted chi-square distribution and (2) they have the same first-order moment. As a result, one may expect that $\sigma_*^2\chi^2(2m(n-1))$ can provide a good approximation to the true distribution of SSE. This idea was first proposed by Rao and Scott (1981) and subsequently studied by Wang (2001). Consequently, the distribution of T can also be approximated by

$$T \sim \frac{(m\sigma_S^2 + m\bar{t}^2\sigma_b^2 + \sigma^2)\chi^2(1)}{\sum_{k=1}^{m} \sigma_k^2\chi^2(2n-2)} \approx \frac{(m\sigma_S^2 + m\bar{t}^2\sigma_b^2 + \sigma^2)\chi^2(1)}{\sigma_*^2\chi^2(2m(n-1))} = \kappa F_{1,2m(n-1)},$$

where

$$\kappa = \frac{m\sigma_S^2 + mt^{-2}\sigma_b^2 + \sigma^2}{\sigma_S^2 + \sigma_b^2\left(\sum t_k^2\right)\bigg/m + \sigma^2}.$$

In what follows, we prove that κ is a positive coefficient, which is always larger than 1 except for some extreme cases. In order to show $\kappa > 1$, consider the following quantity:

$$\Delta = (m\sigma_S^2 + m\bar{t}^2\sigma_b^2 + \sigma^2) - \left(\sigma_S^2 + \frac{\sigma_b^2\left(\sum t_k^2\right)}{m+\sigma^2}\right) = (m-1)\sigma_S^2 + \frac{\sigma_b^2\left(m^2\bar{t}^2 - \sum t_k^2\right)}{m}.$$

Note that

$$m^2\bar{t}^2 - \sum t_k^2 = \left(\sum t_k\right)^2 - \sum t_k^2 \geq 0,$$

unless $t_k \equiv t$ for some t. As a result, Δ will always be positive except for some extreme cases, which implies that κ will be larger than 1 except for some extreme cases. For example, $m = 1$ or $\sigma_S = \sigma_b = 0$. As a result, we can conclude that by applying a standard two-way ANOVA to a one-way ANOVA with repeated measures model, the type I error tends to be inflated.

Recall that the null hypothesis of no treatment effect should be rejected if $T > F_{1-\alpha,1,2m(n-1)}$ under model (13.2).

13.5 Simulation Study

A simulation study was conducted to confirm the conclusions drawn in the previous section. More specifically, the simulation was carried out by using SAS. The number of iterations was chosen to be 1000. The sample size per treatment group is set to be 15. It is assumed that $t_k = t; k = 1, \ldots, m$ for different

$m = 2, 4,$ or 8. For simplicity, we considered $\alpha_i = b_k = 0$ for $\forall i$. For different σ_S and σ_b values, data are generated according to model (13.1) and are analyzed by using a standard two-way ANOVA model. The significance level is chosen to be 5%. The empirical type I error rate is estimated by the proportion of the 1000 iterations, which mistakenly rejected the null hypothesis of no treatment difference (Figures 13.1 through 13.3). The results are summarized in Tables 13.1 through 13.3. The p-values were also plotted in

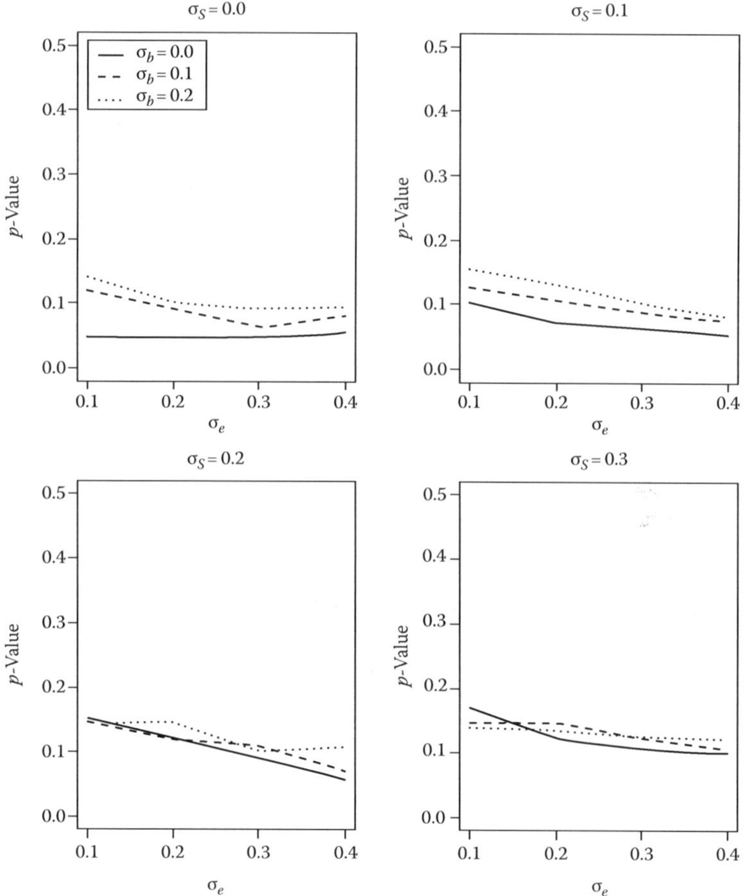

FIGURE 13.1
Empirical type I error ($m = 2$).

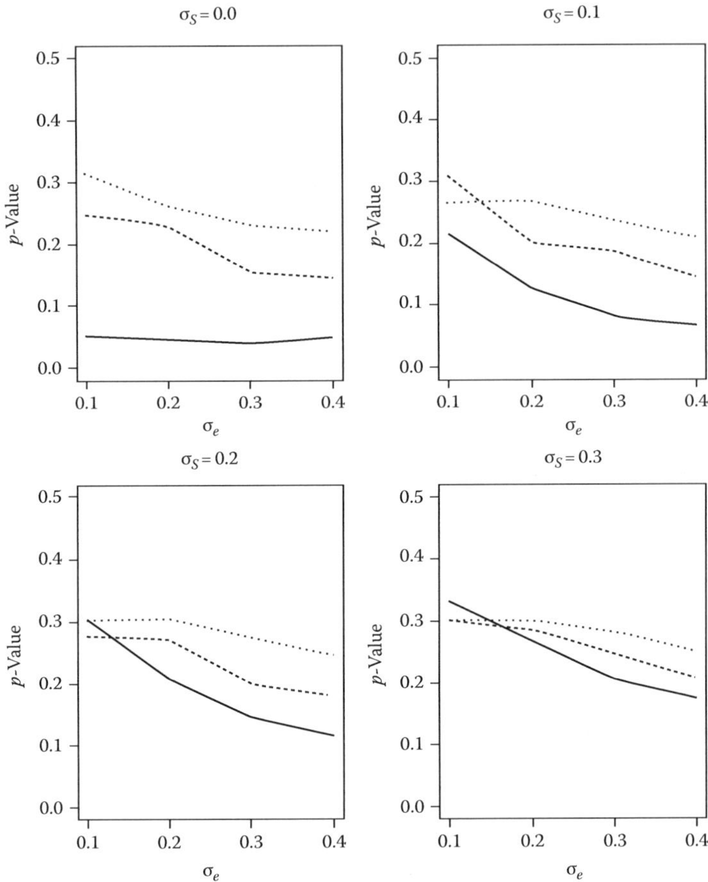

FIGURE 13.2
Empirical type I error ($m = 4$).

Tables 13.1 through 13.3. Based on the results, the following conclusions can be made:

1. When $\sigma_S = \sigma_b = 0$, then the empirical type I error rate is very close to the nominal level 5%. This is because under such a situation, the observations from the same subject but at different time points are independent with each other, which makes the two-way ANOVA a valid analysis.

2. When σ_S or σ_b increases, the type I error rate increases. This can be explained by noting that the variance of these two random variables implies how much dependence there is among the responses from the same subject. When σ_S and σ_b are small, then those responses

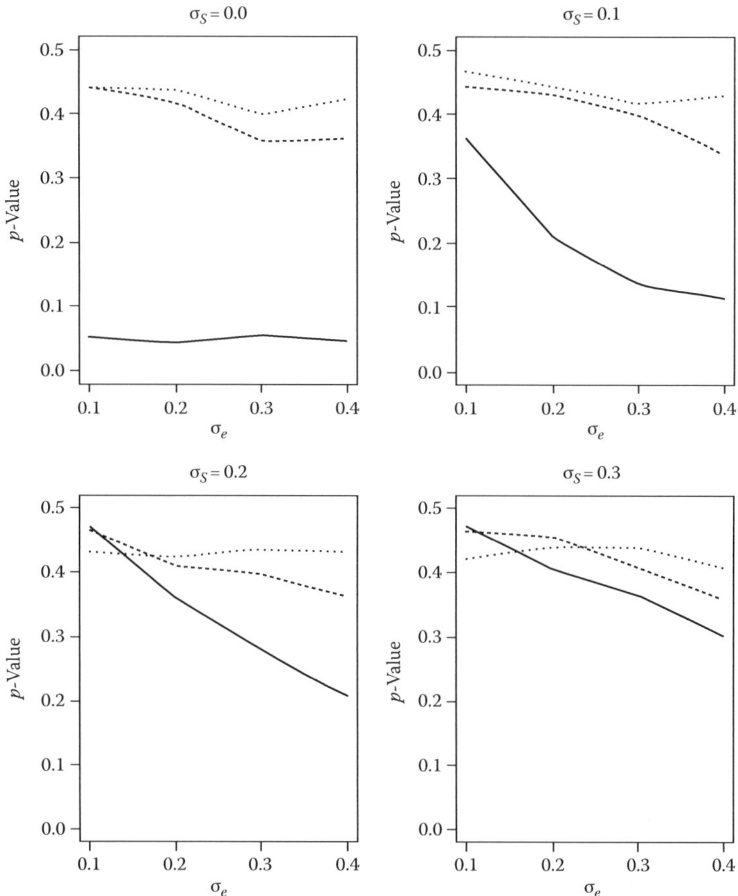

FIGURE 13.3
Empirical type I error ($m = 8$).

from the same subject may seem more like "independent," which makes the naive two-way ANOVA analysis approximately valid. On the other hand, the larger those variances are, the more dependencies there are among the responses from the same subject, which makes the empirical results more far away from the expected.

3. When σ_e increases, the type I error rate decreases toward the nominal level. This can be explained by noting the fact that if σ_e is very large, then σ_S and σ_b become relatively smaller, which implies that the observation from the same subject "looks" more independent. As a result, the empirical type I error rate becomes closer to the nominal level.

TABLE 13.1

Type I Error Rate with $m = 2$

σ_s		σ_e	p-Value	σ_s	σ_b	σ_e	p-Value
0.0	0.0	0.1	0.048	0.2	0.0	0.1	0.151
0.0	0.0	0.2	0.047	0.2	0.0	0.2	0.123
0.0	0.0	0.3	0.049	0.2	0.0	0.3	0.092
0.0	0.0	0.4	0.056	0.2	0.0	0.4	0.057
0.0	0.1	0.1	0.121	0.2	0.1	0.1	0.147
0.0	0.1	0.2	0.092	0.2	0.1	0.2	0.119
0.0	0.1	0.3	0.064	0.2	0.1	0.3	0.110
0.0	0.1	0.4	0.082	0.2	0.1	0.4	0.072
0.0	0.2	0.1	0.141	0.2	0.2	0.1	0.145
0.0	0.2	0.2	0.102	0.2	0.2	0.2	0.146
0.0	0.2	0.3	0.091	0.2	0.2	0.3	0.102
0.0	0.2	0.4	0.096	0.2	0.2	0.4	0.108
0.1	0.0	0.1	0.103	0.3	0.0	0.1	0.170
0.1	0.0	0.2	0.072	0.3	0.0	0.2	0.123
0.1	0.0	0.3	0.065	0.3	0.0	0.3	0.107
0.1	0.0	0.4	0.052	0.3	0.0	0.4	0.100
0.1	0.1	0.1	0.127	0.3	0.1	0.1	0.146
0.1	0.1	0.2	0.106	0.3	0.1	0.2	0.147
0.1	0.1	0.3	0.088	0.3	0.1	0.3	0.123
0.1	0.1	0.4	0.074	0.3	0.1	0.4	0.106
0.1	0.2	0.1	0.155	0.3	0.2	0.1	0.139
0.1	0.2	0.2	0.131	0.3	0.2	0.2	0.135
0.1	0.2	0.3	0.102	0.3	0.2	0.3	0.125
0.1	0.2	0.4	0.081	0.3	0.2	0.4	0.126

13.6 An Example

A two-arm parallel design with 10 repeated measures at equally spaced
time points was conducted to compare two compounds (a test treatment and
an active control) in terms of a clinical endpoint—illness score. A total of
30 patients (15 patients in each treatment group) were enrolled and completed
the study. The data are given in Table 13.4. The illness scores are plotted against
time by each patient in Figure 13.4. As it can be seen, for each patient, the
score follows approximately a straight line. As a result, model (13.1) becomes
a model of choice for this data set. Both tests for comparing intercepts and
slopes are carried out with no significant difference found (see Table 13.5).
However, if we ignore the fact that the observations from the same subject are
actually correlated and naively apply a standard two-way ANOVA analysis,
we can see a highly significant difference with p-value <0.001 (see Table 13.6).

TABLE 13.2

Type I Error Rate with $m = 4$

σ_s	σ_b	σ_e	p-Value	σ_s	σ_b	σ_e	p-Value
0.0	0.0	0.1	0.052	0.2	0.0	0.1	0.303
0.0	0.0	0.2	0.047	0.2	0.0	0.2	0.208
0.0	0.0	0.3	0.041	0.2	0.0	0.3	0.145
0.0	0.0	0.4	0.049	0.2	0.0	0.4	0.115
0.0	0.1	0.1	0.249	0.2	0.1	0.1	0.276
0.0	0.1	0.2	0.229	0.2	0.1	0.2	0.271
0.0	0.1	0.3	0.155	0.2	0.1	0.3	0.199
0.0	0.1	0.4	0.145	0.2	0.1	0.4	0.179
0.0	0.2	0.1	0.313	0.2	0.2	0.1	0.302
0.0	0.2	0.2	0.261	0.2	0.2	0.2	0.304
0.0	0.2	0.3	0.231	0.2	0.2	0.3	0.274
0.0	0.2	0.4	0.221	0.2	0.2	0.4	0.244
0.1	0.0	0.1	0.216	0.3	0.0	0.1	0.331
0.1	0.0	0.2	0.128	0.3	0.0	0.2	0.265
0.1	0.0	0.3	0.082	0.3	0.0	0.3	0.205
0.1	0.0	0.4	0.066	0.3	0.0	0.4	0.173
0.1	0.1	0.1	0.308	0.3	0.1	0.1	0.300
0.1	0.1	0.2	0.202	0.3	0.1	0.2	0.284
0.1	0.1	0.3	0.187	0.3	0.1	0.3	0.246
0.1	0.1	0.4	0.144	0.3	0.1	0.4	0.205
0.1	0.2	0.1	0.265	0.3	0.2	0.1	0.301
0.1	0.2	0.2	0.267	0.3	0.2	0.2	0.300
0.1	0.2	0.3	0.236	0.3	0.2	0.3	0.281
0.1	0.2	0.4	0.208	0.3	0.2	0.4	0.250

13.7 Discussion

In this chapter, we evaluated the consequences of the wrong use of a two-way ANOVA model when in fact the true model is a one-way ANOVA model with repeated measures in terms of type I error probability. It is found that the wrongly use of the two-way ANOVA model when the true model is a one-way ANOVA with repeated measures will inflate the type I error rate. The magnitude of the inflation depends not only on the variability of the subject-specific random effects and the random error but also on the number of the prespecified time points. In practice, it is strongly recommended that a correct statistical model be used for assessment of treatment difference under the valid study design.

TABLE 13.3

Type I Error Rate with $m = 8$

σ_s	σ_b	σ_e	p-Value	σ_s	σ_b	σ_e	p-Value
0.0	0.0	0.1	0.053	0.2	0.0	0.1	0.471
0.0	0.0	0.2	0.043	0.2	0.0	0.2	0.360
0.0	0.0	0.3	0.054	0.2	0.0	0.3	0.278
0.0	0.0	0.4	0.047	0.2	0.0	0.4	0.208
0.0	0.1	0.1	0.442	0.2	0.1	0.1	0.466
0.0	0.1	0.2	0.417	0.2	0.1	0.2	0.410
0.0	0.1	0.3	0.358	0.2	0.1	0.3	0.397
0.0	0.1	0.4	0.360	0.2	0.1	0.4	0.361
0.0	0.2	0.1	0.440	0.2	0.2	0.1	0.432
0.0	0.2	0.2	0.438	0.2	0.2	0.2	0.423
0.0	0.2	0.3	0.401	0.2	0.2	0.3	0.435
0.0	0.2	0.4	0.422	0.2	0.2	0.4	0.432
0.1	0.0	0.1	0.363	0.3	0.0	0.1	0.472
0.1	0.0	0.2	0.210	0.3	0.0	0.2	0.406
0.1	0.0	0.3	0.136	0.3	0.0	0.3	0.365
0.1	0.0	0.4	0.113	0.3	0.0	0.4	0.302
0.1	0.1	0.1	0.443	0.3	0.1	0.1	0.465
0.1	0.1	0.2	0.432	0.3	0.1	0.2	0.456
0.1	0.1	0.3	0.399	0.3	0.1	0.3	0.408
0.1	0.1	0.4	0.338	0.3	0.1	0.4	0.359
0.1	0.2	0.1	0.467	0.3	0.2	0.1	0.421
0.1	0.2	0.2	0.445	0.3	0.2	0.2	0.441
0.1	0.2	0.3	0.417	0.3	0.2	0.3	0.439
0.1	0.2	0.4	0.429	0.3	0.2	0.4	0.408

TABLE 13.4

One-Way ANOVA with Repeated Measures

Parameter	Treatment	N	Estimate	SD	p-Value
Intercept	Test	15	−0.1900	0.35087	0.3785
	Control	15	−0.0611	0.43365	
Slope	Test	15	−0.1638	0.60637	0.4707
	Control	15	0.0053	0.65840	

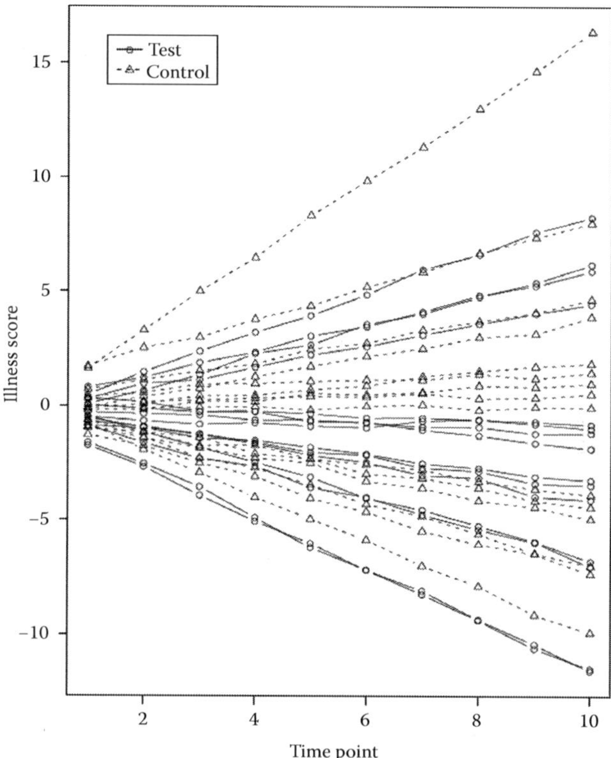

FIGURE 13.4

Illness score by time point.

TABLE 13.5

Naive Two-Way ANOVA

Source	DF	SS	*p*-Value
Treatment	1	84.02	0.0233
Time	9	15.63	0.9994
Treatment * time	9	17.76	0.9991
Error	280	4520.87	

TABLE 13.6

Data Listing for Example

Treatment	Subject	Time Point									
		1	2	3	4	5	6	7	8	9	10
Test	1	−0.884	−0.982	−1.420	−1.515	−1.820	−2.109	−2.521	−2.716	−3.089	−3.247
	2	−0.496	−0.673	−0.822	−0.755	−0.895	−0.961	−0.654	−0.596	−0.796	−1.001
	3	−0.549	−0.951	−1.270	−1.659	−2.171	−2.473	−3.033	−3.150	−3.958	−4.102
	4	0.820	1.228	1.885	2.343	3.058	3.465	4.122	4.852	5.259	5.911
	5	0.346	0.165	−0.263	−0.270	−0.695	−0.728	−1.049	−1.291	−1.618	−1.865
	6	−0.694	−0.918	−1.237	−1.595	−2.042	−2.155	−2.710	−2.850	−3.389	−3.470
	7	−0.293	−0.366	−0.422	−0.638	−0.625	−0.739	−0.918	−0.922	−1.206	−1.194
	8	0.083	0.662	1.355	2.286	2.659	3.571	4.039	4.773	5.391	6.202
	9	0.295	0.971	1.130	1.688	2.231	2.592	3.112	3.595	4.053	4.452
	10	0.032	−0.132	−0.284	−0.243	−0.356	−0.541	−0.532	−0.615	−0.694	−0.817
	11	−1.618	−2.534	−3.542	−4.886	−6.196	−7.176	−8.244	−9.380	−10.608	−11.491
	12	−1.739	−2.676	−3.919	−5.049	−6.01	−7.191	−8.090	−9.348	−10.423	−11.587
	13	0.481	1.476	2.381	3.213	3.945	4.863	5.982	6.640	7.612	8.262
	14	−0.446	−1.147	−1.827	−2.477	−3.118	−4.035	−4.538	−5.236	−5.924	−6.804
	15	−0.878	−1.622	−2.311	−2.635	−3.571	−3.996	−4.749	−5.402	−5.957	−7.023

Control											
	1	0.715	1.128	1.529	1.834	2.484	2.743	3.322	3.705	4.089	4.666
	2	1.731	2.539	2.993	3.787	4.370	5.204	5.863	6.692	7.381	8.014
	3	0.409	0.569	0.942	0.976	1.077	1.144	1.143	1.401	1.208	1.468
	4	−0.964	−1.179	−1.751	−1.728	−2.331	−2.566	−2.781	−3.304	−3.622	−3.888
	5	−0.660	−1.373	−2.296	−2.712	−3.454	−4.294	−4.817	−5.595	−6.449	−6.994
	6	−0.629	−1.181	−1.381	−2.101	−2.320	−2.976	−3.156	−3.576	−4.143	−4.429
	7	−0.092	0.404	0.760	1.280	1.729	2.171	2.514	3.011	3.160	3.898
	8	−0.930	−1.416	−1.847	−2.288	−2.503	−3.295	−3.587	−4.138	−4.414	−4.945
	9	1.639	3.301	5.008	6.489	8.333	9.842	11.322	13.021	14.654	16.375
	10	0.309	0.177	0.242	0.342	0.415	0.391	0.556	0.336	0.370	0.533
	11	−0.125	0.090	0.437	0.457	0.712	0.883	1.294	1.533	1.736	1.858
	12	−0.790	−1.925	−2.931	−3.998	−4.947	−5.859	−7.000	−7.887	−9.136	−9.914
	13	−0.202	−0.195	0.206	0.183	0.547	0.493	0.604	0.900	0.836	0.985
	14	0.048	−0.093	−0.072	−0.037	−0.191	−0.015	0.061	−0.188	−0.046	−0.071
	15	−1.276	−1.760	−2.504	−3.087	−4.048	−4.628	−5.468	−6.045	−6.481	−7.350

14

Validation of QOL Instruments

14.1 Introduction

In clinical research, an instrument (or questionnaire) is often used to provide a standardized and objective means of collecting data on subjective states or events across various therapeutic areas. In practice, although there exist many instruments such as Hamilton-D (Hamilton scale for depression) and Hamilton-A (Hamilton scale for anxiety) for central nervous system (CNS) and quality of life (QOL) assessment in cancer trials, the investigators frequently face the need to develop new ones. This need arises because a proper development and validation of the existing instruments was achieved for a specific purpose and relative to a specific target patient population. While the existing pool of instruments may contain one that has been developed for the target patient population and the desired purpose, new research questions often require new instruments for measurement. Validation of the developed instrument is important to ensure a proper sampling and a valid measurement of the content of the subjective state, behavior, or disease to be measured (Testa, 1987). For illustration purpose, in this chapter we will focus on the validation of QOL instruments. The performance characteristics for the validation of QOL instruments can be applied to other instruments for other purposes across therapeutic areas.

In cancer clinical trials, it has been a concern that the treatment of disease or survival may not be as important as the improvement of QOL, especially for patients with chronic disease. Enhancement of life beyond absence of illness to enjoyment of life is considered more important than the extension of life. In general, there exists no universal definition for QOL. It may vary from one patient population to another and from one therapeutic area to another. For example, Williams (1987) defined QOL as a collective term that encompasses multiple components of a person's social and medical status. However, Smith (1992) interpreted QOL as the way a person feels and how he or she functions in day-to-day activities. The concept of QOL can be traced back to the mid-1920s. Peabody (1927) pointed out that the clinical picture of a patient is an impressionistic painting of the patient surrounded by his or her home, work, friends, joys, sorrows, hopes, and fears. In 1947, the World Health Organization stated that health is a state of complete physical, medical, and social well-being and not merely the absence of disease or infirmity. In 1948, Karnofsky published his performance status index to assess the usefulness of chemotherapy for cancer patients. The New York Heart Association proposed a refined version of its functional classification to assess the effects of cardiovascular symptoms on the performance of physical activities in 1964. In the past several decades, QOL has attracted much attention. Since 1970, several research groups have been actively working on the assessment of QOL in clinical trials. For example, Kaplan et al. (1976) developed the Index of Well-Being to provide a comprehensive measure of QOL. Torrance (1976) and Torrance and Feeny (1989) introduced the concept of utility theory to measure the health state preferences of individuals and quality-adjusted life year to summarize both QOL and quantity of life. Bergner et al. (1981) developed the Sickness Impact Profile to study perceived health and sickness-related QOL. Ware (1987) proposed a set of widely used scales for the Rand Health Insurance Experiment and Williams (1987) studied the effects of QOL on hypertensive patients.

QOL not only can provide information as to how patients feel about drug therapies but it also appeals to the physician's desire for the best clinical practice. It can be used as a predictor of compliance of the patient. In addition, it may be used to distinguish between therapies that appear to be equally efficacious and equally safe at the stage of marketing strategy planning. The information can be potentially used in advertising for the promotion of the drug therapy. However, unlike the analytic instrument, there exist no known standards that can be used as reference. In addition, the QOL instrument is a very subjective tool, which is expected to have a large variation. It is then a concern as to whether the adopted QOL instrument can accurately and reliably quantify patients' QOL. To ensure the accuracy and reliability of QOL assessment in clinical trials, the adopted QOL instrument is necessarily validated in terms of some performance characteristics. In practice, a QOL instrument is usually validated based on some classic validation parameters such as validity, reliability, test–retest reproducibility, responsiveness, and sensitivity. However, it is not clear whether the classic validation can actually verify the instrument. In other words, can the classic validation address whether the questions are the right ones for the assessment of QOL?

In the next section, we briefly review statistical methods for QOL assessment. In Section 14.3, we provide statistical evaluation for the validation of a QOL instrument in terms of performance characteristics of validity, reliability, and test–retest reproducibility. Responsiveness and sensitivity are discussed in Section 14.4. The validation of utility analysis and calibration is discussed in Section 14.5. The controversial issue concerning the use of a parallel questionnaire for the assessment of QOL is discussed in Section 14.6. A brief discussion concerning some statistical tests that may occur in QOL assessment is given in the last section.

14.2 QOL Assessment

In clinical trials, QOL is usually assessed by means of a global physician's assessment or a QOL instrument that consists of a number of questions. The global physician's assessment, such as an analog scale ranging from 0 to 10, is easy to apply by simply asking the question "How is your QOL?" However, it cannot capture the whole spectrum of QOL. In addition, if the drug therapy does improve QOL, no information as to which domain of QOL has improved is provided. The global physician's assessment generally produces large variability and low reproducibility. For a QOL instrument, the questionnaire may be assessed by patients, their spouses/significant others, reviewers (e.g., nurses or social workers), and/or physicians through direct observation or face-to-face or teleconference interview. It can be self-administered or supervised self-administered. Based on the collected data, the health-related QOL can be quantified. Generally, health-related QOL may be described by a number of major domains (or dimensions). The most commonly considered QOL domains include physical functioning and morbidity, emotional or psychological status and well-being, disease-specific symptoms and somatic discomfort, and cognitive function. Other domains, such as intimacy and sexual functioning, economic status and personal productivity, employment, and laboratory test values, are less often used.

In the QOL instrument, each patient score associated with each question is usually referred to as an item. In practice, there may be a larger number of items and it is not practical to analyze the data by item. Thus, items are usually grouped to form subscales, which are often used to evaluate different components of QOL. However, analysis of individual subscales often produces inconsistent results across subscales; consequently, no overall conclusion can be made. As an alternative, these subscales may be combined to form the so-called composite scores, which can be used to assess major domains of QOL.

As a result, QOL may be assessed by analyzing items, subscales, composite scores, and/or the total score. Tandon (1990) applied global statistics to combine the results of a univariate analysis of each subscale. His approach is useful, yet it does not reveal the underlying correlation structure of subscales. As an alternative approach, Olschewski and Schumacher (1990) proposed the use of aggregated measures to reduce the dimension of the measurements. Their method uses the standardized scoring coefficients from factor analysis as data-oriented weights for combining subscales, which neglects small coefficients. The disadvantage of their method is that the selected coefficients are neither unique nor have optimal properties. To overcome these problems, Ki and Chow (1995) suggest the use of factor analysis in conjunction with the analysis of principal components for combining subscales. The proposed method provides statistical justification for the use of composite scores.

14.3 Performance Characteristics

In practice, commonly considered performance characteristics for the validation of an instrument include, but are not limited to, accuracy (or validity), precision (or reliability), and reproducibility (see, USP/NF, 2000; NCCLS, 2001), which are briefly described below.

14.3.1 Validity

The validity of a QOL instrument is defined as the extent to which the QOL instrument measures what it is designed to measure. In other words, it is a measure of biasedness of the instrument. The biasedness of an instrument can reflect the accuracy of the instrument.

In clinical trials, as indicated earlier, the QOL of a patient is usually quantified based on responses to a number of questions related to several components or dimensions of QOL. It is a concern that the questions may not be the right questions to assess the components or domains of QOL of interest. To address this concern, consider a specific component (or domain) of QOL that consists of K items (or subscales), i.e., X_i, $i = 1, \ldots, K$. Also, let Y be the QOL component (or domain) of interest that is unobservable. Suppose that Y follows a normal distribution with mean θ and variance τ^2 and can be quantified by X_i, $i = 1, \ldots, K$. In other words, there exists a function f such that

$$f(X) = f(X_1, X_2, \ldots, X_K) = Y,$$

where $X = (X_1, X_2, \ldots, X_K)'$. Suppose X follows a distribution with mean $\mu = (\mu_1, \ldots, \mu_K)$ and variance Σ. Thus, θ can be estimated by

$$\hat{\theta} = \hat{f}(X_1, X_2, \ldots, X_K),$$

and the bias is given by

$$Bias(\hat{\theta}) = E(\hat{\theta} - \theta) = E(\hat{f}(X_1, X_2, \ldots, X_K)) - \theta.$$

In practice, for convenience, the unknown function f is usually assumed to be the mean of X_i, i.e.,

$$f(X) = f(X_1, X_2, \ldots, X_K) = \frac{1}{K}\sum_{i=1}^{K} X_i = Y.$$

Thus, it is desired to have the mean of X_i, $i = 1,...,K$ close to θ and $\bar{\mu} = 1/K \sum_{i=1}^{K} \mu_i = \theta$. As a result, we may claim that the instrument is validated in terms of its validity if

$$|\mu_i - \bar{\mu}| < \delta \quad \forall i = 1, ..., K.$$

To verify this, we may consider a simultaneous confidence interval for $\mu_i - \bar{\mu}$, $i = 1,...,K$. Let $\mu_i - \bar{\mu} = a_i'\mu$, where $a_i' = (-(1/K)1_{i-1}, 1-(1/K), -(1/K)1_{K-i})$. Suppose the QOL questionnaire is administered to N patients, let

$$\hat{\mu} = \frac{1}{N} \sum_{j=1}^{N} X_j = \bar{X}.$$

Then, the $(1 - \alpha)100\%$ simultaneous confidence intervals for $\mu_i - \bar{\mu}$, $i = 1,...,K$ are given by

$$a_i'\hat{\mu} - \sqrt{\frac{1}{N} a_i'Sa_i} T(\alpha, K, N-K) \le a_i'\mu \le a_i'\hat{\mu} + \sqrt{\frac{1}{N} a_i'Sa_i} T(\alpha, K, N-K), \quad i = 1, ..., K,$$

where

$$S = \frac{1}{N-1} \sum_{j=1}^{N} (X_j - \bar{X})(X_j - \bar{X})',$$

$$T^2(\alpha, K, N-K) = \frac{(N-1)K}{N-K} F(\alpha, K, N-K),$$

$$P(T^2(K, N-K) \le T^2(\alpha, K, N-K)) = 1 - \alpha.$$

We may also consider a Bonferroni adjustment of an overall α level as follows:

$$a_i'\hat{\mu} - \sqrt{\frac{1}{N} a_i'Sa_i} T\left(\frac{\alpha}{2K}, N-1\right) \le a_i'\mu \le a_i'\hat{\mu} + \sqrt{\frac{1}{N} a_i'Sa_i} T\left(\frac{\alpha}{2K}, N-1\right).$$

We then compare the confidence interval with $(-\delta, \delta)$ and reject the null hypothesis that

$$H_0 : |\mu_i - \bar{\mu}| < \delta \quad \forall i = 1, ..., K$$

if any confidence interval falls completely outside $(-\delta, \delta)$.

Note that it is also important to establish concurrent validity in practice. Concurrent validity is established for a new instrument by demonstrating a good correlation with an already existing tool that is widely accepted as measuring the same construct(s). For example, a clinical evaluation by a physician might be considered the *gold standard* diagnosis. Or a well-recognized and accepted diagnostic criterion that is widely used by practitioners in the area may exist. In the case that the existing tool is continuous, a Pearson or Spearman's correlation coefficient can be computed. In the case of a dichotomous diagnostic tool, the area under the receiver operating characteristics curve can be used to establish a good relationship. If no such *gold standard* exists, one option is to compare the new instrument to existing instruments that measure similar or related constructs. This is referred to as convergent validity (discussed later). For example, all instruments in the same general area of general well-being should show some degree of relationship. Furthermore, all instruments or domains designed to measure general well-being should show a somewhat closer relationship to one another than they do to other domains, even if they have been developed on varying populations.

14.3.2 Reliability

The reliability of a QOL instrument reflects the other part of measurements and refers to the freedom from random error. The reliability of an instrument measures the variability of the instrument, which directly relates to the precision of the instrument. Therefore, the items are considered reliable if the variance of Y is small. To verify the reliability of estimating θ by Y, we consider the following hypothesis:

$$H_0: \mathrm{Var}(Y) < \Delta \quad \text{for some fixed } \Delta.$$

The variance of Y is given below:

$$\mathrm{Var}(Y) = \mathrm{Var}\left(\frac{1}{K}\sum_{i=1}^{K} X_i\right) = \frac{1}{K^2} 1'\Sigma 1.$$

The sample distribution of

$$\sum_{j=1}^{N} \frac{(Y_j - \bar{Y})}{\mathrm{Var}(Y)}$$

has a chi-square distribution with $N - 1$ degrees of freedom. Thus, a $(1 - \alpha)100\%$ one-sided confidence interval for $\mathrm{Var}(Y)$ is as follows:

$$\mathrm{Var}(Y) \geq \frac{\sum_{j=1}^{N}(Y_j - \bar{Y})^2}{\chi^2(\alpha, N-1)} = \xi(Y).$$

If $\xi(Y) > \Delta$, then we reject the null hypothesis and conclude that the items are not reliable in estimating θ. As indicated earlier, patients' response to a QOL instrument may vary from one patient population to another and from one therapy to another. Therefore, it is recommended that the variability of QOL scores be studied before and after medication intervention.

Since the items $X_1, X_2, \ldots, X_K$ are relevant to a QOL component, they are expected to be correlated. In classical validation, a group of items with high intercorrelation between items are considered to be internal-consistent. Cronbach's α defined below is often used to measure the intercorrelations between items:

$$\alpha_C = \frac{K}{K-1}\left(1 - \frac{\sum_{i=1}^{K}\sigma_i^2}{\sum_{i=1}^{K}\sigma_i^2 + 2\sum\sum_{i<l}\sigma_{il}}\right),$$

where
$$\sigma_i^2 = \mathrm{Var}(X_i)$$
$$\sigma_{il} = \mathrm{Cov}(X_i, X_l)$$

When the covariance between items is high compared to the variance of each item, α_C is large. To ensure that the items are measuring the same component of QOL, the items under the component should be positively correlated, i.e., $\alpha_C \geq 50\%$. However, if the intercorrelations between items are too high, i.e., α_C is close to 1, it suggests that some of the items are redundant. Note that the variance of Y given below:

$$\mathrm{Var}(Y) = \left(\frac{1}{K-(K-1)\alpha_C}\right)\frac{1}{K}\sum_{i=1}^{K}\mathrm{Var}(X_i)$$

increases with α_C for fixed K and $\mathrm{Var}(X_i)$, $i = 1, \ldots, K$. By including redundant items we cannot improve the precision of the result. It is desired to have independent items reflect the QOL component at different perspectives. However, in that case, it is hard to validate whether the items are measuring the same targeted component of QOL. Therefore, we suggest using items with moderate α_C, i.e., α_C is somewhere between 50% and 80%.

14.3.3 Reproducibility

Reproducibility is defined as the extent to which repeat administrations of the same QOL measure yield the same result, assuming no underlying changes have occurred. The assessment of reproducibility involves expected and/or unexpected variabilities that might occur in the assessment of QOL. It includes inter-time (between time) point and inter-rater (between rater) reproducibility.

For the assessment of reproducibility, the technique of test–retest is often employed. The same QOL instrument is administered to patients who have reached stable conditions at two different time points. These two time points are generally separated by a sufficient length of time that is long enough to wear off the memory of the previous evaluation but not long enough to allow any change in environment. Pearson's product moment correlation coefficient, ρ, of the two repeated results is then studied. In practice, a test–retest correlation of 80% or higher is considered acceptable. To verify this, the sample correlation between test–retest, denoted by r, is calculated from a sample of N patients. The following hypotheses are then tested:

$$H_0 : \rho \geq \rho_0 \quad \text{versus} \quad H_a : \rho < \rho_0.$$

When N is large,

$$Z(r) = \frac{1}{2} \ln\left(\frac{1+r}{1-r}\right)$$

is approximately normally distributed with mean $Z(\rho)$ and variance $1/(N-3)$. The null hypothesis is rejected if

$$\sqrt{N-3}\,(Z(r) - Z(\rho_0)) < z(1-\alpha),$$

where $z(1 - \alpha)$ is the αth quantile of the standard normal distribution. Note that a shift in mean of the score at test–retest may be detected by using a simple paired t-test. The inter-rater reproducibility can be verified by the same method.

14.4 Responsiveness and Sensitivity

The responsiveness of a QOL instrument is usually referred to as the ability of the instrument to detect a difference of clinical significance within a treatment. The sensitivity is a measure of the ability of the instrument to detect a clinically significant difference between treatments. A validated instrument should be able to detect a difference if there is indeed a difference and should not wrongly detect a difference if there is no difference. Chow and Ki (1994) proposed precision and power indices to assess the responsiveness and sensitivity of a QOL instrument when comparing the effect of a drug on QOL between treatments. The precision index measures the probability of not detecting a false difference and the power index reflects the probability of detecting a meaningful difference. The precision and power indices for measuring responsiveness and sensitivity under a time series model proposed by Chow and Ki (1994) are described below.

14.4.1 Statistical Model

For a given QOL index, let X_{ijt} be the response of the ith subject to the jth question (item) at time t, where $i = 1, \ldots, N$, $j = 1, \ldots, J$, and $t = 1, \ldots, T$. Consider the average score over J questions:

$$y_{it} = \bar{X}_{it} = \frac{1}{J} \sum_{j=1}^{J} X_{ijt}.$$

Since the average scores $y_{i1}, \ldots, y_{iT}$ are correlated, the following autoregressive time series model is an appropriate statistical model for y_{it}.

$$y_{it} = \mu + \psi(y_{i,t-1} - \mu) + e_{it}, \quad i = 1, \ldots, N, \quad t = 1, \ldots, T, \tag{14.1}$$

where

μ is the overall mean

$|\psi| < 1$ is the autoregressive parameter

e_{it} are independent identically distributed random errors with mean 0 and variance σ_e^2

It can be verified that $E(e_{it}, e_{jt'}) = 0$ for all i, j and $t \neq t'$, and $E(e_{it}, y_{it'}) = 0$ for all $t' < t$. The autoregressive parameter ψ can be used to assess the correlation of consecutive responses y_{it} and $y_{i,t+1}$. From the above model, it can be shown that the autocorrelation of response with k lag times is ψ^k, which is negligible when k is large. Based on the observed average scores on the ith subject, i.e., $y_{i1}, y_{i2}, \ldots, y_{iT}$, we can estimate the overall mean μ and the autoregressive parameter ψ. The ordinary least squares estimators of μ and ψ can be approximated by

$$\hat{\mu}_i = \bar{y}_{i\cdot},$$

$$\hat{\psi}_i = \frac{\sum_{t=2}^{T} (y_{it} - \bar{y}_{i\cdot})(y_{i,t-1} - \bar{y}_{i\cdot})}{\sum_{t=2}^{T} (y_{it} - \bar{y}_{i\cdot})^2} = r_{it},$$

which are the sample mean and sample autocorrelation of consecutive observations. Under model (14.1), it can be verified that the variance of $\hat{\mu}_i$ is

$$\text{Var}(\bar{y}_{i\cdot}) = \frac{\gamma_{i0}}{T} \left[1 + 2 \sum_{k=1}^{T-1} \frac{T-k}{T} \psi^k \right],$$

where $\gamma_{i0} = \mathrm{Var}(y_{it})$. The standard error of $\hat{\mu}_i$ is then given by

$$s(\bar{y}_{i\cdot}) = \left[\frac{c_{i0}}{T}\left(1 + 2\sum_{k=1}^{T-1}\frac{T-k}{T}r_{i1}^{k}\right)\right]^{1/2},$$

in which

$$c_{i0} = \sum_{i=1}^{T}\frac{(y_{it}-\bar{y}_{i\cdot})^2}{T-1}.$$

Suppose that the N subjects are from the same population with the same variability and autocorrelation. The QOL measurements of these subjects can be used to estimate the mean average scores μ. An intuitive estimator of μ is the sample mean

$$\hat{\mu} = \bar{y}_{\cdot\cdot} = \frac{1}{N}\sum_{i=1}^{N}\bar{y}_{i\cdot}.$$

Under model (14.1), the variance and standard error of $\hat{\mu}$ are given by, respectively,

$$\mathrm{Var}(\bar{y}_{\cdot\cdot}) = \frac{1}{N^2}\sum_{i=1}^{N}\mathrm{Var}(\bar{y}_{i\cdot}),$$

and

$$s(\bar{y}_{\cdot\cdot}) = \frac{1}{N}\left\{\sum_{i=1}^{N}[s(\bar{y}_{i\cdot})]^2\right\}^{1/2} = \frac{1}{N}\left\{\frac{c_0}{T}\sum_{i=1}^{N}\left[1 + 2\sum_{k=1}^{T-1}\frac{T-k}{T}r_1^{k}\right]\right\}^{1/2}$$

$$= \left\{\frac{c_0}{NT}\left[1 + 2\sum_{k=1}^{T-1}\frac{T-k}{T}r_1^{k}\right]\right\}^{1/2}, \tag{14.2}$$

where

$$c_0 = \frac{1}{N(T-1)}\sum_{i=1}^{N}\left[\sum_{t=1}^{T}(y_{it}-\bar{y}_{i\cdot})^2\right],$$

$$r_1 = \frac{1}{N}\sum_{i=1}^{N}r_{i1}.$$

As a result, an approximate $(1 - \alpha)100\%$ confidence interval for μ in (14.1) can then be constructed as follows:

$$\bar{y}_{..} \pm z_{1-1/2\alpha}\, s(\bar{y}_{..}), \tag{14.3}$$

where $z_{1-1/2\alpha}$ is the $100(1 - \alpha/2)$th percentile of a standard normal distribution and $s(\bar{y}_{..})$ is given in (14.2). As a matter of fact, the assumption that all the QOL measurements over time are independent is a special case of model (14.1) with $\psi = 0$. In practice, it is suggested that model (14.1) be used to account for the possible positive correlation between measurements over a test time period. Under model (14.1), it can be seen that the confidence interval given in (14.3) with $r_1 > 0$ is wider than when $\psi = 0$.

Note that, for the statistical validation of an instrument, at a fixed confidence level, the width of the confidence interval in (14.3) is inversely proportional to the prevision of the estimator $\bar{y}_{..}$ and may be used as an indicator of the validity of the instrument. For example, if the width of a confidence interval is too wide, the instrument may not be sensitive due to low power for detecting a positive difference or an equivalence. In what follows, the precision and power indices of QOL instruments will be evaluated under model (14.1).

14.4.2 Precision Index

Suppose that a homogeneous group is divided into two independent groups A and B that are known to have the same QOL. A *good* QOL instrument should have a small chance of *wrongly* detecting a difference. Let $y_i = (y_{i1}, y_{i2}, \ldots, y_{it}, \ldots, y_{iT})'$ be the average scores observed on the ith subject in group A at different time points over a fixed time period. Similarly, denote the average scores for the jth subject in group B over a time period by $w_j = (w_{j1}, w_{j2}, \ldots, w_{jt}, \ldots, w_{jT})'$. The objective is to compare mean average scores between groups to see whether the instrument reflects the expected result statistically. Based on y_i, $i = 1, \ldots, N$ and w_j, $j = 1, \ldots, M$, the difference in mean average scores between groups A and B can be assessed by testing the following hypotheses that $H_0 : \mu_y = \mu_w$ versus $H_a : \mu_y \neq \mu_w$, where μ_y and μ_w are the mean average scores for groups A and B, respectively. Under the null hypothesis, the following test statistic

$$Z = \frac{\bar{y}_{..} - \bar{w}_{..}}{\left[s^2(\bar{y}_{..}) + s^2(\bar{w}_{..})\right]^{1/2}}$$

is approximately distributed as a standard normal distribution when N and M are both large. Therefore, we would reject the null hypothesis if $|Z| > z_{1-\alpha/2}$.

Note that the above test is a uniform most powerful test. The level of significance of the test is α. The confidence interval for $\mu_y - \mu_w$ and the rejection region are given, respectively, by

$$(L, U) = \bar{y}_{..} - \bar{w}_{..} \pm d_\alpha \quad \text{and} \quad \left|\bar{y}_{..} - \bar{w}_{..}\right| > d_\alpha,$$

where

$$d_\alpha = z_{1-\alpha/2}\left[s^2(\bar{y}_{..}) + s^2(\bar{w}_{..})\right]^{1/2}.$$

In general, an interval estimator of $\mu_y - \mu_w$ given by

$$(\bar{y}_{..} - \bar{w}_{..}) \pm d \tag{14.4}$$

is used for detecting a difference in means. A difference is detected if zero lies outside the interval, i.e.,

$$\left|\bar{y}_{..} - \bar{w}_{..}\right| > d.$$

The precision index, denoted by P_d, of an instrument is defined as the probability of the interval (14.4) not detecting a difference when there is no difference between groups, i.e.,

$$P_d = P\left\{\left|\bar{y}_{..} - \bar{w}_{..}\right| \le d \,|\, \mu_y = \mu_w\right\} = P\left\{|Z| \le d[\sigma^2(\bar{y}_{..}) + \sigma^2(\bar{w}_{..})]^{-1/2}\right\}, \tag{14.5}$$

where

$$Z = \frac{(\bar{y}_{..} - \bar{w}_{..}) - (\mu_y - \mu_w)}{\left[\sigma^2(\bar{y}_{..}) + \sigma^2(\bar{w}_{..})\right]^{1/2}}$$

is the standardized random variable that is approximately distributed as a standard normal when N and M are large. It can be seen that the precision index of an instrument is $(1 - \alpha)$ at $d = d_\alpha$. Note that P_d is the confidence level of the interval estimator given in (14.4), which increases as d increases. When d is too big, although the interval has a very high probability to capture the true difference, it may not have a sufficient power for detecting a positive difference.

14.4.3 Power Index

On the other hand, if the QOL instrument is administered to two groups of subjects who are known to have different QOL, then the QOL instrument should be able to correctly detect such a difference with a high probability. The power index of an instrument for detecting a meaningful difference, denoted by $\delta_d(\varepsilon)$, is defined as the probability of detecting a meaningful difference ε. That is,

$$\delta_d(\varepsilon) = P\left\{\left|\bar{y}_{..} - \bar{w}_{..}\right| > d \,|\, \left|\mu_y - \mu_w\right| = \varepsilon\right\}$$

$$= P\left\{Z > (d-\varepsilon)[\sigma^2(\bar{y}_{..}) + \sigma^2(\bar{w}_{..})]^{-1/2}\right\} + P\left\{Z < -(d+\varepsilon)[\sigma^2(\bar{y}_{..}) + \sigma^2(\bar{w}_{..})]^{-1/2}\right\}.$$

$$\tag{14.6}$$

For $d = d_\alpha$, $\delta_d(\varepsilon)$ is the power, which can be calculated as follows:

$$\delta_d(\varepsilon) = P\left\{|\bar{y}_{..} - \bar{w}_{..}| > z_{1-\alpha/2}[s^2(\bar{y}_{..}) + s^2(\bar{w}_{..})]^{1/2} \,\big|\, |\mu_y - \mu_w| = \varepsilon\right\}$$

$$\doteq P\left\{Z < -z_{1-\alpha/2} - \frac{\varepsilon}{\left[s^2(\bar{y}_{..}) + s^2(\bar{w}_{..})\right]^{1/2}}\right\} + P\left\{Z > z_{1-\alpha/2} - \frac{\varepsilon}{\left[s^2(\bar{y}_{..}) + s^2(\bar{w}_{..})\right]^{1/2}}\right\}.$$

$$(14.7)$$

Note that for a fixed ε, $\delta_d(\varepsilon)$ decreases as d increases. We consider an instrument to be responsive in detecting a difference if both P_d and $\delta_d(\varepsilon)$ are above some reasonable limits for a given ε.

In practice, two groups are considered to have equivalent QOL if their mean QOL measurements only differ by less than a meaningful difference η. In this case, it is of interest to detect equivalence rather than a difference. Denote the acceptable limits for the difference between two group means by $(-\Delta, \Delta)$. When the confidence interval of $\mu_y - \mu_w$ given in (14.4) is within the acceptable limits, we conclude that the two groups have equivalent effect on QOL. We will refer to the probability of detecting an equivalence as the power index of an instrument for detecting an equivalence when the true group means differ by less than a meaningful difference η. The power index is then defined as

$$\phi_\Delta(\eta) = \mathop{\mathrm{Inf}}_{|\mu_y - \mu_w|} P\left\{(L, U) \subset (-\Delta, \Delta) \,\big|\, |\mu_y - \mu_w| < \eta\right\}$$

$$= P\left\{(L, U) \subset (-\Delta, \Delta) \,\big|\, \mu_y - \mu_w = \eta\right\},$$

where (L, U) is a confidence interval of $\mu_y - \mu_w$ as given in (14.4). Note that $\phi_\Delta(\eta)$ can be obtained as follows:

$$\phi_\Delta(\eta) = P\left\{(L, U) \subset (-\Delta, \Delta) \,\big|\, \mu_y - \mu_w = \eta\right\}$$

$$= P\left\{(\bar{y}_{..} - \bar{w}_{..} - d, \bar{y}_{..} - \bar{w}_{..} + d) \subset (-\Delta, \Delta) \,\big|\, \mu_y - \mu_w = \eta\right\}$$

$$= P\left\{\frac{(\bar{y}_{..} - \bar{w}_{..}) - \eta}{[\sigma^2(\bar{y}_{..}) + \sigma^2(\bar{w}_{..})]^{1/2}} > \frac{-(\Delta - d) - \eta}{[\sigma^2(\bar{y}_{..}) + \sigma^2(\bar{w}_{..})]^{1/2}} \quad \text{and}\right.$$

$$\left.\frac{(\bar{y}_{..} - \bar{w}_{..}) - \eta}{[\sigma^2(\bar{y}_{..}) + \sigma^2(\bar{w}_{..})]^{1/2}} < \frac{(\Delta - d) - \eta}{[\sigma^2(\bar{y}_{..}) + \sigma^2(\bar{w}_{..})]^{1/2}}\right\}$$

$$= P\left\{\frac{-(\Delta - d) - \eta}{[\sigma^2(\bar{y}_{..}) + \sigma^2(\bar{w}_{..})]^{1/2}} < Z < \frac{(\Delta - d) - \eta}{[\sigma^2(\bar{y}_{..}) + \sigma^2(\bar{w}_{..})]^{1/2}}\right\}, \qquad (14.8)$$

which can be approximated by

$$\phi_\Delta(\eta) \doteq \Phi\left[\frac{(\Delta-d)-\eta}{[\sigma^2(\bar{y}_{..})+\sigma^2(\bar{w}_{..})]^{1/2}}\right] - \Phi\left[\frac{-(\Delta-d)-\eta}{[\sigma^2(\bar{y}_{..})+\sigma^2(\bar{w}_{..})]^{1/2}}\right],$$

where Φ is the cumulative distribution function of a standard normal distribution.

14.4.4 Sample Size Determination

Since the QOL response may vary widely from patient to patient, a large sample size is usually required to attain a reasonable precision and power. Under model (14.1) and the setting as described above, some useful formulae for determination of sample size can be derived based on normal approximation. The formulae can also be applied to many clinical research studies with time-correlated outcome measurements, for example, 24 h monitoring of blood pressure, heart rates, hormone levels, and body temperature.

For a fixed precision index (e.g., $1 - \alpha$), to ensure a reasonably high power index δ for detecting a meaningful difference ε, the sample size per treatment group should not be less than

$$N_\delta = \frac{c[z_{1-\alpha/2}+z_\delta]^2}{\varepsilon^2} \quad \text{for } \delta > 0.5, \tag{14.9}$$

where

$$c = \frac{\gamma_y}{T}\left[1+2\sum_{k=1}^{T-1}\frac{T-k}{T}\psi_y^k\right] + \frac{\gamma_w}{T}\left[1+2\sum_{k=1}^{T-1}\frac{T-k}{T}\psi_w^k\right].$$

For a fixed precision index (e.g., $1 - \alpha$), if the acceptable limit for detecting an equivalence between two treatment means is $(-\Delta, \Delta)$, to ensure a reasonably high power ϕ for detecting an equivalence when the true difference in treatment means is less than a small constant η, the sample size for each treatment group should be at least

$$N_\phi = \frac{c}{(\Delta-\eta)^2}[z_{1/2+1/2\phi}+z_{1-\alpha/2}]^2. \tag{14.10}$$

If both treatment groups are assumed to have some variability and autocorrelation coefficient, the constant c in (14.9) and (14.10) can be simplified as

$$c = \frac{2\gamma}{T}\left[1+2\sum_{k=1}^{T-1}\frac{T-k}{T}\psi^k\right].$$

When $N = \max(N_\phi, N_\delta)$, it ensures that the QOL instrument will have a precision index $1 - \alpha$ and power of no less than δ and ϕ in detecting a difference and an equivalence, respectively. It should be noted that the required sample size is proportional to the variability of the average scores considered. The higher the variability, the larger is the sample size that would be required.

As an example, suppose that there are two independent groups A and B. A QOL index containing 11 questions is administered to subjects at Weeks 4, 8, 12, and 16. The mean scores are analyzed to assess group difference. Denote the mean of QOL score of the subjects in group A and B by Y_{it} and W_{jt}, respectively, where $i, j = 1, ..., N$ and $t = 1, 2, 3, 4$. We assume that Y_{it} and W_{jt} have distributions that follow the time series model described in model (14.1) with common variance $\gamma = 0.5$ square units and have moderate autocorrelation between scores at consecutive time points, say $\psi = 0.5$. For a fixed 95% precision index, by formula (14.9), 87 subjects per group will provide a 90% power for detection of a difference of 0.25 units in means. If the chosen acceptable limits are $(-0.35, 0.35)$, by (14.10), 108 subjects per group will have a power of 90% that the 95% confidence interval of difference in group means will correctly detect an equivalence with $\eta = 0.1$ units. If the sample size is chosen to be 108 per group, it ensures that the power indices for detecting a difference of 0.25 units or equivalence are not less than 90%.

14.5 Utility Analysis and Calibration

14.5.1 Utility Analysis

Gains in quantity of life can be measured in terms of life years gained, while gains in QOL should be measured by an instrument that incorporates a broad spectrum of health status, including physical/mobility function, psychological function, cognitive function, social function, and so forth. Feeny and Torrance (1989) used a utility approach to measure the health-related QOL. Utility is a single summary score, which ranges from zero (for dead) to one (for perfect health). Torrance and Feeny (1989) used QOL utility as quantity-adjustment weights for quality-adjusted life years, which are highly used in cost-effectiveness analysis.

The utility of hypothetical or actual health states may be evaluated by an individual. Utility is the preference of an individual for a health state. The preference of health state can be measured by some standard technique, such as rating scale, standard gamble, and time tradeoff. However, the utility measurements are not very precise. The within-subject variability is around 0.13 and the intersubject variability is approximately 0.3 for the general public and 0.2 for patients experiencing the health state (Feeny and Torrance, 1989). An individual either is experiencing the disease state or understands the hypothetical description of the disease state. A rating scale consists of a line with the least preferred state (e.g., death) on one end and the most preferred state (perfect health) on the other end. An individual will rate the disease state on the line between these two extreme states. Usually, the utility value obtained by this technique has high variability. A utility value of a disease state can be assigned by the standard gamble technique. An individual is given the choice of remaining at the disease state for an additional t years or the alternative, which consists of perfect health for an additional t years with probability p and immediate death with probability $(1 - p)$. The probability p is varied until the individual is indifferent between the two alternatives. Then the preference/utility of that disease state is p. The preference value of a disease state can also be assigned by using a time tradeoff technique. An individual is offered two alternatives: (1) a disease state with a life expectancy of t years or (2) perfect health for x years. Then x is varied until the individual is indifferent regarding the two alternatives. Then, the preference value of the disease state is x/t. The time tradeoff technique is easier for an individual to understand; however, the preference value is the true utility provided that the individual's utility function for additional healthy years is linear in time. If the utility function for additional healthy years is concave, the preference value by the time tradeoff method will underestimate the true utility value of the disease state. For more details regarding the performance of the above utility measuring techniques, the readers should refer to Torrance (1987).

The utility values should be validated for test–retest reproducibility before they are used to measure any change in health state. For the interpretation of improvement in utility, Torrance and Feeny (1989) related the utility values of some marker states. If there are utility values for some marker states, A, B, and C at 0.8, 0.7, and 0.4, respectively, an average improvement of 0.1 in utility of outcome health state from a trial may be described as equivalent to improving from outcome B to A average over all patients in the trial.

Although aggregation of utilities across individuals is commonly used in the analysis of data, it should be done with caution. The utility function may not be the same across subjects. The anchor states, perfect health and death, should be well defined for the same understanding across all subjects. To evaluate the effect of a therapy, the life years gained should be adjusted by the QOL. The quality-adjusted life years is the area under the profile of quality of like utility over time. The quality-adjusted life years gained is usually used in the evaluation of the effectiveness of therapy.

14.5.2 Calibration

Besides the validation of a QOL instrument, another issue of particular interest is the interpretation of an identified significant change in the QOL score. For this purpose, Testa et al. (1993) considered the calibration of change in QOL against the change in life events. A linear calibration curve was used to predict the relationship between the change in QOL index and the change in life events index. Only negative life events were considered. The study was not designed for calibration purposes and the changes in life events were collected as auxiliary information. The effect of change of life events was confounded with the effect of medication. If we want to use calibration to interpret the impact of change in QOL score, further research in the design and analysis method is necessary. Since the impact of life events is subjective and varies from person to person, it is difficult to assign numerical scores/ indices to life events. The relationship between QOL score and life events may not be linear. More complicated calibration functions or transformations may be required. We expect that the QOL score has positive correlation with the life events score; however, the correlation may not be strong enough to give a precise calibration curve. Besides the calibration of the QOL score with the life events score, changes in the QOL score may be related to changes in disease status.

14.6 Analysis of Parallel Questionnaire

Jachuck et al. (1982) indicated that QOL may be assessed in parallel by patients, their relatives, and physicians. The variability of the patient's rating is expected to be larger than those of the relatives' ratings and physicians' ratings. Although QOL scores can be analyzed separately based on individual ratings, they may lead to different conclusions. In this case, determining which rating should be used to assess the treatment effect on QOL has become a controversial issue. On the one hand, it is suggested that patients' ratings should be considered as the primary analysis because only patients' ratings can reflect exactly how patients feel. On the other hand, it is suggested that ratings of patients' relatives (e.g., spouses or significant others) should be considered because patients' ratings may not be accurate and reliable due to their illness. This is probably true especially for sensitive QOL components such as sexual function.

In practice, a typical approach is to analyze each rating separately. This approach, however, may cause the loss of some important information from the responses provided by the different perspectives. Jachuck et al. (1982) pointed out that QOL assessment based on each rating alone may lead to a totally different conclusion. To fully use the information contained in the two ratings, as an alternative, it is suggested that a composite index that combines both patients' ratings and parallel ratings (by their spouses or significant others) be considered. In this case, "Should the individual ratings carry the same weights as the parallel ratings?" has become an interesting question. If the patient's rating is considered to be more reliable than others, it should carry more weight in the assessment of QOL; otherwise, it should carry less weight in the analysis. Ki and Chow (1994) considered the following weighted score function:

$$Z = aX + bY,$$

where

   X and Y denote the ratings of a patient and his or her spouse, respectively
   a and b are the corresponding weights assigned to X and Y

Note that if $a = 1$ and $b = 0$, then the score function reduces to the patient's rating. On the other hand, when $a = 0$ and $b = 1$, the score function represents the spouse's rating. When $a = b = 1/2$, the score function is the average of the two ratings, that is, the patient's rating and his or her spouse's rating are considered equally important.

If one believes that one rating is more reliable than the other, then the more reliable one should carry more weight for the assessment of QOL. The choice of a and b in the above score function determines the relative importance of the ratings in the assessment of QOL. Ki and Chow (1994) proposed using the technique of principal components to determine a and b based on the observed data. The idea is to derive a one-dimensional function of both ratings, which can retain as much information as possible compared to the two-dimensional vector $W = (X, Y)'$. Assume that W follows a bivariate joint distribution with mean $\mu = (\mu_X, \mu_Y)'$ and covariance matrix:

$$\Sigma = \begin{pmatrix} \sigma_X^2 & \rho\sigma_X\sigma_Y \\ \rho\sigma_X\sigma_Y & \sigma_Y^2 \end{pmatrix},$$

where

   σ_X and σ_Y are the standard deviation of X and Y, respectively
   ρ is the linear correlation coefficient between X and Y

Suppose that N patients and their spouses (or significant others) from the same population are administered the QOL questionnaire simultaneously. Then, the mean and covariance matrix of W can be estimated based on observed ratings $W_i = (X_i, Y_i)'$, $i = 1, \ldots, N$, as follows:

$$\hat{\mu} = \bar{W} = (\bar{X}, \bar{Y}),$$

where

$$\bar{X} = \frac{1}{N}\sum_{i=1}^{N} X_i \quad \text{and} \quad \bar{Y} = \frac{1}{N}\sum_{i=1}^{N} Y_i,$$

and

$$\hat{\Sigma} = S = \frac{1}{N-1} \sum_{i=1}^{N} (W_i - \bar{W})(W_i - \bar{W})' = \begin{pmatrix} S_X^2 & rS_XS_Y \\ rS_XS_Y & S_Y^2 \end{pmatrix}.$$

The above sample covariance matrix contains not only the information about the variations of the patients' and spouses' ratings but also the correlation between the two ratings. For the determination of a and b, one approach is to employ the technique of principal components based on both ratings. The first principal component of the observed data $\{W_i, i = 1, \ldots, N\}$ possesses the maximum sample variance, that is,

$$A'SA = a^2S_X^2 + b^2S_Y^2 + 2abrS_XS_Y$$

among all coefficient vectors satisfying

$$A'A = a^2 + b^2 = 1.$$

It can be shown that the numbers in the characteristic vector A associated with the largest characteristic root of S are the coefficients of the first principal component. The characteristic roots of S can be obtained from the characteristic equation

$$|S - \lambda I| = 0.$$

This leads to

$$\begin{vmatrix} S_X^2 - \lambda & rS_XS_Y \\ rS_XS_Y & S_Y^2 - \lambda \end{vmatrix} = 0.$$

Therefore,

$$\lambda = \frac{1}{2}(S_X^2 + S_Y^2) \pm \frac{1}{2}\Delta_{XY},$$

where

$$\Delta_{XY} = \sqrt{(S_X^2 + S_Y^2)^2 - 4S_X^2S_Y^2(1-r^2)}.$$

The largest root is then given by

$$\lambda_1 = \frac{1}{2}(S_X^2 + S_Y^2) + \frac{1}{2}\Delta_{XY}.$$

The first principal component can be obtained by solving the following equations:

$$(S_X^2 - \lambda_1)a + brS_X S_Y = 0,$$

$$a^2 + b^2 = 1.$$

This leads to

$$a = \left(1 + \frac{(\lambda_1 - S_X^2)^2}{r^2 S_X^2 S_Y^2}\right)^{-1/2},$$

and

$$b = \frac{(\lambda_1 - S_X^2)a}{rS_X S_Y}.$$

The sample covariance of the first principal component $y = A'W$ is the largest characteristic root $\lambda_1 = A'SA$ and the percentage of variation expressed by this component is

$$\frac{\lambda_1}{tr(S)},$$

where $tr(S)$ is the trace of S which is given by

$$tr(S) = S_X^2 + S_Y^2.$$

Note that if the sample covariance matrix S is singular, then there is only one nonzero characteristic root. The first principal component explains all the variation in the observations. The percentage of sample variation presented by the first principal component reflects how much information from the observations is retained by the first principal component and the usefulness of the component in representing the observations in a one-dimensional setting. If a large proportion of the variation of the observations can be accounted for by a single principal component, then most of the variation generated by the observations in a two-dimensional space can be expressed along a one-dimensional vector. This appeals to dimensional reduction and the coefficients (a, b) indicate the direction and relative importance of each rating toward QOL assessment.

14.7 An Example

Suppose that the sample covariance matrix of X and Y is

$$S = \begin{pmatrix} 1 & r \\ r & 1 \end{pmatrix},$$

where $r > 0$. The largest characteristic root of S is $1 + r$ and its corresponding characteristic vector is $A = \left(\sqrt{2}/2, \sqrt{2}/2\right)$. The score function is then given by

$$Z = \frac{\sqrt{2}}{2} X + \frac{\sqrt{2}}{2} Y,$$

which gives equal weight to both ratings. The percentage of variation retained by Z is $100(1 + r)/2$. The amount of variation expressed by Z for different values of the linear correlation coefficient r is summarized in Table 14.1. When the two ratings X and Y are highly correlated, the score function retains a very high percentage of variation. When the correlation is moderate, say 0.7, the score function can still retain 85% of the variation of the data. As can be seen from Table 14.1, the score function proposed in this section is simple and easy to use. It reduces a two-dimensional problem to a univariate problem. It duly uses the information from both ratings and gives a better power for statistical tests.

Suppose QOL assessment is administered before drug therapy (at baseline) and at the end of the therapy (endpoint) to patients and their spouses. The hypothesis of interest is one of no drug effect on QOL. Denote the endpoint change from baseline in the patient's rating by X and that of the spouse's rating by Y. When X and Y are analyzed separately, the probabilities of all possible conclusions are summarized in Table 14.2.

TABLE 14.1

Percentage of Variation
Expressed by Z for Various r

r	Percentage of Variation Expressed by Z
0.9	95
0.7	85
0.5	75
0.0	50

TABLE 14.2

Probabilities of All Possible
Conclusions

		Y	
		Accept H_0	**Reject H_0**
X	Accept H_0	P_{AA}	P_{AR}
	Reject H_0	P_{RA}	P_{RR}

As can be seen from Table 14.2, the probability of observing an inconsistent conclusion is given by $P = P_{AR} + P_{RA}$. For a particular case, when X and Y are bivariate normal with linear correlation coefficient ρ, the probabilities of observing inconsistent conclusions can be calculated and are presented in Table 14.3. The analysis of treatment effect can be done on the score function Z to avoid the potential problem of inconsistent results which may occur when the ratings are analyzed separately.

TABLE 14.3

Probability of Inconsistent Conclusions

ρ	$P = P_{AR} + P_{RA}$
−0.9	0.0407
−0.8	0.0561
−0.7	0.0669
−0.6	0.0751
−0.5	0.0815
−0.4	0.0865
−0.3	0.0902
−0.2	0.0929
−0.1	0.0945
−0.0	0.0950
0.1	0.0945
0.2	0.0929
0.3	0.0902
0.4	0.0865
0.5	0.0815
0.6	0.0751
0.7	0.0669
0.8	0.0561
0.9	0.0407

Note: X and Y are bivariate normal with correlation ρ.

14.8 Concluding Remarks

As discussed above, a QOL instrument needs to be validated in terms of its validity, reliability, reproducibility, responsiveness, and sensitivity before it can be applied to assess QOL in clinical trials. However, in current practice an instrument is usually validated either concurrently or retrospectively. If an instrument is to be validated prospectively, an appropriate validation study should be carefully designed. Chow and Ki (1994) discussed statistical characteristics of QOL under a time series model, which may be useful for prospective validation. Appropriate statistical tests for validity, reliability, and test–retest reproducibility should be derived under such a model.

For the assessment of QOL in clinical trials, subscales or composite scores are often analyzed in order to describe different domains of QOL. One of the controversial issues is whether the developed instrument (questionnaire) asks the right questions for the assessment of each individual domain of the QOL. Chow and Ki (1994) provided statistical justification for the use of a composite score in QOL assessment using factor analysis to group relevant questions to form individual domains as suggested by the data. Another controversial issue regarding the use of subscales or composite scores is α adjustment for multiple comparisons. "How to adjust for α?" and "how to interpret the results?" have become important issues in QOL assessment.

In practice, missing data are commonly encountered in QOL assessment. Thus, statistical procedures for handling missing values play an important role for the validity of QOL assessment. A typical approach is to exclude subjects whose missing values have exceeded a prespecified percentage. For those subjects included in the analysis, their missing values will be imputed. Commonly considered procedures for missing value imputation include (1) mean imputation, (2) median imputation, and (3) regression analysis. These methods may not be useful when there is a significant proportion of subjects with missing values.

The interpretation of improvement in the QOL score is always a challenge for the investigator. For example, suppose QOL can be assessed by a mean overall QOL score with the categories presented in Table 14.4.

TABLE 14.4

QOL Categories

Status	QOL Score
Very poor	$0 \leq QOL < 1$
Poor	$1 \leq QOL < 2$
Fair	$2 \leq QOL < 3$
Good	$3 \leq QOL < 4$
Excellent	$4 \leq QOL < 5$

Suppose the mean QOL score at baseline is 1.2, which is considered to be in the *Poor* category. After the treatment, the mean QOL score has improved from 1.2 to 1.9 (an improvement of 0.7), which still falls in the category of *Poor*. In this case, we may conclude that there is no improvement in QOL. However, if we take a close look, some patients with baseline QOL scores close to the boundary may have *significant* improvement (i.e., jump from one category to the next category) even with a small improvement in QOL score. Thus, the analysis of improvement in mean QOL score may not be appropriate. Alternatively, we may consider the so-called shift analysis to capture the information regarding how many subjects are improving and how many subjects are worsening in terms of their QOL status change from category to category. This analysis may provide a good statistical interpretation of the collected data. However, it does not provide any insight of the QOL clinically. Thus, it is suggested that calibration with life events (e.g., promotion, salary raise, losing job, and losing love ones) or health care status (outpatient, emergency, hospitalization, and intensive care) be considered. The approach of calibration against life events and/or health care status could be *a* solution; however, the validation of the calibration has raised another controversial issue in QOL assessment.

Finally, another controversial issue is that whether QOL should be treated as a safety endpoint, an efficacy endpoint, both safety and efficacy, or neither. Unlike the hard clinical endpoint such as survival, different individuals have different perceptions regarding QOL. QOL may not serve as a clinical endpoint for the evaluation of clinical efficacy and/or safety. But it does provide clinical benefit to the patient with the disease under study.

15

Missing Data Imputation

15.1 Introduction

Missing values or incomplete data are commonly encountered in clinical trials. One of the primary causes of missing data is the dropout. Reasons for dropout include, but are limited to, refusal to continue in the study (e.g., withdrawal of informed consent), perceived lack of efficacy, relocation, adverse events, unpleasant study procedures, worsening of disease, unrelated disease, noncompliance with the study, need to use prohibited medication, and death (DeSouza et al., 2009). Following the idea of Little and Rubin (1987, 2002), DeSouza et al. (2009) provided an overview of three types of missingness mechanisms for dropouts. These three types of missingness mechanisms include (1) missing completely at random (MCAR), (2) missing at random (MAR), and (3) missing not at random (MNAR). MCAR refers to the dropout process that is independent of the observed data and the missing data. MAR indicates that the dropout process is dependent on the observed data but is independent of the missing data. For MNAR, the dropout process is dependent on the missing data and possibly the observed data. Depending upon the missingness mechanisms, appropriate missing data analysis strategies can then be considered based on existing analysis methods in the literature. For example, commonly considered methods under MAR include (1) discard incomplete cases and analyze complete cases only, (2) impute or fill in missing values and then analyze the filled-in data, (3) analyze the incomplete data by a method such as likelihood-based method (e.g., maximum likelihood, restricted maximum likelihood, and Bayesian approach), moment-based method (e.g., generalized estimating equations [GEEs] and their variants), and survival analysis method (e.g., Cox proportional hazards model) that does not require a complete data set. On the other hand, under MNAR, commonly considered methods are derived under pattern mixture models (Little, 1994) which can be divided into two types: parametric (see Diggle and Kenward, 1994) and semi-parametric (Rotnitzky et al., 1998).

In practice, the possible causes of missing values in a study can generally be classified into two categories. The first category includes the reasons that are not directly related to the study. For example, a patient may be lost to follow-up because relocation out of the area. This category of missing values can be considered as MCAR. The second category includes the reasons that are related to the study. For example, a patient may withdraw from the study due to treatment-emergent adverse events. In clinical research, it is not uncommon to have multiple assessments from each subject. Subjects with all observations missing are called unit nonrespondents. Because unit nonrespondents do not provide any useful information, these subjects are usually excluded from the analysis. On the other hand, the subjects with some, but not all, observations missing are referred to as item nonrespondents. In practice, excluding item nonrespondents from the analysis is considered against the intent-to-treat (ITT) principle and, hence, is not acceptable. In clinical trials, the primary analysis is usually conducted based on ITT population, which includes all randomized subjects with at least posttreatment evaluation. As a result, most item nonrespondents may be included in the ITT population. Excluding item nonrespondents may seriously decrease power/ efficiency of the study. Statistical methods for missing values imputation have been studied by many authors (see Kalton and Kasprzyk, 1986; Little and Rubin, 1987; Schafer, 1997).

To account for item nonrespondents, two methods are commonly considered. The first method is the so-called likelihood-based method. Under a parametric model, the marginal likelihood function for the observed responses is obtained by integrating the missing responses. The parameter of interest can then be estimated by the maximum likelihood estimator (MLE). Consequently, a corresponding test (e.g., likelihood ratio test) can be constructed. The merit of this method is that the resulting statistical procedures are usually efficient. The drawback is that the calculation of the marginal likelihood could be difficult. As a result, some special statistical or numerical algorithms are commonly applied for obtaining the MLE. For example, the expectation–maximization (EM) algorithm is one of the most popular methods for obtaining the MLE when there are missing data. The other method for item nonrespondents is imputation. Compared with the likelihood-based method, the method of imputation is relatively simple and easy to apply. The idea of imputation is to treat the imputed values as the observed values and then apply the standard statistical software for obtaining consistent estimators. However, it should be noted that the variability of the estimator obtained by imputation is usually different from the estimator obtained from the complete data. In this case, the formulas designed to estimate the variance of the complete data set cannot be used to estimate the variance of estimator produced by the imputed data. As an alternative, two methods are considered for the estimation of its variability. One is based on Taylor's expansion. This method is referred to as the linearization method. The merit of the linearization method is that it requires less computation. However, the drawback is that its formula could be very complicated and/or not trackable. The other approach is based on the resampling method (e.g., bootstrap and jackknife).

The drawback of the resampling method is that it requires an intensive computation. The merit is that it is very easy to apply. With the help of a fast-speed computer, the resampling method has become much more attractive in practice.

Note that imputation is very popular in clinical research. The simple imputation method of last observation carry forward (LOCF) at endpoint is probably the most commonly used imputation method in clinical trials. Although the LOCF is simple and easy for implementation in clinical trials, its validity has been challenged by many researchers. As a result, the search for alternative valid statistical methods for missing values imputation has received much attention in the past decade. In practice, the imputation methods in clinical trials are more diversified due to the complexity of the study design relative to the sample survey. As a result, statistical properties of many commonly used imputation methods in clinical trials are still unknown, while most imputation methods used in the sample survey are well studied. Hence, the imputation methods in clinical trials provide a unique challenge and also an opportunity for the statisticians in the area of clinical research.

In the next section, statistical properties and the validity of the commonly used LOCF method are studied. Some commonly considered statistical methods for missing values imputation are described in the subsequent sections of this chapter. Some recent development and a brief concluding remark are given in the last two sections of this chapter.

15.2 Last Observation Carry Forward

As indicated earlier, LOCF analysis at endpoint is probably the most commonly used imputation method in clinical trials. For illustration purpose, one example is described in the following. Consider a randomized, parallel-group clinical trial comparing r treatments. Each patient is randomly assigned to one of the treatments. According to the protocol, each patient should undergo s consecutive visits. Let y_{ijk} be the observation from the kth subject in the ith treatment group at visit j. The following statistical model is usually considered.

$$y_{ijk} = \mu_{ij} + \varepsilon_{ijk}, \quad \varepsilon_{ijk} \sim N(0, \sigma^2), \tag{15.1}$$

where μ_{ij} represents the fixed effect of the ith treatment at visit j. If there are no missing values, the primary comparison between treatments will be based on the observations from the last visit ($j = s$) because this reflects the treatment difference at the end of the treatment period. However, it is not necessary that every subject completes the study. Suppose that the last evaluable visit is $j^* < s$ for the kth subject in the ith treatment group. Then the value of y_{ij^*k} can be used to impute y_{isk}. After imputation, the data at endpoint are analyzed by the usual analysis of variance (ANOVA) model. We will refer to the procedure described above as LOCF. Note that the method of LOCF is usually applied according to the ITT principle. The ITT population includes all randomized subjects. In clinical research, although the LOCF is commonly employed, it lacks statistical justification. In what follows, its statistical properties and justification are studied.

15.2.1 Bias–Variance Trade-Off

The objective of a clinical study is usually to assess the safety and efficacy of a test treatment under investigation. Statistical inferences on the efficacy parameters are usually obtained. In practice, a sufficiently large number of sample size is required to obtain a reliable estimate and to achieve a desired power for the establishment of the efficacy of treatment. The reliability of an estimator can be evaluated by bias and by variability. A reliable estimator should have a small or zero bias with small variability. Hence, the estimator based on LOCF and the estimator based on completers are compared in terms of their bias and variability. For illustration purpose, we focus on only one treatment group with two visits. Assume that there are a total of $n = n_1 + n_2$ randomized subjects, where n_1 subjects complete the trial, while the remaining n_2 subjects only have observations at visit 1. Let y_{ik} be the response from the kth subject at the ith visit and $\mu_i = E(y_{ik})$. The parameter of interest is μ_2. The estimator based on completers is given by

$$\bar{y}_c = \frac{1}{n_1} \sum_{k=1}^{n_1} y_{i2k}.$$

On the other hand, the estimator based on LOCF can be obtained as

$$\bar{y}_{\text{LOCF}} = \frac{1}{n} \left(\sum_{i=1}^{n_1} y_{i2k} + \sum_{i=n_1+1}^{n} y_{i1k} \right).$$

It can be verified that the bias of $\bar{y}_c$ is 0 with variance σ^2/n_1, while the bias of $\bar{y}_{\text{LOCF}}$ is $n_2(\mu_1 - \mu_2)/n$ with variance $\sigma^2/(n_1 + n_2)$. As noted, although LOCF may introduce some bias, it decreases the variability. In a clinical trial with multiple visits, usually, $\mu_j \approx \mu_s$ if $j \approx s$. This implies that the LOCF is recommended if the patients withdraw from the study at the end of the study. However, if a patient drops out of the study at the very beginning, the bias of the LOCF could be substantial. As a result, it is recommended that the results from the analysis based on LOCF be interpreted with caution.

15.2.2 Hypothesis Testing

In practice, the LOCF is viewed as a pure imputation method for testing the null hypothesis of

$$H_0 : \mu_{1s} = \cdots = \mu_{rs},$$

where μ_{ij} are as defined in (15.1). Shao and Zhong (2003) provided another look of statistical properties of the LOCF under the above null hypothesis. More specifically, they partitioned the total patient population into s subpopulations according to the time when the number of patients drop out from the study. Note that in their definition, the patients who complete the study are considered a special case of "dropout" at the end of the study. Then μ_{ij} represents the population mean of the jth subpopulation under treatment i. Assume that the jth subpopulation under the ith treatment accounts for $p_i \times 100\%$ of the overall population under the ith treatment. They argued that the objective of the ITT analysis is to test the following hypothesis:

$$H_0 : \mu_1 = \cdots = \mu_r, \tag{15.2}$$

where

$$\mu_i = \sum_{j=1}^{s} p_{ij}\mu_{ij}.$$

Based on the above hypothesis, Shao and Zhong (2003) indicated that the LOCF bears the following properties:

1. In the special case of $r = 2$, the asymptotic ($n_i \to \infty$) size of the LOCF under H_0 is $\leq \alpha$ if and only if

$$\lim\left(\frac{n_2\tau_1^2}{n} + \frac{n_1\tau_2^2}{n}\right) \leq \lim\left(\frac{n_1\tau_1^2}{n} + \frac{n_2\tau_2^2}{n}\right),$$

where

$$\tau_i^2 = \sum_{j=1}^{s} p_{ij}(\mu_{ij} - \mu_i)^2.$$

The LOCF is robust in the sense that its asymptotic size is α if $\lim(n_1/n) = n_2/n$ or $\tau_1^2 = \tau_2^2$. Note that, in reality, $\tau_1^2 = \tau_2^2$ is impractical unless $\mu_{ij} = \mu_i$ for all j. However, $n_1 = n_2$ (as a result $\lim(n_1/n) = n_2/n$ is very typical, in practice). The above observation indicates in such a situation $n_1 = n_2$ that LOCF is still valid.

2. When $r = 2$, $\tau_1^2 \neq \tau_2^2$, and $n_1 \neq n_2$, the LOCF has an asymptotic size smaller than α if

$$(n_2 - n_1)\tau_1^2 < (n_2 - n_1)\tau_2^2 \tag{15.3}$$

   or larger than α if the inequality sign in (15.3) is reversed.

3. When $r \geq 3$, the asymptotic size of the LOCF is generally not α except for some special case (e.g., $\tau_1^2 = \tau_2^2 = \cdots = \tau_r^2 = 0$).

Because the LOCF usually does not produce a test with asymptotic significance level α when $r \geq 3$, Zhong and Shao (2002) proposed the following testing procedure based on the idea of post-stratification. The null hypothesis H_0 should be rejected if $T > \chi_{1-\alpha,r-1}^2$, where $\chi_{1-\alpha,r-1}^2$ is a chi-square random variable with $r - 1$ degrees of freedom and

$$T = \sum_{i=1}^{r} \frac{1}{\hat{V}_i}\left(\bar{y}_{i\cdot\cdot} - \frac{\sum_{i=1}^{r}\bar{y}_{i\cdot\cdot}/\hat{V}_i}{\sum_{i=1}^{r}1/\hat{V}_i}\right)^2,$$

$$\hat{V}_i = \frac{1}{n_i(n_i-1)}\sum_{j=1}^{s}\sum_{k=1}^{n_{ij}}(y_{ijk} - \bar{y}_{i\cdot\cdot})^2.$$

Under model (15.1) and the null hypothesis of (15.3), this procedure has the exact type I error α.

15.3 Mean/Median Imputation

Missing ordinal responses are also commonly encountered in clinical research. For those types of missing data, mean or median imputation is commonly considered. Let x_i be the ordinal response from the ith subject, where $i = 1, \ldots, n$. The parameter of interest is $\mu = E(x_i)$. Assume that x_i for $i = 1, \ldots, n_1 < n$ are observed and the rest are missing. Median imputation will impute the missing response by the median of the observed response (i.e., x_i, $i = 1, \ldots, n_1$). The merit of median imputation is that it can keep the imputed response within the sample space as the original response by appropriately defining the median. The sample mean of the imputed data set will be used as an estimator for the population mean. However, as the parameter of interest is population mean, the median imputation may lead to biased estimates.

As an alternative, mean imputation will impute the missing value by the sample mean of the observed units, i.e., $(1/n_1)\sum_{i=1}^{n_1} x_i$. The disadvantage of the mean imputation is that the imputed value may be out of the original response sample space. However, it can be shown that the sample mean of the imputed data set is a consistent estimator of population mean. Its variability can be assessed by the jackknife method proposed by Rao and Shao (1992).

In practice, usually, each subject will provide more than one ordinal response. The summation of those ordinal responses (total score) is usually considered as the primary efficacy parameter. The parameter of interest is the population mean of the total score. In such a situation, mean/median imputation can be carried out for each ordinal response within each treatment group.

15.4 Regression Imputation

The method of regression imputation is usually considered when covariates are available. Regression imputation assumes a linear model between the response and the covariates. The method of regression imputation has been studied by various authors (see Srivastava and Carter, 1986; Shao and Wang, 2002).

Let y_{ijk} be the response from the kth subject in the ith treatment group at the jth visit. The following regression model is considered:

$$y_{ijk} = \mu_i + \beta_i x_{ij} + \varepsilon_{ijk}, \tag{15.4}$$

where x_{ij} is the covariate of the kth subject in the ith treatment group. In practice, the covariates x_{ij} could be demographic variables (e.g., age, sex, and race) or the patient's baseline characteristics (e.g., medical history or disease severity). Model (15.4) suggests a regression imputation method. Let $\hat{\mu}_i$ and $\hat{\beta}_i$ denote the estimators of μ_i and β_i based on the complete data set, respectively. If y_{ijk} is missing, its predicted mean value $y_{ijk}^* = \hat{\mu}_i + \hat{\beta}_i x_{ij}$ is used for imputation. The imputed values are treated as true responses and the usual ANOVA is used to perform the analysis.

15.5 Marginal/Conditional Imputation for Contingency

In an observational study, two-way contingency tables can be used to summarize two-dimensional categorical data. Each cell (category) in a two-way contingency table is defined by a two-dimensional categorical variable (A, B), where A and B take values in $\{1, \ldots, a\}$ and $\{1, \ldots, b\}$, respectively. Sample cell frequencies can be computed based on the observed responses of (A, B) from a sample of units (subjects). Statistical interest includes the estimation of cell probabilities and testing hypotheses of goodness of fit or the independence of the two components A and B. In an observational study, there can be more than one stratum. It is assumed that within a stratum, sampled units independently have the same probability π_A to have missing B and observed A, π_B to have missing A and observed B, and π_C to have observed A and B. (The probabilities π_A, π_B, and π_C may be different in different imputation classes.) As units with both A and B missing are considered as unit nonrespondent, they are excluded in the analysis. As a result, without loss of generality, it is assumed that $\pi_A + \pi_B + \pi_C = 1$. For a two-way contingency table, it is very important for an appropriate imputation method to keep imputed values in the appropriate sample space. Whether in calculating the cell probability or in testing hypotheses (e.g., testing independence or goodness of fit), the corresponding statistical procedures are all based on the frequency counts of a contingency table. If the imputed value is out of the sample space, additional categories will be produced, which is of no practical meaning. As a result, two hot deck imputation methods are thoroughly studied by Shao and Wang (2002).

15.5.1 Simple Random Sampling

Consider a sampled unit with observed $A = i$ and missing B. Two imputation methods were studied by Shao and Wang (2002). The marginal (or unconditional) random hot deck imputation method imputes B by the value of B of a unit randomly selected from all units with observed B. The conditional hot deck imputation method imputes B by the value of B of a unit randomly selected from all units with observed B and $A = i$. All nonrespondents are imputed independently.

After imputation, the cell probabilities p_{ij} can be estimated using the standard formulas in the analysis of data from a two-way contingency table by treating imputed values as observed data. Denote these estimators by $\hat{p}_{ij}^I$, where $i = 1, \ldots, a$ and $j = 1, \ldots, b$. Let

$$\hat{p}^I = (\hat{p}_{11}^I, \ldots, \hat{p}_{1b}^I, \ldots, \hat{p}_{a1}^I, \ldots, \hat{p}_{ab}^I)',$$

and

$$p = (p_{11}, \ldots, p_{1b}, \ldots, p_{a1}, \ldots, p_{ab})',$$

where $p_{ij} = P(A = i, B = j)$. Intuitively, marginal random hot deck imputation leads to consistent estimators of $p_{i.} = P(A = i)$ and $p_{.j} = P(B = j)$, but not p_{ij}. Shao and Wang (2002) showed that $\hat{p}^I$ under conditional hot deck imputation are consistent, asymptotically unbiased, and asymptotically normal.

Theorem 15.1

Assume that $p_C > 0$. Under conditional hot deck imputation,

$$\sqrt{n}(\hat{p}^I - p) \to_d N(0, MPM' + (1 - \pi_C)P),$$

where

$$P = diag\{p\} - pp'$$

and

$$M = \frac{1}{\sqrt{\pi_C}} (I_{axb} - \pi_A diag\{p_{B|A}\}I_a \otimes U_b - \pi_B diag\{p_{A|B}\}U_a \otimes I_b,$$

$$p_{A|B} = \left(\frac{p_{11}}{p_{.1}}, \ldots, \frac{p_{1b}}{p_{.b}}, \ldots, \frac{p_{a1}}{p_{.1}}, \ldots, \frac{p_{ab}}{p_{.b}} \right)',$$

$$p_{B|A} = \left(\frac{p_{11}}{p_{1.}}, \ldots, \frac{p_{1b}}{p_{1.}}, \ldots, \frac{p_{a1}}{p_{a.}}, \ldots, \frac{p_{ab}}{p_{a.}} \right)',$$

where

I_a denotes an a-dimensional identity matrix,
U_b denotes a b-dimensional square matrix with all components being 1,
$\otimes$ is the Kronecker product.

15.5.2 Goodness-of-Fit Test

A direct application of Theorem 15.1 is to obtain a Wald-type test for goodness of fit. Consider the null hypothesis of the form $H_0 : p = p_0$, where p_0 is a known vector. Under H_0,

$$X_W^2 = n(\hat{p}^* - p_0^*)'\hat{\Sigma}^{*-1}(\hat{p}^* - p_0^*) \to_d \chi_{ab-1}^2,$$

where

χ_v^2 denotes a random variable having the chi-square distribution with v degrees of freedom,
$\hat{p}^*(p_0^*)$ is obtained by dropping the last component of $\hat{p}^I$ (p_0),
$\hat{\Sigma}^*$ is the estimated asymptotic covariance matrix of $\hat{p}^*$, which can be obtained by dropping the last row and column of $\hat{\Sigma}$, the estimated asymptotic covariance matrix of $\hat{p}^I$.

Note that the computation of $\hat{\Sigma}^{*-1}$ is complicated. Shao and Wang (2002) proposed a simple correction of the standard Pearson chi-square statistic by matching the first-order moment, an approach developed by Rao and Scott (1987). Let

$$X_G^2 = n \sum_{j=1}^{b} \sum_{i=1}^{a} \frac{(\hat{p}_{ij}^I - p_{ij})^2}{p_{ij}}.$$

It is noted that under conditional imputation the asymptotic expectation of X_G^2 is given by

$$D = \frac{1}{\pi_C}(ab + \pi_A^2 a + \pi_B^2 b - 2\pi_A a - 2\pi_B b + 2\pi_A \pi_B + 2\pi_A \pi_B \delta) - \pi_C ab + (ab - 1).$$

Let $\lambda = D/(ab - 1)$. Then the asymptotic expectation of X_G^2/λ is $ab - 1$, which is the first-order moment of a standard chi-square variable with $ab - 1$ degrees of freedom. Thus, X_G^2/λ can be used just like a normal chi-square statistic to test the goodness of fit. However, it should be noted that this is just an approximated test procedure which is not asymptotically correct. According to Shao and Wang's simulation study, this test performs reasonably well with moderate sample sizes.

15.6 Testing for Independence

Testing for the independence between A and B can be performed by the following chi-square statistic when there is no missing data:

$$X^2 = n \sum_{j=1}^{b} \sum_{i=1}^{a} \frac{(\hat{p}_{ij}^I - \hat{p}_{i \cdot} \hat{p}_{\cdot j})^2}{\hat{p}_{i \cdot} \hat{p}_{\cdot j}} \to_d \chi_{(a-1)(b-1)}^2.$$

It is of interest to know what the asymptotic behavior of the above chi-square statistic is under both marginal and conditional imputation. It is found that under the null hypothesis of A and B is the independent and conditional hot deck imputation

$$X^2 \to_d (\pi_C^{-1} + 1 - \pi_C)\chi_{(a-1)(b-1)}^2$$

and under the marginal hot deck imputation

$$X^2 \to_d \chi^2_{(a-1)(b-1)}.$$

15.6.1 Results under Stratified Simple Random Sampling

When the number of strata is small, stratified samplings are also commonly used in medical study. For example, a large epidemiology study is usually conducted by several large centers. Those centers are usually considered as strata. For those types of study, the number of strata is not very large; however, the sample size within each stratum is very large. As a result, imputation is usually carried out within each stratum. Within the hth stratum, we assume that a simple random sample of size n_h is obtained and samples across strata are obtained independently. The total sample size is $n = \sum_{h=1}^{H} n_h$, where H is the number of strata and n_h is the sample size in stratum h. The parameter of interest is the overall cell probability vector $p = \sum_{h=1}^{H} w_h p_h$, where w_h is the hth stratum weight. The estimator of p based on conditional imputation is given by $\hat{p}^I = \sum_{h=1}^{H} w_h \hat{p}_h^I$. Assume that $n_h = n \to p$ as $n \to \infty, h = 1, ..., H$. Then a direct application of Theorem 15.1 leads to

$$\sqrt{n}(\hat{p}^I - p) \to_d N(0, \Sigma),$$

where

$$\Sigma = \sum_{h=1}^{H} \frac{w_h^2}{p_h} \Sigma_h$$

and Σ_h is the Σ in Theorem 15.1 but restricted to the hth stratum.

15.6.2 When Number of Strata Is Large

In a medical survey, it is also possible to have the number of strata (H) very large, while the sample size within each stratum is small. A typical example is if a medical survey is conducted by the family, then the family can be considered as a stratum and all the members within the family become the samples from this stratum. In such a situation, the method of imputation within the stratum is impractical because it is possible that within a stratum, there are no completers. As an alternative, Shao and Wang (2002) proposed the method of imputation across strata under the assumption that $(\pi_{h,A}, \pi_{h,B}, \pi_{h,C})$, where $h = 1, \ldots, H$, is constant. More specifically, let $n^C_{h,ij}$ denote the number of completers in the hth stratum such that $A = i$ and $B = j$. For a sampled unit in the kth imputation class with observed $B = j$ and missing A, the missing value is imputed by i according to the conditional probability

$$p_{ij} \mid B, k = \frac{\sum_h w_h n^C_{h,ij}/n_h}{\sum_h w_h n^C_{h,\cdot j}/n_h}.$$

Similarly, the missing value of a sampled unit in the kth imputation class with observed $A = i$ and missing B can be imputed by j according to the conditional probability

$$p_{ij} \mid A, k = \frac{\sum_h w_h n^C_{h,ij}/n_h}{\sum_h w_h n^C_{h,i\cdot}/n_h}.$$

Note that $\hat{p}^I$ can be computed by ignoring the imputation classes and treating the imputed values as observed data. The following result establishes the asymptotic normality of $\hat{p}^I$ based on the method of conditional hot deck imputation across strata.

Theorem 15.2

Let $(\pi_{h,A}, \pi_{h,B}, \pi_{h,C}) = (\pi_A, \pi_B, \pi_C)$ for all h. Assume further that $H \to \infty$ and that there are constants c_j, for $j = 1, \ldots, 4$, such that $n_h \le c_1$, $c_2 \le H w_h \le c_3$, and $p_{h,ij} \ge c_4$ for all h. Then

$$\sqrt{n}(\hat{p}^I - p) \to_d N(0, \Sigma),$$

where Σ is the limit of

$$n\left(\sum_{h=1}^{H} \frac{w_h^2}{n_h} \Sigma_h + \Sigma_A + \Sigma_B \right).$$

15.7 Controversial Issues

One of the most controversial issues in missing data imputation is the possible decrease in power. In practice, it is often considered that the most worrisome impact of missing values on the inference for clinical trials is biased on the estimation of the treatment effect. As a result, little attention was paid to the possible loss of power. In clinical trials, it is recognized that missing data imputation may inflate variability and consequently decrease the power. If there is a significant decrease in power, the intended clinical trial will not be able to achieve the study objectives as planned. This would be a major concern during the regulatory review and approval process.

In addition to the issue of the possible loss of power, the following is a summary of controversial issues that present a challenge to clinical scientists when applying missing data imputation in clinical trials:

1. When the data are missing, the data are missing. How can we make up data for the missing data?

2. The validity of the method of LOCF for missing data imputation in clinical trials.

3. When there is a high percentage of missing values, missing data imputation could be biased and misleading.

For the first question, from a clinical scientist's point of view, if the data are missing, they are missing. One should not impute (or make up) data in any way whenever possible—it is always difficult, if not impossible, to verify the assumptions behind the method/model for missing data imputation. However, from a statistician's point of view, we may be able to estimate the missing data based on the information surrounding the missing data under certain statistical assumptions/models. Dropping subjects with incomplete data may not be a good statistics practice (GSP).

For the second question, the method of LOCF for missing values has been widely used in clinical trials for years in practice although its validity has been challenged by many researchers and the regulatory agency such as the United States Food and Drug Administration (FDA). It is suggested that the method of LOCF for missing values should not be considered as the primary analysis for missing data imputation.

For the third question, in practice, if the percentage of missing values exceeds a prespecified number, it is suggested that missing data imputation should not be applied. This raises a controversial issue for the selection of the criterion of the cutoff value for the percentage of the missing value, which will preserve good statistical properties of the statistical inference derived based on the incomplete data set and imputed data.

15.8 Recent Development

As indicated earlier, depending upon the mechanisms of missing data, different approaches may be selected in order to address the medical questions asked. In addition to the methods described in the previous sections of this chapter, the methods that are commonly considered include the mixed effects model for repeated measures (MMRMs), weighted and unweighted GEEs, multiple-imputation–based generalized estimating equations (MI-GEE), and complete-case (CC) analysis of covariance (ANCOVA). For recent developments in missing data imputation, the *Journal of Biopharmaceutical Statistics* (*JBS*) has published a special issue on Missing Data—Prevention and Analysis (Soon, 2009). These recent developments in missing data imputation are briefly summarized in the following.

For a time-saturated treatment effect model and an informative dropout scheme that depends on the unobserved outcomes only through the random coefficients, Kong et al. (2009) proposed a grouping method to correct the biases in the estimation of the treatment effect. Their proposed method could improve the current methods (e.g., the LOCF and the MMRM) and give more stable results in the treatment efficacy inferences. Zhang and Paik (2009) proposed a class of unbiased estimating equations using a pairwise conditional technique to deal with the generalized linear mixed model under benign non-ignorable missingness where specification of the missing model is not needed. The proposed estimator was shown to be consistent and asymptotically normal under certain conditions.

Moore and van der Laan (2009) applied targeted maximum likelihood methodology to provide a test that makes use of the covariate data that are commonly collected in randomized trials. The proposed methodology does not require assumptions beyond those of the log-rank test when censoring is uninformative. Two approaches based on this methodology are provided: (1) a substitution-based approach that targets treatment and time-specific survival from which the log-rank parameter is estimated, and (2) directly targeting the log-rank parameter. Shardell and El-Kamary (2009), on the other hand, used the framework of coarsened data to motivate performing sensitivity analysis in the presence of incomplete data. The proposed method (under pattern-mixture models) allows departures from the assumption of coarsening at random, a generalization of MAR, and independent censoring.

Alosh (2009) studied the missing data problem for count data by investigating the impact of missing data on a transition model, i.e., the generalized autoregressive model of order 1 for longitudinal count data. Rothmann et al. (2009) evaluated the loss to follow-up with respect to the ITT principle on the most important efficacy endpoints for clinical trials of anticancer biologic products submitted to the U.S. FDA from August 2005 to October 2008 and provided recommendations in light of the results.

DeSouza et al. (2009) studied the relative performances of these methods for the analysis of clinical trial data with dropouts via an extensive Monte Carlo study. The results indicate that the MMRM analysis method provides the best solution for minimizing the bias arising from missing longitudinal normal continuous data for small to moderate sample sizes under MAR dropout. For the nonnormal data, the MI-GEE may be a good candidate as it outperforms the weighted GEE method.

Yan et al. (2009) discussed methods used to handle missing data in medical device clinical trials, focusing on the tipping-point analysis as a general approach for the assessment of missing data impact. Wang et al. (2009) studied the performance of a biomarker predicting clinical outcome in a large prospective study under the framework of outcome- and auxiliary-dependent subsampling and proposed a semi-parametric empirical likelihood method to estimate the association between biomarker and clinical outcome. Nie et al. (2009) dealt with censored laboratory data due to assay limits by comparing a marginal approach and a variance-component mixed effects model approach.

15.9 Concluding Remarks

In summary, missing values or incomplete data are commonly encountered in clinical research. How to handle the incomplete data is always a challenge to the statisticians in practice. Imputation as one very popular methodology to compensate for the missing data is widely used in biopharmaceutical research. Compared to its popularity, however, its theoretical properties are far from well understood.

As indicated by Soon (2009), addressing missing data in clinical trials involves missing data prevention and missing data analysis (see also, NRC, 2010). Missing data prevention is usually done through the enforcement of good clinical practices during protocol development and clinical operations personnel training for data collection. This will lead to reduced biases, increased efficiency, less reliance on modeling assumption, and less need for sensitivity analysis. However, in practice, missing data cannot be totally avoided. Missing data often occur due to factors beyond the control of patients, investigators, and the clinical project team.

Note that the Panel on Handling Missing Data in Clinical Trials, Committee on National Statistics at the Division of Behavioral and Social Sciences and Education of National Research Council of the National Academies published a monograph on the Prevention and Treatment of Missing Data in Clinical Trials to assist the FDA in development of regulatory guidance on the issue of missing value in clinical trials (NRC, 2010).

16

Center Grouping

16.1 Introduction

For the approval of a new drug, the United States Food and Drug Administration (FDA) requires that substantial evidence of the effectiveness of the drug be provided through the conduct of adequate and well-controlled clinical trials. In clinical development, multicenter trials are usually considered adequate and well-controlled clinical trials. A multicenter trial is a single study that is conducted simultaneously at more than one study center according to a common protocol. A multicenter trial is often conducted to expedite the patient recruitment process to accrue sufficient number of patients in order to achieve a desired power within a predetermined time frame. The purpose of a multicenter clinical trial is not only to show that the clinical results are reproducible from center to center, but also to establish generalizability of the clinical results from one patient population to another patient population in different geographic locations (Ho and Chow, 1998). As indicated by Chow and Liu (1998b), a multicenter trial is not equivalent to separate single-site trials. The data collected from different centers are intended to be analyzed as a whole. To pool the data for an overall assessment of the effectiveness and safety of the study drug, however, both the FDA and the International Conference on Harmonization (ICH) guidelines require statistical tests for homogeneity across centers in order to detect possible quantitative or qualitative treatment-by-center interaction. A quantitative interaction indicates that the treatment differences are in the same direction across centers, but the magnitude differs from center to center, while a qualitative interaction reveals that substantial treatment differences occur in different directions in different centers (Gail and Simon, 1985). As pointed out by Gail and Simon (1985), no overall statistical inference regarding the treatment effect can be made if there is a significant qualitative interaction between treatment and center. In this case, it is suggested that treatment effect be assessed by the study center.

Lewis (1995) posted some commonly asked questions regarding issues related to design and analysis of multicenter trials (see also, Ho and Chow, 1998). These questions, which are helpful in planning stages of a multicenter trial, include the following:

1. Are some of the centers too small for reliable separate interpretation of the results?
2. Are some of the centers so big that they dominate the results?
3. Do the results at one or more centers look out of line with the others even if not significantly so?
4. Do any of the centers show a trend in the wrong direction?
5. If a treatment-by-center interaction is detected, is the trial valid?

One of the controversial issues in multicenter trials is whether the observed treatment-by-center interaction at the end of the study has any statistical meaning if the study ends up with a number of small centers. Should these small centers be grouped into a larger "dummy" center? What criteria should be considered for grouping? In practice, it is not a good idea to group all small centers into a single large dummy center. Thus, it is of interest to the investigator as to how many dummy centers should be created based on the number of small centers. In this chapter, we will examine the above issues and provide recommendations whenever possible.

In the next section, a rule of thumb for selection of the number of centers proposed by Shao and Chow (1993) is briefly outlined. Section 16.3 discusses the impact of treatment imbalance on statistical power for testing treatment effect. In Section 16.4, some methods for center grouping are introduced. Also included in this section is a proposed treatment for small centers with only patients in one treatment group. In Section 16.5, a valid randomization procedure is discussed. An example concerning a multicenter trial is presented to illustrate the use of the center grouping methodology proposed by Lin et al. (2010) in the last section.

16.2 Selection of the Number of Centers

One purpose of multicenter trials is to expedite the patient recruitment process to accrue sufficient number of patients within a relatively short period of time. The more centers used, the sooner the study would be completed. However, more centers would result in fewer patients in each center. For comparative clinical trials, the comparison between treatments is usually made between patients within centers. Statistically it is undesirable to have too few patients in each center for a valid and unbiased assessment of the treatment effect. As indicated in both the FDA and ICH guidelines, statistical tests for homogeneity across centers should be performed to detect a potential interaction between treatment and center. If a significant qualitative treatment-by-center interaction is observed, regulatory agencies require that the treatment effect be examined by study center. In this case, an overall assessment of the treatment effect by pooling data across centers is not statistically valid. In practice, more centers may increase the chance of observing a significant qualitative treatment-by-center interaction. The significance may be due to (1) heterogeneity across centers or some centers do not constitute a representative sample of the target patient population and (2) heterogeneity among centers or some centers exhibit relatively large variabilities. As a result, how to select an appropriate number of centers from a pool of representative centers is of great concern to the investigator. In multicenter trials, however, the centers are usually selected based on convenience and availability.

Shao and Chow (1993) proposed a rule of thumb suggesting that the number of patients in each center should not be less than the number of centers. For example, if the intended clinical trial calls for 100 patients, the sponsor may choose upto 10 study centers with 10 patients in each. Statistical justification of this rule of thumb will be provided and further discussed in Chapter 19.

16.3 Impact of Treatment Imbalance on Power

For a multicenter clinical trial comparing two treatments, sample size is usually selected to achieve a desired power for detection of a clinically meaningful difference at a prespecified significance level. Under the assumption of normality, the sample size calculation for a balance trial (i.e., each treatment group has the same number of patients) is usually given by

$$n = \frac{2\sigma^2 (z_\alpha + z_\beta)^2}{\Delta^2}$$

with a power of

$$1 - \Phi\left(z_{\alpha/2} - \frac{\Delta}{\sigma\sqrt{2/n}} \right), \tag{16.1}$$

where
σ is the standard deviation of the random error
z_α is the α th quantile of a standard normal distribution
Δ is the difference of clinical importance

Note that the above formula is derived by ignoring the center effect and the effect due to treatment-by-center interaction. In practice, a clinical trial may experience treatment imbalance (i.e., each treatment group has a different number of patients) despite plans to have an equal number of patients in each treatment group. Consequently, it may result in differences among centers. In this case, the power becomes

$$1 - \Phi\left(z_{\alpha/2} - \frac{\Delta}{\sigma/\sqrt{2}(\sqrt{1/n_1 + 1/n_2})} \right).$$ (16.2)

The power given in (16.2) is clearly less than the power given in (16.1). In order to achieve the same power as planned in (16.1), we set (16.2) to be equal to (16.1), which leads to

$$\frac{\Delta}{\sigma\sqrt{2/n}} = \frac{\Delta}{\sigma/\sqrt{2}(\sqrt{1/n_1 + 1/n_2})}.$$ (16.3)

For a fixed sample size N, a practical issue of (16.3) is that n_i, $i = 1, 2$, are not fixed. Before the conduct of the clinical trial, we cannot predict how many patients will be in each center after the completion of the trial. As a result, the only choice to achieve the sample power as planned in (16.1) is to increase the sample size N if we assume that the variance remains the same. It should be noted that when the number of patients are equal across centers the variance of the test statistic will be equal to the variance of the test statistic derived from the single center trial. Hence, the test statistic has the minimum variance.

16.4 Center Grouping

Without loss of generality, consider the following two-way classification random effects model:

$$Y_{ijk} = \mu + A_i + B_j + \varepsilon_{ijk}, \quad i = 1, \ldots, I, \quad j = 1, \ldots, J, \quad k = 1, \ldots, K,$$

where
    A_i is the fixed effect due to the ith treatment (factor A)
    B_j is the random effect due to the jth center (factor B)
    $(AB)_{ij}$ is the random effect due to the interaction between the ith treatment and the jth center
    ε_{ijk} is the random error in observing Y_{ijk}

It is assumed that $A_i = \mu + \alpha_i$, $i = 1, \ldots, I$, where $\sum_i \alpha_i = 0$, B_j, $j = 1, \ldots, J$ are independent and identically distributed (i.i.d.) as a normal random variable with mean 0 and variance σ_B^2, and ε_{ijk} are i.i.d. normal with mean 0 and variance σ^2. In addition, $\{B_j\}$ and $\{\varepsilon_{ijk}\}$ are mutually independent. Then, as indicated in Scheffé (1959),

$$E(\text{SSA}) = (I-1)\sigma^2,$$

$$E(\text{SSB}) = (J-1)(\sigma^2 + IK\sigma_B^2),$$

$$E(\text{SSAB}) = (I-1)(J-1)\sigma^2,$$

$$E(\text{SSE}) = IJ(K-1)\sigma^2,$$

where

$$\text{SSA} = JK\sum_{i=1}^{I}(\bar{Y}_{i\cdot\cdot} - \bar{Y}_{\cdot\cdot\cdot})^2,$$

$$\text{SSB} = IK\sum_{j=1}^{J}(\bar{Y}_{\cdot j\cdot} - \bar{Y}_{\cdot\cdot\cdot})^2,$$

$$\text{SSAB} = \sum_{i=1}^{I}\sum_{j=1}^{J}\sum_{k=1}^{K}(\bar{Y}_{ij\cdot} - \bar{Y}_{i\cdot\cdot} - \bar{Y}_{\cdot j\cdot} + \bar{Y}_{\cdot\cdot\cdot})^2,$$

$$\text{SSE} = \sum_{i=1}^{I}\sum_{j=1}^{J}\sum_{k=1}^{K}(Y_{ijk} - \bar{Y}_{ij\cdot})^2,$$

and

$$\bar{Y}_{i\cdot\cdot} = \frac{1}{JK}\sum_{j=1}^{J}\sum_{k=1}^{K}Y_{ijk},$$

$$\bar{Y}_{\cdot j\cdot} = \frac{1}{IK}\sum_{i=1}^{I}\sum_{k=1}^{K}Y_{ijk},$$

$$\bar{Y}_{ij\cdot} = \frac{1}{K} \sum_{k=1}^{K} Y_{ijk},$$

$$\bar{Y}_{\cdots} = \frac{1}{IJK} \sum_{i=1}^{I} \sum_{j=1}^{J} \sum_{k=1}^{K} Y_{ijk}.$$

If centers (factor B) are combined, then the new SSE becomes

$$\text{New SSE} = \text{SSB} + \text{SSAB} + \text{SSE}.$$

Therefore,

$$E(\text{New SSE}) = [(J-1) + (I-1)(J-1) + IJ(K-1)]\sigma^2 + (J-1)IK\sigma_B^2$$

$$= I(JK-1)\sigma^2 + IK(J-1)\sigma_B^2.$$

This implies that

$$E(\text{New MSE}) = \sigma^2(1+\delta),$$

where

$$\delta = \frac{IK(J-1)\sigma_B^2}{I(JK-1)\sigma^2} = \frac{K(J-1)\sigma_B^2}{(JK-1)\sigma^2}. \tag{16.4}$$

From the above expression, we conclude the following observations. First, $\delta > 0$ if σ_B^2 is not 0. Second, δ is an increasing function in J, the number of centers being combined. Thus, δ is smaller if fewer centers are combined. Finally, δ depends upon the similarity of centers being combined and the number of centers combined. The increases of δ for various I, J, and K and σ_B^2 and σ_{AB}^2 are given in Table 16.1.

Before combining the centers, the treatment effect is tested by

$$\frac{\text{SSA}/(I-1)}{(\text{SSE} + \text{SSAB})/((I-1)(J-1) + IJ(K-1))},$$

TABLE 16.1

δ Under Various Combinations of K, J, and σ_B^2/σ^2

	$K = 2$			$K = 4$		
s_B^2/s^2	$J = 2$	$J = 3$	$J = 4$	$J = 2$	$J = 3$	$J = 4$
0.1	0.10	0.10	0.10	0.07	0.08	0.09
0.2	0.20	0.20	0.20	0.13	0.16	0.17
0.3	0.30	0.30	0.30	0.20	0.24	0.26
0.4	0.40	0.40	0.40	0.27	0.32	0.34
0.5	0.50	0.50	0.50	0.33	0.40	0.43
0.6	0.60	0.60	0.60	0.40	0.48	0.51
0.7	0.70	0.70	0.70	0.47	0.56	0.60
0.8	0.80	0.80	0.80	0.53	0.64	0.69
0.9	0.90	0.90	0.90	0.60	0.72	0.77
1.0	1.00	1.00	1.00	0.67	0.80	0.86

Source: Chow, S.C. and Shao, J., *Stat. Med.*, 25, 1101, 2006. © 2006 by John Wiley & Sons, Ltd. With permission.

which follows a noncentral χ^2 distribution with $(I-1)(J-1) + IJ(K-1)$ degrees of freedom, and the noncentrality parameter of $\Delta = (JK/\sigma^2)\sum_i \alpha_i^2 = JK$.

After the grouping, the treatment effect can be tested by

$$\frac{\text{SSA}/(I-1)}{(\text{SSB}+\text{SSAB}+\text{SSE})/((J-1)+(I-1)(J-1)+IJ(K-1))}$$

$$\approx \left[\frac{(J-1)+(I-1)(J-1)+IJ(K-1)}{I-1}\right]$$

$$\times \left[\frac{\chi^2(I-1,\Delta)}{(1+\sigma_B^2/\sigma^2)\chi^2(J-1)+\chi^2(I-1)(J-1)+\chi^2(IJ(K-1))}\right].$$

The power before and after grouping is obtained by simulation based on 10,000 iterations. The results are summarized in Tables 16.2 through 16.4 for various choices of σ_B^2/σ^2, $\sum_i \alpha_i^2/\sigma^2$, and I, J, K. In these tables, p_1 is used to denote the power after grouping J centers into a dummy center, and p_2 to denote the power before grouping. The relative improvement by grouping is indicated by $\Delta \times 100\% = (p_1 - p_2)/p_2 \times 100\%$. For a proper interpretation, it should be pointed out that all the power listed in the table is the power within each dummy center. Thus, the overall power for all dummy centers combined would be significantly higher than the power within each dummy center. The relative improvement in power within a given dummy center, however, could serve as a sensible criterion for the evaluation of the effect of center grouping. It is suggested that small centers should be combined in such

TABLE 16.2

Power Comparison for $\sigma_B^2/\sigma^2 = 0.01$, $I = 2$

$\dfrac{1}{s^2}\sum a_i^2$		$J = 2$			$J = 3$			$J = 4$		
		p_1	p_2	Δ	p_1	p_2	Δ	p_1	p_2	Δ
$k = 1$	0.00	0.050	0.050	0.00	0.050	0.050	0.00	0.047	0.050	−0.058
	0.1	0.057	0.054	0.05	0.069	0.064	0.08	0.069	0.064	0.08
	0.2	0.072	0.060	0.21	0.099	0.077	0.28	0.099	0.077	0.28
	0.3	0.080	0.064	0.26	0.118	0.091	0.30	0.118	0.091	0.30
	0.4	0.090	0.069	0.31	0.131	0.104	0.26	0.131	0.104	0.26
	0.5	0.094	0.073	0.29	0.156	0.117	0.34	0.157	0.117	0.33
	0.6	0.101	0.077	0.31	0.182	0.130	0.40	0.182	0.130	0.40
	0.7	0.114	0.081	0.40	0.207	0.142	0.46	0.207	0.142	0.46
	0.8	0.118	0.085	0.38	0.217	0.155	0.40	0.217	0.155	0.40
	0.9	0.128	0.089	0.43	0.237	0.167	042	0.237	0.167	0.42
	1.0	0.132	0.009	0.42	0.273	0.179	0.52	0.272	0.178	0.52
$k = 2$	0.0	0.053	0.050	0.05	0.049	0.050	−0.03	0.051	0.050	0.02
	0.1	0.069	0.064	0.08	0.109	0.105	0.04	0.128	0.130	−0.01
	0.2	0.099	0.077	0.28	0.168	0.162	0.03	0.216	0.212	0.02
	0.3	0.118	0.091	0.30	0.229	0.220	0.04	0.299	0.293	0.02
	0.4	0.131	0.104	0.26	0.282	0.277	0.02	0.380	0.372	0.02
	0.5	0.157	0.117	0.34	0.345	0.333	0.04	0.461	0.446	0.03
	0.6	0.182	0.130	0.40	0.403	0.386	0.04	0.522	0.515	0.01
	0.7	0.207	0.142	0.45	0.458	0.438	0.05	0.592	0.578	0.02
	0.8	0.217	0.155	0.40	0.505	0.487	0.04	0.656	0.635	0.03
	0.9	0.237	0.167	0.42	0.555	0.533	0.04	0.701	0.686	0.02
	1.0	0.272	0.179	0.52	0.593	0.576	0.03	0.745	0.731	0.02

a way that the maximum of Δ is reached. As can be seen from Tables 16.2 through 16.4, we conclude the following statements:

1. By properly grouping small centers into a larger dummy center, power can be increased significantly.
2. Under certain conditions, however, center grouping can also decrease power significantly. According to the simulation results by Lin et al. (2010), it is mainly determined by the relative ratio of between-center (or center-to-center) variability versus between-subject (or subject-to-subject) variability, σ_B^2/σ^2. When $\sigma_B^2/\sigma^2 \approx 0.01$, center grouping will generally increase power. However, when $\sigma_B^2/\sigma^2 \approx 1$, center grouping may not help in increasing power in most cases.
3. For a fixed K sample size per arm with a small center, the maximum Δ can be reached by a proper choice J, the number of centers within each dummy center.

TABLE 16.3

Power Comparison for $\sigma_B^2/\sigma^2 = 0.1$, $I = 2$

$\frac{1}{s^2}\sum a_i^2$		$J = 2$			$J = 3$			$J = 4$		
		p_1	p_2	Δ	p_1	p_2	Δ	p_1	p_2	Δ
$k = 1$	0.0	0.043	0.050	−0.15	0.046	0.050	−0.08	0.042	0.050	−0.05
	0.1	0.056	0.055	0.03	0.070	0.064	0.10	0.086	0.074	0.15
	0.2	0.061	0.059	0.03	0.084	0.077	0.08	0.103	0.099	0.04
	0.3	0.076	0.064	0.18	0.110	0.091	0.21	0.143	0.124	0.16
	0.4	0.083	0.069	0.20	0.128	0.104	0.23	0.179	0.148	0.21
	0.5	0.090	0.073	0.23	0.149	0.117	0.28	0.212	0.172	0.23
	0.6	0.105	0.077	0.36	0.173	0.130	0.33	0.239	0.196	0.22
	0.7	0.109	0.081	0.34	0.190	0.142	0.33	0.285	0.220	0.30
	0.8	0.121	0.085	0.41	0.211	0.155	0.36	0.300	0.243	0.23
	0.9	0.125	0.089	0.40	0.235	0.167	0.40	0.342	0.266	0.28
	1.0	0.133	0.093	0.43	0.255	0.179	0.42	0.371	0.288	0.28
$k = 2$	0.0	0.046	0.050	−0.08	0.050	0.050	0.00	0.048	0.050	−0.05
	0.1	0.083	0.082	0.01	0.103	0.105	−0.02	0.131	0.130	0.01
	0.2	0.115	0.114	0.01	0.162	0.162	0.00	0.206	0.212	−0.02
	0.3	0.153	0.146	0.04	0.225	0.220	0.02	0.297	0.293	0.01
	0.4	0.184	0.179	0.03	0.281	0.277	0.01	0.370	0.372	−0.01
	0.5	0.217	0.211	0.02	0.344	0.333	0.03	0.444	0.446	−0.01
	0.6	0.254	0.244	0.04	0.393	0.386	0.02	0.523	0.515	0.01
	0.7	0.291	0.276	0.05	0.459	0.438	0.05	0.594	0.578	0.03
	0.8	0.319	0.307	0.04	0.506	0.487	0.04	0.644	0.635	0.01
	0.9	0.354	0.338	0.05	0.551	0.533	0.03	0.696	0.686	0.02
	1.0	0.382	0.368	0.04	0.594	0.576	0.03	0.744	0.731	0.02

16.5 Procedure for Center Grouping

As discussed above, a dummy center could have a higher power if the smaller centers within the dummy center have smaller between-center variability. In practice, it is then suggested that these smaller centers be grouped into a dummy center for the purpose of increasing power. However, the results may not be valid if the grouping is not done at random. Hence, it is recommended that small centers should be grouped randomly if they are to be grouped into dummy centers. Since the between-center variability is generally assessed by considering

$$E\left(\frac{SSB}{J-1}\right) = \sigma^2 + IK\sigma_B^2,$$

  Controversial Statistical Issues in Clinical Trials

TABLE 16.4

Power Comparison for $\sigma_B^2/\sigma^2 = 1, I = 2$

$\frac{1}{s^2}\sum a_i^2$		$J = 2$			$J = 3$			$J = 4$		
		p_1	p_2	Δ	p_1	p_2	Δ	p_1	p_2	Δ
$k = 1$	0.0	0.043	0.050	−0.14	0.029	0.050	−0.41	0.025	0.050	−0.50
	0.1	0.068	0.082	−0.17	0.044	0.064	−0.31	0.049	0.074	−0.34
	0.2	0.099	0.114	−0.13	0.057	0.077	−0.26	0.068	0.099	−0.31
	0.3	0.124	0.146	−0.15	0.073	0.091	−0.20	0.094	0.124	−0.24
	0.4	0.162	0.179	−0.10	0.080	0.104	−0.23	0.117	0.148	−0.21
	0.5	0.200	0.211	−0.06	0.101	0.117	−0.13	0.141	0.172	−0.18
	0.6	0.217	0.244	−0.11	0.117	0.130	−0.10	0.161	0.196	−0.18
	0.7	0.250	0.276	−0.09	0.131	0.142	−0.08	0.181	0.220	−0.18
	0.8	0.291	0.307	−0.05	0.150	0.155	−0.03	0.210	0.243	−0.14
	0.9	0.308	0.338	−0.09	0.164	0.167	−0.02	0.240	0.266	−0.10
	1.0	0.347	0.368	−0.06	0.182	0.179	−0.01	0.256	0.289	−0.11
$k = 2$	0.0	0.04	0.050	−0.20	0.037	0.050	−0.25	0.048	0.050	−0.05
	0.1	0.068	0.082	−0.17	0.080	0.105	−0.24	0.100	0.130	−0.23
	0.2	0.097	0.114	−0.15	0.133	0.162	−0.18	0.170	0.212	−0.20
	0.3	0.124	0.146	−0.15	0.191	0.220	−0.13	0.251	0.293	−0.14
	0.4	0.155	0.179	−0.13	0.238	0.277	−0.14	0.310	0.372	−0.17
	0.5	0.189	0.211	−0.11	0.287	0.333	−0.14	0.392	0.446	−0.12
	0.6	0.223	0.244	−0.09	0.338	0.386	−0.12	0.469	0.515	−0.09
	0.7	0.250	0.276	−0.09	0.397	0.438	−0.09	0.511	0.578	−0.12
	0.8	0.280	0.307	−0.09	0.440	0.487	−0.10	0.574	0.635	−0.10
	0.9	0.308	0.338	−0.09	0.486	0.533	−0.09	0.630	0.686	−0.08
	1.0	0.343	0.368	−0.07	0.523	0.576	−0.09	0.682	0.731	−0.07

we argue that a valid randomization is to achieve an unbiased estimate for $\sigma^2 + IK\sigma_B^2$. Suppose there are J small centers, which are to be grouped into some dummy centers with n small centers in each. Suppose $i_1, i_2, \ldots, i_J$ is a random permutation of index $i = 1, \ldots, J$. A valid randomization rule is to assign the centers with the same index as the first n indices in the sequence $i_1, i_2, \ldots, i_J$ to the first dummy center. The centers with the same index as the second n indices in the sequence $i_1, i_2, \ldots, i_J$ to the second dummy center, and so on.

For any given dummy center, it can be shown that the centers grouped into this dummy center are in fact chosen by the simple random sampling without replacement (SRSWR) from a population of J small centers. Hence, $\bar{Y}_{\cdot i_t}, t = 1, \ldots, n$, the overall mean of each center in this dummy center, can be considered as a SRSWR from a population of J small centers with overall mean $\bar{Y}_{\cdot i}, i = 1, \ldots, J$. Then the new sum of squares within this dummy center is given by

$$\text{SSB}^* = IK \sum_{t=1}^{n} \left(\overline{Y}_{\cdot i_t \cdot} - \frac{1}{n} \sum_{t=1}^{n} \overline{Y}_{\cdot i_t \cdot} \right)^2$$

So, an estimate for $\sigma^2 + IK\sigma_B^2$ can be obtained by $\text{SSB}^*/(n-1)$. Lohr (1999) showed that given $\overline{Y}_{\cdot j\cdot}, j = 1, \ldots, J$, $\text{SSB}^*/(n-1)$ is an unbiased estimate for

$$\frac{\text{SSB}}{J-1} = \frac{JK}{J-1} \sum_{j=1}^{N} (\overline{Y}_{\cdot j\cdot} - \overline{Y}_{\cdots})^2.$$

Thus,

$$E\left(\frac{\text{SSB}^*}{n-1} \right) = E\left[E\left(\frac{\text{SSB}^*}{n-1} \bigg| \overline{Y}_{\cdot j\cdot}, j = 1, \ldots, J \right) \right] = E\left(\frac{\text{SSB}}{J-1} \right) = \sigma^2 + JK\sigma_B^2.$$

Hence, the proposed randomization procedure for center grouping is valid.

16.6 An Example

To illustrate the proposed concept for center grouping in multicenter trials with a large number of small centers, consider a clinical trial for comparing a test compound with a standard therapy in treating patients with metastatic breast cancer. This study was a comparative, parallel-group, randomized, and double-blind multicenter clinical trial. The study protocol called for 288 patients in approximately 43 centers to achieve a desired statistical power for the evaluation of therapeutic equivalence. It was expected that each center would enroll 6–7 patients. As discussed earlier, the selection of 43 centers did expedite the patient recruitment to achieve the desired number of patients. However, due to a significant variation among centers, seven centers enrolled more than 10 patients in each. The other 36 centers had less than 10 patients in each. As a result, these small centers are necessarily grouped into comparable dummy centers not only to address Lewis' questions (Lewis, 1995) but also to provide an unbiased and fair assessment of the efficacy and safety of the study drug. Firstly, a center with more than 10 patients will stand alone as a single center. Secondly, since there are 36 centers with less than 10 patients in each, grouping these small centers into comparable dummy centers must be considered. Among these 36 small centers, 24 centers have patients in both treatment groups. Twelve centers have patients in only one treatment group.

TABLE 16.5

An Example of Center Grouping

Center Characteristic	After the Trial	After Center Grouping
Number of centers	43	15
Number of centers with more than 10 patients in each	7	7
Number of centers with less than 10 patients in each	36	0
Number of dummy centers	0	8
Number of centers with patients in one group	12	0

Suppose $\sigma_B^2/\sigma^2 = 0.01$ and $\sum_i \alpha_i^2/\sigma^2 = 0.5$. As can be seen from Table 16.2, if the size of the dummy center is selected to be $J = 2$, the relative improvement in power will be $\Delta = 0.29$. On the other hand, if we choose $J = 3$ and $J = 4$, $\Delta = 0.34$ and $\Delta = 0.33$, respectively. In order to achieve the maximum improvement in power, the size of the dummy center should be chosen as 3. It is suggested to group these 24 centers into 8 dummy centers at random. In this case, each dummy center consists of 3 randomly selected centers from the 24 small centers. Applying the above proposed procedure would result in a total of 15 centers. Finally, for the 12 centers with patient(s) in only one treatment group, randomly assign the patient(s) to the 15 centers. The summary of this example is given in Table 16.5.

17

Non-Inferiority Margin

17.1 Introduction

In cancer trials, it is unethical to use a placebo control when approved and effective therapies are available. A response to this problem in the investigation of a new test therapy is to replace the placebo control by an established therapy (which is referred to as active control agent or standard therapy) and to demonstrate that the test therapy is not inferior to the active control agent in the sense that the effect of the test therapy, when compared with the efficacy of the active control agent, is not below some non-inferiority margin. In practice, there may be a need to develop a new therapy that is non-inferior (but not necessarily superior) to an established efficacious therapy; for example, the new therapy is less toxic, easier to administer, and/or less expensive. As a result, a clinical trial for the establishment of non-inferiority of a test therapy as compared to an active control agent has become very popular in drug research and development. Clinical trials of this kind are referred to as active-controlled trials and statistical tests for establishing non-inferiority are called non-inferiority tests. An overview of design concepts and important issues in these trials is provided by D'Agostino et al. (2003).

One of the major considerations in a non-inferiority test is the selection of the non-inferiority margin. A different choice of non-inferiority margin may affect sample size calculation and consequently alter the conclusion of the clinical study. As pointed out in the guideline by the International Conference on Harmonization (ICH), the determination of non-inferiority margins should be based on both statistical reasoning and clinical judgment (ICH, 2000). The focus of this chapter is on the statistical considerations for the determination of non-inferiority margins. Despite the existence of some studies (Tsong et al., 1999; Hung et al., 2003; Laster and Johnson, 2003; Phillips, 2003; Chow and Shao, 2006), there is no established rule or gold standard for the determination of non-inferiority margins in active-controlled trials until a recent draft guidance distributed by the FDA for comments (FDA, 2010a).

According to the ICH E10 guidance on *Choice of Control Group and Related Issues in Clinical Trials* (ICH, 2000), a non-inferiority margin may be selected based on past experience in placebo-controlled trials with valid design under conditions similar to those planned for the new trial, and the determination of a non-inferiority margin should not only reflect uncertainties in the evidence on which the choice is based but also be suitably conservative. Furthermore, as a basic frequentist statistical principle, the hypothesis of non-inferiority should be formulated with population parameters, not estimates from historical trials (Hung et al., 2003). Along these lines, we propose in this chapter a method by Chow and Shao (2006) for selecting non-inferiority margins with some statistical justification. Chow and Shao (2006) proposed non-inferiority margin depends on population parameters including parameters related to the placebo control if it were not replaced by the active control. Unless a fixed (constant) non-inferiority margin can be chosen based on clinical judgment, a fixed non-inferiority margin not depending on population parameters is rarely suitable. Intuitively, the non-inferiority margin should be small when the effect of the active control agent relative to placebo is small or the variation in the population under investigation is large. Chow and Shao's approach ensures that the efficacy of the test therapy is superior to placebo when non-inferiority is concluded. When it is necessary/desired, Chow and Shao's approach can produce a non-inferiority margin that ensures that the efficacy of the test therapy relative to placebo can be established with great confidence.

Because Chow and Shao's proposed non-inferiority margin depends on population parameters, the non-inferiority test designed for the situation where the non-inferiority margin is fixed has to be modified in order to apply it to the case where the non-inferiority margin is a parameter. This is studied in Section 17.3. Sample size calculation, an important issue in the planning stage of a clinical trial, is also discussed. An example concerning a cancer trial for testing non-inferiority of a test therapy for treating patients with a specific cancer is presented in Section 17.4 to illustrate Chow and Shao's proposed method. The determination of non-inferiority margin based on the concept of mixed null hypothesis (Tsou et al., 2007) is discussed on Section 17.5. Recent development such as the 2010 FDA draft guidance on non-inferiority trials and some concluding remarks are given in Sections 17.6 and 17.7, respectively.

17.2 Non-Inferiority Margin

Let θ_T, θ_A, and θ_P be the unknown population efficacy parameters associated with the test therapy, the active control agent, and the placebo, respectively. Also, let $\Delta \geq 0$ be a non-inferiority margin. Without loss of generality, we assume that a large value of population efficacy parameter is desired. The hypotheses for non-inferiority can be formulated as

$$H_0 : \theta_T - \theta_A \leq -\Delta \quad \text{versus} \quad H_a : \theta_T - \theta_A > -\Delta. \tag{17.1}$$

If Δ is a fixed prespecified value, then standard statistical methods can be applied to testing hypotheses (17.1). In practice, however, Δ is often unknown. There exists an approach that constructs the value of Δ based on a placebo-controlled historical trial. For example, $\Delta = a$ fraction of the lower limit of the 95% confidence interval for $\theta_A - \theta_P$ based on some historical trial data (see CBER/FDA, 1999). Although this approach is intuitively conservative, it is not statistically valid because (1) if the lower confidence limit is treated as a fixed value, then the variability in historical data is ignored, and (2) if the lower confidence limit is treated as a statistic, then this approach violates the basic frequentist statistical principle, i.e., the hypotheses being tested should not involve any estimates from current or past trials (Hung et al., 2003).

From a statistical point of view, the ICH E10 guideline suggests that the non-inferiority margin Δ should be chosen to satisfy at least the following two criteria:

Criterion 1: The test therapy is non-inferior to the active control agent and is superior to the placebo (even though the placebo is not considered in the active-controlled trial).

Criterion 2: The non-inferiority margin should be suitably conservative, i.e., variability should be taken into account.

A fixed Δ (i.e., it does not depend on any parameter) is rarely suitable under criterion 1. Let $\delta > 0$ be a superiority margin if a placebo-controlled trial is conducted to establish the superiority of the test therapy over a placebo control. Since the active control is an established therapy, we may assume that $\theta_A - \theta_P > \delta$. However, when $\theta_T - \theta_A > -\Delta$ (i.e., the test therapy is non-inferior to the active control) for a fixed Δ, we cannot ensure that $\theta_T - \theta_P > \delta$ (i.e., the test therapy is superior to the placebo) unless $\Delta = 0$. Thus, it is reasonable to consider non-inferiority margins depending on unknown parameters. Hung et al. (2003) summarized the approach of using the non-inferiority margin of the form:

$$\Delta = \gamma(\theta_A - \theta_P), \tag{17.2}$$

where γ is a fixed constant between 0 and 1. This is based on the idea of preserving a certain fraction of the active control effect $\theta_A - \theta_P$. The smaller $\theta_A - \theta_P$ is, the smaller Δ is. How to select the proportion of γ, however, was not discussed.

Chow and Shao (2006) derived a non-inferiority margin satisfying criterion 1. Let $\delta > 0$ be a superiority margin if a placebo control is added to the trial. Suppose that the non-inferiority margin Δ is proportional to δ, i.e., $\Delta = r\delta$, where r is a known value chosen in the beginning of the trial. To be conservative, r should be ≤ 1. If the test therapy is not inferior to the active control agent but is superior to the placebo, then both

$$\theta_T - \theta_A > -\Delta \quad \text{and} \quad \theta_T - \theta_P > \delta \tag{17.3}$$

should hold. Under the worst scenario, i.e., $\theta_T - \theta_A$ achieves its lower bound $-\Delta$, the largest possible Δ satisfying (17.3) is given by

$$\Delta = \theta_A - \theta_P - \delta,$$

which leads to

$$\Delta = \frac{r}{1+r}(\theta_A - \theta_P). \tag{17.4}$$

From (17.2) and (17.4), $\gamma = r/(r + 1)$. If $0 < r \leq 1$, then $0 < \gamma \leq 1/2$.

The above argument in determining Δ takes criterion 1 into account, but is not conservative enough, since it does not take the variability into consideration. Let $\hat{\theta}_T$ and $\hat{\theta}_P$ be sample estimators of θ_T and θ_P, respectively, based on data from a placebo-controlled trial. Assume that $\hat{\theta}_T - \hat{\theta}_P$ is normally distributed with mean $\theta_T - \theta_P$ and standard error SE_{T-P} (which is true under certain conditions or approximately true under the central limit theorem for large sample sizes). When $\theta_T = \theta_A - \Delta$,

$$P(\hat{\theta}_T - \hat{\theta}_P < \delta) = \Phi\left(\frac{\delta + \Delta - (\theta_A - \theta_P)}{SE_{T-P}}\right), \tag{17.5}$$

where Φ denotes the standard normal distribution function. If Δ is chosen according to (17.4) and $\theta_T = \theta_A - \Delta$, then the probability that $\hat{\theta}_T - \hat{\theta}_P$ is less than δ is equal to 1/2. In view of criterion 2, a value much smaller than 1/2 for this probability is desired, because it is the probability that the estimated test therapy effect is not superior to that of the placebo. Since the probability in (17.5) is an increasing function of Δ, the smaller Δ (the more conservative choice of the non-inferiority margin) is, the smaller the chance that $\hat{\theta}_T - \hat{\theta}_P$ is less than δ. Setting the probability on the left-hand side of (17.5) to ε with $0 < \varepsilon \leq 1/2$, we obtain

$$\Delta = \theta_A - \theta_P - \delta - z_{1-\varepsilon}SE_{T-P},$$

where $z_a = \Phi^{-1}(a)$. Since $\delta = \Delta/r$, we obtain

$$\Delta = \frac{r}{1+r}(\theta_A - \theta_P - z_{1-\varepsilon}SE_{T-P}). \tag{17.6}$$

Figure 17.1 provides an illustration for the selection of the non-inferiority margin according to this idea.

Comparing (17.2) and (17.6), we obtain

$$\gamma = \frac{r}{1+r}\left(1 - \frac{z_{1-\varepsilon}SE_{T-P}}{\theta_A - \theta_P}\right),$$

i.e., the proportion γ in (17.2) is a decreasing function of a type of noise-to-signal ratio (or coefficient of variation).

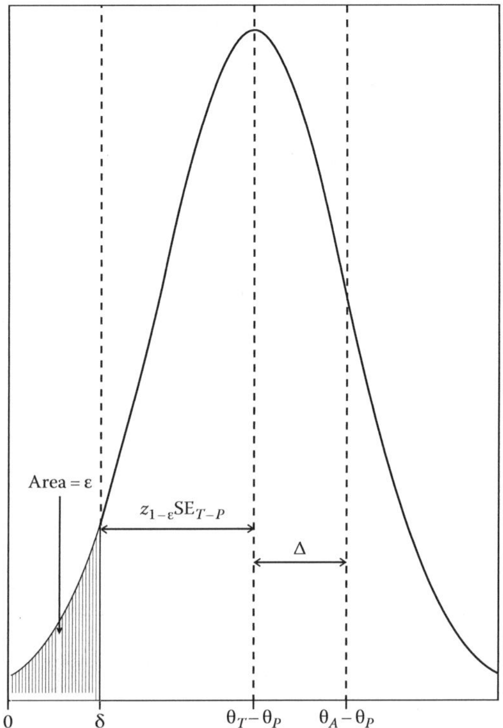

FIGURE 17.1
Selection of non-inferiority margin Δ (the solid curve is the probability density of $\hat{\theta}_T - \hat{\theta}_P$).

The proposed non-inferiority margin (17.6) can also be derived from a slightly different point of view. Suppose that we actually conduct a placebo-controlled trial with a superiority margin δ to establish the superiority of the test therapy over the placebo. Then the power of the large sample t-test for hypotheses $\theta_T - \theta_P \leq \delta$ versus $\theta_T - \theta_P > \delta$ is approximately equal to

$$\Phi\left(\frac{\theta_T - \theta_P - \delta}{\mathrm{SE}_{T-P}} - z_{1-\alpha}\right),$$

where α is the level of significance. Assume the worst scenario $\hat{\theta}_T = \hat{\theta}_A - \Delta$ and that β is a given desired level of power. Then, setting the power to β leads to

$$\frac{\theta_A - \theta_P - \delta - \Delta}{\mathrm{SE}_{T-P}} - z_{1-\alpha} = z_{\beta},$$

i.e.,

$$\Delta = \frac{r}{1+r}[\theta_A - \theta_P - (z_{1-\alpha} + z_\beta)\mathrm{SE}_{T-P}].\tag{17.7}$$

Comparing (17.6) with (17.7), we have $z_{1-\beta} = z_{1-\alpha} + z_\beta$. For $\alpha = 0.05$, the following table gives some examples of values of β, ε, and $z_{1-\varepsilon}$.

β	ε	$z_{1-\varepsilon}$
0.36	0.1000	1.282
0.50	0.0500	1.645
0.60	0.0290	1.897
0.70	0.0150	2.170
0.75	0.0101	2.320
0.80	0.0064	2.486

We now summarize the above discussions as follows:

1. The non-inferiority margin proposed by Chow and Shao (2006) given in (17.6) takes variability into consideration, i.e., Δ is a decreasing function of the standard error of $\hat{\theta}_T - \hat{\theta}_P$. It is an increasing function of the sample sizes, since SE_{T-P} decreases as sample sizes increase. Choosing a non-inferiority margin depending on the sample sizes does not violate the basic frequentist statistical principle. In fact, it cannot be avoided when the variability of sample estimators is considered. Statistical analysis, including sample size calculation at the trial planning stage, can still be performed. In the limiting case ($\mathrm{SE}_{T-P} \to 0$), the non-inferiority margin in (17.6) is the same as that in (17.4).

2. The ε value in (17.6) represents a degree of conservativeness. An arbitrarily chosen ε may lead to highly conservative tests. When sample sizes are large (SE_{T-P} is small), one can afford a small ε. A reasonable value of ε and sample sizes can be determined in the planning stage of the trial.

3. The non-inferiority margin in (17.6) is non-negative if and only if $\theta_A - \theta_P \geq z_{1-\varepsilon} \mathrm{SE}_{T-P}$, i.e., the active control effect is substantial or the sample sizes are large. We might take our non-inferiority margin to be the larger of the quantities in (17.6) and 0 to force the non-inferiority margin to be non-negative. However, it may be wise not to do so. Note that if θ_A is not substantially larger than θ_P, then non-inferiority testing is not justifiable since, even if $\Delta = 0$ in (17.1), concluding H_a in (17.1) does not imply the test therapy is superior over the placebo. Using Δ in (17.6), testing hypotheses (17.1) converts to testing the superiority of the test therapy over the active control agent when Δ is actually negative. In other words, when $\theta_A - \theta_P$ is smaller than a certain margin,

our test automatically becomes a superiority test and the property $P(\hat{\theta}_T - \hat{\theta}_P < \delta) = \varepsilon$ (with $\delta = |\Delta|/r$ still holds).

4. In many applications, there are no historical data. In such cases parameters related to placebo are not estimable and, hence, a non-inferiority margin not depending on these parameters is desired. Since the active control agent is a well-established therapy, let us assume that the power of the level α test showing that the active control agent is superior to placebo by the margin δ is at the level η. This means that approximately

$$\theta_A - \theta_P \geq \delta + (z_{1-\alpha} + z_\eta)\mathrm{SE}_{A-P}.$$

Replacing $\theta_A - \theta_P - \delta$ in (17.6) by its lower bound given in the previous expression we obtain the non-inferiority margin:

$$\Delta = (z_{1-\alpha} + z_\eta)\mathrm{SE}_{A-P} - z_{1-\varepsilon}\mathrm{SE}_{T-P}.$$

To use this non-inferiority margin, we need some information about the population variance of the placebo group. As an example, consider the parallel design with two treatments, the test therapy and the active control agent. Assume that the same two-group parallel design would have been used if a placebo-controlled trial had been conducted. Then $\mathrm{SE}_{A-P} = \sqrt{\sigma_A^2/n_A + \sigma_P^2/n_P}$ and $\mathrm{SE}_{T-P} = \sqrt{\sigma_T^2/n_T + \sigma_P^2/n_P}$, where σ_k^2 is the asymptotic variance for $\sqrt{n_k}(\hat{\theta}_k - \theta_k)$ and n_k is the sample size under treatment k. If we assume $\sigma_P/\sqrt{n_P} = c$, then

$$\Delta = (z_{1-\alpha} + z_\eta)\sqrt{\frac{\sigma_A^2}{n_A} + c^2} - z_{1-\varepsilon}\sqrt{\frac{\sigma_T^2}{n_T} + c^2}. \tag{17.8}$$

Formula (17.8) can be used in two ways. One way is to replace c in (17.8) by an estimate. When no information from the placebo control is available, a suggested estimate of c is the smaller of the estimates of $\sigma_T/\sqrt{n_T}$ and $\sigma_A/\sqrt{n_A}$. The other way is to carry out a sensitivity analysis by using Δ in (17.8) for a number of c values.

17.3 Statistical Test Based on Treatment Difference

When the non-inferiority margin depends on unknown population parameters, statistical tests designed for the case of constant non-inferiority margin may not be appropriate. Valid statistical tests for hypotheses (17.1) with Δ given by (17.2) can be found in Hung et al. (2003), Holmgren (1999), and

Wang et al. (2002b), assuming that (1) γ is known and (2) historical data from a placebo-controlled trial are available and the so-called *constancy condition* holds, i.e., the active control effects are equal in the current and historical patient populations.

Chow and Shao (2006) derived valid statistical tests for the non-inferiority margin given in (17.6) or (17.8), which are summarized below.

17.3.1 Tests Based on Historical Data under Constancy Condition

We first consider tests involving the non-inferiority margin (17.6) in the case where historical data for a placebo-controlled trial assessing the effect of the active control agent are available and the constancy condition holds, i.e., the effect $\theta_{A0} - \theta_{P0}$ in the historical trial is the same as $\theta_A - \theta_P$ in the current active-controlled trial, if a placebo control is added to the current trial. It should be emphasized that the constancy condition is a crucial assumption for the validity of the results in this subsection. A discussion of how to check the constancy condition is given in the next subsection.

Assume that the two-group parallel design is adopted in both the historical and current trials and that the sample sizes are, respectively, n_{A0} and n_{P0} for the active control and placebo in the historical trial and n_T and n_A for the test therapy and active control in the current trial. Without the normality assumption on the data, we adopt the large sample inference approach. Let $k = T, A, A0,$ and $P0$ be the indexes, respectively, for the test and active control in the current trial and the active control and placebo in the historical trial. Assume that $n_k = l_k n$ for some fixed l_k and that, under appropriate conditions, estimators $\hat{\theta}_k$ for parameters θ_k satisfy

$$\sqrt{n_k}(\hat{\theta}_k - \theta_k) \to_d N(0, \sigma_k^2) \tag{17.9}$$

as $n \to \infty$, where $\to_d$ denotes convergence in distribution. Also, assume that consistent estimators $\hat{\sigma}_k^2$ for σ_k^2 are obtained. From (17.9), the independence of data from different groups, and the constancy condition,

$$\frac{\hat{\theta}_T - \hat{\theta}_A + [r/(1+r)](\hat{\theta}_{A0} - \hat{\theta}_{P0}) - \{\theta_T - \theta_A[r/(1+r)](\theta_A - \theta_P)\}}{\text{SE}_{T-C}} \to_d N(0,1). \tag{17.10}$$

From the consistency of $\hat{\sigma}_k^2$ and the fact that $\sqrt{n}\text{SE}_{T-C}$ is a fixed constant, we have

$$\frac{\hat{\text{SE}}_{T-P} - \text{SE}_{T-P}}{\text{SE}_{T-C}} = \frac{\sqrt{n}(\hat{\text{SE}}_{T-P} - \text{SE}_{T-P})}{\sqrt{n}\text{SE}_{T-C}} = o_p(1)$$

and

$$\frac{\hat{SE}_{T-C}}{SE_{T-C}} - 1 = \frac{\sqrt{n}(\hat{SE}_{T-C} - SE_{T-P})}{\sqrt{n}SE_{T-C}} = o_p(1)$$

where $o_p(1)$ denotes a quantity converging to 0 in probability. Then

$$\frac{\hat{\theta}_T - \hat{\theta}_A + [r/(1+r)](\hat{\theta}_{A0} - \hat{\theta}_{P0} - z_{1-\varepsilon}\,\hat{SE}_{T-P}) - (\theta_T - \theta_A + \Delta)}{\hat{SE}_{T-C}}$$

$$= \left[\frac{\hat{\theta}_T - \hat{\theta}_A + [r/(1+r)](\hat{\theta}_{A0} - \hat{\theta}_{P0}) - \{\theta_T - \theta_A + [r/(1+r)](\theta_A - \theta_P)\}}{SE_{T-C}}\right.$$

$$\left. - \frac{r}{1+r}\frac{\hat{SE}_{T-P} - SE_{T-P}}{SE_{T-C}}\right]\frac{SE_{T-C}}{\hat{SE}_{T-C}}$$

$$= \left[\frac{\hat{\theta}_T - \hat{\theta}_A + [r/(1+r)](\hat{\theta}_{A0} - \hat{\theta}_{P0}) - \{\theta_T - \theta_A + [r/(1+r)](\theta_A - \theta_P)\}}{SE_{T-C}} - o_p(1)\right]$$

$$\times\,[1 + o_p(1)].$$

By Slutsky's theorem, we have

$$\frac{\hat{\theta}_T - \hat{\theta}_A + [r/(1+r)](\hat{\theta}_{A0} - \hat{\theta}_{P0} - z_{1-\varepsilon}\,\hat{SE}_{T-P}) - (\theta_T - \theta_A + \Delta)}{\hat{SE}_{T-C}} \to_d N(0,1), \quad (17.11)$$

where

$\hat{SE}_{T-P} = \sqrt{\hat{\sigma}_T^2/n_T + \hat{\sigma}_{P0}^2/n_{P0}}$ is an estimator of $SE_{T-P} = \sqrt{\sigma_T^2/n_T + \sigma_{P0}^2/n_{P0}}$
$\hat{SE}_{T-C}$ is an estimate of SE_{T-C}, the standard deviation of $\hat{\theta}_T - \hat{\theta}_A + [r/(1+r)](\hat{\theta}_{A0} - \hat{\theta}_{P0})$, which is given by

$$\hat{SE}_{T-C} = \sqrt{\frac{\hat{\sigma}_T^2}{n_T} + \frac{\hat{\sigma}_A^2}{n_A} + \left(\frac{r}{1+r}\right)^2\left(\frac{\hat{\sigma}_{A0}^2}{n_{A0}} + \frac{\hat{\sigma}_{P0}^2}{n_{P0}}\right)}.$$

Then, when the non-inferiority margin in (17.6) is adopted, the null hypothesis H_0 in (17.1) is rejected at approximately level α if

$$\hat{\theta}_T - \hat{\theta}_A + \frac{r}{1+r}(\hat{\theta}_{A0} - \hat{\theta}_{P0} - z_{1-\varepsilon}\hat{SE}_{T-P}) - z_{1-\alpha}\hat{SE}_{T-C} > 0.$$

Using result (17.11), we can approximate the power of this test by

$$\Phi\left(\frac{\theta_T - \theta_A + \Delta}{SE_{T-C}} - z_{1-\alpha}\right).$$

Using this formula, we can select the sample sizes n_T and n_A to achieve a desired power, say β, assuming that n_{A0} and n_{P0} are given (in the historical trial). Assume that $n_T/n_A = \lambda$ is chosen. Then n_T should be selected as a solution of

$$\theta_T - \theta_A + \frac{r}{1+r}\left(\theta_A - \theta_P - z_{1-\varepsilon}\sqrt{\frac{\sigma_T^2}{n_T} + \frac{\sigma_{P0}^2}{n_{P0}}}\right)$$

$$= (z_{1-\alpha} + z_\beta)\sqrt{\frac{\sigma_T^2}{n_T} + \frac{\lambda\sigma_A^2}{n_T} + \left(\frac{r}{1+r}\right)^2\left(\frac{\sigma_{A0}^2}{n_{A0}} + \frac{\sigma_{P0}^2}{n_{P0}}\right)}. \tag{17.12}$$

Although (17.12) does not have an explicit solution in terms of n_T, its solution can be numerically obtained once initial values for all parameters are given.

17.3.2 Constancy Condition

Using the historical data usually increases the power of the test for hypotheses with a non-inferiority margin depending on the parameters in the historical trial. On the other hand, using historical data without the constancy condition may lead to invalid conclusions. As indicated by Hung et al. (2003), checking the constancy condition is difficult. In this subsection, we discuss a method of checking the constancy condition under an assumption much weaker than the constancy condition.

Note that the key is to check whether the active control effect $\theta_A - \theta_P$ in the current trial is the same as $\theta_{A0} - \theta_{P0}$ in the historical trial. If we assume that the placebo effects θ_P and θ_{P0} are the same (which is much weaker than the constancy condition), then we can check whether $\theta_A = \theta_{A0}$ using the data under the active control in the current and historical trials.

17.3.3 Tests without Historical Data

We now consider tests where the non-inferiority margin (17.8) is chosen. Following (17.10) and (17.11), we can establish that

$$\frac{\hat{\theta}_T - \hat{\theta}_A + (z_{1-\alpha} + z_\eta)\hat{SE}_{A-P} - z_{1-\varepsilon}\hat{SE}_{T-P} - (\theta_T - \theta_A + \Delta)}{\hat{SE}_{T-A}} \to_d N(0,1),$$

where $\hat{\text{SE}}_{k-l} = \sqrt{\hat{\sigma}_k^2/n_k + \hat{\sigma}_l^2/n_l}$. Hence, when the non-inferiority margin in (17.8) is adopted, the null hypothesis H_0 in (17.1) is rejected at approximately level α if

$$\hat{\theta}_T - \hat{\theta}_A + (z_{1-\alpha} + z_\eta)\hat{\text{SE}}_{A-P} - z_{1-\varepsilon}\hat{\text{SE}}_{T-A} - z_{1-\alpha}\sqrt{\frac{\hat{\sigma}_T^2}{n_T} + \frac{\hat{\sigma}_A^2}{n_A}} > 0.$$

The power of this test is approximately

$$\Phi\left(\frac{\theta_T - \theta_A + \Delta}{\text{SE}_{T-A}} - z_{1-\alpha}\right).$$

If $n_T/n_A = \lambda$, then we can select the sample sizes n_T and n_A to achieve a desired power, say β, by solving

$$\theta_T - \theta_A + (z_{1-\alpha} + z_\eta)\sqrt{\frac{\lambda\sigma_A^2}{n_T} + \frac{\sigma_P^2}{n_P}} - z_{1-\varepsilon}\sqrt{\frac{\sigma_T^2}{n_T} + \frac{\sigma_P^2}{n_P}} = (z_{1-\alpha} + z_\beta)\sqrt{\frac{\lambda\sigma_A^2}{n_T} + \frac{\sigma_T^2}{n_T}}.$$

17.3.4 An Example

A clinical trial was conducted to compare the efficacy of a test therapy for treating patients with a specific cancer who had relapsed following first-line therapy and were refractory to their most recent therapy. A total of 103 patients were included in this study and were randomly assigned into two groups, 51 in the test therapy group and 52 in the active control group. All patients received treatments as a rapid intravenous bolus twice per week for 2 weeks followed by a 10 day rest period. Then, all patients received a maximum of 16 three weeks of treatment cycles. Therefore, the maximum duration of treatment in this study was 48 weeks. The actual number of cycles administered for each patient was based on the response to therapy. One of the primary study endpoints is time-to-disease progression (TTP). Observed TTP data are time-to-event data with right random censoring.

Applying the Kaplan–Meier estimation method to each treatment group, we obtain the estimated probability of TTP. The results are plotted in Figure 17.2. The parameter of interest in this example is θ_k = the median TTP. The sample median under the test therapy is $\hat{\theta}_T = 243$ days with estimated standard error $\hat{\sigma}_T/\sqrt{n_T} = 13.5$ days. The standard error estimate is calculated according to the results in Brookmeyer and Crowley (1982) and Emerson (1982). Similarly, the sample median under the active control is $\hat{\theta}_A = 235$ days with estimated standard error $\hat{\sigma}_A/\sqrt{n_A} = 14.5$ days. The estimate $\hat{\text{SE}}_{T-A} = 19.81$ days. For $\alpha = 0.05$, $z_{1-\alpha}\hat{\text{SE}}_{T-A} = 32.59$. Although $\hat{\theta}_T - \hat{\theta}_A = 8 > 0$, the hypothesis $\theta_T - \theta_A \le 0$ cannot be rejected at the 5% level, possibly due to the large variability in the data set.

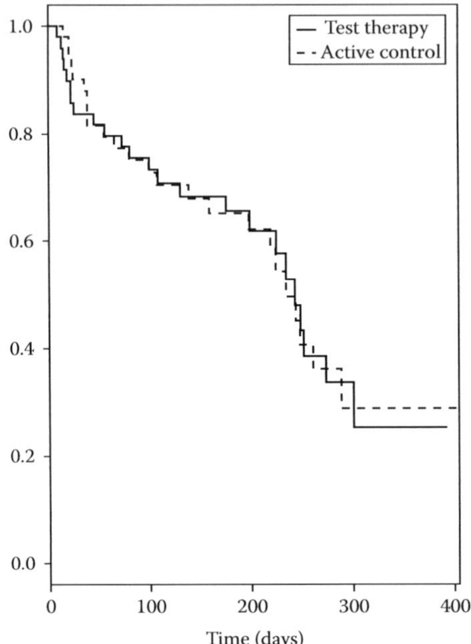

Time (days)

FIGURE 17.2
Kaplan–Meier plot of TTP.

Since we do not have historical data, we apply the test procedure as described in Section 17.3.3. For any $c > 0$, define the statistic

$$W = \hat{\theta}_T - \hat{\theta}_A + (z_{1-\alpha} + z_\eta)\sqrt{\frac{\hat{\sigma}_A^2}{n_A} + c^2} - z_{1-\varepsilon}\sqrt{\frac{\hat{\sigma}_T^2}{n_T} + c^2} - z_{1-\alpha}\sqrt{\frac{\hat{\sigma}_T^2}{n_T} + \frac{\hat{\sigma}_A^2}{n_A}}.$$

If c is an estimate of $\sigma_P/\sqrt{n_P}$, then the test procedure described in Section 17.3.3 rejects the null hypothesis $\theta_T - \theta_A + \Delta \leq 0$ if and only if $W > 0$. As discussed in Section 17.2, the test can be carried out in two ways. In the first method, we estimate $\sigma_P/\sqrt{n_P}$ by $\min(\hat{\sigma}_T/\sqrt{n_T}, \hat{\sigma}_A/\sqrt{n_A})$. Values of the statistic W and estimates of the non-inferiority margin Δ (denoted by $\hat{\Delta}$) for $\alpha = 0.05$, $\eta = 0.8$, and some ε values are given in Table 17.1. It can be seen that if ε is chosen to be 0.05, then an estimate of Δ is 19.37 days and we cannot reject the null hypothesis that $\theta_T - \theta_A + \Delta \leq 0$; if ε is chosen to be 0.1, then an estimate of Δ is 26.52 days and we reject the null hypothesis at the level $\alpha = 0.05$.

In the second method we compute the statistic W and $\hat{\Delta}$ for a set of c values. Results for $\alpha = 0.05$, $\eta = 0.8$, and $\varepsilon = 0.1$ and 0.2 are given in Table 17.2. The results indicate that if $\varepsilon = 0.2$, then the null hypothesis can be rejected for all values of c; if $\varepsilon = 0.1$, then the null hypothesis can be rejected when $c > 13$.

TABLE 17.1

Values of Statistic W and $\hat{\Delta}$ When
$c = \min(\hat{\sigma}_A/\sqrt{n_A}, \hat{\sigma}_T/\sqrt{n_T})$, $\alpha = 0.05$, $\eta = 0.80$

ε	0.50	0.30	0.25	0.20	0.15	0.10	0.05
W	26.40	16.39	13.52	10.33	6.61	1.93	−5:22
$\hat{\Delta}$	50.99	40.98	38.11	34.92	31.20	26.52	19.37

Source: Chow, S.C. and Shao, J., *Stat. Med.*, 25, 1101, 2006. © 2006 by John Wiley & Sons, Ltd. With permission.

TABLE 17.2

Values of Statistic W and $\hat{\Delta}$ When $\varepsilon = 0.1$
and 0.2, $\alpha = 0.05$, $\eta = 0.80$

c	$\varepsilon = 0.1$		$\varepsilon = 0.2$	
	W	$\hat{D}$	W	$\hat{D}$
1	−5:80	18.79	0.16	24.75
2	−5:68	18.91	0.32	24.91
3	−5:49	19.09	0.59	25.18
4	−5:23	19.36	0.96	25.55
5	−4:90	19.69	1.43	26.02
6	−4:50	20.09	2.00	26.59
7	−4:04	20.55	2.65	27.24
8	−3:52	21.07	3.38	27.97
9	−2:95	21.64	4.19	28.78
10	−2:32	22.27	5.07	29.66
11	−1:65	23.65	6.01	30.60
12	−0:94	24.40	7.01	31.60
13	−0:18	25.19	8.06	32.65
14	0.60	26.01	9.16	33.75
15	1.42	26.86	10.30	34.89
16	2.27	26.86	11.48	36.07
17	3.15	27.74	12.70	37.29
18	4.05	28.64	13.95	38.54
19	4.97	29.56	15.22	39.81
20	5.91	30.50	16.53	41.12

Source: Chow, S.C. and Shao, J., *Stat. Med.*, 25, 1101, 2006. © 2006 by John Wiley & Sons, Ltd. With permission.

17.4 Statistical Tests Based on Relative Risk

As discussed in the previous section, the method proposed by Chow and Shao (2006) is a method for selecting non-inferiority margins based on treatment difference. On the other hand, Hung et al. (2003) proposed a margin selection based on relative risk. However, with relative risk, it is difficult to adjust for covariates. In this section, we will outline a statistical method for selecting a non-inferiority margin based on relative risk.

17.4.1 Hypotheses for Non-Inferiority Margin

If the treatment effect is expressed in terms of relative risk, the hypotheses for non-inferiority testing can be formulated as follows:

$$H_0 : \frac{p_T}{p_C} \geq \Delta_2 \quad \text{versus} \quad H_1 : \frac{p_T}{p_C} < \Delta_2, \tag{17.13}$$

where Δ_2 is the non-inferiority margin. The effect retention can be considered on log relative risk scale because the statistics on this scale are better approximately by a normal distribution. The hypotheses in (17.13) can be reexpressed as

$$H_0 : \log(p_C) - \log(p_T) \leq -\log(\Delta_2) \quad \text{versus} \quad H_1 : \log(p_C) - \log(p_T) > -\log(\Delta_2) \tag{17.14}$$

Here $\log(\Delta_2)$ is the new non-inferiority margin. Again, the non-inferiority margin satisfying the two criteria from ICH E10 is given by

$$\log(\Delta_2) = \frac{r}{1+r}\left(\log(p_P) - \log(p_C) - z_{1-\varepsilon}\text{SE}_{\log(P/T)}\right). \tag{17.15}$$

In many applications, there are no historical data. In such cases, we can assume that the power of the level α test showing that the active control agent is more effective than a placebo by the margin $\log(\zeta)$ is at the level η, since the active control agent is a well-established therapy. Consider the parallel design with two treatments, the test product and the active control agent, and assume that the same two-group parallel design would have been used if a placebo-controlled trial had been conducted. Consequently, following a similar idea as described by Chow and Shao (2006), the non-inferiority margin can be obtained by

$$\log(\Delta_2) = (z_{1-\alpha} + z_\eta)\text{SE}_{\log(P/C)} - z_{1-\varepsilon}\text{SE}_{\log(P/T)}, \tag{17.16}$$

where

$$\text{SE}_{\log(P/C)} \approx \sqrt{\frac{(1-p_P)}{p_P n_P} + \frac{(1-p_C)}{p_C n_C}} \quad \text{and} \quad \text{SE}_{\log(P/T)} \approx \sqrt{\frac{(1-p_P)}{p_P n_P} + \frac{(1-p_T)}{p_T n_T}}.$$

The approximation is established as follows.

Let p_T and p_C denote the incidence rates of a clinical event associated with the experimental treatment and the control treatment, respectively, in the patient population targeted by the active control study. Then, the observed incidence rate $\hat{p}_k = O_k/n_k$ is the number of events observed in the group k divided by n_k, for $k = T, C$. The notations O_T and O_C denote the number of events observed in the treatment group and the active control group, respectively. Thus, O_k has a binomial distribution with parameters n_k and p_k, for $k = T, C$. Hence, the variance of the observed incidence rate $\hat{p}_k$ is

$$\text{var}(\hat{p}_k) = \frac{\text{var}(O_k)}{n_k^2} = \frac{p_k(1-p_k)}{n_k} \quad \text{for } k = T, C,$$

and the standard error

$$\text{SE}_{C-T} = \sqrt{\frac{p_C(1-p_C)}{n_C} + \frac{p_T(1-p_T)}{n_T}}.$$

By the delta method, $\text{var}(g(X)) \approx (\partial g/\partial X)^2 \, \text{var}(X)$. Thus, the variance of the log incidence rate ratio is

$$\text{var}\left[\log\left(\frac{p_T}{p_C}\right)\right] = \text{var}[\log(p_T)] + \text{var}[\log(p_C)]$$

$$\approx \frac{\text{var}(p_T)}{p_T^2} + \frac{\text{var}(p_C)}{p_C^2} = \frac{1-p_T}{n_T p_T} + \frac{1-p_C}{n_C p_C}$$

and the standard error $\text{SE}_{\log(C/T)} \approx \sqrt{(1-p_T)/n_T p_T + (1-p_C)/n_C p_C}$.

Here, $(1-p_k)/p_k n_k$ is the asymptotic variance for $\sqrt{n_k}(\log(\hat{p}_k) - \log(p_k))$ and n_k is the sample size under treatment k, for $k = P, C, T$. If we assume that $\text{var}(\log(p_P)) = a^2$, then we have

$$\log(\Delta_2) \approx (z_{1-\alpha} + z_\eta)\sqrt{\frac{(1-p_C)}{p_C n_C} + a^2} - z_{1-\varepsilon}\sqrt{\frac{(1-p_T)}{p_T n_T} + a^2}. \tag{17.17}$$

Formula (17.17) can be used if a in (17.17) is replaced by an estimate. Again, when no information from the placebo control is available, a suggested estimate of a is the smaller of the estimates of $\sqrt{(1-p_C)/p_C n_C}$ and $\sqrt{(1-p_T)/p_T n_T}$.

17.4.2 Tests Based on Historical Data under Constancy Condition

Again we first consider tests involving the non-inferiority margin in the case where historical data for a placebo-controlled trial assessing the effect of the active control agent are available and the constancy condition holds. The definition of constancy condition is similar to that described earlier. It should be emphasized that the constancy condition is a crucial assumption for the validity of the result in this subsection. A discussion on how to check the constancy condition is given in Chow and Shao (2006).

Assume that the two-group parallel design is adopted in both the historical and current trials and that the sample sizes are, respectively, n_{C0} and n_{P0} for the active control and the placebo in the historical trial, and n_T and n_C for the test product and active control in the current trial. Let $k = T, C, C_0,$ and P_0 be the indexes, respectively, for the test and active control in the current trial, and the active control and placebo in the historical trial. Assume that $n_k = l_k n$ for some fixed l_k, and, under appropriate conditions, estimators $\log(\hat{p}_k)$ for parameters $\log(p_k)$ satisfy

$$\sqrt{n_k}\,(\log(\hat{p}_k) - \log(p_k)) \to_d N\left(0, \frac{(1-p_k)}{p_k n_k}\right)$$

as $n \to \infty$. Also, assume that consistent estimators $\hat{\mathrm{var}}(\log(p_k))$ for $\mathrm{var}(\log(p_k))$ are obtained. As in Chow and Shao (2005), we can derive that

$$\frac{\log\left(\dfrac{\hat{p}_C}{\hat{p}_T}\right) + \left(\dfrac{r}{1+r}\right)\left(\log\left(\dfrac{\hat{p}_{P0}}{\hat{p}_{C0}}\right) - z_{1-\varepsilon}\mathrm{SE}_{\log(P/T)}\right) - \left(\log\left(\dfrac{p_C}{p_T}\right) + \log(\Delta_2)\right)}{\hat{\mathrm{SE}}_{\log(C/T)}} \to_d N(0,1)$$

$$(17.18)$$

where

$$\hat{\mathrm{SE}}_{\log(P/T)} = \sqrt{\frac{(1-\hat{p}_P)}{\hat{p}_P n_P} + \frac{(1-\hat{p}_T)}{\hat{p}_T n_T}} \quad \text{is an estimator of } \mathrm{SE}_{\log(P/T)}$$

$\hat{\mathrm{SE}}_{\log(C/T)}$ is an estimator of $\mathrm{SE}_{\log(C/T)}$, the standard deviation of $\log(\hat{p}_C/\hat{p}_T) + (r/1+r)\log(\hat{p}_{P0}/\hat{p}_{C0})$, i.e., $\hat{\mathrm{SE}}_{\log(C/T)} =$

$$\sqrt{\frac{(1-\hat{p}_C)}{\hat{p}_C n_C} + \frac{(1-\hat{p}_T)}{\hat{p}_T n_T} + \left(\frac{r}{1+r}\right)^2\left(\frac{(1-\hat{p}_{P0})}{\hat{p}_{P0} n_{P0}} + \frac{(1-\hat{p}_{C0})}{\hat{p}_{C0} n_{C0}}\right)}.$$

Then, when the non-inferiority margin in (17.15) is adopted, the null hypothesis H_0 in (17.13) is rejected at approximately level α if

$$\log\left(\frac{\hat{p}_C}{\hat{p}_T}\right) + \left(\frac{r}{1+r}\right)\left(\log\left(\frac{\hat{p}_{P0}}{\hat{p}_{C0}}\right) - z_{1-\varepsilon}\,\hat{\mathrm{SE}}_{\log(P/T)}\right) - z_{1-\alpha}\,\hat{\mathrm{SE}}_{\log(C/T)} > 0.$$

Thus we can approximate the power of this test by

$$\Phi\left(\frac{\log(p_C) - \log(p_T) + \log(\Delta_2)}{\mathrm{SE}_{\log(C/T)}} - z_{1-\alpha}\right).$$

Using this formula, we can select the sample size n_T and n_C to achieve a desired power level (say $1-\beta$), assuming that n_{C0} and n_{P0} are given in the historical trial. Suppose that $n_T/n_C = \lambda$ is chosen. Then n_T should be selected as a solution of

$$\log\left(\frac{\hat{p}_C}{\hat{p}_T}\right) + \frac{r}{1+r}\left(\log\left(\frac{\hat{p}_P}{\hat{p}_C}\right) - z_{1-\varepsilon}\sqrt{\frac{(1-\hat{p}_{P0})}{\hat{p}_{P0}n_{P0}} + \frac{(1-\hat{p}_T)}{\hat{p}_T n_T}}\right)$$

$$= (z_{1-\alpha} + z_\beta)\sqrt{\frac{(1-\hat{p}_C)}{\hat{p}_C n_C} + \frac{(1-\hat{p}_T)}{\hat{p}_T n_T} + \left(\frac{r}{1+r}\right)^2\left(\frac{(1-\hat{p}_{P0})}{\hat{p}_{P0}n_{P0}} + \frac{(1-\hat{p}_{T0})}{\hat{p}_{T0}n_{T0}}\right)}. \quad (17.19)$$

Although (17.19) does not have an explicit solution in terms of n_T, its solution can be numerically obtained once initial values for all parameters are given.

17.4.3 Tests without Historical Data

We now consider tests in which a non-inferiority margin (17.14) is chosen. Following the same argument as of Chow and Shao (2006), we can establish that

$$\frac{\log\left(\frac{\hat{p}_C}{\hat{p}_T}\right) + (z_{1-\alpha} + z_\eta)\widehat{SE}_{\log(P/C)} - z_{1-\varepsilon}\,\widehat{SE}_{\log(P/T)} - \left(\log\left(\frac{p_C}{p_T}\right) + \log(\Delta_2)\right)}{\widehat{SE}_{\log(C/T)}} \to_d N(0,1).$$

Hence, when the non-inferiority margin in (17.14) is adopted, the null hypothesis H_0 is rejected at approximately level α if

$$\log\left(\frac{\hat{p}_C}{\hat{p}_T}\right) + (z_{1-\alpha} + z_\eta)\widehat{SE}_{\log(P/C)} - z_{1-\varepsilon}\,\widehat{SE}_{\log(P/T)} - z_{1-\alpha}\sqrt{\frac{(1-\hat{p}_T)}{\hat{p}_T n_T} + \frac{(1-\hat{p}_C)}{\hat{p}_C n_C}} > 0.$$

The power of this test is approximately

$$\Phi\left(\frac{\log(p_C) - \log(p_T) + \log(\Delta_2)}{SE_{\log(C/T)}} - z_{1-\alpha}\right).$$

If $n_T/n_C = \lambda$, then we can select the sample size n_T and n_C to achieve a desired power level (say $1-\beta$) by solving

$$\log\left(\frac{p_C}{p_T}\right) + (z_{1-\alpha} + z_\eta)\sqrt{\frac{(1-p_P)}{p_P n_P} + \frac{\lambda(1-p_C)}{p_C n_T}} - z_{1-\varepsilon}\sqrt{\frac{(1-p_T)}{p_T n_T} + \frac{(1-p_P)}{p_P n_P}}$$

$$= (z_{1-\alpha} + z_\beta)\sqrt{\frac{(1-p_T)}{p_T n_T} + \frac{\lambda(1-p_C)}{p_C n_T}}.$$

Again, since we do not have historical data, for any $a > 0$, we can define the statistic

$$W_{\text{ratio}} = \log\left(\frac{\hat{p}_C}{\hat{p}_T}\right) + (z_{1-\alpha} + z_\eta)\sqrt{\frac{(1-\hat{p}_C)}{\hat{p}_C n_C} + b^2}$$

$$- z_{1-\varepsilon}\sqrt{\frac{(1-\hat{p}_T)}{\hat{p}_T n_T} + b^2} - z_{1-\alpha}\sqrt{\frac{(1-\hat{p}_T)}{\hat{p}_T n_T} + \frac{(1-\hat{p}_C)}{\hat{p}_C n_C}},$$

where b is an estimate of $\sqrt{(1-p_P)/p_P n_P}$. Consequently, the test procedure rejects the null hypothesis (17.13) if and only if $W_{\text{ratio}} > 0$.

17.4.4 An Example

Suppose that a clinical trial was conducted to compare the efficacy of a test treatment to an active control on a clinical adverse event in a target patient population with cardiovascular disease. Suppose that the estimated 5 year event rates in the active control and the treatment group are $\hat{p}_C = 21.2\%$ and $\hat{p}_T = 19.4\%$, respectively, based on a total of 500 patients per group. Consider the scenario where historical data for a placebo-controlled trial assessing the effect of the active control agent are available and the constancy condition holds, i.e., the effect $p_{P0} - p_{C0}$ (p_{P0}/p_{C0}) in the historical trial is the same as $p_P - p_C$ (p_P/p_C) in the current active-controlled trial, if a placebo control is added to the current trial. For the following work, the selections of η and ε are based on Chow and Shao (2006).

For the same data set, the estimated relative risk of the test product relative to the active control is $\hat{p}_T/\hat{p}_C = 0.9151$. We can also derive that the estimated standard errors in the active control and the treatment group are $\sqrt{(1-\hat{p}_C)/\hat{p}_C n_C} = 0.0862$ and $\sqrt{(1-\hat{p}_T)/\hat{p}_T n_T} = 0.0912$, respectively. Also the estimate $\hat{SE}_{\log(C/T)} = 0.0157$. For $\alpha = 0.05$, $z_{1-\alpha}\hat{SE}_{\log(C/T)} = 0.0258$. The estimated relative risk $\hat{p}_T/\hat{p}_C = 0.9151 < 1$, i.e., $\log(\hat{p}_C) - \log(\hat{p}_T) = 0.0887 > 0$, and thus the hypothesis $p_T/p_C \geq 1$ can be rejected at the 5% level. This shows the superiority of the test therapy to the active control agent.

Applying the test procedure described above, the statistic W_{ratio} is defined as

$$W_{\text{ratio}} = \log\left(\frac{\hat{p}_C}{\hat{p}_T}\right) + (z_{1-\alpha} + z_\eta)\sqrt{\frac{(1-\hat{p}_C)}{\hat{p}_C n_C} + b^2}$$

$$- z_{1-\varepsilon}\sqrt{\frac{(1-\hat{p}_T)}{\hat{p}_T n_T} + b^2} - z_{1-\alpha}\sqrt{\frac{(1-\hat{p}_T)}{\hat{p}_T n_T} + \frac{(1-\hat{p}_C)}{\hat{p}_C n_C}},$$

where b is an estimate of $\sqrt{(1-p_P)/p_P n_P}$. Then the test procedure rejects the null hypothesis (17.13) if and only if $W_{\text{ratio}} > 0$. If ε is chosen to be 0.1 and $b = 0.0747$, the value of the statistic $W_{\text{ratio}} = 0.0149 > 0$ and the estimated non-inferiority margin $\hat{\Delta}_2 = 1.1418$ for $\alpha = 0.05$, $\eta = 0.80$, then the null hypothesis $p_T/p_C \geq \Delta_2$ can be rejected at the level $\alpha = 0.05$.

17.5 Mixed Non-Inferiority Margin

In practice, the determination of a non-inferiority margin based on either a test for treatment difference or a test for relative risk would be critical. In this section, a statistical method for selecting a non-inferiority margin with the use of a mixed null hypothesis is described (Tsou et al., 2007). The mixed null hypothesis consists of a margin based on treatment difference and a margin based on relative risk. Both non-inferiority margins will simultaneously satisfy the principles as described in the ICH E10 guideline. Statistical tests for mixed non-inferiority margin are also derived.

17.5.1 Hypotheses for Mixed Non-Inferiority Margin

If the treatment effect is expressed in terms of the mixture of a rate difference and a rate ratio, the mixed null hypothesis is

$$
\begin{aligned}
H_0 : P_T/P_C \geq \Delta_2 \quad &\text{if } p_C \leq \pi^* \\
p_T - p_C \geq \Delta_1 \quad &\text{if } p_C > \pi^*
\end{aligned}
\quad \text{versus} \quad
\begin{aligned}
H_1 : P_T/P_C < \Delta_2 \quad &\text{if } p_C \leq \pi^* \\
p_T - p_C < \Delta_1 \quad &\text{if } p_C > \pi^*
\end{aligned}
$$

$$(17.20)$$

where Δ_1 and Δ_2 are the margins and both satisfy the two criteria stated in Section 17.2. Here, $\pi^* = \Delta_1/(\Delta_2 - 1)$ is the bent point. Thus, the mixed null hypothesis in (17.20) will be the same as a null hypothesis based on relative risk in (17.13) when $p_C \leq \pi^*$, and will be the same as a null hypothesis based

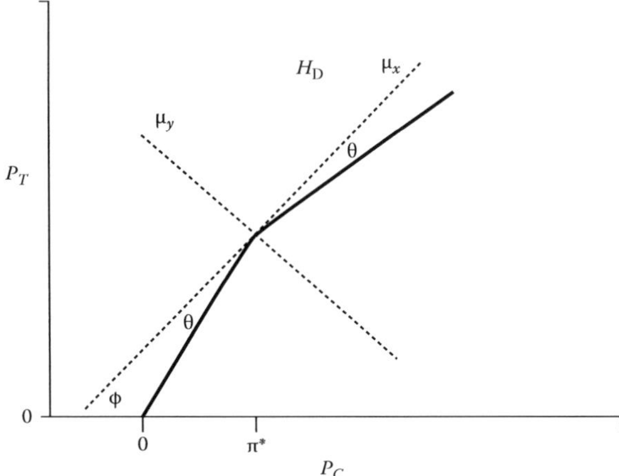

FIGURE 17.3
The original hypothesis: the area above and on the bent line shows the null hypothesis and the area under the bent line shows the alternative hypothesis.

on treatment difference when $p_C > \pi^*$. Assume that $n = n_C = n_T$. Following the approach developed by Wei and Chappel (2005), the mixed hypotheses in (17.20) can be transformed as follows:

$$H_0 : \frac{-\mu_y}{|\mu_x|} \geq \tan(\theta) \quad \text{versus} \quad H_A : \frac{-\mu_y}{|\mu_x|} < \tan(\theta), \tag{17.21}$$

where

$$\begin{pmatrix} \mu_x \\ \mu_y \end{pmatrix} = B \begin{pmatrix} p_C - \pi^* \\ p_T - \pi^* \Delta_2 \end{pmatrix}, \quad B = \begin{pmatrix} \cos(\phi) & \sin(\phi) \\ -\sin(\phi) & \cos(\phi) \end{pmatrix} \quad \text{and} \quad \begin{cases} \theta = \dfrac{1}{2}\tan^{-1}(\Delta_2) - \dfrac{\pi}{8} \\ \phi = \dfrac{1}{2}\tan^{-1}(\Delta_2) + \dfrac{\pi}{8} \end{cases}.$$

$$\tag{17.22}$$

The parameters μ_x and μ_y and the angles θ and ϕ are shown in Figure 17.1. The matrix B, called the rotation matrix, rotates the original bent line by a clockwise angle ϕ. Figures 17.3 and 17.4 provide an illustration for the original hypothesis and the rotated hypothesis.

17.5.2 Non-Inferiority Tests

When the non-inferiority margin depends on unknown population parameters, statistical tests designed for the case of the constant non-inferiority margin may not be appropriate. Chow and Shao (2006) developed valid statistical

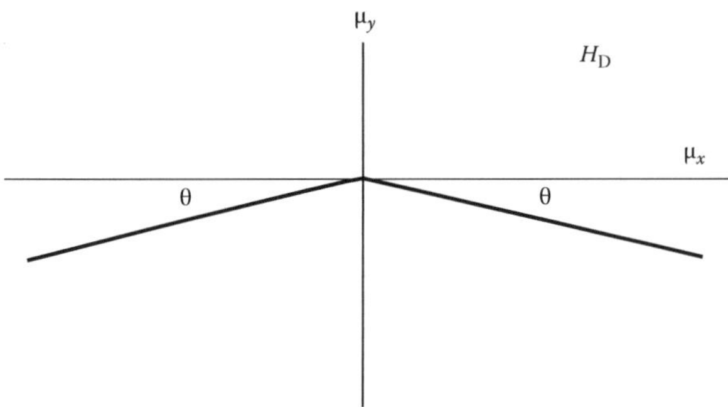

FIGURE 17.4
The rotated null hypothesis.

tests for non-inferiority tests with non-constant non-inferiority margin. Tsou et al. (2007) extended and developed statistical tests for the mixed hypotheses in (17.21). Their method is briefly outlined below.

Let (x_n, y_n) be the estimators of (μ_x, μ_y) such that

$$\begin{pmatrix} x_n \\ y_n \end{pmatrix} = B \begin{pmatrix} \hat{p}_C - \pi^* \\ \hat{p}_T - \pi^* \Delta_2 \end{pmatrix},$$

where the matrix B is defined in (17.22), and $\hat{p}_C$ and $\hat{p}_T$ are the estimated incidence rates in the active control group and treatment group, respectively. Consequently, we can have

$$x_n = \cos(\phi)(\hat{p}_C - \pi^*) + \sin(\phi)(\hat{p}_T - \pi^* \Delta_2) \quad \text{and}$$

$$y_n = -\sin(\phi)(\hat{p}_C - \pi^*) + \cos(\phi)(\hat{p}_T - \pi^* \Delta_2).$$

Consider the following test statistics under H_0:

$$M = \frac{Y_n + \tan(\theta)|X_n|}{\sqrt{\hat{v}(Y_n + \tan(\theta)|X_n|)}},$$

where $\hat{v}(Y_n + \tan(\theta)|X_n|)$ is the estimator of the variance of $Y_n + \tan(\theta)|X_n|$. Let v_{boot} denote the bootstrap variance estimator of the statistic $Y_n + \tan(\theta)|X_n|$.

Let σ_n^2 denote the sampling variance of the parameter $\mu_y + \tan(\theta)|\mu_x|$. Since

$$\frac{Y_n + \tan(\theta)|X_n|}{\sigma_n} \to_d N(0,1), \tag{17.23}$$

if we can prove that the bootstrap estimate of variance v_{boot} is a consistent estimator of the sampling variance σ_n^2, then the above result is established. To show that $v_{\text{boot}}/\sigma_n^2 \to_{a.s.} 1$, we verify the conditions in Theorem 3.8 in Shao and Tu (1995) as follows.

Let $C_1, \ldots, C_n$ be the independent and identically distributed (i.i.d.) random variables with distribution *Bernoulli*(p_C). Then

$$C_i = \begin{cases} 1 & \text{if the event is observed in the active control group with probability } p_C, \\ 0 & \text{otherwise}. \end{cases}$$

Similarly, let $T_1, T_2, \ldots, T_n$ be the i.i.d. random variables with distribution *Bernoulli*(p_T). Then $C_1 + \cdots + C_n$ is the random variable with distribution *Binomial*(n_C, p_C) and $\hat{p}_C = \bar{C}_n$; $T_1 + \cdots + T_n$ is the random variable with distribution *Binomial*(n_T, p_T) and $\hat{p}_T = \bar{T}_n$. Denote $U_i \equiv (C_i, T_i)'$ and the population mean $\mu = (p_C, p_T)'$, then $U_1, \ldots, U_n$ are i.i.d. random vectors. Define a function f to be

$$f\begin{pmatrix} x_n \\ y_n \end{pmatrix} = y_n + \tan(\theta)\,|\,x_n\,|.$$

Then a composite function $f \circ B$ is defined by

$$f\left(B\begin{pmatrix} \hat{p}_C - \pi^* \\ \hat{p}_T - \pi^*\Delta_2 \end{pmatrix}\right) = -\sin(\phi)(\hat{p}_C - \pi^*) + \cos(\phi)(\hat{p}_T - \pi^*\Delta_2)$$

$$+ \tan(\theta)\,|\cos(\phi)(\hat{p}_C - \pi^*) + \sin(\phi)(\hat{p}_T - \pi^*\Delta_2)|.$$

Let $W_n = f \circ B(\bar{U}_n)$. The conditions to be verified in Theorem 3.8 are

1. $E\|U_1\|^2 < \infty$.
2. A sufficient condition $\max_{i_1,\ldots,i_n} |W_n(U_{i_1}, \ldots, U_{i_n}) - W_n|/\tau_n \to_{a.s.} 0$, where the maximum is taken over all integers $i_1, \ldots, i_n$ satisfying $1 \le i_1 \le \cdots \le i_n \le n$, the notation $W_n(U_{i_1}, \ldots, U_{i_n})$ is the statistic W_n based on the data sets $\{U_{i_j}, i = 1, \ldots, n\}, j = 1, \ldots, n$, which are randomly selected from the original data set, and $\{\tau_n\}$ is a sequence of positive numbers satisfying $\liminf_n \tau_n > 0$ and $\tau_n = O(e^{n^q})$ with a $q \in (0, 1/2)$.
3. $f \circ B$ is continuously differentiable in a neighborhood of the population mean with $\nabla(f \circ B) \ne 0$.

We now verify the three conditions as follows.

First, condition (1) is verified since we simply have $E \frown C_1 \frown^2 < \infty$ and $E \frown T_1 \frown^2 < \infty$. We now verify condition (2). When $0 < \phi, \theta < \pi/4$, we have $0 < \sin(\phi), \cos(\phi), \tan(\theta) < 1$. Therefore,

$$W_n = f \circ B(\bar{U}_n)$$

$$= y_n + \tan(\theta) \, | \, x_n \, | \leq | \, \hat{p}_C - \pi^* \, | + | \, \hat{p}_T - \pi^* \Delta_2 \, | \leq | \, \hat{p}_C \, | + | \, \pi^* \, | + | \, \hat{p}_T \, | + | \, \pi^* \Delta_2 \, |.$$

Since the values of $\hat{p}_C$ and $\hat{p}_T$ are both between 0 and 1, the clinical meaningful values of π^* and $\pi^* \Delta_2$ are also between 0 and 1. Thus, the values of $y_n + \tan(\theta)|x_n| \leq 4$. Similarly, the values of $W_n(U_{i_1}, \ldots, U_{i_n}) \leq 4$ for all integers $i_1, \ldots, i_n$ satisfying $1 \leq i_1 \leq \cdots \leq i_n \leq n$. Consequently, we have

$$\max_{i_1, \ldots, i_n} | \, W_n(U_{i_1}, \ldots, U_{i_n}) - W_n \, | \leq 8 \quad \text{and} \quad \max_{i_1, \ldots, i_n} \frac{| \, W_n(U_{i_1}, \ldots, U_{i_n}) - W_n \, |}{\tau_n} \to 0,$$

as $n \to 0$ when we simply choose $\tau_n = e^{n^q}$ with $q = 1/3$. Thus, condition (2) is also verified.

Finally, we verify condition (3). The function $f \circ B$ is continuously differentiable except at the bent point $(\pi^*, \pi^* \Delta_2)$. Thus, if the population mean $\mu = (p_C, p_T)'$ is not equal to the bent point $(\pi^*, \pi^* \Delta_2)'$, then $f \circ B$ is continuously differentiable in a neighborhood of the population mean with $\nabla(f \circ B) \neq 0$. Although the equation $(\pi^*, \pi^* \Delta_2) = (p_C, p_T)$ with $\Delta_1 \geq 0, \Delta_2 \geq 1, 0 \leq p_C, p_T \leq 1$. Since there is no solution that satisfies all the constraints, we claim that the population mean μ is not equal to the bent point $(\pi^*, \pi^* \Delta_2)$. Thus, $f \circ B$ is continuously differentiable in a neighborhood of the population mean. The values of the gradient $\nabla(f \circ B)(\mu)$ are

$$\nabla(f \circ B)(\mu) = \begin{cases} (-\sin(\phi) - \tan(\theta)\cos(\phi), & \cos(\phi) - \tan(\theta)\sin(\phi) \, ' \text{ if } p_C < \pi^*, p_T < \pi^* \Delta_2 \\ (-\sin(\phi) - \tan(\theta)\cos(\phi), & \cos(\phi) + \tan(\theta)\sin(\phi) \, ' \text{ if } p_C < \pi^*, p_T > \pi^* \Delta_2 \\ (-\sin(\phi) + \tan(\theta)\cos(\phi), & \cos(\phi) - \tan(\theta)\sin(\phi) \, ' \text{ if } p_C > \pi^*, p_T < \pi^* \Delta_2 \\ (-\sin(\phi) + \tan(\theta)\cos(\phi), & \cos(\phi) + \tan(\theta)\sin(\phi) \, ' \text{ if } p_C > \pi^*, p_T > \pi^* \Delta_2 \end{cases}.$$

Since $0 < \sin(\phi), \cos(\phi), \tan(\theta) < 1$ for $0 < \phi, \theta < \pi/4$, we have $\nabla(f \circ B)(\mu) \neq 0$. Consequently, condition (3) is verified. This proved (17.23). Then when the non-inferiority margins Δ_1 and Δ_2 in (17.20) are adopted, the null hypothesis H_0 is rejected at approximately level α if

$$Y_n + \tan(\theta) \, | \, X_n \, | + z_{1-\alpha} v_{\text{boot}} < 0.$$

17.5.3 An Example

Consider the same data set in the test procedure for non-inferiority with a mixed null hypothesis. From the data, the estimated value of $\mu_y + \tan(\theta) \cdot |\mu_x|$ is

$$y_n + \tan(\theta)\,|\,x_n\,| = -0.0325 < 0,$$

and the estimated standard error is $\sqrt{v_{\text{boot}}} = 0.0188$ based on 10,000 replications. For $\alpha = 0.05$, $z_{1-\alpha}\sqrt{v_{\text{boot}}} = 0.0309$. The estimated value of $\mu_y + \tan(\theta)|\mu_x|$ is $-0.0325 < 0$, and thus the hypothesis $\mu_y + \tan(\theta)|\mu_x| \geq 0$ can be rejected at the 5% level of significance. Using the test procedure described in this section, the statistic W_{mix} is defined as

$$W_{\text{mix}} = -y_n - \tan(\theta)\,|\,x_n\,| - z_{1-\alpha}v_{\text{boot}}.$$

Then the test procedure rejects the null hypothesis (17.20) if and only if $W_{\text{mix}} > 0$. If ε is chosen to be 0.1, the value of the statistic $W_{\text{mix}} = 0.0834 > 0$ and the estimated non-inferiority margins $\hat{\Delta}_1 = 0.0329$ and $\hat{\Delta}_2 = 1.1418$ for $\alpha = 0.05$, $\eta = 0.80$, then the null hypothesis (17.20) can be rejected at the level of $\alpha = 0.05$.

17.6 Recent Developments

17.6.1 A Special Issue of the *Journal of Biopharmaceutical Statistics*

To reflect and explosive growth of research on non-inferiority trials, the Journal of Biopharmaceutical Statistics (JBS) published a special issue on Active Controlled Clinical Trials (JBS, Vol. 17, No. 2, pp. 197–365, 2007). In this special issue, Hung et al. (2007) discussed the issues of controlling type I error rate of the non-inferiority test using two approaches by defining two types of type I error rates: the within non-inferiority trial type I error rate and the cross-trial type I error rate. Hung et al. (2007) suggested considering both type I error rates when determining the inferiority margin. Koti (2007a,b) focused on the estimation methods of the non-inferiority measurement in the forms of the ratio of parameters.

Following the discussion of simultaneously testing superiority and non-inferiority hypotheses in active controlled clinical trials by Tsong and Zhang (2005, 2007), further compared the type I error rate of superiority test using only the test and active control, historical active control and historical placebo arms. On the other hand, Ng (2007) raised his concerns regarding the increased discovery rate when using simultaneous test routinely in general practices.

As there is a concern regarding the consistency and independency of the non-inferiority from multiple clinical trials, Yan et al. (2007) proposed a method to test for the consistency of non-inferiority from multiple clinical trials, while Tsong et al. (2007) examined the relationship between the choice of non-inferiority margin and the dependency of the non-inferiority test in multiple clinical trials.

Liao et al. (2007) and Tsong and Shen (2007) dealt with nonconventional non-inferiority application. Liao et al. (2007) proposed to use concordance correlation coefficient and the concept of non-inferiority testing for the assessment of agreement on microarray experiments. Tsong and Shen (2007), on the other hand, proposed to use tolerance interval and the two one-sided non-inferiority tests concept for the assessment of exchangeability of test and reference active control treatments.

17.6.2 FDA Draft Guidance

After a series of internal discussions, a draft guidance on non-inferiority clinical trials is currently being distributed by the U.S. Food and Drug Administration (FDA) for comments (FDA, 2010a). Basically, this draft guidance consists of four parts, which are (1) a general discussion of regulatory, study design, scientific, and statistical issues associated with the use of non-inferiority studies when these are used to establish the effectiveness of a new drug; (2) details of some of the issues such as the quantitative analytical and statistical approaches used to determine the non-inferiority margin for use in non-inferiority studies; (3) Q&A of some commonly asked questions; and (4) five examples of successful and unsuccessful efforts for determining non-inferiority margins and the conduct of non-inferiority studies.

In principle, the 2010 FDA draft guidance is very similar to the ICH E10 guideline. However, the 2010 FDA draft guidance provides more details regarding study design and statistical issues. For example, the 2010 FDA draft guidance defines two non-inferiority margins, namely M_1 and M_2, where M_1 is defined as the entire effect of the active control assumed to be present in the non-inferiority study and M_2 is referred to as the largest clinically acceptable difference (degree of inferiority) of the test drug compared to the active control. As indicated in the 2010 FDA draft guidance, M_1 is based on (1) the treatment effect estimated from the historical experience with the active control drug, (2) the assessment of the likelihood that the current effect of the active control is similar to the past effect (the constancy assumption), and (3) the assessment of the quality of the non-inferiority trial, particularly looking for defects that could reduce a difference between the active control and the new drug. On the other hand, M_2 is a clinical judgment which is never greater than M_1, even if for active control drugs with small effects, a clinical judgment might argue that a larger difference is not clinically important. Ruling out a difference between the active control and the test drug that is larger than M_1 is a critical finding that supports the conclusion of effectiveness.

As indicated in the draft guidance, there are essentially two different approaches to the analysis of the non-inferiority study: one is the fixed margin method (or the two confidence intervals method) and the other one is the synthesis method. In the fixed margin method, the margin M_1 is based on estimates of the effect of the active comparator in previously conducted studies, making any needed adjustment for changes in trial circumstances. The non-inferiority margin is then prespecified and it is usually chosen as a margin smaller than M_1 (i.e., M_2). The synthesis method combines (or synthesizes) the estimate of treatment effect relative to the control from the non-inferiority trial with the estimate of the control effect from a meta-analysis of historical trials. This method treats both sources of data as if they came from the same randomized trial to project what the placebo effect would have been had the placebo been present in the non-inferiority trial.

17.7 Concluding Remarks

To assess the type I error rate and the power, a number of simulation studies were performed. The true event rates associated with the active control and the new treatment were given in Tables 17.3 through 17.5. The results were based on 10,000 replications for each simulation run under the assumption that the constancy condition holds. As seen in Table 17.3, the type I error rate is close to 0.05 when the sample size is greater than 100. We may expect that the type I error rate can be preserved when the sample size is large enough. Tables 17.4 and 17.5 display the actual power and simulated powers for different testing hypotheses for different combinations of parameters. The simulation study shows that the mixed test gives a similar result as that in the ratio test when $p_C \leq \pi^*$ and that in the difference test when $p_C > \pi^*$.

Although the ICH E10 guideline and the 2010 FDA draft guidance provide a general framework for the selection of appropriate non-inferiority margins, there is so far no established rule or gold standard for the selection of non-inferiority margins in active-controlled trials. Hung et al. (2003) proposed a margin selection based on relative risk. However, with relative risk, it is difficult to do covariate adjustments. On the other hand, Chow and Shao (2006) proposed a method for selecting non-inferiority margins based on treatment difference. From the example in Section 17.4, the difference test shows the non-inferiority of the new therapy to the active control agent, but does not have the evidence of the superiority of the new therapy to the active control agent. On the other hand, the ratio test concludes the non-inferiority of the new therapy to the active control agent and provides the evidence of the superiority of the new therapy to the active control agent. Consequently, the determination of choosing the difference test or ratio test would be critical. Tsou et al. (2007) proposed a non-inferiority test statistic for testing the mixed hypothesis based on treatment difference and relative risk for active-controlled trials. One benefit of the mixed test is that we do

TABLE 17.3

Empirical Significance Level for Mixed Testing Hypotheses (10,000 Replicates), $n = n_C = n_T$, $\alpha = 0.05$, $\eta = 0.8$, $\varepsilon = 0.1$

					Mixed
p_C	p_T	Δ_1	Δ_2	n	Simulated
0.2	0.2927	0.0927	1.6701	50	0.0717
	0.2738	0.0738	1.4938	80	0.0715
	0.2662	0.0662	1.4296	100	0.0556
	0.2543	0.0543	1.3360	150	0.0526
	0.2423	0.0423	1.2494	250	0.0583
0.3	0.4082	0.1082	1.4778	50	0.0653
	0.3858	0.0858	1.3574	80	0.0656
	0.3768	0.0768	1.3127	100	0.0539
	0.3629	0.0629	1.2468	150	0.0437
	0.3488	0.0488	1.1848	250	0.0551
0.4	0.5174	0.1174	1.3677	50	0.0636
	0.4928	0.0928	1.2774	80	0.0625
	0.4830	0.0830	1.2436	100	0.0519
	0.4678	0.0678	1.1933	150	0.0441
	0.4526	0.0526	1.1455	250	0.0580
0.5	0.6215	0.1215	1.2920	50	0.0638
	0.5958	0.0958	1.2216	80	0.0601
	0.5856	0.0856	1.1950	100	0.0594
	0.5698	0.0698	1.1553	150	0.0597
	0.5540	0.0540	1.1173	250	0.0560
0.6	0.7209	0.1209	1.2340	50	0.0600
	0.6949	0.0949	1.1782	80	0.0560
	0.6847	0.0847	1.1570	100	0.0633
	0.6689	0.0689	1.1253	150	0.0604
	0.6532	0.0532	1.0949	250	0.0520
0.7	0.8153	0.1153	1.1856	50	0.0611
	0.7901	0.0901	1.1415	80	0.0597
	0.7802	0.0802	1.1248	100	0.0600
	0.7651	0.0651	1.0998	150	0.0580
	0.7502	0.0502	1.0757	250	0.0518
0.8	0.9037	0.1037	1.1419	50	0.0602
	0.8803	0.0803	1.1079	80	0.0604
	0.8713	0.0713	1.0951	100	0.0575
	0.8577	0.0577	1.0760	150	0.0593
	0.8443	0.0443	1.0577	250	0.0536
0.9	0.9848	0.0848	1.0996	50	0.0665
	0.9634	0.0634	1.0738	80	0.0645
	0.9558	0.0558	1.0647	100	0.0582
	0.9446	0.0446	1.0513	150	0.0562
	0.9339	0.0339	1.0387	250	0.0475

TABLE 17.4

Actual Powers and Simulated Powers for Different Testing Hypotheses (10,000 Replicates), $n = n_C = n_T$, $\alpha = 0.05$, $\eta = 0.8$, $\varepsilon = 0.1$

					Difference		Mixed
p_C	p_T	Δ_1	Δ_2	n	Simulated	Actual	Simulated
0.20	0.15	0.0993	1.5789	50	0.6064	0.6273	0.5582
		0.0785	1.4348	80	0.6710	0.6912	0.6487
		0.0702	1.3812	100	0.7058	0.7250	0.6909
		0.0574	1.3017	150	0.8198	0.7903	0.8146
		0.0444	1.2266	250	0.8578	0.8728	0.8555
0.30	0.25	0.1122	1.4254	50	0.5802	0.5690	0.5775
		0.0887	1.3234	80	0.6072	0.6264	0.6156
		0.0793	1.2848	100	0.6587	0.6577	0.6570
		0.0648	1.2271	150	0.7011	0.7205	0.7081
		0.0502	1.1718	250	0.7953	0.8070	0.7974
0.40	0.35	0.1189	1.3309	50	0.5419	0.5407	0.5656
		0.0940	1.2536	80	0.5937	0.5945	0.6041
		0.0841	1.2240	100	0.6186	0.6242	0.6250
		0.0687	1.1795	150	0.6793	0.6847	0.6884
		0.0532	1.1364	250	0.7685	0.7708	0.7730
0.50	0.45	0.1207	1.2635	50	0.4892	0.5263	0.5768
		0.0954	1.2031	80	0.5483	0.5788	0.6114
		0.0853	1.1799	100	0.5855	0.6078	0.6380
		0.0697	1.1446	150	0.6408	0.6675	0.6851
		0.0540	1.1103	250	0.7291	0.7535	0.7618
0.60	0.55	0.1175	1.2103	50	0.4982	0.5207	0.5769
		0.0929	1.1629	80	0.5954	0.5738	0.6261
		0.0831	1.1445	100	0.5935	0.6032	0.6513
		0.0679	1.1165	150	0.6841	0.6637	0.7019
		0.0526	1.0891	250	0.7595	0.7511	0.7729
0.70	0.65	0.1092	1.1647	50	0.5070	0.5226	0.6076
		0.0863	1.1281	80	0.5987	0.5786	0.6467
		0.0772	1.1138	100	0.6050	0.6095	0.6712
		0.0630	1.0920	150	0.6919	0.6729	0.7246
		0.0488	1.0706	250	0.7712	0.7634	0.7938
0.80	0.75	0.0942	1.1222	50	0.5334	0.5337	0.6543
		0.0745	1.0955	80	0.6134	0.5962	0.5962
		0.0666	1.0850	100	0.6495	0.6304	0.7170
		0.0544	1.0689	150	0.7172	0.6997	0.7712
		0.0421	1.0529	250	0.8040	0.7956	0.8358
0.90	0.85	0.0685	1.0775	50	0.5859	0.5604	0.7692
		0.0542	1.0608	80	0.6612	0.6379	0.7907
		0.0484	1.0542	100	0.7142	0.6794	0.8089
		0.0396	1.0440	150	0.7899	0.7602	0.8559
		0.0306	1.0339	250	0.8860	0.8619	0.9197

TABLE 17.5

Actual Powers and Simulated Powers for Different Testing Hypotheses (10,000 Replicates), $n = n_C = n_T$, $\alpha = 0.05$, $\eta = 0.8$, $\varepsilon = 0.1$

| | | | | | Mixed | Ratio | |
| | | | | | Simulated | Simulated | Actual |
p_C	p_T	Δ_1	Δ_2	n			
0.20	0.05	0.1092	1.4092	50	0.9196	0.9517	0.8172
		0.0863	1.3115	80	0.9921	0.9927	0.9260
		0.0772	1.2745	100	0.9976	0.9976	0.9601
		0.0630	1.2190	150	0.9999	0.9999	0.9918
0.30	0.15	0.1175	1.3689	50	0.9091	0.8908	0.8087
		0.0929	1.2817	80	0.9693	0.9633	0.9086
		0.0831	1.2486	100	0.9859	0.9816	0.9442
		0.0678	1.1987	150	0.9965	0.9957	0.9839
0.40	0.25	0.1220	1.3000	50	0.8863	0.8521	0.7871
		0.0964	1.2305	80	0.9516	0.9383	0.8871
		0.0862	1.2038	100	0.9677	0.9601	0.9258
		0.0704	1.1636	150	0.9897	0.9879	0.9740
0.50	0.35	0.1221	1.2428	50	0.8712	0.8313	0.7754
		0.0965	1.1875	80	0.9354	0.9143	0.8753
		0.0863	1.1662	100	0.9617	0.9505	0.9154
		0.0705	1.1337	150	0.9874	0.9829	0.9678
0.60	0.45	0.1175	1.1948	50	0.8850	0.8402	0.7736
		0.0929	1.1511	80	0.9441	0.9213	0.8735
		0.0831	1.1341	100	0.9596	0.9450	0.9137
		0.0679	1.1082	150	0.9880	0.9835	0.9667
0.70	0.55	0.1078	1.1521	50	0.8959	0.8451	0.7819
		0.0852	1.1184	80	0.9499	0.9261	0.8817
		0.0762	1.1053	100	0.9691	0.9571	0.9209
		0.0622	1.0852	150	0.9914	0.9866	0.9711
0.80	0.65	0.0913	1.1112	50	0.9238	0.8706	0.8020
		0.0722	1.0870	80	0.9681	0.9477	0.9009
		0.0645	1.0774	100	0.9814	0.9705	0.9374
		0.0527	1.0628	150	0.9953	0.9922	0.9803
0.90	0.75	0.0637	1.0673	50	0.9742	0.9286	0.8365
		0.0504	1.0529	80	0.9901	0.9797	0.9322
		0.0451	1.0472	100	0.9966	0.9922	0.9626
		0.0368	1.0383	150	0.9994	0.9993	0.9917

not need to choose between difference test and ratio test in advance. In particular, this mixed null hypothesis consists of a margin based on treatment difference and a margin based on relative risk. From Tables 17.3 through 17.5, the proposed mixed non-inferiority test not only preserves the type I error rate at the desired level but also gives a similar power as that from the difference test or as that from the ratio test.

For regulatory recommendations, the ICH E10 guideline recommends that the non-inferiority margin should be chosen to satisfy at least two criteria summarized in Section 17.2. In other words, the non-inferiority margin should be chosen in such a way that if the non-inferiority of the test product to the active control therapy is claimed, the test product is not only non-inferior to the active control therapy but also superior to the placebo. In addition, the variability should be taken into account. On the other hand, the 2010 FDA draft guidance recommends two non-inferiority margins, namely M_1 and M_2, where M_1 is the entire effect of the active control assumed to be present in the non-inferiority study and M_2 is the largest clinically acceptable difference of the test drug compared to the active control. As indicated by the FDA, M_2 is a clinical judgment which is never greater than M_1, even if for active control drugs with small effects, a clinical judgment might argue that a larger difference is not clinically important. Ruling out a difference between the active control and the test drug that is larger than M_1 is a critical finding that supports the conclusion of effectiveness.

18

QT Studies with Recording Replicates

18.1 Introduction

As indicated by Tsong and Zhang (2008), delay in cardiac repolarization creates an electrophysiological environment that may set off cardiac arrhythmias, particularly a polymorphic ventricular tachycardia. This condition can degenerate into ventricular fibrillation, leading to sudden cardiac death. The QT interval represents the duration of ventricular depolarization and subsequent repolarization and is typically measured on a 12-lead surface electrocardiogram (ECG) from the beginning of the QRS complex to the end of the T wave (see Figure 18.1). The RR interval, which is the distance between two consecutive R waves, is the inverse of the heart rate. In pharmaceutical research and development, drug-induced prolongation of the QT interval has been used as an indicator of possible cardiac safety problems. The QT interval is often used to indirectly assess the delay in cardiac repolarization, which can predispose to the development of life-threatening cardiac arrhythmias such as torsade depointes (Moss, 1993). The QTc interval is referred to as the QT interval corrected by heart rate. In clinical practice, it is recognized that the prolongation of the QT/QTc interval is related to the increased risk of cardiotoxicity, such as a life-threatening arrhythmia (Temple, 2003). Thus, it is suggested that a careful evaluation of potential QT/QTc prolongation be assessed for potential drug-induced cardiotoxicity.

For the development of a new pharmaceutical entity, most regulatory agencies such as the U.S. FDA require the evaluation of pro-arrhythmic potential (see CPMP, 1997; FDA/TPD, 2003). In recent years, after several drugs were removed from the market because of deaths due to ventricular tachycardia resulting from drug-induced QT prolongation (Pratt et al., 1994; Khongphatthanayothin et al., 1998; Wysowski et al., 2001; Lasser et al., 2002), the International Conference on Harmonization (ICH) issued a guideline on the evaluation of QT/QTc interval prolongation and pro-arrhythmic potential for non-antiarrhythmic drugs (ICH, 2005a) and requested all sponsors submitting new drug applications to conduct at least one thorough QT (TQT) study, normally early in the clinical development with some information about the pharmacokinetics of the drug.

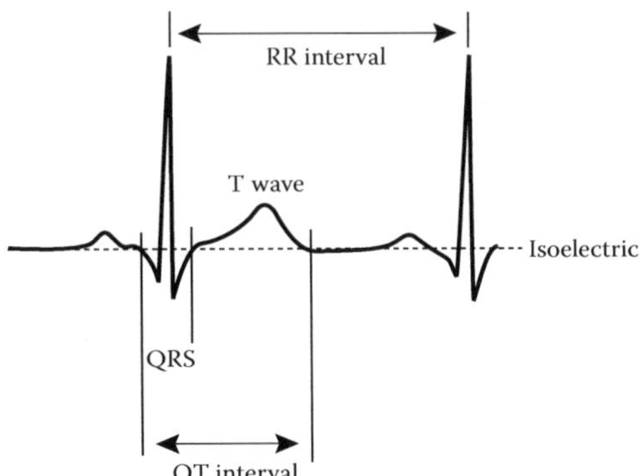

FIGURE 18.1
QT and RR intervals of the surface ECG.

The ICH E14 guideline also provides the basic recommendations on the regulatory requirements on the assessment of drug-induced prolongation of the QT interval. The ICH E14 guideline calls for a placebo-controlled study in normal healthy volunteers with a positive control to assess cardiotoxicity by examining QT/QTc prolongation. Under a valid study design (e.g., a parallel-group design or a crossover design), ECGs will be collected at baseline and at several time points posttreatment for each subject. Malik and Camm (2001) recommend that it would be worthwhile to consider 3–5 replicate ECGs at each time point within 2–5 min periods. Replicate ECGs are then defined as single ECGs recorded within several minutes of a nominal time (PhRMA QT Statistics Expert Working Team, 2003). Along this line, Strieter et al. (2003) studied the effect of replicate ECGs on QT variability in health subjects. In practice, it is then of interest to investigate the impact of recording replicates on power and sample size calculation in routine QT studies.

In clinical trials, a pre-study power analysis for sample size calculation is usually performed to ensure that the study will achieve a desired power (or the probability of correctly detecting a clinically meaningful difference if such a difference truly exists). For QT studies, the following information is necessarily obtained prior to the conduct of the pre-study power analysis for sample size calculation. The information includes (1) the variability associated with the primary study endpoint such as the QT intervals (or the QT interval endpoint change from baseline), (2) the maximal difference in QT interval between treatment groups, and (3) the number of time points where QT measurements are taken. Under the above assumptions, the procedures as described by Longford (1993) and Chow et al. (2003) can then be applied for sample size calculation under the study design (e.g., a parallel-group design or a crossover design). Although QT/QTc studies involve multiple time points, we will consider in this chapter the simplified case with only one time point. And we argue that considering one time point, though conservative, is reasonable for sample size determination purpose. This is particularly true if we focus on the time point where the maximal QT difference between treatments is expected.

The ICH E14 guidance recommends a *thorough* QT/QTc study to decide whether the drug induces QT/QTc prolongation as is evidenced if the upper bound of the 95% confidence interval of the mean drug effect on QTc exceeds 10 ms. Statistical Methods for TQT/QTc study have been proposed by Patterson et al. (2005b) under a linear mixed model and by Eaton et al. (2006) using a confidence interval approach. Hosmane and Locke (2005) examined the power in TQT/QTc studies via a simulation study, while Wang, Pan, and Balch (2008) investigated bias and variance evaluation of QT interval correlation methods. For a review of the statistical design and analysis in QT/QTc studies, see Patterson et al. (2005). The testing method proposed in Patterson et al. (2005b) was essentially an intersection-union method which is typically conservative. To address this issue, Eaton et al. (2006) constructed a confidence interval, via delta-method, for a parameter which sufficiently approximates the parameter of interest. However, this method technically assumes that mean QT/QTc differences between drug and placebo are positive at all time intervals, which is too restrictive and unverifiable in reality. Furthermore, when applying to a function (although smooth) which is presumably close to a non-smooth function (i.e., maximum function), the delta-method may yield a confidence interval whose actual coverage considerably differs from the nominal one, particularly when the sample size is moderate in size. To address these restrictions, Cheng et al. (2008) proposed a new testing method based on the maximum of correlated normal random variables.

The remainder of this chapter is organized as follows. In the next section, commonly used study designs such as a parallel-group design or a crossover design for routine QT studies with recording replicates are briefly described. Power analyses and the corresponding sample size calculations under a parallel-group design and a crossover design are derived in Section 18.3. Extensions to the designs with covariates such as pharmacokinetic (PK) responses are considered in Section 18.4. The sample size allocation optimization is discussed in Section 18.5. Some tests for QT/QTc prolongation are discussed in Section 18.6. Recent developments are given in Section 18.7. Section 18.8 provides some concluding remarks.

18.2 Study Designs and Models

As indicated by Zhang and Machado (2008), for a typical TQT, a randomized four-treatment group design is usually considered. The four treatment arms are (1) drug with therapeutic dose, (2) drug with supratherapeutic dose, (3) positive control, and (4) placebo. A typical study design for TQT studies could be either a parallel-group design or a crossover design. In this section, simple statistical models under a parallel-group design and a crossover design are briefly outlined.

Under a parallel-group design, qualified subjects will be randomly assigned to receive either treatment A or treatment B. ECGs will be collected at baseline and at several time points post treatment. Subjects will fast at least 3 h and rest at least 10 min prior to the scheduled ECG measurements. Identical lead placement and the same ECG machine will be used for all measurements. As recommended by Malik and Camm (2001), 3–5 recording replicate ECGs at each time point will be obtained within 2–5 min periods.

Let y_{ijk} be the QT interval observed from the kth recording replicate of the jth subject who receives treatment i, where $i = 1, 2, j = 1, \ldots, n$, and $k = 1, \ldots, m$. Consider the following model:

$$y_{ijk} = \mu_i + e_{ij} + \varepsilon_{ijk}, \tag{18.1}$$

where

e_{ij} are independent and identically distributed as normal random variables with mean 0 and variance σ_S^2 (between subject or intersubject variability)

ε_{ijk} are independent and identically distributed as normal random variables with mean 0 and variance σ_e^2 (within subject or intra-subject variability or measurement error variance)

Thus, we have $\mathrm{Var}(y_{ijk}) = \sigma_S^2 + \sigma_e^2$.

Under a crossover design, qualified subjects will be randomly assigned to receive one of the two sequences of test treatments under study. In other words, subjects who are randomly assigned to sequence 1 will receive treatment 1 first and then be crossed over to receive treatment 2 after a sufficient period of washout. Let y_{ijkl} be the QT interval observed from the kth recording replicate of the jth subject in the lth sequence who receives the ith treatment, where $i = 1, 2, j = 1, \ldots, n, k = 1, \ldots, m$, and $l = 1, 2$. We consider the following model:

$$y_{ijkl} = \mu_i + \beta_{il} + e_{ijl} + \varepsilon_{ijkl}, \tag{18.2}$$

where

β_{il} are independent and identically distributed normal random period effects (period uniquely determined by sequence l and treatment i) with mean 0 and variance σ_p^2

e_{ijl} are independent and identically distributed normal subject random effects with mean 0 and variance σ_S^2

ε_{ijkl} are independent and identically distributed normal random errors with mean 0 and variance σ_e^2

Thus, $\mathrm{Var}(y_{ijkl}) = \sigma_p^2 + \sigma_S^2 + \sigma_e^2$.

To ensure a valid comparison between the parallel design and the crossover design, we assume that μ_i, σ_S^2, and σ_e^2 are the same as those given in (18.1) and (18.2) and consider an extra variability σ_p^2, which is due to the random period effect for the crossover design.

18.3 Power and Sample Size Calculation

Under models (18.1) and (18.2), Chow et al. (2006) derived formulas for sample size calculation and examined the relationship between a crossover design and a parallel-group design for QT studies with recording replicates. The power analysis for sample size calculations under a parallel-group design and a crossover design are described in the subsequent subsections.

18.3.1 Parallel-Group Design

Under the parallel-group design as described in the previous section, to evaluate the impact of recording replicates on power and sample size calculation, for simplicity, we will only consider one time point post treatment. The results for recording replicates at several posttreatment intervals can be similarly obtained. Under model (18.1), consider sample mean of QT intervals of the jth subject who receives the ith treatment, then $\mathrm{Var}(\bar{y}_{ij\cdot}) = \sigma_S^2 + \sigma_e^2/m$. The hypotheses of interest regarding treatment difference in QT interval are given by

$$H_0 : \mu_1 - \mu_2 \geq 10 \quad \text{versus} \quad H_a : \mu_1 - \mu_2 < 10. \tag{18.3}$$

Under the null hypothesis of no treatment difference, the following statistic can be derived:

$$T = \frac{\bar{y}_{1\cdot\cdot} - \bar{y}_{2\cdot\cdot} - 10}{\sqrt{(2/n)(\hat{\sigma}_S^2 + \hat{\sigma}_e^2/m)}},$$

where

$$\hat{\sigma}_e^2 = \frac{1}{2n(m-1)} \sum_{i=1}^{2} \sum_{j=1}^{n} \sum_{k=1}^{m} (y_{ijk} - \bar{y}_{ij\cdot})^2,$$

and

$$\hat{\sigma}_S^2 = \frac{1}{2(n-1)} \sum_{i=1}^{2} \sum_{j=1}^{n} (\bar{y}_{ij\cdot} - \bar{y}_{i\cdot\cdot})^2 - \frac{1}{2nm(m-1)} \sum_{i=1}^{2} \sum_{j=1}^{n} \sum_{k=1}^{m} (y_{ijk} - \bar{y}_{ij\cdot})^2.$$

Under the null hypothesis in (18.3), T has a central t-distribution with $2n - 2$ degrees of freedom.

Let $\sigma^2 = \mathrm{Var}(y_{ijk}) = \sigma_S^2 + \sigma_e^2$ and $\rho = \sigma_S^2/(\sigma_S^2 + \sigma_e^2)$, then under a given alternative that $H_a: \mu_1 - \mu_2 = d < 10$ in (18.3), the power of the test can be approximated as follows:

$$1 - \beta \approx \Phi\left(-z_\alpha + \frac{\delta}{\sqrt{(2/n)(\rho + (1-\rho)/m)}}\right), \tag{18.4}$$

where

$\delta = (10 - d)/\sigma$ is the relative effect size

Φ is the cumulative distribution of a standard normal

To achieve the desired power of $1 - \beta$ at the α level of significance, the sample size needed per treatment is

$$n = \frac{2(z_\alpha + z_\beta)^2}{\delta^2}\left(\rho + \frac{1-\rho}{m}\right). \tag{18.5}$$

18.3.2 Crossover Design

Under a crossover model (18.2), it can be verified that $y_{i\cdots}$ is an unbiased estimator of μ_i with variance $\sigma_p^2/2 + \sigma_S^2/2n + \sigma_e^2/2nm$. Thus, we used the following test statistic to test the hypotheses in (18.3):

$$T = \frac{\bar{y}_{1\cdots} - \bar{y}_{2\cdots} - 10}{\sqrt{\hat{\sigma}_p^2 + (1/n)\left(\hat{\sigma}_S^2 + \left(\hat{\sigma}_e^2/m\right)\right)}},$$

where

$$\hat{\sigma}_e^2 = \frac{1}{4n(m-1)}\sum_{i=1}^{2}\sum_{j=1}^{n}\sum_{k=1}^{K}\sum_{l=1}^{2}(y_{ijkl} - \bar{y}_{ij\cdot l})^2$$

and

$$\hat{\sigma}_S^2 = \frac{1}{4(n-1)}\sum_{i=1}^{2}\sum_{j=1}^{n}\sum_{l=1}^{2}(\bar{y}_{ij\cdot l} - \bar{y}_{i\cdot\cdot l})^2 - \frac{1}{4nm(m-1)}\sum_{i=1}^{2}\sum_{j=1}^{n}\sum_{k=1}^{m}\sum_{l=1}^{2}(y_{ijkl} - \bar{y}_{ij\cdot l})^2$$

and

$$\hat{\sigma}_p^2 = \frac{1}{2}\sum_{i=1}^{2}\sum_{l=1}^{2}(\bar{y}_{i\cdot\cdot l} - \bar{y}_{\cdots})^2 - \frac{1}{4n(n-1)}\sum_{i=1}^{2}\sum_{j=1}^{n}\sum_{l=1}^{2}(\bar{y}_{ij\cdot l} - \bar{y}_{i\cdot\cdot l})^2.$$

Under the null hypothesis in (18.3), T has a central t-distribution with $2n - 4$ degrees of freedom. Let σ^2 and ρ be defined as in the previous section, and $\gamma = \sigma_p^2/\sigma^2$, then $\mathrm{Var}(y_{ijkl}) = \sigma^2/(1 + \gamma)$.

Under a given alternative that $\mu_1 - \mu_2 = d < 10$ in (18.3), the power of the test can be approximated as follows:

$$1 - \beta \approx \Phi\left(-z_\alpha + \frac{\delta}{\sqrt{\gamma + (1/n)(\rho + (1 - \rho/m))}}\right), \tag{18.6}$$

where $\delta = (10 - d)/\sigma$. To achieve the desired power of $1 - \beta$ at the α level of significance, the sample size needed per treatment group is given by

$$n = \frac{(z_\alpha + z_\beta)^2}{\delta^2 - \gamma(z_\alpha + z_\beta)^2}\left(\rho + \frac{1 - \rho}{m}\right). \tag{18.7}$$

18.3.3 Remarks

Let n_{old} be the sample size with $m = 1$ (i.e., there is a single measure for each subject). Then, we have $n = \rho n_{\mathrm{old}} + (1 - \rho)n_{\mathrm{old}}/m$. Thus, sample size (with recording replicates) required for achieving the desired power is a weighted average of n_{old} and n_{old}/m. Note that this relationship holds under both a parallel and a crossover design. Table 18.1 provides sample sizes required

TABLE 18.1

Sample Size for Achieving the Same Power
with m Recording Replicates

		m	
ρ	1	3	5
1.0	n	$1.00n$	$1.00n$
0.9	n	$0.93n$	$0.92n$
0.8	n	$0.86n$	$0.84n$
0.7	n	$0.80n$	$0.76n$
0.6	n	$0.73n$	$0.68n$
0.5	n	$0.66n$	$0.60n$
0.4	n	$0.60n$	$0.52n$
0.3	n	$0.53n$	$0.44n$
0.2	n	$0.46n$	$0.36n$
0.1	n	$0.40n$	$0.28n$
0.0	n	$0.33n$	$0.20n$

Source: Chow, S.C. et al., *Sample Size Calculation in Clinical Research*, Chapman and Hall/CRC Press, Taylor & Francis, New York, 2008. With permission.

TABLE 18.2

Sample Sizes Required under a Parallel-Group Design

| | Power = 80% | | | | | Power = 90% | | | | |
| | ρ | | | | | ρ | | | | |
(m, δ)	0.2	0.4	0.6	0.8	1.0	0.2	0.4	0.6	0.8	1.0
(3, 0.3)	81	105	128	151	174	109	140	171	202	233
(3, 0.4)	46	59	72	85	98	61	79	96	114	131
(3, 0.5)	29	38	46	54	63	39	50	64	73	84
(5, 0.3)	63	91	119	147	174	84	121	159	196	233
(5, 0.4)	35	51	67	82	98	47	68	89	110	131
(5, 0.5)	23	33	43	53	63	30	44	57	71	84

Source: Chow, S.C. et al., *Sample Size Calculation in Clinical Research,* Chapman and Hall/CRC Press, Taylor & Francis, New York, 2008. With permission.

under a chosen design (either parallel or crossover) for achieving the same power with a single recording ($m = 1$), three recording replicates ($m = 3$), and five recording replicates ($m = 5$).

Note that if ρ closes to 0, then these repeated measures can be treated as independent replicates. As can be seen from the above, if $\rho \approx 0$, then $n \approx n_{\text{old}}/m$. In other words, sample size is indeed reduced when the correlation coefficient between recording replicates is close to 0 (in this case, the recording replicates are almost independent). Table 18.2 shows the sample size reduction for different values of ρ under the parallel design. However, in practice, ρ is expected to be close to 1. In this case, we have $n \approx n_{\text{old}}$. In other words, there is not much gain for considering recording replicates in the study.

In practice, it is of interest to know whether the use of a crossover design can further reduce the sample size when other parameters such as d, σ^2, and ρ remain the same. Comparing formulas (18.5) and (18.7), we conclude that the sample size reduction by using a crossover design depends upon the parameter $\gamma = \sigma_p^2/\sigma^2$, which is a measure of the relative magnitude of period variability with respect to the within-period subject marginal variability. Let $\theta = \gamma/(z_\alpha + z_\beta)^2$, then by (18.5) and (18.7) the sample size n_{cross} under the crossover design and the sample size n_{parallel} under the parallel group design satisfy $n_{\text{cross}} = n_{\text{parallel}}/2(1 - \theta)$. When the random period effect is negligible, that is, $\gamma \approx 0$ and hence $\theta \approx 0$, we have $n_{\text{cross}} = n_{\text{parallel}}/2$. This indicates that the use of a crossover design could further reduce the sample size by half as compared to a parallel-group design when the random period effect is negligible (based on the comparison of the above formula and the formula given in (18.5). However, when the random period effect is not small, the use of a crossover design may not result in sample size reduction. Table 18.3 shows the sample size under different values of γ. It is seen that the possibility of sample size reduction under a crossover design depends upon whether

TABLE 18.3

Sample Sizes Required under a Crossover Design with $\rho = 0.8$

| | Power = 80% | | | | | Power = 90% | | | | |
| | γ | | | | | γ | | | | |
(m, δ)	0.000	0.001	0.002	0.003	0.004	0.000	0.001	0.002	0.003	0.004
(3, 0.3)	76	83	92	102	116	101	115	132	156	190
(3, 0.4)	43	45	47	50	53	57	61	66	71	77
(3, 0.5)	27	28	29	30	31	36	38	40	42	44
(5, 0.3)	73	80	89	99	113	98	111	128	151	184
(5, 0.4)	41	43	46	48	51	55	59	64	69	75
(5, 0.5)	26	27	28	29	30	35	37	39	40	42

Source: Chow, S.C. et al., *Sample Size Calculation in Clinical Research,* Chapman and Hall/CRC Press, Taylor & Francis, New York, 2008. With permission.

the carryover effect of the QT intervals could be avoided. As a result, it is suggested that a sufficient length of washout period be applied between dosing periods to wear off the residual (or carryover) effect from one dosing period to another. For a fixed sample size, the possibility of power increase by crossover design also depends on parameter γ. Figure 18.2 shows that the crossover design results in power increase when γ is close to 0 but may result in considerable power loss when γ is not close to 0.

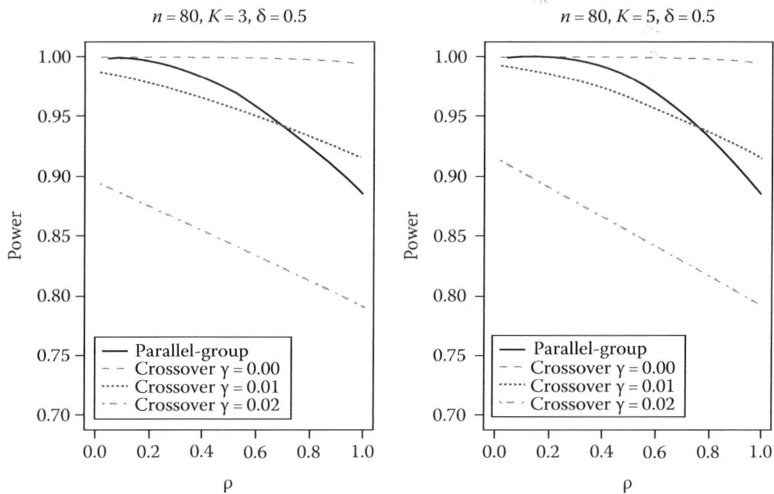

FIGURE 18.2
Power comparison under parallel-group and crossover designs.

18.4 Adjustment for Covariates

In the previous section, we considered models without covariates. In practice, additional information such as some PK responses, for example, area under the blood or plasma concentration time curve and the maximum concentration (C_{max}), which are known to be correlated to the QT intervals, may be available, for example, in an active-controlled QT study. In this case, models (18.1) and (18.2) are necessarily modified to include the PK responses as covariates for a more accurate and reliable assessment of power and sample size calculation (Cheng and Shao, 2007).

18.4.1 Parallel-Group Design

After the inclusion of some relevant covariates such as demographics and/or patient characteristics, model (18.1) becomes

$$y_{ijk} = \mu_i + \eta x_{ij} + e_{ij} + \varepsilon_{ijk},$$

where x_{ij} is some relevant covariate such as PK response for subject j. The least square estimate of η is given by

$$\hat{\eta} = \frac{\sum_{i=1}^{2}\sum_{j=1}^{n}(\bar{y}_{ij\cdot} - \bar{y}_{i\cdot\cdot})(x_{ij} - \bar{x}_{i\cdot})}{\sum_{i=1}^{2}\sum_{j=1}^{n}(x_{ij} - \bar{x}_{i\cdot})^2}.$$

Then $(\bar{y}_{1\cdot\cdot} - \bar{y}_{2\cdot\cdot}) - \hat{\eta}(\bar{x}_{1\cdot} - \bar{x}_{2\cdot})$ is an unbiased estimator of $\mu_1 - \mu_2$ with variance

$$\left[\frac{(\bar{x}_{1\cdot} - \bar{x}_{2\cdot})^2}{\sum_{ij}(x_{ij} - \bar{x}_{i\cdot})^2/n} + 2\right]\left(\rho + \frac{1-\rho}{m}\right)\frac{\sigma^2}{n},$$

which can be approximated by

$$\left[\frac{(v_1 - v_2)^2}{\tau_1^2 + \tau_2^2} + 2\right]\left(\rho + \frac{1-\rho}{m}\right)\frac{\sigma^2}{n},$$

where

$$v_i = \lim_{n\to\infty} \bar{x}_{i\cdot}$$

$$\tau_i^2 = \lim_{n\to\infty}\sum_{j=1}^{n}(x_{ij} - \bar{x}_{i\cdot})^2/n$$

Similarly, to achieve the desired power of $1 - \beta$ at the α level of significance, the sample size needed per treatment group is given by

$$n = \frac{(z_\alpha + z_\beta)^2}{\delta^2} \left[\frac{(v_1 - v_2)^2}{\tau_1^2 + \tau_2^2} + 2 \right] \left(\rho + \frac{1 - \rho}{m} \right). \tag{18.8}$$

In practice, v_i and τ_i^2 are estimated by the corresponding sample mean and sample variance from the pilot data. Note that if there are no covariates or the PK responses are balanced across treatments (i.e., $v_1 = v_2$), then formula (18.8) reduces to (18.5).

18.4.2 Crossover Design

After taking the PK response into consideration as a covariate, model (18.2) becomes

$$y_{ijkl} = \mu_i + \eta x_{ijl} + \beta_{il} + e_{ijl} + \varepsilon_{ijkl}.$$

Then $(\bar{y}_{1\cdots} - \bar{y}_{2\cdots}) - \hat{\eta}\,(\bar{x}_{1\cdot} - \bar{x}_{2\cdot})$ is an unbiased estimator of $\mu_1 - \mu_2$ with variance

$$\left[\gamma + \frac{(\bar{x}_{1\cdot} - \bar{x}_{2\cdot})^2}{\sum_{ijl}(x_{ijl} - \bar{x}_{i\cdot})^2/n} + 1 \right] \left(\rho + \frac{1 - \rho}{m} \right) \sigma^2,$$

which can be approximated by

$$\left[\gamma + \frac{(v_1 - v_2)^2}{\tau_1^2 + \tau_2^2} + 1 \right] \left(\rho + \frac{1 - \rho}{K} \right) \sigma^2,$$

where

$$v_i = \lim_{n \to \infty} \bar{x}_{i\cdot}$$

$$\tau_i^2 = \lim_{n \to \infty} \sum_{jl} (x_{ijl} - \bar{x}_{i\cdot})^2/n$$

Similarly, to achieve the desired power of $1 - \beta$ at the α level of significance, the sample size needed per treatment group is given by

$$n = \frac{(z_\alpha + z_\beta)^2}{\delta^2 - \gamma(z_\alpha + z_\beta)^2} \left[\frac{(v_1 - v_2)^2}{\tau_1^2 + \tau_2^2} + 1 \right] \left(\rho + \frac{1 - \rho}{m} \right). \tag{18.9}$$

When there are no covariates or the PK responses satisfy $v_1 = v_2$, then formula (18.9) reduces to (18.7). Formulas (18.8) and (18.9) indicate that under either a parallel-group or a crossover design, a larger sample size is required to achieve the same power if the covariate information is to be incorporated.

18.5 Optimization for Sample Size Allocation

For optimization of the allocation of n (the number of subjects) and m (the number of recording replicates) in routine QT studies with recording replicates, we may consider two approaches, namely, the fixed power approach and the fixed budget approach. The fixed power approach is to find optimal allocation of n and m for achieving a desired (fixed) power in the way that the total budget is minimized. For the fixed budget approach, the purpose is to find optimal allocation of n and m for achieving maximum power.

In this section, for simplicity, we will only describe the solution under a parallel-group design. The results under a crossover design can be similarly obtained. Let C_1 be the cost for recruiting a subject and C_2 be the associated cost for each QT recording replicate. To find n and K for achieving a desired (fixed) power of $1 - \beta$ under the minimal budget is equivalent to minimizing $C = nC_1 + nmC_2$ under the constraint of $2(z_\alpha + z_\beta)^2(\rho m + 1 - \rho) - nm\delta^2 = 0$. Under the given constraint, the total cost can be expressed as a function of m:

$$C(K) = \frac{2(z_\alpha + z_\beta)^2}{\delta^2}\left(\rho C_2 m + \frac{(1-\rho)C_1}{m} + \rho C_1 + (1-\rho)C_2\right),$$

which attains its minimum at

$$m = \left[\sqrt{\frac{C_1(1-\rho)}{C_2\rho}}\right] + 1,$$

where function $[t]$ denotes the integer part of t. In practice, we may consider choosing the m value among $m = 1, 3$, and 5 that will result in the smallest C.

When the total budget is fixed, say, $nC_1 + nmC_2 = C_0$, where C_0 is a known constant, the power function (18.4) becomes a function of m only:

$$H(m) = \Phi\left(-z_\alpha + \frac{\delta}{\sqrt{\dfrac{2(C_1 + C_2 m)}{C_0}\left(\rho + \dfrac{1-\rho}{m}\right)}}\right),$$

whose maximal value also occurs at $K = \left[\sqrt{C_1(1-\rho)/C_2\rho}\right] + 1$.

Note that for any fixed ρ, both the fixed power approach (for achieving a desired power but minimizing the total budget) and the fixed budget approach (for achieving the minimal power under a fixed total budget) result in the same optimal choice of K (the number of replicates), which is given by $m = \left[\sqrt{C_1(1-\rho)/C_2\rho}\right] + 1$.

18.6 Test for QT/QTc Prolongation

In the previous sections, we focused on statistical tests for mean QT/QTc difference between treatment groups for a given time interval under a parallel-group design and a crossover design. As an alternative, Cheng et al. (2008) proposed to test the maximum of QT/QTc differences between treatment groups across all time intervals for the detection of potential QT/QTc prolongation. Their proposed method under a parallel-group design and a crossover design are described in the following.

18.6.1 Parallel-Group Design

Under model (18.1), define $\delta_k = \mu_{1k} - \mu_{2k}$ and $\theta = \max_{1 \le k \le m} \delta_k$, then a QT/QTc study is equivalent to testing the following hypotheses:

$$H_0 : \theta \ge 10 \quad \text{versus} \quad H_a : \theta < 10. \tag{18.10}$$

Suppose the non-inferiority in QTc prolongation can be claimed via a 95% confidence upper bound based on a statistic U, then according to the ICH E14 guidance this means that $U + z_{0.05}\text{SE}(U) < 10$, or equivalently, $(U - 10)/\text{SE}(U) < -z_{0.05}$, which rejects H_0 (18.10) at the 5% level of significance. Here $\text{SE}(U)$ denotes the estimated standard error of U. Define $W_k = \bar{y}_{1 \cdot k} - \bar{y}_{2 \cdot k}$, where $\bar{y}_{i \cdot k}$ is the sample mean for the ith treatment at the kth time interval, and $W = (W_1, \dots, W_m)'$. Cheng et al. (2008) proved the following asymptotic result.

Theorem 18.1

Let $T = \max_{1 \le k \le m} W_k$, and θ is defined in (18.10), then

$$\sqrt{n}(T - \theta) \to_d N(0, 2(\sigma_1^2 + \sigma^2)),$$

where $\to_d$ means convergence in distribution.

Proof

The random vector W is normally distributed with mean $\delta = (\delta_1, \dots, \delta_m)'$ and variance $\Sigma = (\tau_{kl}) = (2\sigma_1^2/n)U_m + (2\sigma^2/n)I_m$, where U is the $m \times m$ matrix of ones and I_m is the $m \times m$ identity matrix. By Afonja (1972), the moment generating function of T is

$$M_T(t) = \sum_{k=1}^{m} e^{\delta_k t + (\sigma_1^2 + \sigma^2)t^2/n} \, \Phi_{m-1}(d_k; R_{-k}),$$

where

$$d_k = \{d_{kl}\}_{l \neq k}, \ d_{kl} = \frac{(\delta_l - \delta_k)}{2\sigma}\sqrt{n} - \frac{\sigma t}{\sqrt{n}},$$

and $\Phi_{m-1}(d_k; R_{-k})$ is the survival function of an $m - 1$ dimensional mean 0 normal random vector whose variance is the correlation matrix of W_{-k}, the random vector formed by removing the kth component of W. Then the moment generating function of $\sqrt{n}(T - \theta)$ is

$$M_{\sqrt{n}(T-\theta)}(t) = e^{-t\sqrt{n}} \sum_{k=1}^{m} e^{\delta_k t \sqrt{n} + (\sigma_1^2 + \sigma^2)t^2} \Phi_{m-1}(d_k; R_{-k})$$

$$= e^{-t\sqrt{n}} \sum_{k=1}^{m} e^{\delta_k t \sqrt{n} + (\sigma_1^2 + \sigma^2)t^2} I(\delta_k = \theta)(1 + o(1)) = e^{(\sigma_1^2 + \sigma^2)t^2}(1 + o(1)),$$

which implies the claim.

By Theorem 18.1, an asymptotic α level test rejects H_0 in (18.10) if and only if

$$\frac{T - 10}{\sqrt{2(\hat{\sigma}_1^2 + \hat{\sigma}^2)/n}} < -z_\alpha, \tag{18.11}$$

where

$$\hat{\sigma}_1^2 + \hat{\sigma}^2 = \sum_{i=1}^{2} \sum_{k=1}^{m} \frac{(y_{ijk} - \bar{y}_{i \cdot k})^2}{(2m(n-1))}.$$

When the number of patients n for each treatment is small, the normal approximation of distribution of T as suggested in Theorem 18.1 may not work well. Thus, Cheng et al. (2008) proposed a small sample correction of the distribution of T. Let $a_k = \{a_{kl}\}$, where $a_{kl} = \sqrt{n}(\delta_l - \delta_k)/2\sigma$ for $k \neq l$ and $a_{kk} = -\infty$. Let k_0 be such that $\delta_{k_0} = \max_{1 \leq k \leq m} \delta_k = \theta$, then according to Afonja (1972),

$$E(T) = \sum_{k=1}^{m} \delta_k \int_{a_k}^{\infty} \phi_m(z, R_k) + \sum_{k=1}^{m} \sqrt{\frac{2(\sigma_1^2 + \sigma^2)}{n}} \int_{a_k}^{\infty} z_k \phi_m(z, R_k)$$

$$= \theta \int_{a_{k_0}}^{\infty} z_k \phi_m(z, R_{k_0}) + o\left(\frac{1}{\sqrt{n}}\right) = \theta\rho + o\left(\frac{1}{\sqrt{n}}\right),$$

thus

$$E(T) \approx \theta\rho, \quad \rho = \int_{a_{k_0}}^{\infty} \phi_m(z, R_{k_0}). \tag{18.12}$$

Similarly since

$$E(T^2) = \sum_{k=1}^{m} \delta_k^2 \int_{a_k}^{\infty} \phi_m(z, R_k) + \sum_{k=1}^{m} \sqrt{\frac{2(\sigma_1^2 + \sigma^2)}{n}} \int_{a_k}^{\infty} z_k \phi_m(z, R_k)$$

$$+ \sum_{k=1}^{m} \frac{2(\sigma_1^2 + \sigma^2)}{n} \int_{a_k}^{\infty} z_k^2 \phi_m(z, R_k)$$

$$= \theta^2 \int_{a_{k_0}}^{\infty} \phi_m(z, R_{k_0}) + \frac{2(\sigma_1^2 + \sigma^2)}{n} \int_{a_{k_0}}^{\infty} z_{k_0}^2 \phi_m(z, R_{k_0}) + o\left(\frac{1}{\sqrt{n}}\right),$$

we have

$$\mathrm{Var}(T) \approx \frac{2(\sigma_1^2 + \sigma^2)}{n} \gamma, \quad \gamma = \int_{a_{k_0}}^{\infty} z_{k_0}^2 (z, R_{k_0}). \tag{18.13}$$

Now by replacing in (18.12) and (18.13) k_0, a_{k_0}, σ_1^2, and σ^2 with their obvious estimators, we get $\hat{\rho}$ and $\hat{\gamma}$. Then a small sample corrected level α test rejects H_0 in (18.10) if and only if

$$\frac{T - 10\hat{\rho}}{\sqrt{2(\hat{\sigma}_1^2 + \hat{\sigma})\hat{\gamma}/n}} < -z_\alpha. \tag{18.14}$$

18.6.2 Crossover Design

Let y_{ijkl} be the average QTc responses (possibly adjusted for baseline) over the recording replicates at the lth time interval of the kth treating period for the jth subjects in the ith sequence, $i = 1, 2$, $j = 1, \ldots, n$, $k = 1, 2$, and $l = 1, \ldots, m$. Under a crossover design, treatment index u is a function of (i, k), hence denoted as $u = d(i, k)$. Consider the following model:

$$y_{ijkl} = \mu + \alpha_k + \beta_{ul} + a_{ij} + b_{ijk} + \varepsilon_{ijkl}, \tag{18.15}$$

where
  μ is the overall mean
  α_k is the period effect
  β_{ul} is the treatment effect at lth time interval
  a_{ij} is the subject random effect
  b_{ijk} is the period random effect nested in the jth subject in the ith sequence
  ε_{ijkl} is the random error

We assume that $a_{ij} \sim N(0, \sigma_2^2)$, $b_{ijk} \sim N(0, \sigma_1^2)$, $\varepsilon_{ijkl} \sim N(0, \sigma^2)$, a_{ij}, b_{ijk}, and ε_{ijkl}'s are independent. Under model (18.15), the treatment effect at the lth time interval is $\delta_l = \beta_{1l} - \beta_{2l}$. Let $\theta = \max_{1 \le l \le m} \delta_l$, then the hypotheses of QTc prolongation in a TQT/QTc study under the crossover design is the same as (18.10). Define $W_l = (\bar{y}_{1 \cdot 1l} - \bar{y}_{1 \cdot 2l} + \bar{y}_{2 \cdot 2l} - \bar{y}_{2 \cdot 1l})/2$, $l = 1, \ldots, m$, then it is straightforward to show that $W = (W_1, \ldots, W_m)'$ has the same distribution as described earlier. A test similar to the one derived in the previous section can therefore be constructed.

18.6.3 Numerical Study

A simulation was conducted to evaluate the performance of the asymptotic test described in Section 18.6.1 (Cheng et al., 2008). For ease of comparison, Cheng et al. (2008) considered a similar setup as that given in Eaton et al. (2006). In other words, six time intervals (i.e., $m = 6$) and $\sigma_1^2 + \sigma^2$ was chosen to be 100. In addition, $\rho = \sigma_1^2/(\sigma_1^2 + \sigma^2) = 0.2, 0.4, 0.6, 0.8$, and $n = 40, 60, 80, 100$. The estimated size for $(\delta_1, \delta_2, \delta_3, \delta_4, \delta_5, \delta_6) = (1, 1, 10, 1, 1, 1)$ is given in Table 18.4. The estimated power for $(\delta_1, \delta_2, \delta_3, \delta_4, \delta_5, \delta_6) = (1, 2, 5, 1, 4, 1)$ is given in Table 18.5. All estimations were obtained based on 5000 simulation runs.

To illustrate the proposed test procedure, consider an example concerning a TQTc study with time-dependent recording replicates. Under the parallel-group design, 380 qualified subjects were randomly assigned to either a test treatment or an active control agent ($n = 190$). Subjects were at rest prior to the scheduled ECG. QT measurements were taken in recordings of five replicates within 2 min of one another. Five time intervals ($m = 5$) were considered 2 h apart. The vector W was calculated as

$$W = (8.98, 8.47, 7.96, 8.78, 10.05)', \quad T = 10.05.$$

TABLE 18.4

Estimated Size under $(\delta_1, \delta_2, \delta_3, \delta_4, \delta_5, \delta_6) = (1, 1, 10, 1, 1, 1)$

n	$\rho = 0.2$	$\rho = 0.4$	$\rho = 0.6$	$\rho = 0.8$
40	0.0452	0.0494	0.0482	0.0516
60	0.0524	0.0548	0.0520	0.0528
80	0.0486	0.0502	0.0496	0.0594
100	0.0478	0.0524	0.0514	0.0484

Source: Chow, S.C. et al., *Sample Size Calculation in Clinical Research*, Chapman and Hall/CRC Press, Taylor & Francis, New York, 2008. With permission.

TABLE 18.5

Estimated Power under $(\delta_1, \delta_2, \delta_3, \delta_4, \delta_5, \delta_6) = (1,2,5,1,4,1)$

N	$\rho = 0.2$	$\rho = 0.4$	$\rho = 0.6$	$\rho = 0.8$
40	0.6022	0.6410	0.6684	0.6962
60	0.8234	0.8252	0.8408	0.8514
80	0.9190	0.9206	0.9246	0.9326
100	0.9628	0.9686	0.9664	0.9708

Source: Chow, S.C. et al., *Sample Size Calculation in Clinical Research,* Chapman and Hall/CRC Press, Taylor & Francis, New York, 2008. With permission.

Since $\hat{\sigma}_1^2 + \hat{\sigma}^2 = 229.78$, we have

$$\frac{T-10}{\sqrt{2(\hat{\sigma}_1^2+\hat{\sigma})/n}} = \frac{10.05-10}{\sqrt{2\times229.78/190}} = 0.03 > -1.64 = -z_{0.05}.$$

Hence we do not reject H_0, implying that there was no statistical evidence to claim the test drug's non-inferiority to placebo in QTc prolongation.

18.7 Recent Developments

To discuss some statistical issues that are commonly encountered in TQT studies, Tsong and Zhang (2008) put together a special issue on Statistical Issues in Design and Analysis of Thorough QTc Clinical Trials in the *Journal of Biopharmaceutical Statistics.* These recent developments are briefly summarized in the following.

In an ongoing effort to try to understand the variability of QT/QTc data and determine how that variability would affect the design, analysis, and conclusions drawn from data collected in TQT/QTc studies, five PhRMA companies performed a retrospective analysis of placebo and nondrug resting ECG data (Agin et al., 2008). Based on the variability observed in the placebo and nondrug data, and on the power simulations, the PhRMA QT Statistics Expert Team suggested raising the upper confidence bound to define a negative QT/QTc study from 7.5 ms to at least 10 ms in the final version of the ICH E14 guideline. On the other hand, Ma et al. (2008) examined the performances of several approaches (including individual QT corrections and model-based QT analysis methods) to the analysis of QT changes based on QTc data obtained from a pharmaceutical company. Their simulation results suggested that the mixed effects modeling approach is more powerful than other methods which are commonly used in QT studies.

In their chapter, Zhang and Machado (2008) attempt to address some statistical issues including study design, primary statistical analysis, assay sensitivity analysis, and sample size calculation for a TQT study from regulatory perspectives. Chow et al. (2008a) discussed the strategy of using replicate ECG recordings at each time point to improve the power in the assessment of the drug-induced QT/QTc prolongation. Zhang et al. (2008), on the other hand, discussed the design strategy of assessing the maximum QTc changes using the bootstrap approach. Along this line, Cheng et al. (2008) proposed an asymptotic test based on the maximum differences under both parallel-group and crossover designs.

Wang et al. (2009) investigated the statistical properties of QTc intervals using individual-based correction (IBC), population-based correction (PBC), and fixed correction (FC) methods under both linear and log-linear regression models for the QT–RR relationship where RR is the time elapsing between two consecutive heartbeats. Based on a simulation study, Wang et al. (2009) suggested that in the analysis of QT intervals using PBC or FC methods, the RR interval may be included as a covariate in the model to adjust for the remaining correlation of QTc interval with RR interval. This approach will not only reduce the within-subject variability but also increase the statistical power for the assessment of QT/QTc prolongation.

For the assessment of QT/QTc prolongation, Zhang (2008) proposed two approaches, namely, a multiple local tests approach and a global average test. Zhang (2008) indicated that the type I error rate needs to be adjusted for the multiple local tests procedure, while no type I error rate adjustment is needed for the global average test. Tsong et al. (2008) indicated that the approaches proposed by Zhang are testing seemingly different hypotheses (the two sets of hypotheses are nested). Because of the property of the nested hypotheses, Tsong et al. (2008) suggested that Zhang's proposed methods may be applied to the same study data for assay validation tests.

Tian and Natarajan (2008) raised concerns on the impact of baseline measurement on the change from baseline to QTc intervals. In their chapter, they evaluated the effect of baseline on the change from baseline using the placebo data from several TQT studies. Tsong et al. (2008) pointed out that current QT concentration methods might result in a biased underestimate of the maximum prolongation of the QTc interval.

18.8 Concluding Remarks

Although the ICH E14 guideline provides the basic recommendations on the regulatory requirements on the assessment of drug-induced prolongations of the QT interval, details in measurements and statistics under various study designs (e.g., time-matched design with recording replicates) are yet to be fully developed. For the TQT studies using replicate ECG recordings, one of the controversial issues is whether a recording replicate is truly a replicate. Another controversial issue relates to the validity of the matched time points approach. In other words, is it clinically/statistically justifiable? In addition, the control of inter- and intra-subject variabilities in the assessment of QT/QTc prolongation is another issue of practical interest to the clinical scientists and biostatistician.

Under a parallel-group design, the possibility that the sample size can be reduced depends upon the parameter ρ, the correlation between the QT recording replicates. As indicated earlier, when ρ closes to 0, these recording repeats can be viewed as (almost) independent replicates. As a result, $n \approx n_{\text{old}}/m$. When ρ is close to 1, we have $n \approx n_{\text{old}}$. Thus, there is not much gain for considering recording replicates in the study. On the other hand, assuming that all other parameters remain the same, the possibility of further reducing the sample size by a crossover design depends upon the parameter γ, which is a measure of the magnitude of the relative period effect. When analyzing QT intervals with recording replicates, we may consider change from baseline. It is, however, not clear which baseline should be used when there are also recording replicates at the baseline. Strieter et al. (2003) proposed the use of the so-called time-matched change from the baseline, which is defined as the measurement at a time point on the post-baseline day minus the measurement at the same time point on the baseline. The statistical properties of this approach, however, are not clear. In practice, it may be of interest to investigate relative merits and disadvantages among the approaches using (1) the most recent recording replicates, (2) the mean recording replicates, or (3) the time-matched recording replicates as the baseline. This requires further research.

In the previous section, the test procedure based on maximum of correlated normal random variables proposed by Cheng et al. (2008) was discussed. Although the tests were derived under a balanced design without covariates, they can be easily generalized to allow for unbalance between the two treatment groups and adjustment of important covariates such as baseline QTc measures and/or heart rates. Note that in justifying our method, we did not assume any specific form for the variance structure of the model. This implies that our proposed method will still be valid when covariance structures other than compound symmetric, for example, an AR(1) structure, is more appropriate, or when heteroscedasticity is suspected. It should be noted that our formulation of hypotheses in (2) represents only one of the possible interpretations of QTc prolongation evidence. Other definitions are worth considering. For example, under a parallel-group design, define

$$\vartheta = \max_{1 \le k \le m} \mu_{1k} - \max_{1 \le k \le m} \mu_{2k},$$

then we could propose testing the following hypotheses:

$$H_0 : \vartheta \geq 10 \quad \text{versus} \quad H_a : \vartheta < 10.$$

The above hypotheses are relevant in an active-controlled QT/QTc study where the maximal prolongation of the two drugs occurs at different time intervals where a globe comparison rather than a time-matched comparison is desired. It is seen that our proposed method can be easily modified to test the above hypotheses.

19

Multiregional Clinical Trials

19.1 Introduction

For the approval of a drug product, the United States Food and Drug Administration (FDA) requires that at least two adequate and well-controlled clinical trials be conducted in order to provide substantial evidence of the effectiveness and safety of the drug product. The characteristics of an adequate and well-controlled clinical trial include a valid design and appropriate statistical tests for data analysis. A valid statistical design can not only minimize bias and variability that may be associated with the trial but also help to address the scientific/medical questions and/or hypotheses of the trial. An appropriate statistical test can provide a fair and unbiased assessment of the effectiveness and safety of the study drug with certain assurance. When conducting a clinical trial, it may be desirable to have the study done at a single study site if (1) the study site can provide an adequate number of relatively homogeneous patients that represent the targeted patient population under study and (2) the study site has sufficient capacity, resources, and supporting staff to sponsor the study. The advantage of a single-site study is that it provides consistent assessment for the efficacy and safety in a similar medical environment. However, a single-site study has some limitations and hence may not be feasible in many clinical trials, especially when the intended clinical trials are for relatively rare chronic diseases and the clinical endpoints for the intended clinical trials are relatively rare (Goldberg and Kury, 1990). As an alternative, a multicenter trial is usually considered. A multicenter study is a study conducted at more than one distinct center where the data collected from these centers are intended to be analyzed as a whole. Unlike a single-site study, a multicenter trial is much more complicated. Although, in practice, multicenter trials do expedite the patient recruitment process, some practical issues in design and analysis need to be carefully considered. These design and analysis issues include the selection of centers, the randomization of treatments, the use of a central laboratory for laboratory evaluation, and the existence of treatment-by-center interaction (Chow and Liu, 1998b). Note that the FDA indicates that an a priori division of a single multicenter trial into two studies is acceptable to the FDA for establishing the reproducibility of drug efficacy to new drug application approval.

However, a multicenter trial does not address the question whether the clinical results can be generalized to different patient populations (e.g., different race or same race with different culture) with similar patient characteristics. For this purpose, a multiregional (multinational) multicenter trial is usually considered. A multiregional (multinational) trial is a trial conducted at more than one distinct region (country) where the data collected from these regions (countries) are intended to be analyzed as a whole.

In recent years, multiregional (multinational or global) trials have become increasingly common in clinical development. In addition to the interest of generalizability, the purpose of multiregional (multinational) trials is multifold. First, a multiregional (multinational) trial makes the study drug available to patients from different regions (countries), which will be beneficial to the region (country), especially when no other alternative therapies are available in that region (country). Second, a multiregional (multinational) trial provides physicians from different regions (countries) the opportunity to obtain experience on medical practice of the study drug through the trial. In addition, a multiregional (multinational) trial may be used as a pivotal trial to fulfill the regulatory requirement of drug registration in some regions (countries). Finally, a multiregional (multinational) trial provides an overall assessment of the performance of the study drug across regions (countries) under study (Ho and Chow, 1998).

In the next section, some commonly seen practical issues in the design and analysis of multicenter trials are outlined. Also included are some practical issues and/or difficulties that are commonly encountered in multiregional (multinational) trials. Section 19.3 provides statistical justification for selecting the number of sites in a multicenter trial. Sample size calculation and allocation for a multiregional (multinational) study are discussed in Section 19.4. In Section 19.5, some statistical methods for bridging studies are described. Some concluding remarks are given in the last section.

19.2 Multiregional (Multinational), Multicenter Trials

19.2.1 Multicenter Trials

In a multicenter trial, an identical study protocol is used at each center. A multicenter trial is a trial with a center or site as a natural blocking or stratified variable that provides replications of clinical results. As a result, a multicenter trial should permit an overall estimation of the treatment effect for the targeted patient population across various centers. As was indicated earlier, a multicenter trial with a number of centers is often conducted to expedite the patient recruitment process. Although these centers follow the same study protocol, some design and analysis issues need to be carefully considered when planning a multicenter trial (Suwelack and Weihrauch, 1992; Philipp and Weihrauch, 1993; Ho and Chow, 1998). These design and analysis issues include the selection of centers, the randomization of treatment, the use of a central laboratory for laboratory evaluation, and the evaluation of treatment-by-center interaction. These issues are briefly outlined in the following sections.

19.2.1.1 Site Selection

In multicenter trials, the selection of centers is important to constitute a representative sample for the targeted patient population. In practice, the centers are usually selected based on convenience and availability. When planning a multicenter trial with a fixed sample size, it is important to determine the allocation of the centers and the number of patients in each center. For comparative clinical trials, it is not desirable to have too few patients in each center because the comparison between treatments is usually made between patients within centers. A rule of thumb is that the number of patients in each center should not be less than the number of centers for a reliable evaluation of the effectiveness and safety of the study drug (Shao and Chow, 1993). For example, if the intended clinical trial calls for 100 patients, the selection of not more than 10 sites is preferable. Some statistical justification is provided in the next section. Although a multicenter trial has its advantages, it also suffers from some difficulties in site selection. For example, if the enrollment is too slow, the sponsor may wish to (1) terminate the inefficient study sites, (2) increase the enrollments for the most aggressive sites, or (3) open new sites during the course of the trial. Each action may introduce potential biases to the study. In addition, the sponsor may ship unused portions of the study drugs from the terminated sites to the newly opened sites for cost-effectiveness consideration. This can certainly increase the chance of mixing up the randomization schedules and consequently decrease the reliability of the study.

19.2.1.2 Randomization of Treatments

In multicenter trials, we usually select investigators first and then select patients at each selected investigator's site. At each selected investigator's site, the investigator will usually enroll qualified patients sequentially. A qualified patient is referred to as a patient who meets the inclusion and exclusion criteria and has signed the informed consent form. The primary concern is that neither the selection of investigators nor the recruitment of patients is random. In practice, although the selection of investigators and patients at the selected sites is not random, patients are assigned to treatment groups at random. The collected clinical data are then analyzed as if they were obtained under the assumption that the sample is randomly selected from a homogeneous patient population. This process is referred to as the invoked population model and is currently widely accepted in clinical research. As a result, randomization is usually performed by study sites in multicenter trials. Note that Lachin (1988) provides a comprehensive summary of the randomization basis for statistical tests under various models. To provide a valid statistical evaluation of the effectiveness and safety of the study drug, randomization is important to ensure that patients selected from the intended patient population constitute a representative sample of the intended patient population. Statistical inference can then be drawn based on some probability distribution assumption of the intended patient population. The probability distribution assumption depends on the method of randomization under a randomization model. A study without randomization will result in the violation of the probability distribution assumption, and consequently no accurate and reliable statistical inference on the study drug can be drawn. It should be noted that in multicenter trials, a large number of study sites may increase the chance of making errors in randomization schedules.

19.2.1.3 Central Laboratory

As indicated earlier, a multicenter trial is usually conducted to enroll enough patients within a desired time frame. In this case, a concern may be whether the laboratory tests should be performed by local laboratories or by a central laboratory. The relative advantages and drawbacks between the use of a central laboratory and local laboratories include (1) the combinability of data, (2) timely access to laboratory data, (3) laboratory data management, and (4) cost. A central laboratory provides combinable data with unique normal ranges, while local laboratories may produce uncombinable data due to different equipment, analysts, and normal ranges. As a result, laboratory data obtained from a central laboratory are more accurate and reliable compared with those obtained from local laboratories. In multicenter trials, it is not uncommon that laboratory tests are performed by local laboratories. In this case, it is suggested that laboratory test results be standardized according to the investigator's normal ranges or local laboratories' normal ranges before analysis (see, e.g., Chung-Stein, 1996). Note that before the data from different laboratories can be combined for analysis, it may be of interest to evaluate the repeatability (within-laboratory variability) and reproducibility (between-site variability) of the results, which can be done by sending to each laboratory identical samples that represent a wide range of possible values, and analyze using the method of analysis of variance.

19.2.1.4 Treatment-by-Center Interaction

For a multicenter trial, the FDA guideline suggests that individual center results should be presented. In addition, the FDA suggests that tests for homogeneity across centers (i.e., for detecting treatment-by-center interaction) be done. The significant level used to declare the significance of a given test for a treatment-by-center interaction should be considered in light of the sample size involved. Any extreme or opposite results among centers should be noted and discussed. For the presentation of the data, demographic, baseline, and post-baseline data as well as efficacy data should be presented by center, even though the combined analysis may be the primary one. Gail and Simon (1985) classify the nature of interaction as either quantitative or qualitative. A quantitative interaction between treatment and center indicates that the treatment differences are in the same direction across centers but the magnitude differs from center to center, while a qualitative interaction reveals that substantial treatment differences occur in different directions in different centers. If there is no evidence of treatment-by-center interaction, the data can be pooled for analysis across centers. The analysis with combined data provides an overall estimate of the treatment effect across centers. In practice, however, if there are a large number of centers, we may observe significant treatment-by-center interaction, either quantitative or qualitative. In addition, a multicenter trial with too many centers may end up with a major imbalance among centers, in that some centers may have a few patients and others a large number of patients. If there are too many small centers with a few patients in each center, we may consider the following two approaches. The first approach is to combine these small centers to form a new center based on their geographical locations or some criteria prespecified in the protocol. The data can then be analyzed by treating the created center as a regular center. Another approach is to randomly assign the patients in these small centers to those larger centers and reanalyze the data. This approach is valid under the assumption that each patient in a small center has an equal chance of being treated at a large center.

19.2.2 Multiregional (Multinational), Multicenter Trials

As indicated earlier, a multiregional (multinational) trial is a trial conducted at more than one distinct region (country) where the data collected from these regions (countries) are intended to be analyzed as a whole. Within each region (country), the trial in fact is a multicenter trial. As a result, a multiregional (multinational) trial can be viewed as a trial consisting of a number of multicenter trials conducted at different regions (countries) under the same study protocol. In practice, it is a concern whether a multiregional (multinational) trial can maintain the integrity of the trial due to the complexity which includes difficulties that are already common in multicenter trials within each region (country) as described in the previous section. To maintain the integrity of the trial and to achieve the desired accuracy and reliability for an overall assessment of the effectiveness and safety of the study drug, it is important to identify all possible causes of bias and variability. These possible causes of bias and variability could be classified into four categories of (1) expected and controllable, (2) expected but uncontrollable, (3) unexpected but controllable, and (4) unexpected and uncontrollable. In general, these biases and variabilities are mostly due to confounding and differences in culture, medical culture/ practice, standards, and regulatory, which will be discussed below.

19.2.2.1 Confounding

In a multicenter trial, qualified patients within a particular country (e.g., China or Japan) tend to be of the same race, which may be different than those patients who are from other countries (e.g., the United States and Germany). An immediate concern is what if there is a potential confounding effect between treatment and race. If the confounding effect between treatment and race does exist, it is difficult to evaluate whether the observed treatment difference is due to treatment or race. In addition, the use of concomitant medication is also a concern, especially when the multiregional (multinational) trial involves the third countries. This is because the quality, efficacy, and safety of the concomitant medications may be a concern. Most of these concomitant medications may or may not be approved by regulatory agencies from other countries. The potential drug-to-drug interaction may contaminate the true treatment effect of the study drug. This is very common for those patients from Chinese countries in the Asian Pacific region who are likely to take traditional Chinese medicines (or herbal medicines) during the conduct of the trial even if they are told not to. These confounding effects present great challenges to clinical researchers and biostatisticians as well.

19.2.2.2 Culture

When planning a multiregional (multinational) trial, it is very important to understand and appreciate culture differences from different countries. These culture differences may have an impact on the conduct of the trial. For example, before a multiregional (multinational) trial can be conducted, most regulatory agencies require that the study protocol be submitted to an institutional review board (IRB) for review and approval. The purpose of an IRB review is not only to assess the potential risk of the intended trial for patient protection but also to ensure the validity and integrity of the intended trial. Different countries, however, may assess the potential risk differently due to the difference in culture. In addition, patients are required to sign an informed consent form before they can be enrolled into the study. It is the investigator's responsibility to explain the potential risk/benefit of the study drug to the patients before they sign the informed consent form. However, in some countries such as China, most patients are unlikely and unwilling to sign an informed consent form if they were told that the study medication is a test drug rather than a new drug under investigation. It is a traditional Chinese culture not to take a test drug. Patients are likely to try a new drug. As a result, we may have a problem obtaining signed informed consent forms from patients. For good clinical practice (GCP), it is unethical to tell patients that they will be taking a new drug rather than a test drug under investigation. Therefore, it is suggested that a well-designed educational program be implemented by the health authority to eliminate the difficulties caused by the difference in culture.

19.2.2.3 *Medical Culture/Practice*

In multiregional (multinational) trials, one of the primary concerns is whether the collected clinical data can be combined for the assessment of the effectiveness and safety of the study drug. Although critical information can be captured by a set of standard case report forms (CRFs), it is very likely that we may capture different information due to differences in (1) the translation of the CRF in different languages, (2) the understanding of medical personnel, and (3) medical culture/practice. In different countries, there is certainly a need to translate the CRF to their respective languages so that patients, clinical monitors, and investigators have same knowledge regarding what information the trial is intended to capture. This is important especially for those countries in which English is not a popular language. A poorly translated CRF may mislead patients to provide inaccurate or even wrong information of little value to the intended trial. In many cases, differences in medical culture and/or practice may result in a very different diagnosis of a similar symptom; consequently, the interpretation or assessment of the efficacy and safety parameters may be different. This is always true for reporting adverse events (AEs). For example, an observed rare but severe AE in one country may be coded differently in a different country if the observed AE is commonly seen in the medical community of the particular country. As a result, AE coding may be different from one country to another, which provides a challenge for having a fair and unbiased assessment of safety across different countries. As described earlier in the previous section, it is likely that a local laboratory will be used for laboratory tests in multinational trials. It is expected that different laboratories in different countries will have different laboratory normal ranges due to differences in medical culture and/or practice. In the interest of combining laboratory data for an overall assessment of safety, it is suggested that the laboratory data be standardized according to respective laboratory normal ranges before pooling for analysis.

19.2.2.4 *Regulatory Requirement*

For drug research and development, most regulatory agencies have similar but slightly different regulations to ensure the drug product has the claimed efficacy and safety. In addition, many regulations and guidelines/guidances were also imposed to ensure that the approved drug product meets standards for identity, strength, quality, purity, and stability as specified in the pharmacopedia in the respective countries such as the United States Pharmacopedia (USP) in the United States and the Chinese Pharmacopedia (CP) in the People's Republic of China. It should be noted that the standards for assay development/validation and test procedures, sampling plans, and acceptance criteria for potency, content uniformity, dissolution, and disintegration may differ from one country to another. These differences may result in a potential treatment-by-country interaction. Consequently, it is difficult to combine the collected clinical data for an overall assessment of the efficacy and safety of the study drug.

19.2.2.5 Drug Management

Drug management is a great challenge in multinational trials. Randomization schedules are usually generalized by country with a stratification factor (if desirable) and an appropriate block size for treatment balance. The generalized randomization schedules will then be forwarded to drug management for packaging and shipment. The complication is not the randomization or drug packaging but the shipment to the study sites. In many cases, the study drug may not be available in some countries and need to be imported from other countries. Different countries have different regulations for importing investigational drugs. It may take weeks or months for the processing. If the duration of the intended trial is over a few years, the sponsor may have to take the drug expiration dating period into account to make sure that the study drug will not be expired prior to the end of the study. Another consideration for drug management is to make sure that sufficient drugs will be supplied during the conduct of the study. Any unused drugs need to be returned or disposed depending on specific regulations of individual countries. One solution, which is probably the most cost effective, is to consider the so-called interactive voice randomization system (IVRS) for randomization and drug management. The IVRS is used to ship sufficient drugs to specific sites on time in a more cost-effective way.

19.3 Selection of the Number of Sites

In clinical trials, multiple sites are necessarily considered because one single study site may not have enough resources and/or capacity to handle all the subjects that enter the study. In addition, multiple sites will expedite patient enrollment. In practice, it is not desirable to have too few subjects in each study site. On the other hand, too many study sites may increase the chance of observing so-called treatment-by-center interaction, which makes an overall inference on the treatment effect impossible. Thus, at the planning stage of a clinical trial, how many study sites should be used in order for achieving optimal statistical properties for a given sample size is a commonly asked question.

The question regarding how many study sites should be used is, in fact, a two-stage sampling problem. One first selects a number of study sites and, for each sampled study site, one then selects a number of patients. Shao and Chow (1993) proposed statistical testing procedures in a two-stage sampling problem with large within-class sample sizes. In addition, they derived a two-stage sampling plan by minimizing the expected squared volume (ESV) (or the generalized variance) of the confidence region related to the test. Some results for a two-stage sampling plan are described in the subsequent subsections.

19.3.1 Two-Stage Sampling

For a given clinical trial comparing K treatment groups, we first draw a random sample of n study sites. For each sampled study site, we then recruit M_k subjects, $k = 1, \ldots, K$. Denote by X_{ijk} the random variable for the jth subject from the kth treatment group in the ith study site, $i(\text{site}) = 1, \ldots, n$, $j(\text{subject}) = 1, \ldots, M_k$, and $k(\text{treatment}) = 1, \ldots, K$ and

$$X_i = (X_{ijk}, j = 1, \ldots, M_k, k = 1, \ldots, K).$$

Then, X_i is a random $\left(\sum_k M_k\right)$ vector and $X_1, \ldots, X_n$ are independent and identically distributed. For each i, the components of X_i have the same distribution if they are from the treatment group. Thus, the means and the variances of X_{ijk}, denoted by μ_k and σ_k^2 respectively, are unknown but depend on k only. In the second-stage sampling, for each selected study site, we recruit a simple random sample of m_k subjects without replacement who will receive the kth treatment, where $1 \leq m_k \leq M_k$ and $k = 1, \ldots, K$. The total number of subjects recruited from each selected study site is $\sum_k m_k$ and the total number of subjects in the clinical trial is $\left(\sum_k m_k\right) n$. Now, the question is how to select n and m_k.

Let x_{ijk} denote clinical response observed from the jth subject in the ith study site who receives the kth treatment group, where $i = 1, \ldots, n, j = 1, \ldots, m_k$, and $k = 1, \ldots, K$. Also, let $\bar{x}_k$ and $\hat{\sigma}_k^2$ be the sample mean and sample variance from the kth treatment group, respectively, where

$$\bar{x}_k = \frac{1}{nm_k} \sum_{i=1}^{n} \sum_{j=1}^{m_k} x_{ijk},$$

and

$$\hat{\sigma}_k^2 = \frac{1}{nm_k - 1} \sum_{i=1}^{n} \sum_{j=1}^{m_k} (x_{ijk} - \bar{x}_k)^2.$$

Using the techniques described by Cochran (1977), we have $E(\bar{x}_k) = \mu_k$ and

$$\text{Var}(\bar{x}_k) = \frac{1}{nm_k} \sigma_k^2 [1 + (m_k - 1)\rho_k],$$

where ρ_k is the correlation coefficient between x_{ijk} and $x_{ij'k}$ with $j \neq j'$. In many pharmaceutical problems, $\rho_k = 0$ and hence

$$\mathrm{Var}(\bar{x}_k) = \frac{1}{nm_k}\,\sigma_k^2. \tag{19.1}$$

Under (19.1),

$$s_k^2 = \frac{\hat{\sigma}_k^2}{nm_k} \tag{19.2}$$

is an unbiased estimator of $\mathrm{Var}(\bar{x}_k)$. In the case where $\rho_k \neq 0$, the variance estimator in (19.2) is not valid. For each fixed k,

$$\bar{x}_{ik} = \frac{1}{m_k}\sum_{j=1}^{m_k} x_{ijk}, \quad i=1,\ldots,n,$$

are independent and identically distributed. Therefore,

$$\mathrm{Var}(\bar{x}_k) = \frac{1}{n}\mathrm{Var}(\bar{x}_{ik}), \quad i=1,\ldots,n.$$

An unbiased estimator of $\mathrm{Var}(\bar{x}_k)$ is the sample variance of $\{\bar{x}_{ik}, i=1,\ldots,n\}$:

$$s_k^2 = \frac{1}{n(n-1)}\sum_{i=1}^{n}(\bar{x}_{ik} - \bar{x}_k)^2, \tag{19.3}$$

which we can use to replace the estimator in (19.2) when $\rho \neq 0$. Note that the estimator in (19.2) is more efficient than that in (19.3) when $\rho_k = 0$, and (19.2) and (19.3) are equivalent when $m_k = 1$.

Assume that nm_k is large so that approximately $100(1-\alpha)\%$ lower and upper confidence bounds for μ_k are given by

$$L_k = \bar{x}_k - z_\alpha s_k \quad \text{and} \quad U_k = \bar{x}_k + z_\alpha s_k, \tag{19.4}$$

respectively, where z_α is the $(1-\alpha)$th quantile of the standard normal distribution. An approximately $100(1-\alpha)\%$ joint confidence region for the vector $\mu = (\mu_1,\ldots,\mu_K)$ is

$$\left\{\mu : \sum_k \left[\frac{(\bar{x}_k - \mu_k)}{s_k}\right]^2 \leq \chi_\alpha^2(K)\right\}, \tag{19.5}$$

where $\chi_\alpha^2(K)$ is the $(1-\alpha)$th quantile of the chi-square distribution with K degrees of freedom.

19.3.2 Testing Procedure

Shao and Chow (1993) proposed a testing procedure in a two-stage sampling problem with large within-class (i.e., within treatment in our case) sample sizes and derived a two-stage sampling plan by minimizing the ESV (or the generalized variance) of the confidence region related to the test assuming that there is an increasing order of mean across treatment groups, that is,

$$\mu_1 < \mu_2 < \cdots < \mu_K, \tag{19.6}$$

where μ_k's satisfy

$$a_k < \mu_k < b_k, \quad k = 1,\dots,K, \tag{19.7}$$

in which (a_k, b_k) are in-house acceptance limits or release targets used for quality assurance of the manufactured products. The basis for construction of a_k's and b_k's is information obtained from previous studies. Note that if we choose the a_k's and b_k's so that $b_k \leq a_{k+1}$, $k = 1, \dots, K - 1$, then (19.7) implies (19.6). Since the μ_k's are unknown, we need to make a decision based on x_{ijk}'s. Let H_0 denote the null hypothesis that (19.6) (or (19.7)) does not hold and H_a the alternative hypothesis that (19.6) (or (19.7)) is true. Then our problem becomes a statistical testing problem of H_0 versus H_a. The form of the null hypothesis, however, is so complicated that there is no simple testing procedure available in the literature. When we test (19.7), we can express H_0 as

$$H_0 : \mu_k < a_k \quad \text{or} \quad \mu_k > b_k \quad \text{for at least one } k. \tag{19.8}$$

In the special case of $K = 1$, we may adopt the two one-sided α level tests approach in the assessment of bioequivalence (see, e.g., Westlake, 1976; Hauck and Anderson, 1984; Schuirmann, 1987). That is, we reject $H_0 : \mu_1 < a_1$ or $\mu_1 > b_1$ if and only if

$$a_1 < L_1 \quad \text{or} \quad U_1 < b_1,$$

where L_1 and U_1 are given in (19.4). Generalizing this idea to the case of $K \geq 3$, Shao and Chow (1993) proposed the following testing procedure for (19.7): H_0 in (19.8) is rejected if and only if

$$a_k < L_k \quad \text{and} \quad U_k < b_k, \quad k = 1,\dots,K, \tag{19.9}$$

where L_k and U_k are given in (19.4). A geometric interpretation of this test procedure is that we reject H_0 whenever

$$C \subset R, \tag{19.10}$$

where
$$C = (L_1, U_1) \times \cdots \times (L_K, U_K),$$
$$R = (a_1, b_1) \times \cdots \times (a_K, b_K).$$

Since the (L_k, U_k)'s are independent, C is actually a confidence region for μ with an approximate level $(1 - \alpha)^K$. It can be shown that

$$\sup_{H_0} \lim_{nm_k \to \infty, k=1, \ldots, K} P(C \subset R \mid H_0) = \alpha. \tag{19.11}$$

For example, when $K = 1$, the left-hand side of (19.11) is greater than or equal to

$$\lim_{nm_1 \to \infty} P(a_1 < L_1 \text{ and } U_1 < b_1) \geq \alpha - \lim_{nm_1 \to \infty} P(U_1 < b_1 \mid \mu_1 = a_1)$$

$$= \alpha - \lim_{nm_1 \to \infty} \Phi\left(z_\alpha - \frac{b_1 - a_1}{s_1}\right) = \alpha,$$

since $s_1 \to 0$. Hence (19.11) holds. We now turn to the test of H_0 that (19.6) does not hold. Let $\delta_k = \mu_{k+1} - \mu_k$, $k = 1, \ldots, K - 1$. Then we can express H_0 as

$$H_0 : \delta_k < 0 \quad \text{for at least one } k. \tag{19.12}$$

Note that (19.12) is a special case of (19.8) with $a_k = 0$ and $b_k = \infty$. Hence we can test (19.12) based on a procedure similar to (19.9): we reject H_0 in (19.12) if and only if

$$0 < \bar{x}_{k+1} - \bar{x}_k - z_\alpha[s_{k+1}^2 + s_k^2]^{1/2}, \quad k = 1, \ldots, K - 1. \tag{19.13}$$

19.3.3 Optimal Selection

As indicated above, although we are able to control the type I error rate, we are unable to control the other type of error rate, that is,

$$P(H_0 \text{ is not rejected} \mid H_a) = P(C \not\subset R \mid H_a),$$

where C and R are given in (19.10). One way to reduce this statistical error is to minimize the *size* of the region C. The K-dimensional volume of C is

$$v = (U_1 - L_1) \cdots (U_K - L_K) = (2z_\alpha)^K (s_1 \cdots s_K).$$

Since we cannot minimize υ by selecting sample sizes before the samples are drawn, we propose to select n and m_k by minimizing the ESV

$$\text{ESV} = E(\upsilon^2) = (2z_\alpha)^{2K}(\sigma_1^2\sigma_2^2 \cdots \sigma_K^2)\frac{1}{n^K(m_1m_2 \cdots m_K)}, \qquad (19.14)$$

under the constraint that a study site can handle only a limited number of subjects. Motivation for this approach is also the fact that the ESV in (19.14) is proportional to the generalized variance, which is the K-dimensional volume of the confidence region defined by (19.5) and is a measure of the asymptotic relative efficiency (see, e.g., Serfling, 1980); hence, minimizing the ESV is equivalent to minimizing the generalized variance.

From (19.14), minimizing ESV is equivalent to minimizing the function

$$J(n,m_1,\ldots,m_K) = \frac{1}{n^K(m_1m_2 \cdots m_K)}.$$

Note that although the σ_k's affect the ESV, they do not affect the selection of sample sizes according to the criterion of minimizing the ESV.

Let c_0 denote the cost of each subject. The total cost is then $c_0 n\left(\sum_k m_k\right)$ and the cost constraint is

$$c_0 n\left(\sum_k m_k\right) \le c,$$

where c is a given upper limit for the total cost. Suppose that a given study site can handle only N subjects owing to limited availability of resources. The resources constraint is then

$$n\left(\sum_k m_k\right) \le N.$$

When there is no cost constraint (e.g., resources constraint), we simply take $C = \infty$ ($N = \infty$). Let L be the integer part of $\min(N, c/c_0)$. We then minimize $J(n,m_1,\ldots,m_K)$ subject to $n\left(\sum_k m_k\right) \le L$, $1 \le m_k \le M_k$, and n, m_k's are integers, $k = 1,\ldots, K$.

Consider the problem of minimizing the function $J(n,m_1,\ldots,m_K)$ over the region

$$A = \left\{(n,m_1,\ldots,m_K): 1 \le m_k \le M_k, k = 1,\ldots,K, n\left(\sum_k m_k\right) \le L\right\}.$$

Clearly, the derivative of the function J does not vanish on the set

$$A_0 = \left\{ (n, m_1, \ldots, m_K) : 1 \le m_k \le M_k, \, k = 1, \ldots, K, n \left(\sum_k m_k \right) \le L \right\}.$$

Hence, the minimum of J is on the set

$$A_1 = \left\{ (n, m_1, \ldots, m_K) : 1 \le m_k \le M_k, \, k = 1, \ldots, K, n \left(\sum_k m_k \right) = L \right\}.$$

On the set A_1, $n = L / \left(\sum_k m_k \right)$ and

$$J(n, m_1, \ldots, m_K) = J_1(m_1, \ldots, m_K) = L^{-K} \frac{m}{w},$$

where $m = \sum_k m_k$ and $w = m_1 \ldots m_K$. Then

$$\frac{\partial J_1}{\partial m_k} = L^{-K} \left(\frac{K m^{K-1}}{w} - \frac{m^K}{m_k w} \right), \quad k = 1, \ldots, K.$$

Setting

$$\frac{\partial J_1}{\partial m_k} = 0, \quad k = 1, \ldots, K,$$

we obtain

$$m_k = \frac{m}{K}, \quad k = 1, \ldots, K,$$

that is, J has a minimum on A_1 as long as $m_1 = m_2 = \cdots = m_K$. If there is an integer m^* such that $1 \le m^* \le M_k$ for all k and L/Km^* is an integer, then J has a minimum at $m_1 = \cdots = m_k = m^*$ and $n = L/Km^*$. If L/Km^* is not an integer for all possible m^*, then we should select m^* in the set $\{1, 2, \ldots, \min(M_1, M_2, \ldots, M_K)\}$ such that $Km^*[L/Km^*]$ is as large as possible. Thus, a solution is given by

$$m_1 = m_2 = \cdots = m_K = m^*, \tag{19.15}$$

$$n = \left[\frac{L}{Km^*}\right], \tag{19.16}$$

where $[L/Km^*]$ is the integer part of L/Km^* and we choose m^* from the set of integers $\{1, 2, \ldots, \min(M_1, M_2, \ldots, M_K)\}$ such that

$$Km^*\left[\frac{L}{Km^*}\right] \text{ is as large as possible.} \tag{19.17}$$

In particular, if there is an integer $m^* \leq \min(M_1, \ldots, M_K)$ such that L/Km^* is an integer, then $m_1 = \cdots = m_K = m^*$ and $n = L/Km^*$ is a solution. There may be several sampling plans that satisfy (19.15) through (19.17).

A sampling plan that satisfies (19.15) through (19.17) is optimal in terms of the ESV only. We would have to use other criteria to choose a sampling plan when there are several plans that satisfy. As an example, consider the situation where $K = 4$, $M_1 = 2$, $M_2 = 4$, $M_3 = 6$, $M_4 = 8$, and $L = 100$. Since $\min(M_1, M_2, M_3, M_4) = 2$, possible values of m^* are 1 and 2. For $m^* = 2$, $Km^* = 4 \times 2 = 8$, the largest n we can take is 12, which gives the total sample size $96 < L$. Similarly, for $m^* = 1$, $Km^* = 4$, the largest n we can use is 25, which gives the total sample size $100 = L$. Hence $m^* = 1$ and $n = 25$ is the unique plan that satisfies (19.15) through (19.17). To compare this plan with other sampling plans, consider the single-stage sampling plan with $m_k = M_k$ for all k and $n = 5$ (which also gives the total sample size 100). A simple calculation shows that the ESV of the single-stage sampling plan over the ESV of the plan that satisfies (19.15) through (19.17) is 162.8%. Therefore, the single-stage sampling plan is not efficient. Note that the sampling plan that takes $\{m_k\}$ in proportion to $\{M_k\}$ produces the same ESV as the single-stage sampling.

In case of $\rho_k \neq 0$, although the testing procedures described above are valid regardless of whether $\rho_k = 0$ (assuming we use the variance estimator (19.3)), the sampling plan given by (19.15) through (19.17) is not necessarily good when $\rho_k \neq 0$. In fact, when $\rho_k \neq 0$ the optimal sampling plan, if it exists, depends on the ρ_k's and, therefore, the problem may be unsolvable since the ρ_k's are unknown. This difficulty is not a serious concern for many problems in the pharmaceutical industry, since $\rho_k = 0$ for all k is a reasonable assumption. Furthermore, in many cases $\rho_k \neq 0$ but is relatively small. We then expect that the sampling plan given by (19.15) through (19.17) is nearly optimal.

19.3.4 An Example

A study protocol for a clinical trial usually includes a statement regarding sample size determination to justify the selected sample size based on a pre-study power analysis. Suppose a placebo-controlled clinical trial entails the selection of a sample size of 200 patients to achieve the desired power for the detection of a clinically meaningful difference. The question then is: "How many study sites should one use?" Suppose that each study site can handle only a maximum of 40 patients. The study director needs to decide the number of study sites (n) and the number of patients at each study site (m_1 for the control group and m_2 for the treatment group) under the following constraints:

$$m_1 \le 40, \quad m_2 \le 40, \quad m_1 + m_2 \le 40, \quad n(m_1 + m_2) \le 200.$$

If we use the ESV criterion described earlier, we obtain the following plans:

Plan	$m_1 = m_2$	$m_1 + m_2$	n
1	1	2	100
2	2	4	50
3	4	8	25
4	5	10	20
5	10	20	10
6	20	40	5

Note that the plans 1–6 produce the same ESV and are all optimal in terms of the ESV. Hence we need to use some other criterion to choose a plan from plans 1 to 6. Note that, for a multicenter study, the FDA requires that one examines the treatment-by-study-site interaction before one pools the data for analysis. An increase in the number of study sites may increase the chance of a treatment-by-study-site interaction. As a rule of thumb, it is preferred that the number of study sites be less than the number of patients in each study site, that is, $n < m_1 + m_2$. Only plans 5 and 6 satisfy $n < m_1 + m_2$. If one expects a treatment-by-study-site interaction, then sampling plan 6 is preferred because the comparison between treatments occurs within each study site.

19.4 Sample Size Calculation and Allocation

19.4.1 Some Background

As indicated by Uesaka (2009), the primary objective of a multiregional bridging trial is to show the efficacy of a drug in all participating regions while also evaluating the possibility of applying the overall trial results to each region. To apply the overall results to a specific region, the results in that region should be consistent with either the overall results or the results from other regions. A typical approach is to show consistency among regions by demonstrating that there exists no treatment-by-region interaction. Recently, the Ministry of Health, Labor and Welfare (MHLW) of Japan published a guidance on *Basic Principles on Global Clinical Trials* that outlines the basic concepts for planning and implementing multiregional trials in a Q&A format (MHLW, 2007). In this guidance, special consideration was placed on the determination of the number of Japanese subjects required in a multiregional trial. As indicated, the selected sample size should be able to establish the consistency of treatment effects between the Japanese group and the entire group.

To establish the consistency of the treatment effects between the Japanese group and the entire group, it is suggested that the selected size should satisfy

$$P\left(\frac{D_J}{D_{All}} > \rho\right) \geq 1 - \gamma,\tag{19.18}$$

where D_J and D_{All} are the treatment effects for the Japanese group and the entire group, respectively. Along this line, Quan et al. (2010) derived closed form formulas for the sample size calculation/allocation for normal, binary, and survival endpoints. As an example, the formula for continuous endpoint assuming that $D_J = D_{NJ} = D_{All} = D$, where D_{NJ} is the treatment effect for the non-Japanese subjects, is as follows:

$$N_J \geq \frac{z_{1-\gamma}^2 N}{(z_{1-\alpha/2} + z_{1-\beta})^2 (1-\rho)^2 + z_{1-\gamma}^2 (2\rho - \rho^2)},\tag{19.19}$$

where N and N_J are the sample size for the entire group and the Japanese group, respectively. Note that the MHLW of Japan recommends that ρ should be chosen to be either 0.5 or greater and γ should be chosen to be either 0.8 or greater in (19.18). As an example, if we choose $\rho = 0.5$, $\gamma = 0.8$, $\alpha = 0.05$, and $\beta = 0.9$, then $N_J/N = 0.224$. In other words, the sample size for the Japanese group has to be at least 22.4% of the overall sample size for the multiregional trial.

In practice, $1 - \rho$ is often considered a non-inferiority margin. If ρ is chosen to be greater than 0.5, the Japanese sample size will increase substantially. It should be noted that the sample size formulas given by Quan et al. (2010) are derived under the assumption that there are no differences in treatment effects for the Japanese group and non-Japanese group. In practice, it is expected that there will be a difference in treatment effect due to ethnic differences. Thus, the formulas for sample size calculation/ allocation derived by Quan et al. (2010) are necessarily modified in order to take into consideration the effect due to ethnic differences.

As an alternative, Kawai et al. (2008) proposed an approach to rationalize partitioning the total sample size among the regions so that a high probability of observing a consistent trend under the assumed treatment effect across regions can be derived, if the treatment effect is positive and uniform across regions in a multiregional trial. Uesaka (2009) proposed new statistical criteria for testing consistency between regional and overall results which do not require impractical sample sizes, and discussed several methods of sample size allocation to regions. Basically, three rules of sample size allocation in multiregional clinical trials are discussed. These rules include (1) allocating equal size to all regions, (2) minimizing total sample size, and (3) minimizing the sample size of a specific region. It should be noted that the sample size of a multiregional trial may become very large when one wishes to ensure consistent results between region of interest and the other regions or between the regional results and the overall results regardless of which rules of sample size allocation are used.

19.4.2 Proposals of Statistical Guidance—Asian Perspective

As indicated earlier, based on the MHLW guidance, several methods for the determination of sample size in a specific region have been proposed (see, e.g., Quan et al., 2010; Uesaka, 2009). In addition, Ko et al. (2010) focus on a specific region and establish four statistical criteria for consistency between the region of interest and overall results. More specifically, two criteria are to assess whether the treatment effect in the region of interest is as large as that of the other regions or of the regions overall, while the other two criteria are to assess the consistency of the treatment effect of the specific region with other regions or the regions overall.

The global drug development plays an important role in a scientific manner to pharmaceutical research. However, the statistical work to draw a statistical inference with regard to translational medicine research is still in a preliminary stage. To provide a comprehensive understanding of statistical design and methodology that are commonly employed in global drug development, under the support of the Bureau of Pharmaceutical Affairs, Department of Health, Taiwan, the National Health Research Institutes and Formosa Cancer Foundation organized one symposium on "Current Advanced Statistical Issues in Clinical Trials—Flexibility and Globalization" held on November 21, 2008, and a closed-door meeting on "Designs of Clinical Trials in New Drug Developments" held on November 22, 2008 in Taipei, Taiwan. As a result, a proposal of statistical guidance to multiregional trials was developed. This proposal is briefly described in the following section. We first give a definition of the so-called *Asian region*.

19.4.2.1 Definition of the Asian Region

When planning a multiregional trial, the definition of the Asian region is very critical, since there are many regional countries in Asia. According to the International Conference on Harmonization (ICH) E5 guideline, the ethnic factors are classified into the following two categories: intrinsic and extrinsic factors. Intrinsic ethnic factors are factors that define and identify the population in the new region and may influence the ability to extrapolate clinical data between regions. They are more genetic and physiologic in nature, e.g., genetic polymorphism, age, and gender. On the other hand, extrinsic ethnic factors are factors associated with the environment and culture. Extrinsic ethnic factors are more social and cultural in nature, e.g., medical practice, diet, and practices in clinical trials and conduct.

For example, the increasing evidence that genetic determinants may mediate variability among persons in response to a drug implies that the patients' responses to therapeutics may vary among racial and ethnic groups. In other words, after the intake of identical doses of a given agent, some ethnic groups may have clinically significant side effects, whereas others may show no therapeutic response. An example of such a situation can be seen in the study by Caraco (2004). Caraco pointed out that some of this diversity in rates of response can be ascribed to differences in the rate of drug metabolism, particularly by the cytochrome P-450 superfamily of enzymes. While 10 isoforms of cytochrome P-450 are responsible for the oxidative metabolism of most drugs, the effect of genetic polymorphisms on catalytic activity is most prominent for 3 isoforms—CYP2C9, CYP2C19, and CYP2D6. Among these three, CYP2D6 has been most extensively studied and is involved in the metabolism of about 100 drugs, including β-blockers, and antiarrhythmic, antidepressant, neuroleptic, and opioid agents. Several studies revealed that some patients are classified as having "poor metabolism" of certain drugs owing to the lack of CYP2D6 activity. On the other hand, patients having some enzyme activity are classified into three subgroups: those with "normal" activity (or extensive metabolism), those with reduced activity (intermediate metabolism), and those with markedly enhanced activity (ultrarapid metabolism). Most importantly, the distribution of CYP2D6 phenotypes varies with race. However, the frequency of the phenotype associated with poor metabolism is 1% in both the Chinese and Japanese populations. Another study also showed that there exist no ethnic differences in CYP2C19 among Chinese, Japanese, and Korean populations (Myrand et al., 2008). Considering genetic polymorphism, the International HapMap Project also shows that the Chinese and Japanese genome look alike. All these data may reasonably support that the countries of China, Hong Kong, Japan, Korea, and Taiwan can be regarded as the *Asian region*.

On the other hand, the frequency of HLA alleles is associated with Stevens–Johnson syndrome (Chung et al., 2004). However, the prevalence rates of HLA-B*1502 for Chinese, Japanese, and Korean populations are, respectively, 1.9%–7.1%, <0.3%, and 0.2% (see, e.g., Ueta et al., 2007). That is, there exist differences within Asian populations in this regard. Consequently, the definition of a region may possibly vary from disease to disease. In fact, all differences and similarities in both intrinsic and extrinsic ethnical factors should be considered for the definition of the *Asian region*.

Within the Asian region, each country may consider accepting all the data derived from other countries in the "Asian region." For example, Taiwan accepts all Asian data. A study by Lin et al. (2001) found that the so-called *Taiwanese*, accounting for 91% of the total population in Taiwan, are comprised of Minnan and Hakka people who are closely related to the southern Han, and are clustered with other southern Asian populations such as Thai and Malaysian in terms of HLA typing. Those who are the descendants of northern Han are separated from the southern Asian cluster, and form a cluster with the other northern Asian populations such as Korean and Japanese. The Taiwanese regulatory authority, therefore, accepts data from trials conducted in Taiwan as well as in other Asian countries, if those trials meet Taiwanese regulatory standards and are conducted in compliance with GCP requirements.

19.4.2.2 Bridging the Results to the Asian Region

The aim of a multiregional trial is to show the efficacy of a drug in various global regions, and concurrently to evaluate the possibility of applying the overall trial results to each region. Therefore, how to bridge the results of the multiregional trial to the "Asian region" is another important issue.

Let D_{Asia} be the observed treatment effect for the Asian region and D_{All} the observed treatment effect from all regions. Given that the overall result is significant at α level, we will judge whether the treatment is effective in the Asian region by the following criterion:

$$D_{\text{Asia}} \geq \rho D_{\text{All}} \quad \text{for some } 0 < \rho < 1. \tag{19.20}$$

Other consistency criteria can be found in Uesaka (2009) and Ko et al. (2010). Selection of the magnitude, ρ, of the consistency trend may be critical. All differences in ethnic factors between the Asian region and other regions should be taken into account. The Japanese MHLW suggests that ρ be 0.5 or greater. However, the determination of ρ will be and should be different from product to product and from therapeutic area to therapeutic area. For example, in a multiregional liver cancer trial, the Asian region can definitely require a larger value of ρ, since it will contribute more subjects than other regions.

In addition to the consistency criterion in (19.20), the following criteria suggested by Uesaka (2009) and Ko et al. (2010) can also be used:

$$D_{\text{Asia}} \geq \rho D_C \quad \text{for some } 0 < \rho < 1,$$

$$\rho \leq \frac{D_{\text{Asia}}}{D_{\text{All}}} \leq \frac{1}{\rho} \quad \text{for some } 0 < \rho < 1,$$

$$\rho \le \frac{D_{\text{Asia}}}{D_C} \le \frac{1}{\rho} \quad \text{for some } 0<\rho<1,$$

where D_C denotes the observed treatment effect from regions other than the Asian region. The first criterion is to assess whether the treatment effect in the Asian region is as large as that of the other regions, while the last two criteria are to assess the consistency of the treatment effect of the Asian region with overall regions or other regions.

19.4.2.3 Sample Size for Multiregional Trials

When planning a multiregional trial, it is suggested that the study objectives should be clearly stated in the study protocol. Once the study objectives are confirmed, a valid study design can be chosen and the primary clinical endpoints can be determined accordingly. Based on the primary clinical endpoint, the sample size required for achieving a desired power can then be calculated. Recent approaches for sample size determination in multiregional trials developed by Kawai et al. (2008), Quan et al. (2010), and Ko et al. (2010) are all based on the assumption that the effect size is uniform across regions. For example, assume that we focus on the multiregional trial for comparing a test product and a placebo control based on a continuous efficacy endpoint. Let X and Y be some efficacy responses for patients receiving the test product and the placebo control, respectively. For convention, both X and Y are normally distributed with variance σ^2. We assume that σ^2 is known, although it can generally be estimated. Let μ_T and μ_P be the population means of the test and placebo, respectively, and let $\Delta = \mu_T - \mu_P$. Assume that effect size (Δ/σ) is uniform across regions. The hypothesis of testing for the overall treatment effect is given as

$$H_0 : \Delta \le 0 \quad \text{versus} \quad H_a : \Delta > 0.$$

Let N denote the total sample size for each group planned for detecting an expected treatment difference $\Delta = \delta$ at the desired significance level α and with power $1 - \beta$. Thus,

$$N = 2\sigma^2 \left\{ \frac{(z_{1-\alpha} + z_{1-\beta})}{\delta} \right\}^2 ,$$

where $z_{1-\alpha}$ is the $(1 - \alpha)$th percentile of the standard normal distribution. Once N is determined, special consideration should be placed on the determination of the number of subjects from the Asian region in the multiregional trial. The selected sample size should be able to establish the consistency of treatment effects between the Asian region and the regions overall. To establish the consistency of treatment effects between the Asian region and the entire group, it is suggested that the selected sample size should satisfy that the assurance probability of the consistency criterion in (19.20), given that $\Delta = \delta$ and the overall result is significant at α level, is maintained at a desired level, say 80%. That is,

$$P_\delta(D_{Asia} \geq \rho D_{All} \mid Z > z_{1-\alpha}) > 1 - \gamma \qquad (19.21)$$

for some prespecified $0 < \gamma \leq 0.2$. Here Z represents the overall test statistic.

Ko et al. (2010) calculated the sample size required for the Asian region based on (19.21). For $\beta = 0.1$, $\alpha = 0.025$, and $\rho = 0.5$, the sample size for the Asian region has to be around 30% of the overall sample size to maintain the assurance probability of (19.21) at 80% level. On the other hand, by considering a two-sided test, Quan et al. (2010) derived closed form formulas for the sample size calculation for normal, binary, and survival endpoints based on the consistency criterion (19.20). For example, if we choose $\rho = 0.5$, $\gamma = 0.2$, $\alpha = 0.025$, and $\beta = 0.9$, then the Asian sample size has to be at least 22.4% of the overall sample size for the multiregional trial.

It should be noted that the sample size determination given in Kawai et al. (2008), Quan et al. (2010), and Ko et al. (2010) are all derived under the assumption that the effect size is uniform across regions. In practice, it might be expected that there is a difference in treatment effect due to ethnic difference. Thus, the sample size calculation derived by Kawai et al. (2008), Quan et al. (2010), and Ko et al. (2010) may not be of practical use. More specifically, some other assumptions addressing the ethnic difference should be explored. For example, we may consider the following assumptions:

1. Δ is the same but σ^2 is different across regions.
2. Δ is different but σ^2 is the same across regions.
3. Δ and σ^2 are both different across regions.

Statistical methods for the sample size determination in multiregional trials should be developed based on the above assumptions.

19.4.2.4 Remarks

A multiregional trial may incorporate subjects from many countries around the world under the same protocol. After showing the overall efficacy of a drug in all global regions, we can simultaneously evaluate the possibility of applying the overall trial results to each region and consequently support registration in each region. In the previous subsections, we described some proposals given by Tsou et al. (2011) regarding statistical guidance to multiregional trials. In Tsou et al.'s proposal, both the MHLW guidance and the 11th Q&A for the ICH E5 guideline can serve as a framework on how to demonstrate the efficacy of a drug in all participating regions while also evaluating the possibility of applying the overall trial results to each region by conducting a multiregional trial. Most importantly, the consistency criterion presented in the Japanese guideline can be used to apply the overall results from the multiregional trial to the Asian region.

In Zhou et al.'s proposal, the sample size calculation for multiregional trials should take the possibility of ethnic differences into account. When planning a multiregional trial, the regions involved are expected to participate in the global development as early as possible. Therefore, the ethnic differences might be detected at any stage of early drug development. On the other hand, the analyses on the Asian data in the multiregional trial may not have enough statistical power. Thus, the number of subjects required for the Asian region in the multiregional trial should be large enough to establish the consistency of treatment effects between the Asian region and the regions overall. Also note that the sample size required in (19.21) is for the entire Asian region with similar ethnicity. Each country in the Asian region may contribute a different size of subjects to the multiregional trial. However, for the evaluation of consistency, each country may consider accepting all the data derived from other countries in the Asian region.

Multiregional trials might have benefits on decreasing Asian patients' exposures on unapproved drugs, reducing drug lag, and increasing available treatment options. From the beginning of the twenty-first century, the trend for clinical development in Asian countries being undertaken simultaneously with clinical trials conducted in Europe and the United States has been speedily rising. In particular, Taiwan, Korea, Hong Kong, and Singapore have already had much experience in planning and conducting the multiregional trials. It should be noted that conducting multiregional trials may require more management skills due to various cultures, languages, religions, and medical practices. This kind of cross-cultural management may be challenging.

19.5 Statistical Methods for Bridging Studies

In recent years, the influence of ethnic factors on clinical outcomes for the evaluation of efficacy and safety of study medications under investigation has attracted much attention from regulatory authorities, especially when the sponsor is interested in bringing an approved drug product from the original region (e.g., the United States or European Union) to a new region (e.g., Asian Pacific region). To determine if clinical data generated from the original region are acceptable in the new region, the ICH issued a guideline on *Ethnic Factors in the Acceptability of Foreign Clinical Data*. The purpose of this guideline is not only to permit adequate evaluation of the influence of ethnic factors, but also to minimize duplication of clinical studies in the new region (ICH, 1998). This guideline is known as ICH E5 guideline.

As indicated in the ICH E5 guideline, a bridging study is defined as a study performed in the new region to provide pharmacokinetic (PK), pharmacodynamic (PD), or clinical data on efficacy, safety, dosage, and dose regimen in the new region that will allow extrapolation of the foreign clinical data to the population in the new region. The ICH E5 guideline suggests that the regulatory authority of the new region assess the ability to extrapolate foreign data based on the bridging data package, which consists of (i) information including PK data and any preliminary PD and dose-response data from the complete clinical data package (CCDP) that is relevant to the population of the new region and, if needed, (ii) a bridging study to extrapolate the foreign efficacy data and/or safety data to the new region. The ICH E5 guideline indicates that bridging studies may not be necessary if the study medicines are insensitive to ethnic factors. For medicines characterized as insensitive to ethnic factors, the type of bridging studies (if needed) will depend upon experience with the drug class and upon the likelihood that extrinsic ethnic factors could affect the medicine's safety, efficacy, and dose response. On the other hand, for medicines that are ethnically sensitive, a bridging study is usually needed since the populations in two regions are different. In the ICH E5 guideline, however, no criteria for assessment of the sensitivity to ethnic factors for determining whether a bridging study is needed are provided. Moreover, when a bridging study is conducted, the ICH guideline indicates that the study is readily interpreted as capable of bridging the foreign data if it shows that dose response, safety, and efficacy in the new region are similar to those in the original region. However, the ICH does not clearly define the similarity.

Shih (2001) interpreted it as consistency among study centers by treating the new region as a new center of multicenter clinical trials. Under this definition, Shih (2001) proposed a method for assessment of consistency to determine whether the study is capable of bridging the foreign data to the new region. Alternatively, Shao and Chow (2002) proposed the concepts of reproducibility and generalizability probabilities for assessment of bridging studies. If the influence of the ethnic factors is negligible, then we may consider the reproducibility probability to determine whether the clinical results observed in the original region are reproducible in the new region. If there is a notable ethnic difference, the generalizability probability can be assessed to determine whether the clinical results in the original region can be generalized in a similar but slightly different patient population due to the difference in ethnic factors. In addition, Chow et al. (2002) proposed to assess bridging studies based on the concept of population (or individual) bioequivalence. Along this line, Hung (2003) and Hung et al. (2003) considered the assessment of similarity based on testing for non-inferiority between a bridging study conducted in the new region and the previous one conducted in the original region. This leads to the argument regarding the selection of non-inferiority margin (Chow and Shao, 2006). Note that other methods such as the use of the Bayesian approach have also been proposed in the literature (see, e.g., Liu et al., 2002a).

19.5.1 Test for Consistency

For the assessment of similarity between a bridging study conducted in a new region and studies conducted in the original region, Shih (2001) considered all of the studies conducted in the original region as a multicenter trial and proposed to test the consistency among study centers by treating the new region as a new center of a multicenter trial.

Suppose there are K reference studies in the CCDP. Let T_i denotes the standardized treatment group difference, i.e.,

$$T_i = \frac{\bar{x}_{T_i} - \bar{x}_{C_i}}{s_i\sqrt{1/m_{T_i} + 1/m_{C_i}}},$$

where

$\bar{x}_{T_i}(\bar{x}_{C_i})$ is the sample mean of $m_{T_i}(m_{C_i})$ observations in the treatment (control) group,

s_i is the pooled sample standard deviation.

Shih (2001) considered the following predictive probability for testing consistency:

$$p(T \mid T_i, i = 1, ..., K) = \left(\frac{2\pi(K+1)}{K}\right)^{-K/2} \exp\left[\frac{-K(T-\bar{T})^2}{2(K+1)}\right]. \tag{19.22}$$

19.5.2 Test for Reproducibility and Generalizability

On the other hand, when the ethnic difference is negligible, Shao and Chow (2002) suggested assessing reproducibility probability for similarity between clinical results from a bridging study and studies conducted in the CCPD. Let x be a clinical response of interest in the original region. Let y be similar to x but a response in a clinical bridging study conducted in the new region. Suppose the hypotheses of interest are

$$H_0 : \mu_1 = \mu_0 \quad \text{versus} \quad H_a : \mu_1 \neq \mu_0.$$

We reject H_0 at the 5% level of significance if and only if $|T| > t_{n-2}$, where t_{n-2} is the $(1 - \alpha/2)$th percentile of the t distribution with $n - 2$ degrees of freedom, $n = n_1 + n_2$

$$T = \frac{\bar{y} - \bar{x}}{\sqrt{\dfrac{(n_1 - 1)s_1^2 + (n_0 - 1)s_0^2}{n - 2}} \sqrt{\dfrac{1}{n_1} + \dfrac{1}{n_0}}},$$

and $\bar{x}$, $\bar{y}$, s_0^2, and s_1^2 are sample means and variances for the original region and the new region, respectively. Thus, the power of T is given by

$$p(\theta) = P(|T| > t_{n-2}) = 1 - \Im_{n-2}(t_{n-2} \mid \theta) + \Im_{n-2}(-t_{n-2} \mid \theta),$$

where

$$\theta = \frac{\mu_1 - \mu_0}{\sigma\sqrt{1/n_1 + 1/n_0}},$$

and $\Im_{n-2}(\cdot \mid \theta)$ denotes the cumulative distribution function of the noncentral t distribution with $n - 2$ degrees of freedom and the noncentrality parameter θ. Replacing θ in the power function with its estimate $T(x)$, the estimated power

$$\hat{p} = P(T(x)) = 1 - \Im_{n-2}(t_{n-2} \mid T(x)) + \Im_{n-2}(-t_{n-2} \mid T(x)) \tag{19.23}$$

is defined as a reproducibility probability for a future clinical trial with the same patient population. Note that when the ethnic difference is notable, Shao and Chow (2002) recommended assessing the so-called generalizability probability for similarity between clinical results from a bridging study and studies conducted in the CCPD.

19.5.3 Test for Similarity

Using the criterion for assessment of population (individual) bioequivalence, Chow, Shao, and Hu (2002) proposed the following measure of similarity between x and y:

$$\theta = \frac{E(x - y)^2 - E(x - x')^2}{E(x - x')^2/2},$$

where
  x' is an independent replicate of x,
  y, x, and x' are assumed to be independent.

Since a small value of θ indicates that the difference between x and y is small (relative to the difference between x and x'), similarity between the new region and the original region can be claimed if and only if $\theta < \theta_U$, where θ_U is a similarity limit. Thus, the problem of assessing similarity becomes a problem of testing the following hypotheses:

$$H_0 : \theta = \theta_U \quad \text{versus} \quad H_a : \theta \neq \theta_U.$$

Let $k = 0$ indicate the original region and $k = 1$ indicate the new region. Suppose that there are m_k study centers and n_k responses in each center for a given variable of interest. For simplicity, we only consider the balanced case where centers in a given region have the same number of observations. Let z_{ijk} be the ith observation from the jth center of region k, b_{jk} be the between-center random effect, and e_{ijk} be the within-center measurement error. Assume that

$$z_{ijk} = \mu_k + b_{jk} + e_{ijk}, \quad i = 1,\dots,n_k, \quad j = 1,\dots,m_k, \quad k = 0,1,$$

where
μ_k is the population mean in region k,
$b_{jk} \sim N(0, \sigma_{Bk}^2)$,
$e_{ijk} \sim N(0, \sigma_{Wk}^2)$,
$\{b_{jk}\}$ and $\{e_{ijk}\}$ are independent.

Under the above model, the criterion for similarity becomes

$$\theta = \frac{(\mu_0 - \mu_1)^2 + \sigma_{T1}^2 - \sigma_{T0}^2}{\sigma_{T0}^2}, \tag{19.24}$$

where $\sigma_{Tk}^2 = \sigma_{Bk}^2 + \sigma_{Wk}^2$ is the total variance (between-center variance plus within-center variance) in region k. The above hypotheses are equivalent to

$$H_0 : \varsigma \geq 0 \quad \text{versus} \quad H_a : \varsigma < 0,$$

where $\varsigma = (\mu_0 - \mu_1)^2 + \sigma_{T0}^2 - (1 + \theta_U)\sigma_{T0}^2$.

19.6 Concluding Remarks

In multiregional (multinational) multicenter trials, it is important to maintain the integrity of the trial by minimizing or controlling all possible sources (both expected and unexpected) of bias, variability, and/or confounding effects that may occur during the conduct of the trial. For this purpose, it is strongly recommended that a steering committee which consists of key individuals across countries be established. The purpose of this committee is multifold. It monitors the performance of the trial to maintain the integrity of the trial. It provides scientific/medical advice to the medical community from different countries for consistent assessment of the study drug. In addition, it helps to resolve any issues/problems that may be encountered during the conduct of the study. The function of the committee should be independent of the sponsor to maintain the integrity of the trial. Note that the analysis of a multiregional (multinational) trial is different from that of a meta-analysis of independent clinical trials in different countries. The analysis of multiregional (multinational) trials combines data observed from each country; the data are generated based on the methods prospectively specified in the same study protocol with the same method of randomization and probably at the same time. In contrast, a meta-analysis combines data retrospectively observed from a number of independent clinical trials involving different regions (countries), which may be conducted under different study protocols with different randomization schemes at different times. In either case, the treatment-by-region (treatment-by-country) interaction for multiregional (multinational) trials or treatment-by-region (treatment-by-country) for meta-analysis must be carefully evaluated before pooling the data for analysis.

In addition to the controversial issues regarding (1) the selection of the optimal number of study sites, (2) sample size calculation and allocation of specific region, and (3) statistical methods for bridging studies described above, another controversial issue which has a direct impact on the quality and validity of the conduct of multiregional (multinational) clinical trials is the possible lost-in-translation due to ethnic differences among regions. Translation in language refers to possible lost-in-translation of the informed consent form and/or CRFs in multiregional (multinational or global) clinical trials. Lost-in-translation is commonly encountered due to differences not only in language but also in perception, culture, and medical practice. A typical approach for the assessment of the possible lost-in-translation is to first translate the informed consent form and/or the CRFs by an experienced expert in the subject area and then perform a back-translation by a different experienced but independent expert in the subject area. The back-translated version is then compared with the original version for consistency. This can be done through the conduct of a small-scale pilot study. Qualified subjects from the target patient population will be randomly assigned to receive either the original version or the back-translated version. The responses will be collected and analyzed for comparison. If the back-translated version passes the test for consistency as compared to the original version, we then conclude that there is no evidence of lost-in-translation in the translated version and hence the translated version is considered validated. The translated version can then be used in the intended multiregional (multinational) clinical trial.

20

Dose Escalation Trials

20.1 Introduction

As therapeutic agents for cancer treatment can induce severe safety concern even at lower dose levels, phase I trials for new anticancer agents are often conducted on terminal cancer patients for whom the test cytotoxic drugs may be the last hope. The primary scientific objective of the evaluation of new chemotherapeutic agents in cancer patients during phase I clinical development is to employ an efficient, reliable, and yet practical dose-finding design to search the maximum dose with an acceptable and manageable safety profile for use in subsequent phase II trials (Koyfman et al., 2007). The dose with an acceptable and manageable safety profile is usually referred to as the maximum tolerable dose (MTD). The unacceptable or unmanageable safety profile is generally called the dose-limiting toxicity (DLT), which is predefined by some criteria such as grade 3 or greater hematological toxicity according to the United Sates National Cancer Institute's Common Toxicity Criteria. Thus, MTD is the highest possible but still tolerable dose with respect to some pre-specified DLT (see, e.g., Storer, 1993; Babb et al., 1998; Korn et al., 1999). Hence, an identified MTD is often considered as the optimal dose for subsequent clinical trials conducted at a later phase of clinical development.

The main purpose of phase I cancer trials is to establish the MTD with an adequate precision. The following considerations are important for the selection of an appropriate design in phase I trials for estimation of the MTD:

1. The patients are critically ill. Some of them are even in the terminal stage of the disease and the test anticancer agent may be the last hope for the patients.

2. The number of patients available for phase I cancer trials is relatively small.

3. The patient population is usually rather heterogeneous because phase I cancer trials might enroll terminal cancer patients with different types of malignant tumors at various disease stages.

4. Phase I cancer trials can be viewed as a screening process where anticancer cytotoxic agents with a tolerable safety profile are selected and their MTDs are determined with a minimal number of patients in a minimal amount of time.

5. Most anticancer agents generally can introduce serious, irreversible, life-threatening or even fatal toxicity. Thus, phase I cancer trials are usually conducted to establish the MTD. In fact, regulatory agencies sometimes dictate the dose from the first patient.

For early-phase cancer trials for dose finding, many useful designs including Bayesian dose-finding designs have been proposed in the literature (see, e.g., Storer, 1989, 1993, 2001; Piantadosi and Liu, 1996; Thall and Russel, 1998; Whitehead and Williamson, 1998; O'Quigley et al., 2001; Chang and Chow, 2005; Loke et al., 2006; Zhou et al., 2006). In practice, however, only two types are commonly used (Dent and Eisenhauer, 1996; Eisenhauer et al., 2000; Le Tourneau et al., 2009). These are the algorithm-based designs which follow a traditional escalation rule (TER), e.g., the "3 + 3" design as well as the model-based designs using the continual reassessment method (CRM) (see, e.g., O'Quigley et al., 1990; O'Quigley and Shen, 1996; Heyd and Carlin, 1999; O'Quigley, 2001; Babb and Rogatko, 2004; Kamp et al., 2007; Paoletti and Kramer, 2009).

The TER has been criticized for resulting in the underestimation of the MTD and for including too many patients at a suboptimal level, among other concerns (see, e.g., Heyd and Carlin, 1999; Chow and Chang, 2006). As a result, the CRM has become very popular. However, it remains unclear as to the relative merits and disadvantages of the CRM compared to the TER design, especially the general "$a + b$" TER design with and without dose de-escalation. Hence clinical trial investigators and statisticians continue to (often quite arbitrarily) choose between the two types of designs, usually without providing any justification for their choice or the planned sample size. Moreover, there are no clear criteria guidelines as to how such designs should be chosen and justified in study protocols for statistical validity. Thus many protocols for phase I dose-finding studies continue to be approved without such justification, resulting in potentially severe consequences as it could mean that the design and sample size eventually used may actually not be sufficient/suitable to adequately answer the research question of interest.

In the next section, standard TER trial design with and without dose de-escalation is briefly described. Also included in this section is the description of the general "$a + b$" TER design without dose de-escalation. In Section 20.3, the model-based CRM trial design is introduced. Also included in this section is the use of the CRM trial design in conjunction with the Bayesian approach for dose finding in cancer trails. Criteria for design selection and statistical justification for sample size calculation are given in Section 20.4. Section 20.5 provides a brief concluding remark.

20.2 Traditional Escalation Rule

In early-phase cancer trial, the TER, which are known as the "3 + 3" rules, are commonly used. The "3 + 3" rule is to enter three patients at a new dose level and then enter another three patients when DLT is observed. The assessment of the six patients is then performed to determine whether the trial should be stopped at that level or to escalate to the next dose level. Basically there are two types of "3 + 3" rules, namely, the TER and strict traditional escalation rule (STER). TER does not allow dose de-escalation but STER does when two of three patients have DLTs.

Note that the "3 + 3" rules can be generalized to the "$a + b$" TER (without dose de-escalation) and STER (with dose de-escalation), which are described in the following section.

For the general "$a + b$" TER design without dose de-escalation, suppose that there are a patients at dose level i. If less than c/a patients have DLTs, then the dose is escalated to the next dose level $i + 1$. If more than d/a ($d \geq c$) patients have DLTs, then the previous dose $i - 1$ will be considered the MTD. If no less than c/a but no more than d/a patients have DLTs, b more patients are treated at this dose level i. If no more than e ($e \geq d$) of the total of $a + b$ patients have DLTs, then the dose is escalated. If more than e of the total of $a + b$ patients have DLT, then the previous dose $i - 1$ will be considered the MTD. It can be seen that the traditional "3 + 3" TER without dose de-escalation is a special case of the general "$a + b$" design with $a = b = 3$ and $c = d = e = 1$.

Basically, the general "$a + b$" TER design with dose de-escalation is similar to the design without dose de-escalation. However, it permits more patients to be treated at a lower dose (i.e., dose de-escalation) when excessive DLT incidences occur at the current dose level. The dose de-escalation occurs when more than d/a ($d \geq c$) or more than $e/(a + b)$ patients have DLTs at dose level i. In this case, b more patients will be treated at dose level $i - 1$ provided that only a patients have been previously treated at this prior dose. If more than a patients have already been treated previously, then dose $i - 1$ is the MTD. The de-escalation may continue to the next dose level $i - 2$ if necessary.

20.3 Continual Reassessment Method

The concept of CRM was first applied in phase I oncology trials by O'Quigley et al. (1990). The primary goal is not only to assess the dose–toxicity relationship, but also to determine MTD. Due to the potential high toxicity of the study drug, in practice usually only a small number of patients (e.g., three to six) are treated at each ascending dose level. The most common approach is the "3 + 3" TER with a prespecified sequence for dose escalation. However, this ad hoc approach is found to be inefficient and often underestimates the MTD, especially when the starting dose is too low. The CRM is developed to overcome these limitations. The estimation or prediction from CRM is weighted by a number of data points. Therefore, if the data points are mostly around the estimated value, then the estimation is more accurate. CRM assigns more patients near MTD; consequently, the estimated MTD is much more precise and reliable. In practice, this is the most desirable operating characteristic of the Bayesian CRM.

20.3.1 Implementation of CRM

For the implementation of the model-based CRM design, the following information is required:

1. *Starting dose*: e.g., the initial dose is usually selected as $1/10$ of LD_{10} in mice.

2. *Dose range and number of dose levels*: Typically, 5–10 dose levels are selected for dose finding. Modified Fibonacci dose escalation factor (sequence) is usually considered within the selected dose range.

3. *Prior information on the MTD*: Any prior knowledge regarding MTD would be helpful. For example, DLT rate at MTD.

4. *Dose–toxicity model*: The following dose–toxicity model is often considered:

$$p(x) = [1 + b\exp(-ax)]^{-1},$$

where $p(x)$ is the probability of toxicity with dose x. The above can be solved for (predicted) MTD as follows:

$$\text{MTD} = \frac{1}{a}\ln\left(\frac{b\theta}{1-\theta}\right),$$

where θ is the probability of DLT (DLT rate) at MTD. Note that for an aggressive tumor and a transient and non-life-threatening DLT, θ could be as high as 0.5. For persistent DLT and less aggressive tumors, θ could be as low as 0.1–0.25. A commonly used value for θ is somewhere between 0 and $1/3 = 0.33$ (see, e.g., Crowley, 2001).

5. *Escalation rule*: e.g., minimum number of patients per dose level before escalation is n.

6. *Stopping rule*: e.g., maximum number of patients at a dose level is 6.

Note that the assignment of patients to the most updated MTD leads to the majority of the patients assigned to the dose levels near MTD, which allows a more precise estimate of MTD with a minimum number of patients. In practice, potential dose jump and delayed response are commonly seen when utilizing CRM in dose escalation trials.

20.3.2 CRM in Conjunction with Bayesian Approach

Chang and Chow (2005) proposed a hybrid frequentist–Bayesian CRM in conjunction with utility-adaptive randomization for clinical trial designs with multiple endpoints. They proposed a hyper-logistic function family with multiple parameters gives users flexibility for probability modeling. Under their proposed method, CRM reassesses a dose–response relationship based on an accrued data of the ongoing trial, which allows investigators to make decisions based on a constantly updated dose–response model. In addition, their proposed utility-adaptive randomization for multiple endpoint trials allows more patients to be assigned to superior treatment groups.

The utility-based CRM adaptive approach proposed by Chang and Chow (2005) can be summarized by the following steps:

Step 1: Construct utility function based on trial objectives.

Step 2: Propose a probability model for dose–response relationship.

Step 3: Construct prior probability distributions of the parameters in the response model.

Step 4: Form the likelihood function based on incremental information on treatment response during the trial.

Step 5: Reassess model parameters or calculate the posterior probability of the model parameters.

Step 6: Update the expected utility function based on dose–response model.

Step 7: Determine next action or make adaptations such as changing the randomization or drop inferior treatment arms.

Step 8: Further collect trial data and repeat Steps 5–7 until stopping criteria are met.

At Step 1, a utility function can be constructed as follows. Let $X = \{x_1, x_2, ..., x_k\}$ be the action space where x_i is a coded value for an action of anything that would affect the outcomes or decision making such as a treatment, a withdrawal of a treatment arm, a protocol amendment, stopping the trial, an investment of advertising for the prospective drug, or any combination of the above. x_i can be either a fixed dose or a variable of a dose given to a patient. If action x_i is not taken, then $x_i = 0$. Let $y = \{y_1, y_2, ..., y_m\}$ be the outcomes of interest, which can be efficacy or toxicity of a test drug, the cost of trial, etc. Each of these outcomes y_i is a function of action $y_i(x)$, $x \in X$. The utility is then defined as

$$U = \sum_{j=1}^{m} w_j = \sum_{j=1}^{m} w(y_j),$$

where
   U is normalized such that $0 \leq U \leq 1$
   w_j are some prespecified weights

For Step 2, each of the outcomes can be modeled by the following generalized probability model:

$$\Gamma_j(p) = \sum_{i=1}^{k} a_{ji} x_i, \quad j = 1, \ldots, m,$$

where $p = (p_1, \ldots, p_m)$, $p_j = P(y_j \geq \tau_j)$, and τ_j is a threshold for the jth outcome. The link function, $\Gamma_j(\bullet)$, is a generalized function of all the probabilities of the outcomes. For a univariate case, a logistic model is commonly used for monotonic response. Note that for utility, however, we usually do not know whether it is monotonic or not. As a result, Chang and Chow (2005) suggested the use of a hyper-logistic function in modeling utility index.

At Step 3, the Bayesian approach requires the specification of prior probability distribution of the unknown parameter tensor a_{ji}. The assessment of the parameters in the model can be carried out in different ways: Bayesian, frequentist, or hybrid approach. Bayesian and hybrid approaches are to assess the probability distribution of the parameter, while the frequentist approach is to provide a point estimate of the parameter. We can then form the likelihood function based on incremental information on treatment response during the trial (Step 4) and reassess model parameters or calculate the posterior probability of the model parameters (Step 5). Then, update the expected utility function based on the dose–response model (Step 6).

At Step 7, we can determine the next action. As mentioned earlier, the actions or adaptations taken should be based on trial objectives or utility function. A typical action is a change of the randomization schedule. From the dose–response model, since each dose is associated with a probability of response, two approaches, namely, deterministic and probabilistic approaches, can be taken. The former refers to the optimal approach where actions can be taken to maximize the expected utility, while the latter refers to adaptive randomization where the treatment assignment to the next patient is not fully determined by the algorithm. The dose level assigned to the next patient based on optimization of the expected utility is given by

$$x_{n+1} = \arg\max_{x_i} \bar{U} = \sum_{j=1}^{m} p_j w_j.$$

It, however, should be noted that the above optimal approach may not be feasible due to its difficulties in practice. As indicated in Chang and Chow (2005), many of the response-adaptive randomizations can be used to increase the expected response. However, these adaptive randomizations are difficult to apply directly in the case of multiple endpoints. As an alternative, Chang and Chow (2005) suggested the use of so-called utility-adaptive randomization algorithm. This utility-adaptive randomization combines the idea from randomized-play-winner (Rosenberger and Lachin, 2003) and Lachin's urn models. More details can be found in Chang and Chow (2005).

20.3.3 Extended CRM Trial Design

The typical CRM can be extended to CRM(n_i), where n_i is the number of patients in each dose level i in conjunction with a Bayesian approach with various prior distributions, and possible dose jump and dose delays in CRM trial designs. In practice, it is of interest to compare the extended CRM trial design (with possible dose jump and dose delays) with the extended "$a + b$" TER trial design (with and without dose escalation) in terms of some performance characteristics such as the probability of correctly identifying the MTD.

20.4 Design Selection and Sample Size

In most protocols of dose escalation trials, little details regarding design selection and/or sample size calculation/justification are provided. Although many simulations have been performed to empirically compare the TER design and the CRM design and its various modifications, little or no empirical evidence is available regarding the relative performance between the TER trial design and the CRM design. In this section, some criteria for design selection and performance characteristics for sample size determination are proposed.

20.4.1 Criteria for Design Selection

For selecting an appropriate study design, two criteria based on a fixed sample size approach and a fixed power approach (i.e., fixed the probability of correctly identifying the MTD) are commonly considered.

For a fixed sample size, the optimal design can be chosen based on one or more of the following:

1. Number of DLT expected
2. Bias and variability of the estimated MTD
3. Probability of observing DLT prior to MTD
4. Probability of correctly identifying the MTD
5. Probability of overdosing

In other words, we may choose the design with the highest probability of correctly identifying the MTD. If it is undesirable to have patients experience the DLT, we may choose the design with the smallest number of DLT expected. In practice, we may compromise the above criteria for choosing the most appropriate design to meet our need.

On the other hand, for a fixed power approach (i.e., fixed the probability of correctly identifying the MTD), the optimal design can be similarly chosen based on one or more of the following:

1. Number of patients expected
2. Number of DLT expected
3. Bias and variability of the estimated MTD
4. Probability of observing DLT prior to MTD
5. Probability of overdosing

Thus, we may choose the design with the smallest number of patients expected. If it is desirable to minimize the exposure of patients prior to MTD, we may choose the design with the smallest probability of observing DLT prior to MTD. Similarly, we may compromise the above criteria for choosing the most appropriate design to meet our need. In some cases, the investigator may want to control potential overdose. In this case, we may choose a design with the minimum number of patients expected to be exposed to the dose beyond MTD.

20.4.2 Sample Size Justification

As indicated above, for most protocols of the dose escalation trials, little or no details regarding sample size justification is provided. When conducting a clinical trial, good statistics practices are necessarily followed for good clinical practice in order to ensure the success of the intended clinical trial. Thus, it is suggested that statistical justification for the selected sample size be provided, which will give statistical assurance for achieving the study objectives of the intended trial. Unlike most clinical trials, the traditional pre-study power analysis for sample size calculation is not applicable for dose escalation trials. For sample size justification of dose escalation trials, the following performance characteristics are useful: (1) the number of DLTs expected prior to MTD, (2) the bias and variability of the estimated MTD, (3) the probability of observing DLT prior to MTD, (4) the probability of correctly identifying the MTD, and (5) the probability of overdosing. In what follows, as an example, sample size calculations for the general "$a + b$" TER without and with dose de-escalation are described in the following section.

20.4.2.1 General TER without Dose De-Escalation

For simplicity, we consider sample size calculation based on the performance characteristic of the probability of correctly identifying the MTD. Under the general "$a + b$" design without dose de-escalation, the probability of concluding that the MTD has been reached at dose i is given by

$$P_i^* = P(\text{MTD} = \text{dose } i)$$

$$= P(\text{escalation at dose } \leq i \text{ and stop escalation at dose } i+1)$$

$$= (1 - P_0^{i+1} - Q_0^{i+1})\left(\prod_{j=1}^{i}(P_0^j + Q_0^j)\right), \quad i \leq i < K,$$

where

$$P_0^j = \sum_{k=1}^{c-1}\binom{a}{k}p_j^k(1-p_j)^{a-k},$$

and

$$Q_0^j = \sum_{k=c}^{d}\sum_{m=0}^{e-k}\binom{a}{k}p_j^k(1-p_j)^{a-k}\binom{b}{m}p_j^m(1-p_j)^{b-m}.$$

The expected number of patients at dose j is then given by

$$n_j = \sum_{i=0}^{K-1} n_{ji}P_i^*, \tag{20.1}$$

where

$$n_{ji} = \begin{cases} \dfrac{aP_0^j + (a+b)Q_0^j}{P_0^j + Q_0^j} & \text{if } j < i+1, \\[1.5em] \dfrac{a(1 - P_0^j - P_1^j) + (a+b)(P_1^j - Q_0^j)}{1 - P_0^j - Q_0^j} & \text{if } j = i+1, \\[1.5em] 0 & \text{if } j > i+1. \end{cases}$$

Note that, without consideration of undershoots (an attempt to de-escalate to a dose level at a lower dose than the starting dose level) and overshoots (an attempt to escalate to a dose level at the highest level planned), the expected number of DLTs at dose i can be obtained as $n_i p_i$. As a result, the total number of DLTs for the trial is given by $\sum_{i=1}^{K} n_i p_i$.

Under the general "$a + b$" design with dose de-escalation, the probability of concluding that the MTD has been reached at dose i is given by

$$P_i^* = P(\text{MTD} = \text{dose } i)$$

$$= P(\text{escalation at dose } \leq i \text{ and stop escalation at dose } i+1)$$

$$= \sum_{k=i+1}^{K} p_{ik},$$

where

$$p_{ik} = (Q_0^i + Q_1^i)(1 - P_0^k - Q_0^k)\prod_{j=1}^{i-1}(P_0^j + Q_0^j)\prod_{j=i+1}^{k-1}Q_2^j,$$

$$Q_1^j = \sum_{k=0}^{c-1}\sum_{m=0}^{e-k}\binom{a}{k}p_j^k(1-p_j)^{a-k}\binom{b}{m}p_j^m(1-p_j)^{b-m},$$

$$Q_2^j = \sum_{k=0}^{c-1}\sum_{m=e+1-k}^{b}\binom{a}{k}p_j^k(1-p_j)^{a-k}\binom{b}{m}p_j^m(1-p_j)^{b-m}.$$

The expected number of patients at dose j is then given by

$$n_j = n_{jK}P_K^* \sum_{i=0}^{K-1}\sum_{k=i+1}^{K} n_{jik}p_{ik}, \tag{20.2}$$

where

$$n_{jK} = \frac{aP_0^j + (a+b)Q_0^j}{P_0^j + Q_0^j},$$

$$n_{jik} = \begin{cases} \dfrac{aP_0^j + (a+b)Q_0^j}{P_0^j + Q_0^j} & \text{if } j < i, \\[2ex] a+b & \text{if } i \le j < k, \\[2ex] \dfrac{a(1-P_0^j-P_1^j)+(a+b)(P_1^j-Q_0^j)}{1-P_0^j-Q_0^j} & \text{if } j = k, \\[2ex] 0 & \text{if } j > k \end{cases}$$

and

$$P_1^j = \sum_{i=c}^{d} \binom{a}{k} p_j^k (1-p_j)^{a-k}.$$

Consequently, the total number of DLTs for the trial is given by $\sum_{i=1}^{K} n_i p_i$.

For the CRM trial design, there exists no closed form for sample size calculation. Thus, a clinical trial simulation is often conducted in order to evaluate the performance characteristics described above for sample size calculation. As an example, consider a dose escalation trial for identifying the MTD of a compound for the treatment of a certain cancer. A simulation with 5000 runs is planned for the evaluation of the above performance characteristics. The simulation was conducted under the following parameter specifications:

1. The initial dose was chosen to be 0.3 mg/kg (e.g., one–tenth of LD_{10} in mice).
2. The dose range considered is from 0.3 to 2.8 mg/kg.
3. The modified Fibonacci sequence is considered. That is, there are six dose levels, which are 0.3, 0.6, 1, 1.5, 2.1, and 2.8 mg/kg.
4. The DLT rate at MTD is assumed to be 1/3 = 33%.

For the algorithm-based trial design, the "3 + 3" TER design and the "3 + 3" STER design with maximum dose de-escalation allowed as 1 are considered. For the CRM method, CRM(n), where n is the number of patients per dose level, $n = 1, 2$, and 3. A logistic toxicity model is assumed. The Bayesian approach with a uniform prior is considered for the estimation of the parameters of the toxicity model. For CRM(n), the dose escalation and stopping rules include the following:

1. The number of doses allowed to skip is 0, i.e., dose jump is not allowed.
2. The minimum number of patients per dose level before escalation is n.
3. The maximum number of patients at a dose level is 6.

TABLE 20.1

Summary of Simulation Results

Design	Number of Patients Expected (N)	Number of DLT Expected	Mean MTD (SD)	Probability of Selecting Correct MTD
"3 + 3" TER	15.96	2.8	1.26 (0.33)	0.526
"3 + 3" STER[a]	17.56	3.2	1.02 (0.30)	0.204
CRM(1)[b]	10.60	3.4	1.51 (0.08)	0.984
CRM(2)[b]	13.57	2.8	1.57 (0.20)	0.884
CRM(3)[b]	16.37	2.7	1.63 (0.26)	0.784

[a] Allows dose de-escalation.
[b] CRM(n) = CRM with n patients per dose level; uniform prior dose was used.

Simulation results are summarized in Table 20.1.

As can be seen from Table 20.1, the "3 + 3" TER without dose de-escalation and CRM(2) have the smallest number of DLTs expected before reaching the MTD. As expected, the "3 + 3" TER design and the "3 + 3" STER design underestimate the MTD with larger standard deviations as compared to the CRM(n) trial design. In terms of the probability of correctly identifying the MTD, CRM(n) with $n = 1$ and $n = 2$ are preferred. Sample sizes required for the trial designs under study range from 11 to 18. Based on the overall comparison in terms of the performance characteristics, CRM(n) with $n = 2$ is recommended for the proposed study.

20.5 Concluding Remarks

Over the past two decades, many simulations have been performed to empirically compare the standard dose escalation design, up-and-down designs, the original CRM, and its various modifications. The results can be found in O'Quigley and Cheveret (1991), Korn et al. (1994, 1999), Goodman et al. (1995), O'Quigley (1999), and Storer (2001). Some of the results are summarized as follows:

1. The standard dose escalation design treats more patients at the subtherapeutic dose levels.

2. The standard dose escalation design underestimates the MTD.

3. The original CRM requires fewer patients than the standard dose escalation design does.

4. The average number of cohorts in the original CRM with a patient per cohort is larger than that of the standard dose escalation design.

Hence, the duration of the trials using the original CRM may be longer than other phase I designs.

5. The average number of cohorts reduces dramatically for the modified CRM with three patients per cohort and is similar to that of the standard dose escalation design.

6. The two-stage (modified) CRM does not provide better performance than the one-stage modified CRM.

7. The CRM is independent of the targeted percentile of some tolerance distribution that is pre-specified for other designs. In addition, it has, theoretically, convergence properties.

8. No design performs uniformly well in all possible dose–response settings.

9. The estimates of MTD generated from the CRM generally have smaller bias although the bias is relatively small.

For the CRM, the toxicity model will be reassessed after the response of the previous patient is observed. The next patient will then be assigned based on the estimated MTD (the patient will be assigned to the closest dose level). It is not efficient to have an independent statistician to reassess the toxicity model and then assign the patient for each level. Alternatively, a clinical trial simulation can be run with respect to all possible scenarios for randomization. Thus, once the response of the previous patient is observed, we can simply check the pregenerated table and assign the next patient to the appropriate dose level.

Note that some SAS codes are available in Chang (2008).

21

Enrichment Process in Target Clinical Trials

21.1 Introduction

As indicated by many researchers (e.g., Simon and Maitournam, 2004; Maitournam and Simon, 2005; Casciano and Woodcock, 2006; Dalton and Friend, 2006; Varmus, 2006), the disease targets at the molecular level can be identified after completion of the Human Genome Project (HGP). As a result, the importance of diagnostic tests for the identification of molecular targets increases as more targeted clinical trials will be conducted for the individualized treatment of patients (personalized medicine). For example, based on the risk of distant recurrence determined by a 21-gene Oncotype DX® breast cancer assay, patients with a recurrence score of 11–25 in the TAILORx (Trial Assigning Individualized Options for Treatment) trial sponsored by the United States National Cancer Institute (NCI) are randomly assigned to receive either adjuvant chemotherapy and hormonal therapy or adjuvant hormonal therapy alone (Sprarano et al., 2006). On the other hand, based on a 70-gene molecular signature, the MINDACT (Microarray in Node-negative Disease may Avoid ChemoTherapy) trial randomizes patients with a low-risk molecular prognosis and a high-risk clinical prognosis to the use of clinicopathologic criteria or gene signature in treatment decisions for the possible avoidance of chemotherapy (MINDACT, 2006). These two trials have an important implication for future individualized treatments for thousands of breast cancer patients (Swain, 2006). The Oncotype DX used in the TAILORx trial is a reverse transcriptase–polymerase chain reaction (RT-PCR) assay based on 21 genes, while the MINDACT trial employs a 70-gene molecular signature derived from the microarray (Van de Vijver et al., 2002; van't Veer, 2002; Paik et al., 2004, 2006).

Despite different technical platforms employed in the diagnostic devices for molecular targets used in the two trials, both assays belong to a group of the *in vitro* diagnostic multivariate index assay (IVDMIA) based on the selected differentially expressed genes for detection of the patients with molecular targets (FDA, 2006a). In addition, to reduce the variation, the IVDMIAs do not usually use all genes during the development stage. Therefore, identification of the differentially expressed genes between different groups of patients is the key to the accuracy and reliability of the devices for molecular targets. Once the differentially expressed genes are identified, the next task is to search an optimal representation or algorithm which provides the best discrimination ability between the patients with molecular targets and those without. The current validation procedure for diagnostic device is for the assay based on one analyte. However, the IVDMIAs are in fact parallel assays based on the intensities of multiple analytes. As a result, the current approach to assay validation for one analyte may not be appropriate and is inadequate for validation of IVDMIAs.

With respect to the enrichment design for the targeted clinical trials, patients with positive diagnosis for molecular targets are randomized to receive the test drug or the control. However, because no IVDMIA can provide a perfectly correct diagnosis, some patients with positive diagnosis may not actually have molecular targets. Consequently, the treatment effect of the test drug for the patients with targets is underestimated. On the other hand, estimation of the treatment effect based on the data from the targeted clinical trials needs to take into consideration the variability associated with the estimates of accuracy of the IVDMIA such as positive predictive value (PPV) and false positive (FP) rate obtained from the clinical effectiveness trials of the IVDMIA.

In the next section, commonly used approaches for identification of differentially expressed genes are reviewed. Also included in this section is the discussion of the relative merit and disadvantages of current methods. A set of interval hypotheses, which takes into consideration the minimal biological meaningful expression level, is proposed. Based on the interval hypotheses, Liu et al. (2007) suggested a two one-sided tests procedure. A discussion of the optimal representation or an algorithm of the IVDMIA based on the expression levels of the selected differentially expressed genes for the best diagnosis of molecular targets is provided in Section 21.3. Also included in this section is a recommendation for determining the number of genes to be included in the IVDMIA. In Section 21.4, the deficiency of the current validation for one analyte used for the IVDMIA is discussed. In addition, the issues and challenges for validation of the IVDMIA are also addressed in this section. Bias in estimation of the treatment effect of the test drug in the targeted clinical trials is discussed in Section 21.5. Approaches for obtaining the unbiased estimator of the treatment effect for patients with molecular targets and their variance are also given in this section. Design and analysis for target clinical trials are given in Sections 21.6 and 21.7, respectively. A discussion is provide in the last section.

____

21.2 Identification of Differentially Expressed Genes

For a given gene, the fold change is defined as the ratio of average expression level of the gene, which is measured by the intensity under one condition (e.g., tested or patients with a certain disease) to that under another condition (e.g., controlled or normal subjects without the disease). A gene is declared to be differentially expressed if the observed fold change either exceeds a prespecified threshold or is below a predetermined lower threshold. We refer to this procedure as the *fixed fold–change rule*. The fixed fold–change rule does not take into consideration the variation in estimation of the average intensity. In addition, it is not in the framework of hypothesis testing and therefore the probability associated with errors for decision making cannot be quantified and/or assessed. On the other hand, most current available statistical methods for identification of differentially expressed genes such as the *t*-test, permutation *t*-test, or significance analysis of microarray are in fact based on the following traditional hypotheses testing for equality (see, e.g., Tusher et al., 2001; Dudoit et al., 2002; Simon et al., 2003; Wang and Ethier, 2004):

$$H_0:\mu_{Di}-\mu_{Ni}=0 \quad \text{versus} \quad H_a:\mu_{Di}-\mu_{Ni}\neq 0, \tag{21.1}$$

where

$i = 1, \ldots, G,$

μ_{Ti} and μ_{Ci} are the true average expression levels on the log-scale (base 2) of gene i of the patients with molecular targets and the normal subjects without molecular targets, respectively.

As pointed out by Liu et al. (2007), the traditional hypotheses testing for equality is only to detect whether the difference in the average expression levels is 0 between the tested and controlled conditions. It fails to take into account the magnitudes of the biologically meaningful fold changes. In addition, due to simultaneously testing thousands of genes at the same time, with a small number of replicated samples, the FP rate for identifying differentially expressed genes is extremely high. Therefore, various methods are proposed to resolve this issue. Basically, they are applications of multiple comparison procedures to use some arbitrarily selected stringent cutoff of *p*-values to control false discovery rate (Hochberg and Tamhane, 1987; Benjamini and Hochberg, 1995) or to apply a combination of less stringent *p*-values for traditional hypotheses testing and the fixed fold–change rule (MAQC Consortium, 2006). However, all of these methods fail to take into account both magnitudes of biologically meaningful fold change and statistical significance simultaneously.

Since the objective is to identify the differentially expressed genes, the hypothesis for identifying differentially expressed genes should be formulated as the alternative hypothesis. On the other hand, gene i is said to be differentially expressed if the difference in average expression levels between the tested and controlled samples is either greater than a minimal biologically meaningful limit C_i (over-expressed) or smaller than a maximal biological meaningful limit $-C_i'$ (under-expressed). As a result, the hypotheses for identifying differential expressed genes between the tested and controlled samples can be formulated as follows (Liu et al., 2007):

$$H_0: -C_i' \le \mu_{iD} - \mu_{iN} \le C_i \quad \text{versus} \quad H_a: \mu_{iD} - \mu_{iN} < -C_i'$$

$$\text{or} \quad \mu_{iD} - \mu_{iN} > C_i, \quad i = 1, \ldots, G. \tag{21.2}$$

The parameter space for H_0 is $[-C_i', C_i]$, which represents the interval of no differential expression. On the other hand, the parameter space of the alternative hypothesis is the union of the intervals of over-expression (C_i, ∞) and under-expression $(-\infty, -C_i')$. In general, each gene should have its own differential expression limits and the differential expression limits do not have to be symmetric about 0. However, for the sake of illustration, without loss of generality, in what follows, we assume that the differential expression limits are the same and are symmetric about 0. The interval hypotheses for differentially expressed genes can be then formulated as

$$H_0: \left| \mu_{iD} - \mu_{iN} \right| \le C \quad \text{versus} \quad H_1: \left| \mu_{iD} - \mu_{iN} \right| > C, \quad i = 1, \ldots, G, \tag{21.3}$$

where C is some biologically meaningful differential expression limit. Furthermore, the interval hypotheses can be decomposed into two sets of one-sided hypotheses:

$$H_{0U}: \mu_{iD} - \mu_{iN} \le C \quad \text{versus} \quad H_{aU}: \mu_{iD} - \mu_{iN} > C$$

or

$$H_{0L}: \mu_{iD} - \mu_{iN} \ge -C \quad \text{versus} \quad H_{aL}: \mu_{iD} - \mu_{iN} < -C, \quad i = 1, \ldots, G. \tag{21.4}$$

The first set of hypotheses is to verify whether the difference in average expression level between the tested and controlled samples for gene i is higher than the prespecified upper differential expression limit for over-expression. The second set of hypotheses is to evaluate whether the difference in average expression levels between the tested and controlled samples for gene i is lower than the predetermined lower differential expression limit for under-expression.

Since the parameter space of the alternative hypothesis in (21.3) is the union of the parameter spaces of the two one-sided hypotheses given in (21.4), H_0 in (21.3) is rejected at the α level of significance if and only if either H_{0U} or H_{0L} is rejected at the $\alpha/2$ level of significance. In other words, under normal assumption, the two one-sided tests procedure proposed by Liu et al. (2007) rejects the null hypothesis of (21.3) and we conclude that gene i is differentially expressed between the tested and controlled samples at the α level of significance if

$$t_{Ui} = \frac{\bar{Y}_{iD} - \bar{Y}_{iN} - C}{\sqrt{s_{pi}^2(1/n_{iD} + 1/n_{iN})}} > t_{(\alpha/2,n_{iD}+n_{iN}-2)}$$

$$\text{or} \quad t_{Li} = \frac{\bar{Y}_{iD} - \bar{Y}_{iN} + C}{\sqrt{s_{pi}^2(1/n_{iD} + 1/n_{iN})}} < -t_{(\alpha/2,n_{iD}+n_{iN}-2)}, \tag{21.5}$$

where

$\bar{Y}_{ik}$ and n_{ik} are the sample mean expression and sample size of gene i under treatment k, respectively,

s_{pi}^2 is the pooled sample variance for gene i, where $i = 1,\dots,G$ and $k = T, C$.

Figure 21.1 gives the rejection region of the two one-sided tests procedure at the α level of significance for $C = 1$, and $n_{iD} = n_{iN} = 5$ together with the rejection region of the conventional two-sample t-test for the hypothesis of equality. From Figure 21.1, an interval of no differential expression is formulated in the acceptance region for the interval hypothesis while the acceptance region for the two-sample t-test contains a single point of 0. In addition, the rejection region of the two one-sided tests procedure is a subset of that of the two-sample t-test. Consequently, the two-sided tests procedure will reduce the probability of falsely identifying unexpressed genes differentially expressed. It is straightforward to verify that under the normality assumption, the power function of the two one-sided tests is symmetric at the average of C_i and C_i' and it is an α-level test.

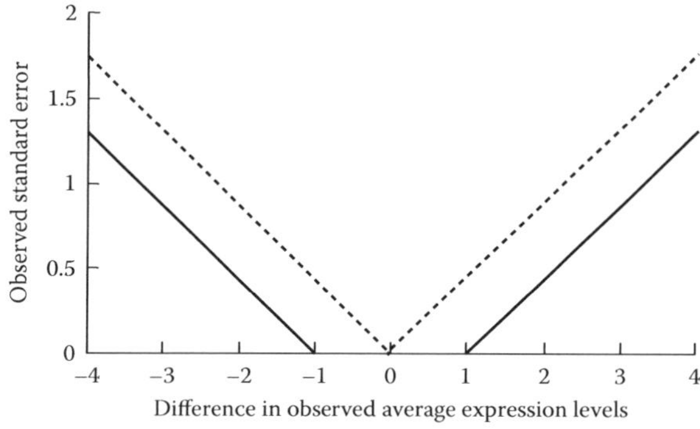

FIGURE 21.1

Rejection regions of the two one-sided tests procedure and the unpaired two-sample t-test (dashed line) for $C = 1$, $n_{iT} = n_{iC} = 5$, and the $\alpha = 0.05$ nominal level.

21.3 Optimal Representation of *in Vitro* Diagnostic Multivariate Index Assays

For an IVDMIA to be clinically meaningful and its validation to be practically feasible, it must be parsimonious with a clinically meaningful threshold that can provide the best diagnostic accuracy for the molecular targets under investigation. In addition, the IVDMIA is in fact some form of parallel assays with many analytes, and hence these analytes can be treated as multiple diagnostic markers with expression levels being the measurements in the same unit. As a result, a linear representation of expression levels of the selected differentially expressed genes presents a reasonable approach to the diagnosis of molecular targets. It follows that the result of any IVDMIA with a linear representation is a continuous variable with a predetermined cut-off for the diagnosis of a molecular target. Therefore, first, we need to determine the coefficients in the linear combination of the multiple markers not only to have the best discrimination ability for the classification of patients with a minimal classification error but also to provide the best diagnostic accuracy. There are many indices for evaluation of diagnostic accuracy such as sensitivity, specificity, FP rate, PPV, and negative predictive value (NPV). However, these indices change when a different threshold is used. On the other hand, the area under the receiver operating characteristic (ROC) curve is a quantitative criterion for the evaluation of the overall performance of diagnostic accuracy. As a result, we recommend using the generalized area under the ROC curve based on multiple diagnostic markers for the evaluation of the diagnostic accuracy of the IVDMIA (Su and Liu, 1993). Then, based on the area under the generalized ROC curve of the IVDMIA, a threshold can be determined to balance between the sensitivity and specificity for clinical application.

Suppose that a total of g differentially expressed genes has been selected for the IVDMIA. Let $\mathbf{Y}_{Dk}(\mathbf{Y}_{Nk})$ be a g-vector of the expression levels of gene i for patient k with (without) molecular targets, $k = 1, \ldots, n_D(n_N)$. Assume that $\mathbf{Y}_{Dk} \sim N(\boldsymbol{\mu}_D, \boldsymbol{\Sigma}_D)$ and $\mathbf{Y}_{Nk} \sim N(\boldsymbol{\mu}_N, \boldsymbol{\Sigma}_N)$, a linear representation of the IVDMIA has the form of $a'\mathbf{Y}_{Dk}$ $(a'\mathbf{Y}_{Nk})$ that has the best diagnostic accuracy if it can provide the maximal area under the ROC curve. In other words, one needs to determine the coefficients in $\mathbf{a}$ such that $P(a'\mathbf{Y}_{Dk} > a'\mathbf{Y}_{Nk})$ is maximized. Su and Liu (1993) showed that the Fisher linear discrimination function provides the coefficients of the best linear combination:

$$\mathbf{a}_0 = (\Sigma_D + \Sigma_N)^{-1}(\mu_D - \mu_N). \tag{21.6}$$

These coefficients can not only minimize the classification error but also provide the largest area under the generalized ROC curve, which is given by

$$A = \Phi\left(\sqrt{(\mathbf{m}_D - \mathbf{m}_N)'(\mathbf{S}_D + \mathbf{S}_N)^{-1}(\mathbf{m}_D - \mathbf{m}_N)}\right), \tag{21.7}$$

where $\Phi(\cdot)$ is the distribution function of the standard normal random variable. A consistent estimate of A can be obtained by replacing the parameters with their unbiased estimators, i.e., sample mean vectors $\overline{\mathbf{Y}}_D$ and $\overline{\mathbf{Y}}_N$ and sample covariance matrices $\mathbf{S}_D$ and $\mathbf{S}_N$ (Su and Liu, 1993). Reiser and Faraggi (1997) provided a confidence interval for A. However, in the case of the IVDMIA derived from microarray experiments, the number of genes usually exceeds tens of thousands and the number of patients is rarely in hundreds. Consequently, unstable estimation of the covariance matrices because of small sample sizes results in very poor prediction for the patient's status of molecular targets (see, e.g., Simon et al., 2003). As a result, from the result of their cross-validation experiments, Simon et al. (2003) recommended the use of diagonal linear discriminate function (DLDF) or the compound covariate predictor (CCP) for their superior performance of correct classification over other methods. For the DLDF, not only the covariances among genes are set to be zero but also the homogeneity is assumed for the variances between the patients and normal subjects.

From (21.6), it can be seen that the estimators of the coefficients in $\mathbf{a}_0$ are proportional to the traditional t-statistic, which are also the coefficients used in the CCP. Therefore, the more differentially expressed the genes are, the more weights of the genes are for the DLDF. In this regard, one could include all genes in the DLDF or CCP for the IVDMIA. However, if a gene is not differentially expressed between the patients with and without molecular targets, it will have a small t-statistic and hence does not contribute to the prediction ability of the resulting DLDF or CCP. Therefore, during the early development stage of the IVDMIA, all possible genes should be included for identification of differentially expressed genes. However, for the construction of the linear representation of the IVDMIA, those genes with no differential expressions should be dropped. Unfortunately, how many and which genes should be included in the linear representation still remain a great challenge to the researchers. One rule of thumb is that the number of genes and the genes to be included in the classifier should reach a balance between the practicality and amount of information required for an accurate diagnosis of molecular targets. If there is unequivocal evidence that a certain biological pathway is involved in the pathogenesis of a disease, then from a viewpoint of biology, all genes affecting this pathway should be included in the classifier. Suppose that the sample sizes are equal for the patients with and without molecular targets. One measure that can be used for possible determination of the number of genes included in the classifier is the partial between-group distance (PBGD) defined as

$$\mathrm{PBGD} = \frac{\sum_{i=1}^{g} (\overline{Y}_{iD} - \overline{Y}_{iN})^2 / s_{pi}^2}{\sum_{i=1}^{G} (\overline{Y}_{iD} - \overline{Y}_{iN}^2) / s_{pi}^2}. \tag{21.8}$$

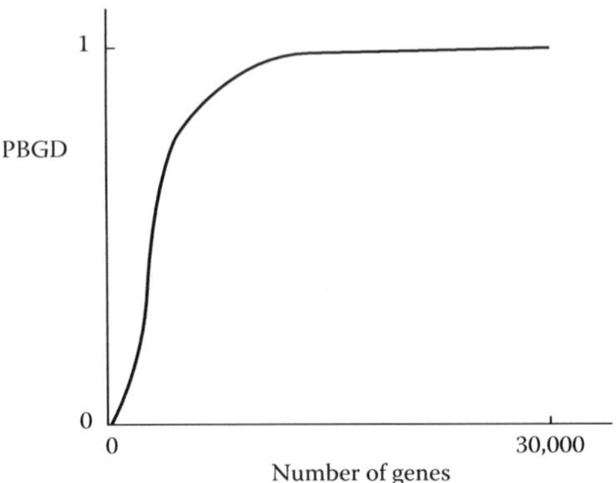

FIGURE 21.2
Number of genes and PBGD.

The range of PBGD is from 0 to 1. Because most of genes tested during the early development stage of the IVDMIA are not differentially expressed, and if we put $(\overline{Y}_{iD} - \overline{Y}_{iN})^2/s_{pi}^2$ into the numerator of PBGD in (21.8) according to its magnitude sequentially, then PBGD is an increasing function of the number of genes. In order to be clinically practical and to be validated feasible, one desirable characteristic of any IVDMIA is to provide a high diagnostic accuracy with a set of a small number of genes. Under this ideal situation, PBGD is very steep and reaches the plateau of 1 very quickly as shown in Figure 21.2. On the other hand, there might be several candidates for the classifier with similar diagnostic accuracy. Due to the principle of parsimony, treating the coefficients in the classifiers as fixed constants, based on the paired areas under the generalized ROC curves, one can apply the non-inferiority test to choose a classifier with the smallest number of genes but with an equivalent diagnostic accuracy (Li et al., 2008; Liu et al., 2006). However, the non-inferiority test based on the difference in paired areas of the generalized ROC curves derived from multiple markers requires further research.

21.4 Validation of *in Vitro* Diagnostic Multivariate Index Assays

As described above, Oncotype DX used in the TAILORx trial is a RT-PCR assay based on 21 genes, while a 70-gene molecular signature derived from the microarray is used in the MINDACT trial. Therefore, IVDMIAs are parallel assays with multiple biomarkers and multiple medical decision points. It follows that validation of IVDMIA should address the performance and assay validation for each component as well as the overall quality performance of the whole IVDMIA (Frueh, 2006; Patterson et al., 2006). The Food and Drug Administration (FDA) draft guidance suggests that for each target or expression pattern, the performance characteristics include assay sensitivity, reproducibility, validation of cut-off, reference range or medical decision point, assay range and specificity (FDA, 2003a). The FDA draft guidance also suggests consulting the guidelines on protocols for assay validation in clinical laboratory published by the Clinical Laboratory Standard Institutes (CLSI). However, these protocols are for a single analyte and are not suitable for complicated assays with multiple markers and a statistical algorithm for diagnosis. As a result, the assay validation of IVDMIA should employ different approaches although the principle of accuracy and precision remains the same (Canales et al., 2006; Ji and Davis, 2006). However, because the overall analytical performance of the IVDMIA is determined by the performance of the individual component markers, at the minimum, the performance of each single gene should be evaluated by the approved guidelines on validation protocols issued by the CLSI.

Traditionally, one key issue for assay validation of the IVDMIA is the reference standards with known concentrations for the establishment of the calibration curve, assessment of accuracy from recovery experiment, and evaluation of linearity and linear range of the IVDMIA. Recently, Shippy et al. (2006) investigated the relationship of the expression measurement of a transcript in a titration sample and the relationship between the signals of a given transcript in the two titration samples and that of each individual sample in the Microarray Quality Control (MAQC) study. They found that differences in normalization, platforms, and laboratory practices can lead to deviations from the mixing ratio expected in traditional assay validation and they proposed empirical measurements to estimate the true mRNA fraction in the titration samples. On the other hand, Tong et al. (2006) also examined the use of external RNA controls for the assessment of the accuracy of the expression ratios between samples with known expression levels in the same MAQC study. They recommended a comprehensive study for modeling concentration response to determine the tolerance ranges for linear fit, slope and y-intercept for assay assessment, specificity in the context of FPs and false negatives. These findings by the investigators of the MAQC study indicate difficulty in obtaining the known concentration reference standards and assay validation for the IVDMIA based on the microarray platforms, and hence more research is needed for the challenges of validation of analytical aspects of the IVDMIA.

On the other hand, for a linear representation, the optimal algorithm to provide the best discrimination ability and diagnosis of a molecular target for the IVDMIA is the diagonal linear discriminant function. Recall that the selected genes in DLDF are differentially expressed between the patients with and without molecular targets and weights are proportional to the t-statistics. Therefore, the DLDF is an aggregate measure of expression levels with weights reflecting their relative contributions to the algorithm. But masking effects may occur while the relative unimportant genes with small weights are differentially expressed more than those with large weights. Once the weights are determined in the development stage, to avoid possible masking effect, the expression levels of each individual gene must exceed a prespecified lower limit for the overall assay results to reach the threshold for a positive diagnosis of the molecular target. Theses prespecified limits should be determined from the biological and clinical knowledge of relative roles of selected genes in the pathway of pathogenesis of the underlying disease.

Agreement and reproducibility are very important performance characteristics of IVDMIA and have recently drawn a lot of attention in the data generated from microarray experiments. For example, Dobbin et al. (2005), Irizarry et al. (2005), Larkin et al. (2005), and Members on Toxicogenomics Research Consortium (2005) examined the agreement on measurements of gene expressions between laboratories and across different platforms. Testing the hypothesis of zero Pearson correlation coefficient (PCC) is one of the most common statistical methods to assess comparability of gene expression levels between technical replicates within and across laboratories. However, to evaluate comparability on gene expressions within and between laboratories is to assess the agreement of the measurements of the technical replicates for the same genes of the same samples. Hence the objective for the evaluation of comparability is to investigate the closeness or equivalence of gene expression levels between technical replicates of the same samples. Although PCC is an excellent statistic for the evaluation of linear association, it is location- and scale-invariant. Hence it cannot detect changes in accuracy and precision and cannot be used for the assessment of agreement of gene expression levels between technical replicates which requires evaluation of equivalence in both accuracy and precision. Therefore, hypothesis of zero linear correlation by PCC is not appropriate for the evaluation of agreement of gene expression levels between technical replicates of the same samples.

On the other hand, the concordance correlation coefficient, proposed by Lin (1989, 1992) and Lin et al. (2002) is a product of PCC and a factor consisting of location and scale shifts. Therefore, it can be employed to evaluate the agreement of gene expression levels between the technical replicates of the same samples. In order to meet the minimal requirement of agreement, the hypothesis for the assessment of the agreement of gene expression levels between technical replicates should be formulated as the non-inferiority hypothesis, where not only does the linear association exceed a prespecified threshold, but the means and variability between technical replicates are also equivalent within some predetermined limits. Both the asymptotic method and the exact procedure based on generalized pivotal quantities are available for an interval estimation of the concordance correlation coefficient for the evaluation of the agreement of gene expression levels between two technical replicates, which exceeds some minimal requirement of agreement (Lin, 1989; Liao et al., 2007).

21.5 Enrichment Process

In clinical research, it is always of particular interest to clinicians to identify patients with disease targets under study who are most likely to respond to the treatment under study. In practice, an enrichment process is often employed to identify such a target patient population. Clinical trials utilizing an enrichment design are referred to as target clinical trials. After completion of an HGP, the disease targets at a certain molecular level can be identified and should be utilized for the treatment of diseases (Maitournam and Simon, 2005; Casciano and Woodcock, 2006). As a result, diagnostic devices for the detection of diseases using biotechnology such microarray, polymerase chain reaction, mRNA transcript profiling, and others become possible in practice (FDA, 2005, 2007). The treatments specific for the molecular targets could then be developed for those patients who are most likely to benefit. Consequently, personalized medicine could become a reality. The clinical development of Herceptin® (trastuzumab), which is targeted at patients with metastatic breast cancer with an over-expression of HER2 (human epidermal growth factor receptor) protein, is a typical example. We will refer to these treatments as the targeted treatments or drugs. Development of targeted treatments involves translation from the accuracy and precision of diagnostic devices for molecular targets to the effectiveness and safety of the treatment modality for the patient population with the targets. Therefore, the evaluation of targeted treatments is much more complicated than that of traditional drugs. To address the issues of development of the targeted drugs, in April 2005, the FDA issued the Drug-Diagnostic Co-Development Concept Paper.

In clinical trials, subjects with and without disease targets may respond to the treatment differently with different effect sizes. In other words, patients with disease targets may show a much larger effect size, while patients without disease targets may exhibit a relatively small effect size. In practice, fewer subjects are required for detecting a bigger effect size. Thus, the traditional clinical trials may conclude that the test treatment is ineffective based on the detection of a combined effect size, while the test treatment is in fact effective for those patients with positive disease targets. Thus, personalized medicine is possible if we can identify those subjects with positive disease targets. As indicated in the FDA Drug-Diagnostic Co-development Concept Paper, one of the useful designs for the evaluation of the targeted treatments is the enrichment design (see also Chow and Liu, 2004). Under the enrichment design, the targeted clinical trials consist of two phases. The first phase is the enrichment phase in which each patient is tested by a diagnostic device for detection of the predefined molecular targets. Then, patients with a positive result by the diagnostic device are randomized to receive either the targeted treatment or a concurrent control. However, in practice, no diagnostic test is perfect with 100% PPV. As a result, some of the patients enrolled in targeted clinical trials under the enrichment design might not have the specific targets and hence the treatment effects of the drug for the molecular targets could be underestimated due to misclassification (Liu and Chow, 2008).

Under the enrichment design, following the idea described in Liu and Chow (2008), Liu et al. (2009) proposed using the expectation-maximization (EM) algorithm (Dempster et al., 1977; McLachlan and Krishnan, 1997) in conjunction with the bootstrap technique (Efron and Tibshirani, 1993) for obtaining the inference of the treatment effects. Their method, however, depends upon the accuracy and reliability of the diagnostic device. A poor (i.e., less accurate and reliable) diagnostic device may result in a large proportion of misclassification which has an impact on the assessment of the true treatment effect. To overcome (correct) the problem of an inaccurate diagnostic device, we propose using the Bayesian approach in conjunction with the EM algorithm and bootstrap technique for obtaining a more accurate and reliable estimate of treatment effect under various study designs recommended by the FDA.

To illustrate the potential impact and significance of the enrichment process, consider the example of Herceptin for treating patients with metastatic breast cancer with and without over-expression of HER2 protein using the gene amplification by fluorescence *in situ* hybridization or clinical trial assay (CTA) which is an investigational immunohistochemical (IHC) assay consisting of four-point ordinal score system (0, 1+, 2+, 3+). Table 21.1 gives the treatment effects of Herceptin plus chemotherapy as a function of HER2 over-expression. As can be seen from Table 21.1, Herceptin plus chemotherapy provides statistically significantly additional clinical benefit in terms of overall survival over chemotherapy alone for patients with a staining score of 3+, while Herceptin plus chemotherapy fails to provide additional

TABLE 21.1

Treatment Effects as a Function of HER2
Over-Expression or Amplification

HER2 Assay Result	Number of Patients	Relative Risk for Mortality (95%)
CTA 2+ or 3+	469	0.80 (0.64, 1.00)
FISH (+)	325	0.70 (0.53, 0.91)
FISH (−)	126	1.06 (0.70, 1.63)
CTA 2+	120	1.26 (0.82, 1.94)
FISH (+)	32	1.31 (0.53, 3.27)
FISH (−)	83	1.11 (0.68, 1.82)
CTA 3+	349	0.70 (0.51, 0.89)
FISH (+)	293	0.67 (0.51, 0.89)
FISH (−)	43	0.88 (0.39, 1.98)

Source: From U.S. FDA Annotated Redlined Draft Package Insert for Herceptin, Rockville, MD, 2006.
FISH, fluorescence *in situ* hybridization.

survival benefit for patients with a CTA score of 2+. However, as indicated in the Decision Summary of HercepTest® (a commercial IHC assay for over-expression of HER2 protein), about 10% of samples have discrepant results between 2+ and 3+ staining intensity. In other words, some patients tested with a score of 3+ may actually have a score of 2+ and vice versa.

The proposed methodology will allow the clinician to identify optimal clinical benefit to patients who are most likely to respond to the treatment under investigation through an enrichment process. Targeted clinical trials under an enrichment design will make personalized medicine a reality. The proposed methodology can be applied not only to different types of study endpoints such as continuous variables, binary responses, and time-to-event data for testing hypotheses of equality, superiority/non-inferiority, and equivalence, but also to various critical diseases across therapeutic areas such as cardiovascular, infectious diseases, and oncology in public health.

21.6 Study Designs of Target Clinical Trials

Under an enrichment design, one of the objectives of targeted clinical trials is to evaluate the treatment effects of the molecular targeted test treatment in the patient population with a molecular target. The diagram in the FDA Concept Paper (FDA, 2005) for demonstration of this design is reproduced in Figure 21.3.

Under the above enrichment design, Liu et al. (2009) considered a two-group parallel design in which patients with a *positive* result by the diagnostic device are randomized in a 1:1 ratio to receive the molecular targeted test treatment (T) or a control treatment (C) (see Figure 21.4). In other words, only patients with positive diagnosed results are included in the study. For simplicity, Liu et al. (2009) assumed that the primary efficacy endpoint is a continuous variable. Let Y_{ij} be the responses of the jth subject in the ith group, where $j = 1, \ldots, n_i$; $i = T, C$. Y_{ij} are assumed to be approximately normally distributed with homogeneous variances between the test and control

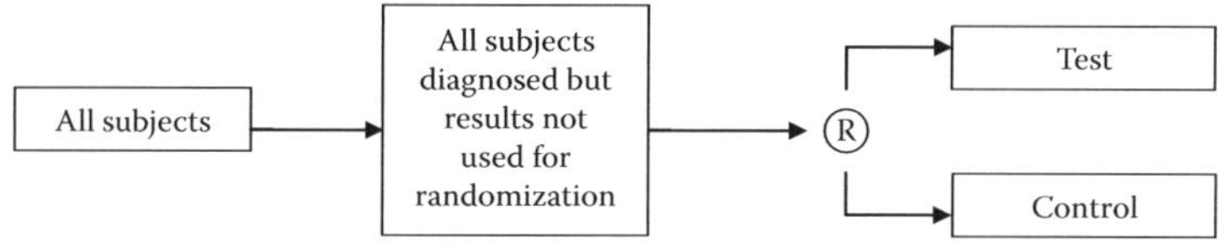

FIGURE 21.3
Targeted clinical trials under an enrichment design.

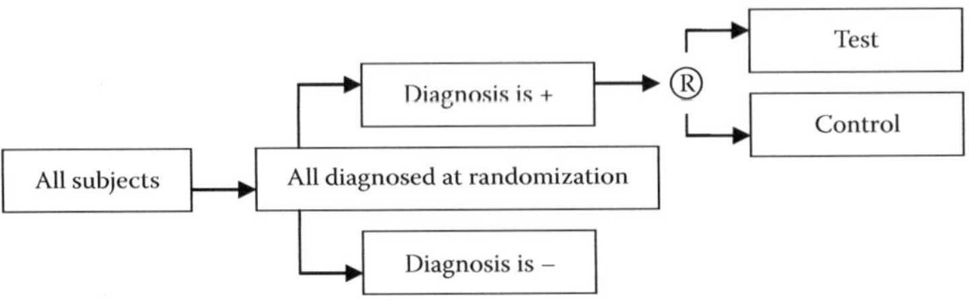

FIGURE 21.4
Enrichment design for patients with positive results.

TABLE 21.2

Population Means by Treatment and Diagnosis

Positive Diagnosis	True Target Condition	Indicator of Diagnostic	Test Group	Control Group	Difference
+	+	γ	μ_{T+}	μ_{C+}	$\mu_{T+} - \mu_{C+}$
	–	$1 - \gamma$	μ_{T-}	μ_{C-}	$\mu_{T-} - \mu_{C-}$

Note: γ is the positive predictive value (PPV).

treatments. Table 21.2 gives the expected values of Y_{ij} by treatment and diagnostic result of the molecular target. In Table 21.1, μ_{T+}, μ_{C+} (μ_{T-}, μ_{C-}) are the means of test and control groups for the patients with (or without) a molecular target. The inference for the treatment effects could be obtained through either estimation or hypothesis testing. For estimation, the parameter of interest is the treatment effects for the patients truly having the molecular target $\theta = \mu_{T+} - \mu_{C+}$. However, this effect may be contaminated due to misclassification, i.e., for those subjects who do not have a molecular target but have positive diagnosed results and those subjects who have a molecular target and negative diagnosed results.

The hypothesis for detection of treatment difference in the patient population truly with a molecular target is the hypothesis of interest:

$$H_0 : \mu_{T+} - \mu_{C+} = 0 \quad \text{versus} \quad H_a : \mu_{T+} - \mu_{C+} \neq 0. \tag{21.9}$$

As indicated above, Liu et al. (2009) proposed statistical methods for assessment of the treatment effect for patients with positive diagnosed results under the enrichment design described in Figure 21.4. Their methods suffer from the lack of information regarding the proportion of subjects who truly have molecule targets in the patient population and the unknown PPV. Consequently, the conclusion drawn from the collected data may be biased and misleading. In addition to the study designs as given in Figures 21.3 and 21.4, the 2005 FDA Concept Paper also recommended the following two study designs for different study objectives (see Figures 21.5 and 21.6).

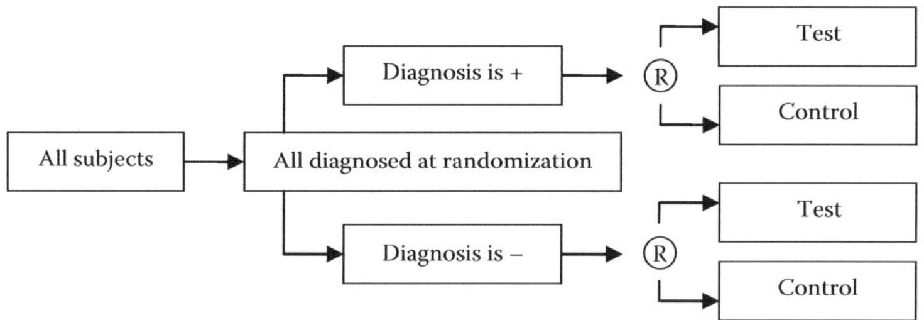

FIGURE 21.5
Enrichment design for patients with and without molecular targets.

This study design allows the evaluation of the treatment effect within subpopulations, i.e., the subpopulation of patients with positive or negative results. Similar to Table 21.1 for the study design given in Figure 21.3, the expected values of Y_{ij} by treatment and diagnostic result of the molecular targets are summarized in Table 21.2. As a result, it may be of interest to estimate the following treatment effects:

$$\theta_1 = \gamma_1(\mu_{T++} - \mu_{C++}) + (1 - \gamma_1)(\mu_{T+-} - \mu_{C+-}),$$

$$\theta_2 = \gamma_2(\mu_{T-+} - \mu_{C-+}) + (1 - \gamma_2)(\mu_{T--} - \mu_{C--}),$$

$$\theta_3 = \delta\gamma_1(\mu_{T++} - \mu_{C++}) + (1 - \delta)\gamma_2(\mu_{T-+} - \mu_{C-+}),$$

$$\theta_4 = \delta\gamma_1(\mu_{T+-} - \mu_{C+-}) + (1 - \delta)\gamma_1(\mu_{T--} - \mu_{C--}),$$

$$\theta_5 = \delta[\gamma_1(\mu_{T+-} - \mu_{C+-}) + (1 - \gamma_1)(\mu_{T+-} - \mu_{C+-})]$$
$$+ (1 - \delta)[\gamma_2(\mu_{T-+} - \mu_{C-+}) + (1 - \gamma_2)(\mu_{T--} - \mu_{C--})],$$

where δ is the proportion of subjects with positive molecule targets. Following a similar idea as described in the previous section, estimates of $\theta_1 - \theta_5$ can be obtained. In other words, estimates of θ_1 and θ_2 can be obtained based on data collected from the subpopulations of subjects with and without positive diagnoses who truly have a molecular target of interest. Similarly, the combined treatment effect θ_5 can be assessed. These estimates, however, depend upon both γ_i, $i = 1, 2$ and δ. To obtain some information regarding γ_i, $i = 1, 2$ and δ, the FDA recommends the following alternative enrichment design which includes a group of subjects without any diagnoses and a subset of subjects who will be diagnosed at the screening stage (Table 21.3, Figure 21.6).

TABLE 21.3

Population Means by Treatment and Diagnosis

Positive Diagnosis	True Target Condition	Indicator of Diagnostic	Test Group	Control Group	Difference
+	+	γ_1	μ_{T++}	μ_{C++}	$\mu_{T++} - \mu_{C++}$
	−	$1 - \gamma_1$	μ_{T+-}	μ_{C+-}	$\mu_{T+-} - \mu_{C+-}$
−	+	γ_2	μ_{T-+}	μ_{C-+}	$\mu_{T-+} - \mu_{C-+}$
	−	$1 - \gamma_2$	μ_{T--}	μ_{C--}	$\mu_{T--} - \mu_{C--}$

Note: γ_i is the PPV, $i = 1$ (positive diagnosis) and $i = 2$ (negative diagnosis); μ_{ijk} is the mean for subjects in the ith group with the kth true target status but with jth diagnosed result.

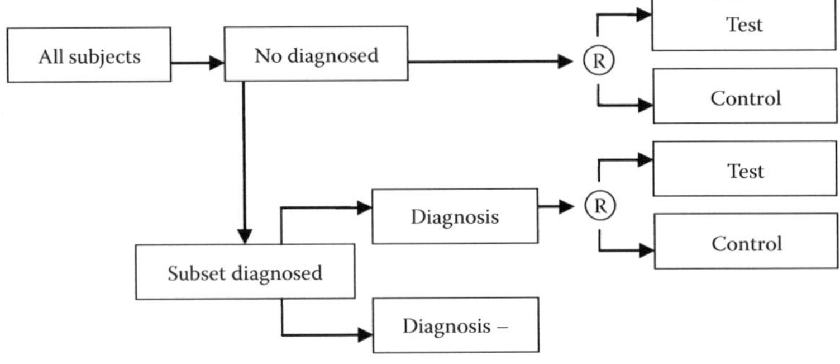

FIGURE 21.6
Alternative enrichment design for targeted clinical trials.

Simon and Maitournam (2004) and Maitournam and Simon (2005) provide sample size determination for the targeted clinical trials for both continuous and binary endpoints. However, variability associated with estimates of PPV, NPV, FP rate, and false negative rate is not considered in the sample size calculation and relative efficiency of the targeted clinical trials to the untargeted ones. On the other hand, for example, getfitnib is the specific inhibitor of the tyrosine kinase of epidermal growth factor receptor (EGFR) that is involved in the pathway of the pathogenesis of non-small cell lung cancer (NSCLC). However, the response rate of getfitnib in patients with NSCLC is only about 10%. In addition, for another EGFR inhibitor, erlotinib, the survival of patients with NSCLC is correlated significantly with the expression, polysomy, amplification, and mutation of EGFR. Therefore, multiple pathways with multiple targets may be involved in most diseases. Consequently, in the foreseeable future, it is very likely that a cocktail of molecularly targeted agents will be employed to treat diseases with multiple targets. Therefore, research on the innovative and novel designs and analyses for targeted clinical trials in evaluation of multiple drugs for multiple molecular targets is urgently needed.

21.7 Analysis of Target Clinical Trials

Liu et al. (2009) considered the situation where a particular molecular target involved with the pathway in the pathogenesis of the disease has been identified and there is a validated diagnostic device available for detection of the identified molecular target. It is assumed that the device is only for detection of the molecular target and is not for prognosis of clinical outcomes of patients. In addition, it is also assumed that the device has been evaluated in the diagnostic effectiveness trial and met the regulatory requirements for diagnostic accuracy.

Let $\bar{y}_T$ and $\bar{y}_C$ be the sample means of test and control treatments, respectively. Since no diagnostic test is perfect for diagnosis of the molecular target of interest without error, some patients with a positive diagnostic result may in fact not have a molecular target. It follows that

$$E(\bar{y}_T - \bar{y}_C) = \gamma(\mu_{T+} - \mu_{C+}) + (1 - \gamma)(\mu_{T-} - \mu_{C-}), \qquad (21.10)$$

where γ is the PPV. Liu and Chow (2008) indicated that the expected value of the difference in sample means consists of two parts. The first part is the treatment effects of the molecularly targeted drug in patients with a positive diagnosis who truly have a molecular target of interest. The second part is the treatment effects of patients with a positive diagnosis but who in fact do not have a molecular target. The reason for developing the targeted treatment is based on the assumption that the efficacy of the targeted treatment is greater in patients truly with a molecular target than in those without a target. In addition, the targeted treatment is also expected to be more efficacious than the untargeted control in the patient population truly with molecular targets. It follows that $\mu_{T+} - \mu_{C+} > \mu_{T-} - \mu_{C-}$. As a result, the difference in sample means obtained under the enrichment design for targeted clinical trials actually underestimated the true treatment effects of the molecularly targeted test drug in the patient population truly with a molecular target of interest. As can be seen from (21.10), the bias of the difference in sample means decreases as the PPV increases. On the other hand, the PPV of a diagnostic test increases as the prevalence of the disease increases (Fleiss et al., 2003). For a disease which is highly prevalent, say greater than 10%, even with a high diagnostic accuracy of 95% sensitivity and specificity for the diagnostic device, the PPV is only about 67.86%. It follows that the downward bias of the traditional difference in sample means could be substantial for the estimation of treatment effects of the molecularly targeted drug in patients truly with the target of interest.

The traditional unpaired two-sample t-test approach is to reject the null hypothesis in (21.9) at the α level of significance if

$$t = \left| \frac{(\bar{y}_T - \bar{y}_C)}{\sqrt{s_p^2(1/n_T + 1/n_C)}} \right| \geq t_{\alpha/2, n_T + n_C - 2},$$

where

s_p^2 is the pooled sample variance,

t_{α,n_T+n_C-2} is the αth upper percentile of a central t-distribution with $n_T + n_C - 2$ degrees of freedom.

Since $\bar{y}_T - \bar{y}_C$ underestimates $\mu_{T+} - \mu_{C+}$, the planned sample size may not be sufficient for achieving the desired power for detecting the true treatment effects in patients truly with a molecular target of interest. Based on the above t-statistic, the corresponding $(1 - \alpha)100\%$ confidence interval can be obtained as follows:

$$(\bar{y}_T - \bar{y}_C) \pm t_{\alpha/2,n_T+n_C-2}\sqrt{s_p^2\left(\frac{1}{n_T} + \frac{1}{n_C}\right)}.$$

Although all patients randomized under the enrichment design have a positive diagnosis, the true status of the molecular target for individual patients in the targeted clinical trials is in fact unknown. It follows that under the assumption of homogeneity of variance, Y_{ij} are independently distributed as a mixture of two normal distributions with mean μ_{i+} and μ_{i-} respectively, and common variance σ^2 (McLachlan and Peel, 2000):

$$\varphi(y_{ij}\,|\,\mu_{i+},\sigma^2)^\gamma\,\varphi(y_{ij}\,|\,\mu_{i-},\sigma^2)^{1-\gamma}, \quad i = T, C; \quad j = 1,\ldots,n_i, \qquad (21.11)$$

where $\varphi(\cdot\,|\,\cdot)$ denotes the density of a normal variable.

However, γ is an unknown PPV, which is usually estimated from the data. Therefore, the data obtained from the targeted clinical trials are incomplete because the true status of the molecular target of the patients is missing. The EM algorithm is one of the methods for obtaining the maximum likelihood estimators of the parameters for an underlying distribution from a given data set when the data are incomplete or have missing values. On the other hand, the diagnostic device for the detection of molecular targets has been validated in diagnostic effectiveness trials for its diagnostic accuracy. Therefore, the estimates of the PPV for the diagnostic device can be obtained from the previously conducted diagnostic effectiveness trials. As a result, we can apply the EM algorithm to estimate the treatment effect for the patients truly with a molecular target by incorporating the estimates of the PPV of the device obtained from the diagnostic effectiveness trials as the initial values.

For each patient, we have a pair of variables (Y_{ij}, X_{ij}), where Y_{ij} is the observed primary efficacy endpoint of patient j in treatment i and X_{ij} is the latent variable indicating the true status of the molecular target of patient j in treatment i; $j = 1,\ldots, n_i$, $i = T, C$. In other words, X_{ij} is an indicator variable with value of 1 for patients truly with a molecular target and with a value of 0 for patients truly without a target. In addition, X_{ij} are assumed to be independent and identically distributed (i.i.d.) Bernoulli random variables with probability γ for the molecular target. Let $\Psi = (\gamma, \mu_{T+},$ $\mu_{T-}, \mu_{C+}, \mu_{C-}, \sigma^2)'$ be the vector containing all unknown parameters and $\mathbf{y}_{obs} = (y_{T1}, \ldots, y_{Tn_T}, y_{C1}, \ldots, y_{Cn_C})'$ be the vector of the observed primary efficacy endpoints from the targeted clinical trials. It follows that the complete-data log-likelihood function is given by

$$\log L_c(\Psi) = \sum_{j=1}^{n_T} x_{Tj} \left[\log \gamma + \log \varphi(y_{Tj} \mid \mu_{T+}, \sigma^2) \right]$$

$$+ \sum_{j=1}^{n_T} (1 - x_{Tj}) \left[\log(1 - \gamma) + \log \varphi(y_{Tj} \mid \mu_{T-}, \sigma^2) \right]$$

$$+ \sum_{j=1}^{n_C} x_{Cj} \left[\log \gamma + \log \varphi(y_{Cj} \mid \mu_{C+}, \sigma^2) \right]$$

$$+ \sum_{j=1}^{n_C} (1 - x_{Cj}) \left[\log(1 - \gamma) + \log \varphi(y_{Cj} \mid \mu_{C-}, \sigma^2) \right]. \tag{21.12}$$

Furthermore, from the previous diagnostic effectiveness trials, an estimate of the PPV of the device is known. Therefore, at the initial step of the EM algorithm for estimating the treatment effects in patients truly with a molecular target, the observed latent variables X_{ij} are generated as i.i.d. Bernoulli random variables with the PPV γ estimated by that obtained from the diagnostic effectiveness trial. The procedures for implementation of the EM algorithm in conjunction with the bootstrap procedure for inference of θ in the patient population truly with a molecular target are briefly described in the following.

At the $(k+1)$st iteration, the E-step requires the calculation of the conditional expectation of the complete-data log-likelihood $L_c(\Psi)$, given the observed data $\mathbf{y}_{obs}$, using currently fitting $\hat{\Psi}^{(k)}$ for Ψ.

$$Q(\Psi; \hat{\Psi}^{(k)}) = E_{\Psi(k)} \left\{ \log L_c(\Psi) \mid \mathbf{y}_{obs} \right\}$$

Since $\log L_c(\Psi)$ is a linear function of the unobservable component labeled variables x_{ij}, the E-step is calculated by replacing x_{ij} by its conditional expectation given by y_{ij}, using $\hat{\Psi}^{(k)}$ for Ψ. That is, x_{ij} is replaced by

$$\hat{x}_{ij}^{(k)} = E_{\Psi(k)} \left\{ x_{ij} \mid y_{ij} \right\} = \frac{\hat{\gamma}_i^{(k)} \varphi\left(y_{ij} \mid \hat{\mu}_{i+}^{(k)}, (\hat{\sigma}_i^2)^{(k)}\right)}{\hat{\gamma}_i^{(k)} \varphi\left(y_{ij} \mid \hat{\mu}_{i+}^{(k)}, (\hat{\sigma}_i^2)^{(k)}\right) + \left(1 - \hat{\gamma}_i^{(k)}\right) \varphi\left(y_{ij} \mid \hat{\mu}_{i-}^{(k)}, (\hat{\sigma}_i^2)^{(k)}\right)}, \quad i = T, C,$$

which is the estimate of the posterior probability of the observation y_{ij} with molecular target after the kth iteration. The M-step requires the computation of $\hat{\gamma}_i^{(k+1)}$, $\hat{\mu}_{i+}^{(k+1)}$, $\hat{\mu}_{i-}^{(k+1)}$, and $(\hat{\sigma}_i^2)^{(k+1)}$; $i = T, C$, by maximizing $\log L_c(\Psi)$. It is equivalent to computing the sample proportion, the weighted sample mean, and sample variance with the weight x_{ij}. Since $\log L_c(\Psi)$ is linear in the x_{ij}, it follows that x_{ij} are replaced by their conditional expectations $\hat{x}_{ij}^{(k)}$. On the $(k + 1)$th iteration, the intent is to choose the value of Ψ, say $\hat{\Psi}^{(k+1)}$, that maximizes $Q(\Psi; \hat{\Psi}^{(k)})$. It follows that on the M-step of the $(k + 1)$st iteration, the current fit for the PPV of the test drug group and control group is given by

$$\hat{\gamma}_i^{(k+1)} = \frac{\sum_{j=1}^{n_i} \hat{x}_{ij}^{(k)}}{n_i}, \quad i = T, C.$$

Under the assumption that $n_T = n_C$, it follows that the overall PPV is estimated by

$$\hat{\gamma}^{(k+1)} = \frac{\left(\hat{\gamma}_T^{(k+1)} + \hat{\gamma}_C^{(k+1)}\right)}{2}.$$

The means of the molecularly targeted test drug and control can then be estimated, respectively, as

$$\hat{\mu}_{T+}^{(k+1)} = \frac{\sum_{j=1}^{n_T} \hat{x}_{Tj}^{(k)} y_{Tj}}{\sum_{j=1}^{n_T} \hat{x}_{Tj}^{(k)}}, \quad \hat{\mu}_{T-}^{(k+1)} = \frac{\sum_{j=1}^{n_T} (1 - \hat{x}_{Tj}^{(k)}) y_{Tj}}{\sum_{j=1}^{n_T} (1 - \hat{x}_{Tj}^{(k)})},$$

$$\hat{\mu}_{C+}^{(k+1)} = \frac{\sum_{j=1}^{n_C} \hat{x}_{Cj}^{(k)} y_{Cj}}{\sum_{j=1}^{n_C} \hat{x}_{Cj}^{(k)}} \quad \text{and} \quad \hat{\mu}_{C-}^{(k+1)} = \frac{\sum_{j=1}^{n_C} (1 - \hat{x}_{Cj}^{(k)}) y_{Cj}}{\sum_{j=1}^{n_C} (1 - \hat{x}_{Cj}^{(k)})},$$

with unbiased estimators for the variances of the molecularly targeted drug and control given respectively by

$$(\hat{\sigma}_T^2)^{(k+1)} = \frac{\left(\sum_{j=1}^{n} \hat{x}_{Tj}^{(k)} (y_{Tj} - \hat{\mu}_{T+}^{(k)})^2 + \sum_{j=1}^{n} (1 - \hat{x}_{Tj}^{(k)})(y_{Tj} - \hat{\mu}_{T-}^{(k)})^2\right)}{(n_T - 2)}$$

and

$$(\hat{\sigma}_C^2)^{(k+1)} = \frac{\left(\sum_{j=1}^{n} \hat{x}_{Cj}^{(k)}(y_{Cj} - \hat{\mu}_{C+}^{(k)})^2 + \sum_{j=1}^{n} (1 - \hat{x}_{Cj}^{(k)})(y_{Cj} - \hat{\mu}_{C-}^{(k)})^2 \right)}{(n_C - 2)}.$$

It follows that an unbiased estimate for the pooled variance is given as

$$(\hat{\sigma}^2)^{(k+1)} = \frac{[(n_T - 2) \times (\hat{\sigma}_T^2)^{(k+1)} + (n_C - 2) \times (\hat{\sigma}_C^2)^{(k+1)}]}{(n_T + n_C - 4)}.$$

Therefore, the estimator for the treatment effects in patients with a molecular target θ obtained from the EM algorithm is given as $\hat{\theta} = \hat{\mu}_{T+} - \hat{\mu}_{C+}$.

Liu et al. (2009) proposed to apply the parametric bootstrap method to estimate the standard error of $\hat{\theta}$.

Step 1: Choose a large bootstrap sample size, say $B = 1000$. For $1 \le b \le B$, generate the bootstrap sample y_{obs}^{b} according to the probability model in (21.11). The parameters in (21.11) for generating bootstrap samples y_{obs}^{b} are substituted by the estimators obtained from the EM algorithm based on the original observations of primary efficacy endpoints from the targeted clinical trials.

Step 2: The EM algorithm is applied to the bootstrap sample y_{obs}^{b} to obtain estimates $\hat{\theta}_b^*$, $b = 1, \ldots, B$.

Step 3: An estimator for the variance of $\hat{\theta}$ by the parametric bootstrap procedure is given as

$$S_B^2 = \sum_{b=1}^{B} \frac{(\hat{\theta}_b^* - \bar{\hat{\theta}}^*)^2}{(B-1)}, \quad \text{where } \bar{\hat{\theta}}^* = \sum_{b=1}^{B} \frac{\hat{\theta}_b^*}{B}.$$

Let $\hat{\theta}$ be the estimator for the treatment effects in patients truly with a molecular target obtained from the EM algorithm. Nityasuddhi and Böhning (2003) show that the estimator obtained under the EM algorithm is asymptotically unbiased. Let S_B^2 denote the estimator of the variance of $\hat{\theta}$ obtained by the bootstrap procedure. It follows that the null hypothesis is rejected and the efficacy of the molecularly targeted test drug is different from that of the control in the patient population truly with a molecular target at the α level if

$$t = \left| \frac{\hat{\theta}}{\sqrt{S_B^2}} \right| \ge z_{\alpha/2}, \tag{21.13}$$

where $z_{\alpha/2}$ is the $\alpha/2$ upper percentile of a standard normal distribution. Thus, the corresponding $(1 - \alpha)100\%$ asymptotic confidence interval for $\theta = \mu_{T+} - \mu_{C+}$ can be constructed as $\hat{\theta} \pm z_{1-\alpha/2}\sqrt{S_B^2}$ (see, e.g., Basford et al., 1997). It should be noted that although the assumption that $\mu_{T+} - \mu_{C+} > \mu_{T-} - \mu_{C-}$ is one of the reasons for developing the targeted treatment, this assumption is not used in the EM algorithm for the estimation of θ. Hence, the inference for θ by the proposed procedure is not biased in favor of the targeted treatment.

As indicated earlier, the method proposed by Liu et al. (2009) suffers from the lack of information regarding the uncertainty in accuracy of the diagnostic device. As an alternative, we propose considering a Bayesian approach to incorporate the uncertainty in accuracy and reliability of the diagnostic device for the molecular target into the inference of treatment effects of the targeted drug. For each patient, we have a pair of variables (y_{ij}, x_{ij}), where y_{ij} is the observed primary efficacy endpoint of patient j in treatment I and x_{ij} is the latent variable indicating the true status of the molecular target of patient j in treatment I; $j = 1, \ldots, n_i$, $i = T, C$. In other words, x_{ij} is an indicator variable with value of 1 for patients with a molecular target and with a value of 0 for patients without a target. x_{ij} are assumed to be i.i.d. Bernoulli random variables with the probability of the molecular target being γ. Thus, $x_{ij} = 1$ if $y_{ij} \sim N(\mu_{i+}, \sigma^2)$ and $x_{ij} = 0$ if $y_{ij} \sim N(\mu_{i-}, \sigma^2)$, $i = T, C$; $j = 1, \ldots, n_i$. The likelihood function is given by

$$L(\Psi \mid Y_{obs}, x_{ij}) = \prod_{j, x_{Tj}=1} \gamma\varphi(y_{Tj} \mid \mu_{T+}, \sigma^2) \times \prod_{j, x_{Tj}=0} (1-\gamma)\varphi(y_{Tj} \mid \mu_{T-}, \sigma^2)$$

$$\times \prod_{j, x_{Cj}=1} \gamma\varphi(y_{Cj} \mid \mu_{C+}, \sigma^2) \times \prod_{j, x_{Cj}=0} (1-\gamma)\varphi(y_{Cj} \mid \mu_{C-}, \sigma^2),$$

where $i = T, C$; $j = 1, \ldots, n_i$ and $\varphi(\cdot \mid \cdot)$ denotes the density of a normal variable.

For the Bayesian approach, a beta distribution can be employed as the prior distribution for γ, while normal prior distributions can be used for μ_{i+} and μ_{i-}. In addition, a gamma distribution can be used as a prior distribution for σ^{-2}. Under the assumptions of these prior distributions, the conditional posterior distributions of γ, μ_{i+}, μ_{i-}, σ^{-2} can be derived. In other words, assuming that $f(\gamma) \sim Beta(\alpha_\gamma, \beta_\gamma)$, $f(\mu_{i+}) \sim N(\lambda_{i+}, \sigma_0^2)$, $f(\mu_{i-}) \sim N(\lambda_{i-}, \sigma_0^2)$, and $f(\sigma^{-2}) \sim Gamma(\alpha_g, \beta_g)$, where μ_{i+}, μ_{i-}, and γ are assumed to be independent and α_γ, β_γ, α_g, β_g, λ_{i+}, λ_{i-} and σ_0^2 are assumed to be known. Thus, the conditional posterior distribution of x_{ij} is given by

$$f(x_{ij} \mid \gamma, \mu_{i+}, \mu_{i-}, Y_{obs}) \sim Bernoulli\left(\frac{\gamma\varphi(y_{ij} \mid \mu_{i+}, \sigma_0^2)}{\gamma\varphi(y_{ij} \mid \mu_{i+}, \sigma_0^2) + (1-\gamma)\varphi(y_{ij} \mid \mu_{i-}, \sigma_0^2)}\right),$$

where $E_\Psi[x_{ij}|\gamma, \mu_{i+}, \mu_{i-}, Y_{obs}] = \gamma\varphi(y_{ij}|\mu_{i+}, \sigma^2)/(\gamma\varphi(y_{ij}|\mu_{i+}, \sigma^2) + (1 - \gamma)\varphi(y_{ij}|\mu_{i-}, \sigma^2))$, $i = T, C; j = 1, \ldots, n_i$ in the EM algorithm. The joint distribution of $\gamma, \mu_{i+}, \mu_{i-}$, and σ^2 is given by

$$f(\gamma, \mu_{i+}, \mu_{i-}, \sigma^2 | Y_{obs}, x_{ij})$$

$$= \prod_{j, x_{Tj}=1} \varphi(y_{Tj} | \mu_{T+}, \sigma^2) \times \prod_{j, x_{Tj}=0} \varphi(y_{Tj} | \mu_{T-}, \sigma^2)$$

$$\times \prod_{j, x_{Cj}=1} \varphi(y_{Cj} | \mu_{C+}, \sigma^2) \times \prod_{j, x_{Cj}=0} \varphi(y_{Cj} | \mu_{C-}, \sigma^2) \times \varphi(\mu_{T+} | \lambda_{T+}, \sigma_0^2)$$

$$\times \varphi(\mu_{T-} | \lambda_{T-}, \sigma_0^2) \times \varphi(\mu_{C+} | \lambda_{C+}, \sigma_0^2) \times \varphi(\mu_{C-} | \lambda_{C-}, \sigma_0^2)$$

$$\times \frac{\Gamma(\alpha_\gamma + \beta_\gamma)}{\Gamma(\alpha_\gamma)\Gamma(\beta_\gamma)} (\gamma)^{\sum\limits_{j=1}^{n_T} x_{Tj} + \sum\limits_{j=1}^{n_C} x_{Cj} + \alpha_\gamma - 1} (1-\gamma)^{\sum\limits_{j=1}^{n_T}(1-x_{Tj}) + \sum\limits_{j=1}^{n_C}(1-x_{Cj}) + \beta_\gamma - 1}.$$

Thus, the conditional posterior distribution of $\gamma, \mu_{i+}, \mu_{i-}$, and σ^{-2} and can be obtained as follows:

$$f(\gamma | \mu_{i+}, \mu_{i-}, \sigma^{-2}, Y_{obs}, x_{ij}) \sim Beta\left(\sum_{j=1}^{n_T} x_{Tj} + \sum_{j=1}^{n_C} x_{Cj} + \alpha_\gamma, \sum_{j=1}^{n_T}(1 - x_{Tj}) + \sum_{j=1}^{n_C}(1 - x_{Cj}) + \beta_\gamma\right),$$

$$f(\mu_{i+} | \gamma, \mu_{i-}, \sigma^{-2}, Y_{obs}, x_{ij}) \sim N\left(\frac{\sigma^{-2}\sum_{j=1}^{n_i} x_{ij}y_{ij} + \sigma_0^{-2}\lambda_{i+}}{\sigma^{-2}\sum_{j=1}^{n_i} x_{ij} + \sigma_0^{-2}}, \frac{1}{\sigma^{-2}\sum_{j=1}^{n_i} x_{ij} + \sigma_0^{-2}}\right),$$

$$f(\mu_{i-} | \gamma, \mu_{i+}, \sigma^{-2}, Y_{obs}, x_{ij}) \sim N\left(\frac{\sigma^{-2}\sum_{j=1}^{n_i} (1 - x_{ij})y_{ij} + \sigma_0^{-2}\lambda_{i-}}{\sigma^{-2}\sum_{j=1}^{n_i} (1 - x_{ij}) + \sigma_0^{-2}}, \frac{1}{\sigma^{-2}\sum_{j=1}^{n_i} (1 - x_{ij}) + \sigma_0^{-2}}\right),$$

$$f(\sigma^{-2} | \gamma, \mu_{i+}, \mu_{i-}, Y_{obs}, x_{ij})$$

$$\sim Gamma\left(\frac{n_T + n_C}{2} + \sigma_g, \frac{1}{2}\sum_{i=T,C}\left[\sum_{j=1}^{n_i} x_{ij}(y_{ij} - \mu_{i+})^2 + \sum_{j=1}^{n_i} (1 - x_{ij})(y_{ij} - \mu_{i-})^2\right] + \beta_g\right),$$

respectively. Consequently, the conditional posterior distribution of $\theta = \mu_{T+} - \mu_{C+}$ can be obtained as follows:

$$f(\hat{\theta} \mid \gamma, \mu_{i+}, \mu_{i-}, \sigma^2, Y_{\text{obs}}, x_{ij})$$

$$\sim N\left(\frac{\sigma^{-2} \sum_{j=1}^{n_T} x_{Tj} y_{Tj} + \sigma_0^{-2} \lambda_{T+}}{\sigma^{-2} \sum_{j=1}^{n_T} x_{Tj} + \sigma_0^{-2}} + \frac{\sigma^{-2} \sum_{j=1}^{n_C} x_{Cj} y_{Cj} + \sigma_0^{-2} \lambda_{C+}}{\sigma^{-2} \sum_{j=1}^{n_C} x_{Cj} + \sigma_0^{-2}} \right.,$$

$$\left. \frac{1}{\sigma^{-2} \sum_{j=1}^{n_T} x_{Tj} + \sigma_0^{-2}} + \frac{1}{\sigma^{-2} \sum_{j=1}^{n_C} x_{Cj} + \sigma_0^{-2}} \right).$$

As a result, statistical inference for $\theta = \mu_{T+} - \mu_{C+}$ can be obtained. Following similar ideas, statistical inferences for the treatment effects (θ_1 through θ_5 as described earlier) can be derived. Note that different prior assumptions for γ, μ_{i+}, μ_{i-}, and σ^2 may be applied depending upon disease targets across different therapeutic areas. However, different prior assumptions will result in different statistical inference for the assessment of the treatment effect under study.

21.8 Discussion

Currently, the inclusion and exclusion criteria for clinical trials are based on some clinical signs and symptoms or their corresponding measurements. However, as more molecular targets of the diseases are identified, the expression profiles of the molecular targets more frequently become inclusion and exclusion criteria, e.g., HercepTest for the diagnosis of the *HER2 neu* gene for the treatment of Herceptin in patients with invasive breast cancer. Microarray platform is the breakthrough technology that can simultaneously measure the genome-wide expression profiles of the pathways involved with the pathogenesis of the disease. But the translation of microarray technology to the diagnostic devices for molecular targets in the treatment of the disease by the molecularly targeted agents still faces many challenges (Simon, 2006, 2008). Because the goal of genomic composite biomarker classifiers or IVDMIA is to treat patients with a molecular target with the molecularly targeted drugs and not to treat patients without a target with ineffective and unnecessary treatments, clinical validation is as equally important as analytical validation of IVDMIA.

One of the critical issues for clinical validation is the definition and availability of the gold standard for the diagnosis of the molecular targets used for the evaluation of sensitivity, specificity, PPV, FP rate, and ROC curve. Some investigators use the classifiers derived from other quantitative gene expression platforms, e.g., RT-PCR, as the gold standard. However, in essence, these platforms are not the gold standard and classification error may also occur using these technology platforms for the diagnosis of the same molecular target. As a result, almost none of the parameters concerning the diagnostic accuracy can be estimated without the gold standard. Under the situation without a gold standard, one can only assess the agreement or equivalence in the diagnosis of the molecular target (Liu et al., 2002b). However, equivalence in diagnosis between the test IVDMIA and the reference classifier based on other technological platforms implies that both are accurate or both are inaccurate in the diagnosis of the molecular target.

For the clinical effectiveness trial of the IVDMIA of the diagnostic accuracy of the molecular target, the inclusion and exclusion criteria for patients should be exactly the same as those for the targeted clinical trials for the evaluation of the efficacy and safety of the molecularly targeted agents. In addition, all procedures of the test IVDMIA evaluated in the clinical effectiveness trials and used for diagnosis in the target clinical (utility) trials should be prespecified in the protocols and should be the same methods derived from the development stage of the classifier such as sample collection, RNA extraction, cDNA/cRNA synthesis, dye labeling, hybridization, scanning, normalization procedures, and thresholds. In addition, reproducibility for the correct diagnosis such as within- and between-laboratory agreement should be also evaluated in the clinical effectiveness trial of the IVDMIA.

For the development of any classifier, the prevalence rate must be taken into consideration. For example, since the misclassification rate of the DLDF is a function of the prevalence rate, determination of thresholds also depends upon the prevalence rate. On the other hand, because the molecularly targeted agents are specific inhibitors of their target and may induce a large treatment effect in patients with a molecular target, targeted clinical trials are in general more efficient than untargeted trials (Maitournam and Simon, 2005; Simon and Maitournam, 2004). Moreover, if the prevalence rate of the target in the patient population is low, the recruitment period of the targeted clinical trials will be much longer than the untargeted ones. In addition, the PPV is proportional to the prevalence rate. Therefore, if the prevalence rate of the target is below 0.01, then the PPV will be below 0.5. From (21.9), the treatment effect of the molecular target will be seriously underestimated. However, when the prevalence rate is 0.1 and above, the FP rate will decrease to below 10%. In this case, bias still exists but with a moderate magnitude. Furthermore, similar to gender or age, the genomic composite biomarker classifier is also another variable with the expression profiles to stratify patients into subgroups with and without molecular targets. If the prevalence rate of a certain target is low, the number of patients in this subgroup will be very low. It follows that it might take a very long time to recruit patients and the trial might not have sufficient power to prove the effectiveness of the molecularly targeted agent even if the targeted clinical trial is more efficient. As a result, prevalence rate is a determining factor for the development of molecularly targeted treatments. But how low is the prevalence rate? Is the personalized medicine for a subgroup of one patient with his or her distinct signature attainable? Does a cocktail of molecularly targeted agents for multiple targets represent a feasible approach to targeted therapy? These are just a few challenges that one must ponder about for the development of diagnostic multivariate assays and molecularly targeted therapy.

22

Clinical Trial Simulation

22.1 Introduction

Clinical trial simulation (CTS) is defined as a process that uses computers to mimic the conduct of a clinical trial by creating *virtual* patients and extrapolating clinical outcomes for each virtual patient based on prespecified assumptions/models. CTS is a powerful tool for designing, monitoring, analyzing, and planning clinical trials. It has been used for several decades (Maxwell and Domenet, 1971; Kimko and Duffull, 2003; Chang, 2011). CTS plays an important role in pharmaceutical/clinical research and development. However, it did not receive much attention and become increasingly popular in the pharmaceutical industry until recently (Parmigiani, 2002; Chang, 2011).

In clinical trials, a complicated trial design may be necessarily employed for achieving study objectives. Under a complicated trial design and/or statistical model, there may exist no closed form for statistical inference (e.g., point estimate or confidence interval) for the study endpoints (e.g., safety or efficacy parameters) of interest. In this case, CTS is often employed to evaluate the performance of the derived statistical inference. A typical approach is to generate virtual clinical data under an assumed model, which is treated as a true model. Based on the generated data, statistical inference can then be obtained. A simulation usually involves a large number of runs. In other words, a simulation will generate virtual clinical data under the same model a large number of times. In each run (sample), statistical inference such as point estimate or confidence interval can then be obtained. Based on the point estimates and confidence intervals, one can evaluate the performance of the statistical inference in terms of (1) bias, (2) standard error, (3) coverage probability, and (4) power of the point estimates and/or confidence intervals.

In clinical research, CTS is a useful tool not only for monitoring the conduct of the trial and its outcomes but also for identifying potential problems and providing recommendations early. In addition, it is helpful in studying the validity and sensitivity of the trial if the study should deviate from the study protocol. Under the prespecified model, CTS can also provide useful information regarding the (predicted) clinical outcomes beyond the scope of the study. CTS can help depicting the relationships between the inputs such as dose, dosing time, patient characteristics, and disease severity and the clinical outcomes such as changes in efficacy and safety parameters (e.g., treatment effects, signs and symptoms, laboratory tests, and adverse events). In practice, a CTS is often conducted to evaluate the performance of clinical outcomes under different assumptions and various design scenarios at the planning stage of the intended clinical trial.

One of the most controversial issues in CTS is the validity of the assumed model and its assumptions. If the assumed model and its assumptions are incorrect and/or (seriously) deviate from the true model and assumptions, the simulation results could be biased and hence misleading. Another controversial issue is that if we can verify the assumed model and its assumptions, then there is no need to conduct clinical trials. Many clinicians are against the concept of drawing conclusions based on simulation results, especially when the assumed model is seriously in doubt. In practice, if it is not impossible, the validity of the assumed model and its assumptions are often difficult to verify, which is the primary reason and motivation for conducting a clinical trial.

In the next section, the process for conducting a CTS including a valid statistical model and its assumptions required for conducting a CTS are briefly outlined. Some commonly considered algorithms and/or procedures in CTS such as the expectation–maximization (EM) algorithm and bootstrap are described in Sections 22.3 and 22.4, respectively. In Section 22.5, some applications such as target clinical trials with enrichment design and dose escalation trials in cancer research are given. Some concluding remarks are discussed in the last section.

22.2 Process for Clinical Trial Simulation

In clinical research, the purpose of CTS is to simulate the behavior of a test treatment in patients with the disease under study. Thus, CTS requires (1) a statistical model with certain assumptions in order to simulate the behavior of the drug in the body of a living organism and (2) a study protocol that provides the dosage and data-collecting schedules for the trial. The dosage schedule indicates when drug treatments are to be given to the individual subjects and how much of the drug is to be administered. The data-collecting schedule describes what observations or measurements of the study endpoints are to be taken of the subject and at what times. Note that there can be multiple dosage schedules and multiple observation schedules even for a single clinical trial.

22.2.1 Model and Assumptions

In CTS, a linear model under a valid study design with certain assumptions is often considered to evaluate the effectiveness and safety of a test treatment under investigation. As an example, consider a randomized, parallel-group, double-blind clinical trial comparing T treatments. Let y_{ij} be the response of the ith subject who receives the jth treatment, $i = 1, \ldots, n_j$; $j = 1, \ldots, T$. The following linear model is usually employed:

$$y_{ij} = \mu + \mu_j + S_i + e_{ij}, \quad i = 1, \ldots, n_j, \quad j = 1, \ldots, T, \tag{22.1}$$

where
  μ is the overall mean
  μ_j is the effect of the jth treatment
  S_i is the random effect due to the ith subject
  e_{ij} are random errors in observing y_{ij}

In practice, S_i are independent and identically distributed with mean 0 and variance σ_S^2, e_{ij} are independent and identically distributed with mean 0 and variance σ_e^2, and S_i and e_{ij} are mutually independent. Note that σ_S^2 and σ_e^2 are usually referred to as between-subject (or inter-subject) and within-subject (or intra-subject) variability, respectively. In most cases, the maximum likelihood estimates (MLEs) or consistent estimates of the study parameters are obtained based on asymptotic results of large samples. Under model (22.1) and its corresponding assumptions, a clinical simulation can be carried out using the following steps:

Step 1: Generate random observations (Gentle, 1998) under model (22.1) and assumptions.

Step 2: Calculate the MLEs or consistent estimates of the parameters of interest, such as treatment effects.

Step 3: Repeat the above two steps a large number of times, say 10,000 times, and obtain statistical inferences such as point estimates and/or confidence intervals of the study parameters of interest.

Step 4: Based on the 10,000 point estimates and/or confidence intervals, evaluate the performance of the statistical inference in terms of some performance characteristics such as bias, standard error, mean squared error (MSE), and/or coverage probability.

In practice, the above steps can be repeated for different combinations of study parameters specifications and distribution assumptions for sensitivity or robustness analysis.

22.2.2 Performance Characteristics

As indicated earlier, CTS is often conducted when there exists no closed form for statistical inference under a complicated trial design. In this case, statistical inference is usually obtained based on asymptotic results. Thus, it is of interest to evaluate the finite sample performance of the obtained statistical inference through a CTS in terms of some performance characteristics.

In practice, commonly considered performance characteristics include, but are not limited to, (1) bias for evaluation of accuracy, (2) variability or MSE for assessment of reliability, (3) coverage probability for controlling type I error rate, and (4) sensitivity for deviations from assumptions. For a given study endpoint, bias and MSE can be obtained. The coverage probability is defined as the number of times the obtained confidence intervals cover the true value divided by the total number of simulation runs.

In some simulations, if the study objective is to detect a clinically meaningful difference, the performance characteristic of power is usually considered. Power is defined as the probability of correctly detecting a clinically meaningful difference if such a difference truly exists.

22.2.3 An Example

For illustration purpose, consider the following analysis of covariance (ANCOVA) model:

$$y_{ij} = \mu + f(x_{ij}) + \mu_j + S_i + e_{ij}, \quad i = 1,\dots,n_j, \quad j = 1,\dots,T, \qquad (22.2)$$

where

y_{ij}, μ, μ_j, S_i, and e_{ij} are as defined in (22.1)

$x_{ij} = (x_{1ij}, x_{2ij}, \dots, x_{Kij})$ is the corresponding vector of covariates that are relevant to the response y_{ij}

f is a function that links y_{ij} and x_{ij}

Under model (22.2), a Monte Carlo simulation can be performed to evaluate the bias, variability, MSE, and coverage probabilities of the parameter estimates using the following steps:

Step 1: Using S/R programming, we generated two sets of correlated values to indicate the measures of response and covariates—y_{ij} (indicates measures of response) and x_{ij} (measures of the corresponding covariates)—with the assumed model above. The data were generated by setting the number of treatment t, number of subjects for each treatment n, value of overall mean mu, corresponding values of $f(x)$ (denoted by f.x), treatment effects (denoted by mu.trt), standard deviations of random effect (denoted by sd.S) and of random error (denoted by sd.e). Sample programs are given in Table 22.1.

Step 2: Using these two sets of variables (y_{ij} and x_{ij}), we can calculate the estimates of the study parameters of interest. Note that if there exist no closed forms for these estimates, the method of EM algorithm can be used, which is given in the next section.

Step 3: We then repeat Steps 1 and 2 a large number of times in order to calculate the bias, variability, mean absolute error (MAE), MSE, and coverage probability (rate) based on asymptotic normality assumption. Sample programs are given in Table 22.2.

TABLE 22.1

Sample Program for Generating Random Numbers

```
data.gen.f=function(t, n, mu, f.x, mu.trt, sd.S, sd.e){
    mu.j=rep(mu.trt,n)
## process of generating a (n1+n2+...nT) vector of random effect
s.random=rnorm(max(n),mean=0, sd=sd.S)
    S.1=s.random[1:n[1]]
    for (j in 2:T){
        S.2=s.random[1:n[j]]
        S.1=c(S.1,S.2)
    }
    S=S.1
## generate a (n1+n2+...nT) vector of independent normals
    e=rnorm(length(f.x), mean=0, sd=sd.e)
    y=mu+f.x+mu.j+S+e
}
```

TABLE 22.2

Sample Programs for Calculation of Bias, Variability, MAE, MSE,
and Coverage Rate

```
performance.est.f=function(theta, theta.est){
Bias=mean(theta.est-theta)
MAE=mean(abs(theta.est-theta))
MSE=mean((theta.est-theta)^2)
se=sd(theta.est)/sqrt(length(theta.est))
lower=theta.est-qnorm(0.975)*se
upper=theta.est+qnorm(0.975)*se
cover.rate=mean(as.numeric((theta<=upper)&(theta>=lower)))
stat=c(Bias,MAE,MSE,cover.rate)
names(stat)=c("Bias", "MAE", "MSE", "coverage rate")
print(stat)
}
```

22.2.4 Remarks

As can be seen, the success of CTS depends upon the validity of the assumed
model. If the model is incorrect, the results which are obtained under the
wrong model would be biased and hence misleading. As a result, one of the
controversial issues regarding the use of CTS in clinical trials for addressing some scientific/medical questions under a complicated study design
and model is the validity of the assumed model. However, if one can show
that the assumed model is correct and almost 100% accurate and reliable,
there is no need to conduct a clinical trial because the model is predictive of
the clinical outcomes of patients who receive the test treatment. In practice,

unfortunately, it is impossible to test the validity of the assumed model until we have conducted the clinical trial. Thus, a sensitivity or robustness study is usually conducted to assess the impact of possible deviations from the unknown true model with respect to study parameters and model assumptions.

22.3 EM Algorithm

As indicated in the previous section, it is important to obtain MLEs or consistent estimates of the parameters of interest such as treatment effects in CTS. In many cases, closed forms for MLEs may not exist. In this case, the method of EM algorithm is a very useful tool for finding the MLEs of parameters of interest under an appropriate statistical model, where the model depends on some unobserved latent variables. The EM algorithm has become very popular in clinical research and development since it was introduced by Dempster et al. (1977). The method of EM algorithm is an iterative method which involves two steps, namely, an expectation (E) step that computes the expectation of the log-likelihood evaluated using the current estimate for the latent variables and a maximization (M) step that computes parameters maximizing the expected log-likelihood found on the E step. These parameter estimates are then used to determine the distribution of the latent variables in the next E step. It, however, should be noted that the convergence analysis of the EM algorithm given by Dempster et al. (1977) was flawed. A correct convergence analysis can be found in Wu (1983).

22.3.1 General Description

Given a likelihood function $L(\theta, x, z)$, where θ is the parameter vector, x is the observed data, and z represents the unobserved latent data or missing values, the MLE can be determined by the marginal likelihood of the observed data $L(\theta, x, z)$. However, this quantity is often intractable in practice. In general, the EM algorithm seeks to find the MLE of the marginal likelihood by iteratively applying the following two steps:

E step: Calculate the expected value of the log-likelihood function, with respect to the conditional distribution of z given x under the current estimate of the parameters $\theta^{(t)}$:

$$Q(\theta \mid \theta^{(t)}) = E_{Z|x,\theta^{(t)}}[\log L(\theta; x, Z)].$$

M step: Find the parameter that maximizes this quantity:

$$\theta^{(t+1)} = \arg\max_{\theta} Q(\theta \mid \theta^{(t)}).$$

Note that the EM algorithm is particularly useful when the likelihood is an exponential family. In such a case, the *E* step becomes the sum of expectations of sufficient statistics, and the *M* step involves maximizing a linear function. Thus, it is possible to derive closed-form updates for each step. In addition, the EM method can be modified to compute maximum a posteriori estimates for Bayesian inference. It should also be noted that there are other methods for finding MLEs. These iterative methods include gradient descent, conjugate gradient, or variations of the Gauss–Newton method. Unlike the method of EM algorithm, such methods typically require the evaluation of first and/or second derivatives of the likelihood function.

22.3.2 An Example

As an example, consider the following simple regression model. Let y_{ij} be the ith subject who receives the jth treatment, where $i = 1, \ldots, n_j$, $j = 1, \ldots, T$, and $\sum_{j=1}^{T} n_j = n$. Let $x_{ij} = (x_{1ij}, x_{2ij}, \ldots, x_{Kij})$ be the corresponding vector of covariates that are relevant to the response y_{ij}. The simple regression model can be expressed as

$$Y = X\beta + \varepsilon,$$

where
   Y is the $n \times 1$ vector
   X is an $n \times K$ fixed matrix
   $\beta = (\beta_1, \beta_2, \ldots, \beta_K)^T$ is a $K \times 1$ matrix of unknown parameters
   the error term $\varepsilon = (e_{11}, \ldots, e_{1n_1}, e_{21}, \ldots, e_{Tn_T})^T$, $\varepsilon \sim N(0, \Sigma)$

where Σ has dimensions $n \times n$ and for $\mathrm{Cov}(e_{ij}, e_{kl}) = 0$, $\forall i \neq k$ or $j \neq l$ else $\mathrm{Var}(e_{ij}) = \sigma^2$ and σ^2 is unknown parameters, actually $\Sigma = \sigma^2 I_n$, I_n being the identity matrix.

   Note that, if we were to observe e_{ij}, we could easily find simple closed-form MLEs of the parameters σ^2 we would use:

$$\hat{\sigma}^2 = \frac{\sum_{j=1}^{T} \sum_{i=1}^{n_j} e_{ij}^T e_{ij}}{\sum_{j=1}^{T} n_j} = \frac{E}{\sum_{j=1}^{T} n_j}.$$

The sufficient statistics for σ^2, because e_{ij} cannot be observed, aid the EM algorithm to calculate estimates of the missing sufficient statistics by setting it equal to its expectation, conditional on the observed data vector Y and the fixed matrix X. It is an iterative algorithm.

E step: Let $\tau = 0, 1, \ldots$ index the iterations number, $\hat{\beta}^{(\tau)}$ and $\hat{\sigma}^{2^{(\tau)}}$ denote the vector β, σ^2 value at the end of the τth iteration respectively, then we have

$$\hat{\beta}^{(\tau)} = (X^T (\hat{\Sigma}^{(\tau)})^{-1} X)^{-1} X^T (\hat{\Sigma}^{(\tau)})^{-1} Y,$$

$$\hat{E} = E\left\{ \sum_{j=1}^{T} \sum_{i=1}^{n_j} e_{ij}^T e_{ij} \bigg| y_{ij}, x_{ij}, \hat{\beta}^{(\tau)}, \hat{\sigma}^{2^{(\tau)}} \right\} = \sum_{j=1}^{T} \sum_{i=1}^{n_j} \left[\hat{e}_{ij}^T \hat{e}_{ij} + tr \mathrm{Cov}\left\{ e_{ij} \big| y_{ij}, x_{ij}, \hat{\beta}^{(\tau)}, \hat{\sigma}^{2^{(\tau)}} \right\} \right],$$

where the two terms are

$$\hat{e}_{ij} = E\left(e_{ij} \big| y_{ij}, x_{ij}, \hat{\beta}^{(\tau)}, \hat{\sigma}^{2^{(\tau)}} \right) = E(y_{ij} - x_{ij}\hat{\beta}^{(\tau)}) = 0$$

and

$$\sum_{j=1}^{T} \sum_{i=1}^{n_j} tr \mathrm{Cov}\left\{ e_{ij} \big| y_{ij}, x_{ij}, \hat{\beta}^{(\tau)}, \hat{\sigma}^{2^{(\tau)}} \right\} = \sum_{j=1}^{T} \sum_{i=1}^{n_j} (y_{ij} - x_{ij}\hat{\beta}^{(\tau)})^2,$$

respectively.

M step: Let E_{ij} be replaced by the appropriate sufficient statistics, then for MLE the iterative equations are

$$\hat{\sigma}^{2^{(\tau+1)}} = \frac{\sum_{j=1}^{T} \sum_{i=1}^{n_j} (y_{ij} - x_{ij}\hat{\beta}^{(\tau)})^2}{n},$$

$$\hat{\beta}^{(\tau+1)} = \left(X^T \left(\hat{\Sigma}^{(\tau+1)} \right)^{-1} X \right)^{-1} X^T \left(\hat{\Sigma}^{(\tau+1)} \right)^{-1} Y,$$

where $\hat{\Sigma}^{(\tau)} = \hat{\sigma}^{2^{(\tau)}} I_n$.

As the iterative original value we can use the identity matrix for $\Sigma^{(0)}$. Theoretically the convergence can be obtained by the EM algorithm at local maximum at least. For illustration purpose, some sample programs for the EM algorithm are given in Table 22.3.

TABLE 22.3

Sample Programs for EM Algorithm

```
n=10
K=3
alpha=0.05
tao<-0
X<-matrix(data,n,K)
y<-matrix(data,n,1)
bslash<-function(X,y)
{
  X<-qr(X)
  qr.coef(X,y)
}
B_hat<-matrix(,K,10000)
E_hat<-matrix(,n,10000)
B1<-matrix(,K,1)
B2<-matrix(,K,1)
z<-matrix(,K,1)
norm<-function(B1,B2)
{
  i=1:K
  z[i]<-B1[i]-B2[i]
  no<-sqrt(sum(z[i]^2))
}
sigma_hat<-c()
Q<-matrix(0,K,n)
P<-matrix(0,n,1)
sigma_hat[0]<-diag(n)
B_hat[,0]<-bslash(X,y)
E_hat[,0]<-y-X%*%B_hat[,0]
sigma_hat[0]<-crossprod(E_hat[,0],E_hat[,0])/n
SIGMA_hat<-diag(sigma_hat[0])
Q<-crosspros(X,sqrt(solve(SIGMA_hat)))
P<-crossprod(sqrt(solve(SIGMA_hat)),Y)
B_hat[,1]<-bslash(Q,P)
while (norm(B_hat[,tao],B_hat[,tao+1])>alpha)
{
  tao<-tao+1
  SIMGMA<-matrix(0,n,n)
  Q<-matrix(0,K,n)
  P<-matrix(0,n,1)
  E_hat[,tao]<-y-X%*%B_hat[,tao]
  sigma_hat[tao]<-crossprod(E_hat[tao],E_hat[tao])/n
  SIGMA_hat<-diag(sigma_hat[tao],n)
  Q<-crosspros(X,sqrt(solve(SIGMA_hat)))
  P<-crossprod(sqrt(solve(SIGMA_hat)),Y)
  B_hat[tao+1]<-bslash(Q,P)
}
print(B_hat[,tao])
print(SIGMA_hat)
```

22.4 Resampling Method: Bootstrapping

In CTS, it is necessary to obtain estimates of summary statistics in order to calculate confidence intervals of the parameters of interest under an assumed statistical model. For this purpose, some resampling methods such as Jackknifing or bootstrapping are commonly considered. In this section, we will introduce the use of bootstrapping in CTS.

22.4.1 General Description

Bootstrapping is a resampling technique used to obtain estimates of summary statistics (such as its variance) by sampling from an empirical distribution of the observed data. In the case where a set of observations can be assumed to be from an independent and identically distributed population, this can be done by randomly drawing a number of samples with equal size from the observed data set. In practice, a simple random sampling with replacement from the original data set is often employed.

Although bootstrapping provides asymptotically consistent estimates under some regularity conditions, it does not guarantee that the resultant estimates will have good finite sample performance. It has a tendency to be overly optimistic. In practice, bootstrapping is often used as an alternative to inference based on parametric assumptions when those assumptions are in doubt or where parametric inference is impossible or requires very complicated formulas for the calculation of standard errors. Bootstrapping is simple, straightforward, and easy to implement for obtaining estimates of standard errors and confidence intervals for complex estimators of complex parameters of the distribution, such as percentile points, proportions, odds ratio, and correlation coefficients.

22.4.2 Types of Bootstrap Scheme

In practice, there are several types of bootstrapping schemes that may be applied depending upon the purpose and/or the need of the study objectives. For example, for univariate problems, the approach of resampling the individual observations with replacement is usually considered. On the other hand, in small samples, a parametric bootstrap approach might be preferred. For other problems, a smooth bootstrap may be considered. These bootstrapping schemes are briefly described below.

Case resampling: Case resampling is an approach of resampling the individual observations (case) with replacement. It can be performed as follows. We first resample the data with replacement. Then the statistic of interest is computed from the resample. We repeat this routine a large number of times in order to obtain a more precise estimate of the bootstrap distribution of the statistic.

Smooth bootstrap: Under this scheme, a small amount of random noise is added on to each resampled observation. This is equivalent to sampling from a kernel density estimate of the data.

Parametric bootstrap: For parametric bootstrap, a parametric model is fitted. Samples of random numbers are then drawn from this fitted model. For each sample, the estimate of interest is calculated. This sampling process is repeated many times as for other bootstrap methods.

Resampling residuals: The method of resampling residuals is often applied in regression problems. The method proceeds as follows. Fit the model and retain the fitted values $\hat{y}_i$ and the residuals $\hat{\varepsilon}_i = y_i - \hat{y}_i$, $i = 1, \ldots, n$. for each pair (x_i, y_i), where x_i is the explanatory variable. Then, add a randomly resampled residual $\hat{\varepsilon}_j$ to the response variable y_i, i.e., $y_i^* = y_i + \hat{\varepsilon}_j$, where j is selected randomly from the list $(1, \ldots, n)$ for every i. Refit the model based on y_i^* and retain the quantities of interest. Repeat the process a large number of times.

22.4.3 Methods for Bootstrap Confidence Intervals

Methods for constructing bootstrap confidence intervals include, but are not limited to, (1) percentile bootstrap, (2) studentized bootstrap, and (3) bias-corrected bootstrap. The percentile bootstrap is probably the simplest method for obtaining confidence intervals. The confidence interval is derived by using the 2.5th and the 97.5th percentiles of the bootstrap distribution as the limits of the 95% confidence interval. This method can be applied to any statistics. It will work well in cases where the bootstrap distribution is symmetrical and centered on the observed statistic (Efron, 1982).

22.5 Clinical Applications

22.5.1 Target Clinical Trials with Enrichment Designs

One of the immediate clinical applications of using the EM algorithm in conjunction with bootstrapping in CTS is the example concerning a target clinical trial with enrichment process as described in Section 21.7. Liu et al. (2009) conducted a simulation study to evaluate finite sample performance of the proposed method of the EM algorithm. In the simulation, μ_{T-}, μ_{C+}, and μ_{C-} are assumed to be equal and set to be a generic value of 100. To investigate the impact of the positive predictive value, sample size, difference in means, and variability, Liu et al. (2009) considered the following specifications of parameters: (1) the positive predicted value is set to be 0.5, 0.7, 0.8, and 0.9, which reflect a range of low, median, and high positive predicted values and (2) the range of the standard deviation σ

is set as 20, 40, or 60. To investigate the finite sample properties, the sample sizes are set as 50, 100, and 200 per group. The mean differences are chosen as a fraction of the standard deviation, from 10% to 60% by 10%; and 75% and 100%. In addition, the size of the proposed testing procedure was investigate at $\mu_{T+} = 100$.

For each of 288 combinations, 5000 random samples were generated and the number of the bootstrap samples was set to be 1000. The simulation results indicate that the absolute relative bias of the estimator for θ by the current method ranges from 10% to more than 50% and increases as the positive predictive value decreases. On the other hand, most of the absolute relative biases of the estimator for θ obtained by the EM algorithm are smaller than 0.05% although it can be as high as 10% for few combinations when the difference in means is 2. The variability has little impact on the bias of both methods. However, for the EM procedure, the relative bias tends to decrease as the sample size increases. The bias of the current method with consideration of the true status of molecular target can be as high as 50% when the positive predictive value is low. Consequently, the empirical coverage probabilities of the corresponding 95% confidence interval can be as low as only 0.28% when the positive predictive value is 50%, mean difference is 20, standard deviation is 20, and n is 200. The coverage probability of the 95% confidence interval by the current method is an increasing function of the positive predictive value. On the other hand, only 36 of the 288 coverage probabilities (12.5%) of the 95% confidence intervals by the current method exceed 0.9449 and 24 of them occur when the positive predictive value is 0.9. On the contrary, only 14.6% of the 288 coverage probabilities of the 95% confidence intervals by the EM method are below 0.9449. However, 277 of the 288 coverage probabilities of the 95% confidence interval constructed by the EM algorithm are above 0.94. No coverage probability of the EM method is below 0.91. Therefore, the proposed procedures for the estimation of the treatment effects in patient populations with a molecular target by the EM algorithm are not only unbiased but also provide sufficient coverage probability.

22.5.2 Dose Escalation Trials

As indicated in Section 20.3, dose escalation trials are usually conducted to identify the maximum tolerable dose (MTD) in cancer research. The identified MTD is often considered as the optimal dose for subsequent clinical trials in the later phase of clinical development. The most commonly employed trial design is the algorithm-based "3 + 3" traditional escalation rule (TER) trial design with a prespecified sequence of dose levels. This approach, however, is found to be inefficient and often underestimates the MTD, especially when the starting dose is too low. Alternatively, the continual reassessment method (CRM) trial design has become very popular since it was introduced by O'Quigley et al. (1990). The CRM trial design is developed to overcome the limitations of the "3 + 3" TER trial design. The CRM trial design assigns more patients near the MTD; consequently, the estimated MTD is more accurate and reliable. Chang and Chow (2005) considered a hybrid Bayesian approach for dose escalation trials. Their approach can be summarized in the following steps:

Step 1: Construct a utility function based on trial objectives.

Step 2: Propose a probability model for dose–response relationship.

Step 3: Construct prior probability distribution of the parameters in the response model.

Step 4: Form the likelihood function based on incremental information on treatment response during the trial.

Step 5: Reassess model parameters or calculate the posterior probability of the model parameters.

Step 6: Update the expected utility function based on the dose–response model.

Step 7: Determine the next action or make adaptations such as changing the randomization or dropping inferior treatment arms.

Step 8: Further collect trial data and repeat Steps 5–7 until stopping criteria are met.

Note that a commonly considered dose–response relationship in cancer trials is the dose-toxicity model of $p(x) = [1 + b \exp(-ax)]^{-1}$, where $p(x)$ is the probability of toxicity with dose x. Under this dose-toxicity model, the (estimated) MTD can be obtained as follows:

$$\text{MTD} = \frac{1}{a} \ln\left(\frac{b\theta}{1-\theta}\right),$$

where θ is the probability of dose-limiting toxicity (DLT) (DLT rate) at MTD. It also should be noted that the assignment of patients to the most updated (predicted) MTD leads to the majority of patients being assigned to dose levels near MTD, which allows a more precise estimate of the MTD with a minimum number of patients. In practice, potential dose jump and delayed response are commonly seen when utilizing CRM in dose escalation trials.

Step-by-step SAS codes for identifying MTD are available in Chang (2008) (see also, TriSaft Intl., 2002). However, many readers have indicated that they were unable to reproduce the numbers as indicated in the book. Thus, we also developed SAS codes following the steps given in Chang's book. We found that the major inconsistency is due to the fact that Chang's program does not initialize the posterior distribution of parameter a at the beginning of each simulation run. Different iterations should be independent. In other words, the 500th iteration should not use the posterior distribution of parameter a obtained from the 499th or previous iteration. Revised SAS codes are provided in Table 22.4. Table 22.5 provides a summary of simulation results obtained using Chang's programs and the revised programs.

TABLE 22.4

Sample SAS Programs for CRM

```
* Simulating the trial using Continual Reassessment Method and
Traditional Escalation Rules;
* This SAS code is developed based on the one given in the book
"Adaptive Design Theory and Implementation Using SAS and R" by
Dr. Mark Chang;
/*Notations:
aMin and aMax = the lower and upper limits for prior on the
parameter a;
AveN = average number of patient treated;
   b = parameter in the dose-response probability model which
       is given as
                      p(x) = 1/(1+b*exp(-a*x))
CorPro = proportion of correctly identifying the MTD;
CohS = cohort size;
doses{i} = dose amount at dose level i;
          dx = the interval width for numerical integration;
          DoseJp = 1/0 means no limited dose jump, and no dose jump;
          g{i} = the prior distribution of the model parameter a;
          MTRate = the maximum tolerated rate defined for MTD;
          nIntPts = number of intervals for numerical integration
            in calculating the posterior;
          nLevels = number of dose levels;
          nLmax = the maximum number of subjects allowed at the
            same dose level;
          nPts = total number of patients;
          nSims = number of simulations;
          RRo{i} = true response rates for dose level i;

          Key output variables:
              AveMTD = average MTD from nSims times of simulation;
              SdMTD = standard deviation of MTDs;
              DLTs = average dose limiting toxicity;
*/
Title "Adaptive Dose Finding Design using CRM";

%Macro CRM(nSims=100, nPts=30,CohS= 1,   nLevels=10, b=100,
   aMin=0.1,
               aMax=0.3,nIntPts=100, MTRate=0.30,nLMax=4, DoseJp=0);
            Data CRM;
                Set D input;
                Keep nPts CohS nLevels AveMTD SdMTD DLTs
                  CorPro AveN nLMax;
                Array nPtsAt{&nLevels};
                Array nRsps{&nLevels};
                Array pog{&nIntPts}; *Posterior distribution of a;
```

TABLE 22.4 (continued)

Sample SAS Programs for CRM

```
                    Array g{&nIntPts};
                       Array Doses{&nLevels};
                       Array RRo{&nLevels};
                       Array RR{&nLevels};
                       seed=2736;
                       nLevels=&nLevels;
                    nPts=&nPts;
                       CohS=&CohS;
                       nLMax = &nLMax;
                       dx = (&aMax-&aMin)/&nIntPts;
                    DLTs=0;
                       AveMTD=0;
                       varMTD=0;
                    TrueMTD =0;
                       CorPro = 0;
                    AveN = 0;

/* determine the maximum tolerated dose(MTD)which is the
maximum dose with
dose-limiting toxicity rate not greater than the maximum
tolerated toxicity rate
          */
            DTRdif = 1;
               Do i = 1 to nLevels;
                 if 0 <= &MTRate- RRo{i} < DTRdif    then
                     Do;
            TrueMTD = Doses{i};
              DTRdif = &MTRate- RRo{i};
                     End;
            End;
               putlog TrueMTD =   ;
   *Simulation begins;
            Do iSim=1 to &nSims;   *number of simulation;
              *begin of the do-loop for simulation;
            Do tt=1 to &nIntPts;
               pog{tt} = g{tt};   /*Initialize the
                  posterior distribution of a using
                            its prior distribution; the
                            program in the
                            book does not do any
                            initialization of the
                            posterior distribution at
                            the beginning of
```

(continued)

TABLE 22.4 (continued)

Sample SAS Programs for CRM

```
                                    each iteration of simulation.
                                     without
                                    re-initialization of the
                                     posterior distribution,
                                    the estimated MTD will
                                     quickly jump to a
                                    level close to the true MTD
                                                       */
                End;

            Do i=1 to nLevels;
                nPtsAt{i}=0; * number of patients
                 at dose level i;
                nRsps{i}=0;  * number of responses
                 at dose level i;
            End;

        iLevel=1;   *the current MTD level;
            iLevel0=1;
            TnPts = 0; *the total number of patients
             who have been treated so far;
            r = 0;

            Do iCoh=1 to nPts;        *begin of the
              do-loop for dose finding using CRM;
                  iLevel=min(iLevel, nLevels);
                  If TnPts + CohS <= nPts then TnPts
                    = TnPts + CohS;
                    Else go to Finisher ; *the study
                      ends if all patients have been
                      treated;
                  if nPtsAt{iLevel} >= nLMax then go
                    to Finisher;  /*the study ends if
                    the number of subjects attains
                                    the maximum allowed
                                      number for a dose
                                      level */
                  Rate=RRo{iLevel};
                  nPtsAt{iLevel}=nPtsAt{iLevel}+CohS;
                    *number of patients at dose
                    level{ilevel};
                  r=Ranbin(seed,CohS, Rate);
                  nRsps{iLevel}=nRsps{iLevel}+r;
                    *number of responses at dose
                    level{iLevel};
```

TABLE 22.4 (continued)

Sample SAS Programs for CRM

```
             ** Posterior distribution of a;
                c=0;
                Do k=1 to &nIntPts;
                * numerical integral of the
                posterior distribution of a;
                ak=&aMin+(k-0.5)*dx;
                Rate=1/(1+&b*Exp(-
                  ak*Doses{iLevel}));
                 L = Rate**r*(1-
                   Rate)**(CohS-r); *likelihood
                   function for the current
                   cohort;
                pog{k}=L*pog{k};
                c=c+pog{k}*dx;
             End;

             Do k=1 to &nIntPts;
                 pog{k}=pog{k}/c;
             End;

           ** Predict response rate and current
             MTD;
             MTD=Doses{iLevel};
             MinDR=1;
             iLevel0 =iLevel;
         Do i=1 to nLevels;                 * begin
           of the do-loop for i;
                 RR{i}=0;
                 Do k=1 to &nIntPts;
                    *calculate the estimated
                    response rate at each dose
                    level using numerical
                    integral;
                    ak=&aMin+k*dx;
                    RR{i}=RR{i}+1/(1+&b*Exp(-
                      ak*Doses{i}))*pog{k}*dx;
                 End;
                 DR=Abs(&MTRate-RR{i});
                 if 0<= DR < MinDR  Then
                    Do;
                       *find the current MTD
                       which is closest to the
                       maximum tolerated rate;
```

(continued)

TABLE 22.4 (continued)

Sample SAS Programs for CRM

```
                              MinDR=DR;
                               iLevel=i;
                               MTD=Doses{i};
                         End;
                    End;
                      * end of the do-loop for i;
            If iSim =1 then putlog RR1-RR8 ;
                 If iLevel > iLevel0 and &DoseJp = 0
                   Then
                    Do;
                    iLevel = iLevel0+1;  *No dose jump
                      is allowed if DoseJp=0;
                          MTD=Doses{iLevel};
                    End;
                 if 0<iSim <=10 then
              putlog iSim= iLevel= ; *output the
                current MTD level for the ith
                iteration;
                         End;   *end of the do-loop
                                  for dose finding using
                                  CRM;

          *accumulate number of toxicity events and
            number of patients treated when the
            study is done;
         Finisher:
            Do i=1 to nLevels;
               DLTs=DLTs+nRsps{i};
              AveN = AveN+nPtsAt{i};
            End;

          If TrueMTD - 0.1 < MTD <=TrueMTD then
         CorPro = CorPro +1;
            AveMTD=AveMTD+MTD;   *sum of MTD;
            VarMTD=VarMTD+MTD**2;
      End;   *end of do-loop for simulation;

         CorPro = CorPro/&nSims;
      AveMTD = AveMTD/&nSims;
         SdMTD=(VarMTD/&nSims-AveMTD**2)**0.5;
         AveN = AveN/&nSims;      * average MTD;
         DLTs = DLTs/&nSims;      * average DLTs;
         Output;
         Run;
         Proc print Data=CRM;
         Run;
   %Mend CRM;
```

TABLE 22.4 (continued)

Sample SAS Programs for CRM

```
Data Dinput;
     Array g{100};
     Array RRo{8} (.01,.02,.03,.05,.12,.17,.22,.4);
      Array Doses{8};
      Do i = 1 to 8;
         Doses(i)=i;
      end;
      Do k=1 to 100;   *uniform distribution as the prior
         distribution;
         g{k}=1;
      End;

Run;


Proc Print Data= Dinput;
        Var Doses1-Doses8 RRo1-RRo8;
Run;



%CRM (nSims=1000, nPts=8,CohS=1, nLevels=8, b=150, aMin=0,
  aMax=3,
        nIntPts=100, MTRate=0.17, nLMax=8, DoseJp=1);
%CRM (nSims=1000, nPts=16,CohS=1, nLevels=8, b=150, aMin=0,
  aMax=3,
        nIntPts=100, MTRate=0.17,nLMax=16, DoseJp=1);
*the macro with the above input has the same function as in
  the example on page 303;
* probability of correctly identifying the true MTD is
  relatively high (0.72)for nPts=8;
* but it is simply an exception—when the sample size changes
  to 7 or 9, the probability
  is quite low;
* it is helpful to have a look at the log for the sequence of
  estimated MTD;
%CRM (nSims=1000, nPts=7,CohS=1, nLevels=8, b=150, aMin=0, aMax=3,
        nIntPts=100, MTRate=0.17, nLMax=7, DoseJp=1);
%CRM (nSims=1000, nPts=9,CohS=1, nLevels=8, b=150, aMin=0, aMax=3,
        nIntPts=100, MTRate=0.17, nLMax=9, DoseJp=1);
%CRM (nSims=1000, nPts=20,CohS=1, nLevels=8, b=150, aMin=0, aMax=3,
        nIntPts=100, MTRate=0.17, nLMax=20, DoseJp=1);
%CRM (nSims=1000, nPts=40,CohS=1, nLevels=8, b=150, aMin=0, aMax=3,
        nIntPts=100, MTRate=0.17, nLMax=40, DoseJp=1);
Run;
```

TABLE 22.5

Simulation Results for CRM

Program	N	Mean DLTs	Mean MTD	Sd MTD	Correct Pr[a]
By Dr. Chang[b]	8	1.24	6.01	0.07	NA
By Dr. Chang	16	2.65	6.00	0.06	NA
Revised version[c]	7	0.18	4.43	1.25	0.0
Revised version	8	0.28	4.97	1.70	0.72
Revised version	9	0.43	5.27	2.12	0.0
Revised version	16	1.51	4.94	1.73	0.16
Revised version	20	2.11	5.08	1.52	0.23
Revised version	40	4.88	5.40	1.14	0.30

[a] Correct Pr is the probability of correctly identifying the true MTD.
[b] The results are from Chang (2008, Table 15.2, p. 303).
[c] The simulation results using the revised programs given in Table 22.4.

As can be seen from Table 22.5, the relatively high probability for the sample size $N = 8$ is a particular case. The relatively high probability is just by chance. If other levels are the true MTD rather than level 6 as specified in the simulation, the probability will not be so high for $N = 8$. It should also be noted that the prior distribution of parameter a is far from accurate. In fact, the estimated toxicity rate at each dose level is normally greater than its true value. The log of SAS shows that the increment of the estimated MTD is not greater than one level at each step.

22.6 Concluding Remarks

In recent years, the use of adaptive design methods in clinical trials has become very popular due to its flexibility for identifying any possible signal (preferably optimal) of safety and efficacy of test treatment under investigation. However, appropriate statistical methods may not be available due to the complexity of adaptive design used. In this case, CTS plays an important role due to the following reasons: (1) the statistical theory of adaptive design is complicated with very limited analytical solutions available under some strong assumptions; (2) the concept of CTS is very intuitive and easy to implement; (3) CTS can be used to model very complicated situations with minimum assumptions and type I error can be strongly controlled; (4) using CTS, we can not only calculate the power of an adaptive design but also generate many other important operating characteristics such as expected sample size, conditional power, and repeated confidence interval, which ultimately leads to the selection of an optimal trial design or clinical development plan; (5) CTS can be used to study the validity and robustness of an adaptive design in different hypothetical clinical settings, or protocol deviations; (6) CTS can be used to monitor trials, project outcomes, anticipate problems, and suggest remedies before it is too late; (7) CTS can also be used to visualize the dynamic trial process from patient recruitment, drug distribution, treatment administration, pharmacokinetic processes, to biomarker and clinical responses; and, finally, (8) CTS has minimal cost and can be done in a short time.

In summary, the use of CTS in adaptive designs has the following advantages: (1) the type I error rate is controlled, (2) sensitivity analysis is easy to carry out for risk assessment, (3) it allows for the identification of an optimal design with various criteria especially when the candidate adaptive designs are less well understood than those described in the Food and Drug Administration (FDA) guidance on adaptive clinical trial designs, and (4) CTS can be used to achieve a better planning, better design, better monitoring, and better execution. However, it should be noted that CTS provides *"a"* solution but not *"the"* solution to the most difficult question in pharmaceutical/clinical research and development. Any misuse or abuse of CTS could be biased in decision making and hence misleading.

23

Traditional Chinese Medicine

23.1 Introduction

In recent years, as more and more innovative drug products are going off patent, the search for new medicines that treat critical and/or life-threatening diseases has become the center of attention of many pharmaceutical companies. As indicated by Chow and Liu (2000b), pharmaceutical research and development is a lengthy and costly process. On average, it may take more than 12 years to bring a promising compound to the market. The probability of success, however, is usually very low. In the past several decades, tremendous effort was put on drug research and development, and yet only a handful of new drug products were approved by the regulatory agencies. As a result, an alternative approach for drug discovery is necessary. This leads to the study of the potential use of promising traditional Chinese medicines (TCMs), especially those intended for treating critical and/or life-threatening diseases. A TCM is defined as a Chinese herbal medicine developed for treating patients with certain diseases as diagnosed by the four major Chinese diagnostic techniques of inspection, auscultation and olfaction, interrogation, and pulse taking and palpation, based on traditional Chinese medical theory of global dynamic balance among the functions/activities of all the organs of the body.

Unlike evidence-based clinical research and development of a Western medicine (WM), clinical research and development of a TCM is usually experience-based with anticipated variability due to a subjective evaluation of the disease under study. The use of TCM in humans for treating various diseases has a history of more than 5000 years and yet no scientific documentation is available regarding clinical evidence of safety and efficacy of these TCMs.

In the past several decades, regulatory agencies of both China and Taiwan have debated which direction the TCM should take—Westernization or modernization. The Westernization of TCM refers to the adoption of the typical (Western) process of pharmaceutical research and development for the scientific evaluation of the safety and effectiveness of the TCM products under investigation, while the modernization of TCM is to evaluate the safety and effectiveness of TCM the Chinese way (i.e., different sets of regulatory requirements and evaluation criteria) scientifically. Although both China and Taiwan do attempt to build up an environment for the modernization of TCM, they seem to adopt the Westernization approach. As a result, in this chapter, we will place our emphasis on the Westernization of TCM.

In practice, it is a concern whether a TCM can be scientifically evaluated the Western way due to some fundamental differences between a WM and a TCM. These fundamental differences include differences in formulation, medical practice, drug administration, diagnostic procedure and criteria for evaluation, and flexibility. Under these differences, it is then of interest to the investigators regarding how to conduct a scientifically valid (i.e., an adequate and well-controlled) clinical trial for the evaluation of the clinical safety and efficacy of the TCM under investigation. In addition, it is also of particular interest to the investigators as to how to translate an observed significant difference detected by the Chinese diagnostic procedure (CDP) to a clinically meaningful difference based on some well-established clinical study endpoint. The purpose of this chapter is to provide some basic considerations regarding practical issues that are commonly encountered when conducting a TCM clinical trial the Western way.

In the next section, some fundamental differences between a WM and a TCM which have an impact on the Westernization of TCM are described. These fundamental differences include the concept of global dynamic balance/harmony among the organs of the body (TCM) versus local site action (WM); subjective diagnostic techniques of inspection, auscultation and olfaction, interrogation, pulse taking and palpation (TCM) versus objectively clinical evaluation (WM); and personalized flexible dose with multiple components (TCM) versus fixed dose of single active ingredient (WM). Section 23.3 provides some basic considerations of TCM clinical trials. These basic considerations include study design, validation of a quantitative instrument developed for the four major TCM diagnostic techniques, the use/preparation of matching placebo, and sample size calculation. Some practical issues that are commonly encountered when conducting a TCM clinical trail are given in Section 23.4. Section 23.5 provides some recent developments for the assessment of TCM such as test for consistency in statistical quality control (QC) of raw material and/or final product, stability analysis, and calibration of CDPs against well-established study endpoints used for the assessment of WM. Some concluding remarks, including future strategy and recommendations in TCM research and development, are given in the last section of this chapter.

23.2 Fundamental Differences

As indicated earlier, the process for pharmaceutical research and development of WMs is well established, and yet it is lengthy and costly. This lengthy and costly process is necessary to ensure the efficacy, safety, quality,

TABLE 23.1

Fundamental Differences between a WM and a TCM

Description	WM	TCM
Active ingredient	Single	Multiple
Dose	Fixed	Flexible
Diagnostic procedure	Objective; validated	Subjective; not validated
Therapeutic index	Well-established	Not well-established
Medical mechanism	Specific organs	Global dynamic balance/harmony among organs
Medical perception	Evidence-based	Experience-based
Statistics	Population	Individual

stability, and reproducibility of the drug product under investigation. For pharmaceutical research and development of a TCM, one may consider directly applying this well-established process to the TCM under investigation. However, this process may not be feasible due to some fundamental differences between a TCM and a WM. Some fundamental differences between a WM and a TCM are summarized in Table 23.1. These fundamental differences are briefly described in the following sections.

23.2.1 Medical Theory/Mechanism and Practice

TCM is more than a 3000-year-old holistic medical system encircling the entire scope of human experience. It combines the use of Chinese herbal medicines, acupuncture, massage, and therapeutic exercise such as Qigong (the practice of internal air) and Taigie for both treatment and prevention of disease. With its unique theories of etiology, diagnostic systems, and abundant historical literature, TCM itself consists of Chinese culture and philosophy, clinical practice experience, and the use of many medical herbs.

Chinese doctors believe that how a TCM functions in the body is based on the eight principles, five-element theory, five Zang and six Fu, and information regarding channels and collaterals. Eight principles consist of Yin and Yang (i.e., negative and positive), cold and hot, external and internal, and Shi and Xu (i.e., weak and strong). The eight principles help Chinese doctors to differentiate syndrome patterns. For instance, people with Yin will develop disease in a negative, passive, and cool way (e.g., diarrhea and back pain), while people with Yang will develop disease in an aggressive, active, progressive, and warm way (e.g., dry eyes, tinnitus, and night sweats). The five elements (earth, metal, water, wood, and fire) correspond to particular organs in the human body. Each element operates in harmony with the others.

The five Zang (or Yin organs) include the heart (including the pericardium), lung, spleen, liver, and kidney, while the six Fu (or Yang organs) include the gall bladder, stomach, large intestine, small intestine, urinary bladder, and three cavities (i.e., chest, epigastrium, and hypogastrium). Zang organs can manufacture and store fundamental substances. These substances are then transformed and transported by Fu organs. TCM treatments involve a thorough understanding of the clinical manifestations of Zang–Fu organ imbalance, and knowledge of appropriate acupuncture points and herbal therapy to rebalance or maintain the balance of the organs. The channels and collaterals are the representation of the organs of the body. They are responsible for conducting the flow of energy and blood through the entire body.

The elements of TCM can also help to describe the etiology of disease including six exogenous factors (i.e., wind, cold, summer, dampness, dryness, and fire), seven emotional factors (i.e., anger, joy, worry, grief, anxiety, fear, and fright), and other pathogenic factors. Once all the information is collected and processed into a logical and workable diagnosis, the traditional Chinese medical doctor can determine the treatment approach.

Under the medical theory and mechanism described above, Chinese doctors believe that all of the organs within a healthy subject should reach the so-called global dynamic balance or harmony among organs. Once the global balance is broken at certain sites such as heart, liver, or kidney, some signs and symptoms then appear to reflect the imbalance at these sites. An experienced Chinese doctor usually assesses the causes of global imbalance before a TCM with flexible doses is prescribed to fix the problem. This approach is sometimes referred to as a personalized (or individualized) medicine approach.

23.2.2 Medical Practice

Different medical perceptions regarding signs and symptoms of certain diseases could lead to a different diagnosis and treatment for the diseases under study. For example, the signs and symptoms of type 2 diabetic subjects could be classified as the disease of thirst reduction by Chinese doctors. The disease of type 2 diabetes is not recognized by Chinese medical literature although they have the same signs and symptoms as the well-known disease of thirst reduction. This difference in medical perception and practice has an impact on the diagnosis and treatment of the disease.

In addition, we tend to see the therapeutic effect of WMs sooner than that of TCMs. TCMs are often considered for patients who have chronic diseases or non-life-threatening diseases. For critical and/or life-threatening diseases such as cancer or stroke, TCMs are often used as the second-line or third-line treatment with no other alternative treatments. In many cases, such as in patients with a later phase of cancer, TCMs are often used in conjunction with WMs without the knowledge of the primary care physicians, which might have contaminated (e.g., due to drug-to-drug interaction) the treatment effect under investigation.

23.2.3 Techniques of Diagnosis

The CDP for patients with certain diseases consists of four major techniques, namely, inspection, auscultation and olfaction, interrogation, and pulse taking and palpation. All these diagnostic techniques aim mainly at providing an objective basis for differentiation of syndromes by collecting symptoms and signs from the patient. Inspection involves observing the patient's general appearance (strong or weak, fat or thin), mind, complexion (skin color), five sense organs (eye, ear, nose, lip, and tongue), secretions, and excretions. Auscultation involves listening to the voice, expression, respiration, vomit, and cough. Olfaction involves smelling the breath and body odor. Interrogation involves asking questions about specific symptoms and the general condition including history of the present disease, past history, personal life history, and family history. Pulse taking and palpation can help to judge the location and nature of a disease according to the changes of the pulse.

The CDP of inspection, auscultation and olfaction, interrogation, and pulse taking and palpation is subjective, with large between-rater variability (i.e., variability from one Chinese doctor to another). This subjectivity and variability will have an impact not only on the patient's evaluability but also on the prescribability of TCM, which will be further discussed in the subsequent sections.

23.2.3.1 Objective versus Subjective Criteria for Evaluability

For the evaluation of a WM, objective criteria based on some well-established clinical study endpoints are usually considered. For example, response rate (i.e., complete response plus partial response based on tumor size) is considered a valid clinical endpoint for evaluating clinical efficacy of oncology drug products. Unlike WMs, CDP for the evaluation of a TCM is very subjective. The use of a subjective CDP has raised the following issues. First, it is a concern whether the subjective CDP can accurately and reliably evaluate clinical efficacy and safety of the TCM under investigation. Thus, it is suggested that the subjective CDP should be validated in terms of its accuracy, precision, and ruggedness before it can be used in TCM clinical trials. A validated CDP should be able to detect a clinically significant difference if the difference truly exists. On the other hand, it is not desirable to wrongly detect a difference when there is no difference.

In clinical trials, evaluation is usually based on some validated tools (instruments) such as laboratory tests. Test results are then evaluated against some normal ranges for abnormality. Thus, it is suggested that the CDP must be validated in terms of validity and reliability, and its false-positive and false-negative rates, before it can be used for the evaluation of clinical efficacy and safety of the TCM under investigation.

23.2.4 Treatment

TCM prescriptions typically consist of a combination of several components. The combination is usually determined based on the medical theory of global dynamic balance (or harmony) among organs, and the observations from the CDP. The use of CDP is to find out what caused the imbalance among these organs. The treatment is to reinstall the balance among these organs. Thus, the dose and treatment duration are flexible in order to achieve the balance. This concept leads to the concept of so-called personalized (or individualized) medicine, which minimizes intra-subject variability.

23.2.4.1 Single Active Ingredient versus Multiple Components

Most WMs contain a single active ingredient. After drug discovery, an appropriate formulation (or dosage form) is necessarily developed so that the drug can be delivered to the site of action in an efficient way. At the same time, an assay is necessarily developed to quantitate the potency of the drug. The drug is then tested on animals for toxicity and on humans (healthy volunteers) for pharmacological activity. Unlike the WMs, TCMs usually consist of multiple components with certain relative proportions among the components. As a result, the typical approach for the evaluation of a single active ingredient for WM is not applicable to TCMs with multiple components.

In practice, one may suggest evaluating the TCM component by component. However, this is not feasible due to the following difficulties. First, analytical methods for quantitation of individual components may not be available or often not tractable. Thus, the pharmacological activities of these components are often not known. It should be noted that in many cases the component which comprises the major proportion of the TCM may not be the most active component. On the other hand, the component that has the least proportion of the TCM may be the most active component of the TCM. In practice, it is not known which relative proportions among these components can lead to the optimal therapeutic effect of the TCM. In addition, the relative component-to-component and/or component-by-food interactions are usually unknown, which may have an impact on the evaluation of clinical efficacy and safety of the TCM.

23.2.4.2 Fixed Dose versus Flexible Dose

Most WMs are usually administered in a fixed dose (say 10 mg tablets or capsules). On the other hand, since a TCM consists of multiple components with possible varied relative proportions among the components, a Chinese doctor usually prescribes the TCM with different relative proportions of the multiple components based on the signs and symptoms of the patient according to his/her best judgment following a subjective evaluation based on the CDP. Thus, unlike a WM which is prescribed as a fixed dose, a TCM is often prescribed as an individualized flexible dose.

The approach of WM with a fixed dose is a population approach to minimize the between-subject (or inter-subject) variability, while the approach to TCM with an individualized flexible dose is to minimize the variability within each individual. In practice, it is a concern whether an individual flexible dose is compatible with a Western evaluation of the TCM. An individualized flexible dose depends heavily upon the Chinese doctor's subjective judgment, which may vary from one Chinese doctor to another. As a result, although an individualized flexible dose does minimize intra-subject variability, the variability from one Chinese doctor to another (i.e., the doctor-to-doctor or rater-to-rater variability) could be huge, and hence nonnegligible.

23.2.5 Remarks

For the research and development of a TCM, before a TCM clinical trial is conducted, the following controversial questions are often asked:

1. Will the TCM clinical trial be conducted by Chinese doctors alone, Western clinicians alone, Western clinicians who have some background of Chinese herbal medicine alone, or both Chinese doctors and Western clinicians?
2. Will traditional Chinese diagnostic and/or trial procedures be used throughout the TCM clinical trial?
3. Upon approval, is the TCM intended for use by Chinese doctors or Western clinicians?

With respect to the first two questions, if the TCM clinical trial is to be conducted by Chinese doctors alone, the following questions arise. First, should the CDP be validated in order to provide an accurate and reliable assessment of the TCM? In addition, it is of interest to determine how an observed difference obtained from the CDP can be translated to the clinical endpoint commonly used in similar WM clinical trials with the same indication. These two questions can be addressed statistically by the calibration and validation of the CDP with respect to some well-established clinical endpoints for the evaluation of WMs. If the TCM clinical trial is to be conducted by Western clinicians or Western clinicians who have some background of Chinese herbal medicine, the standards and consistency of clinical results as compared to those WM clinical trials are ensured. However, the good characteristics of TCM may be lost during the process of the conduct of the TCM clinical trials. On the other hand, if the TCM clinical trial is to be conducted by both Chinese doctors and Western clinicians, differences in medical practice and/or possible disagreement regarding the diagnosis, treatment, and evaluation are major concerns.

For the third question, if the TCM is intended for use by Chinese doctors but it is conducted by Western clinicians, differences in perception regarding how to prescribe the TCM are of great concern. The preparation of a package insert based on the clinical data could be a major issue, not only to the sponsor but also to regulatory authorities. Similar comments apply to the situation where the TCM is intended for use by Western clinicians, but the trial is conducted by Chinese doctors.

As a result, it is suggested that the intention of use (i.e., labeling for the indication) be clearly evaluated when planning a TCM clinical trial. In other words, the sponsor needs to determine whether the TCM is intended for use by Western clinicians only, Chinese doctors only, or both Western clinicians and Chinese doctors at the planning stage of a TCM clinical trial, for an adequate package insert of the target diseases under study.

23.3 Basic Considerations

In this section, we describe some basic considerations that are necessary in order to ensure the success of a TCM clinical trial.

23.3.1 Study Design

To demonstrate clinical efficacy and safety of a TCM under investigation, like WMs, it is suggested that a randomized parallel-group, placebo-controlled clinical trial be conducted. However, it may not be ethical if the disease under study is critical and/or life-threatening provided that a WM is available. Alternatively, a randomized placebo-control crossover clinical trial or a parallel-group design consisting of three arms (i.e., the TCM under study, a WM as an active control, and a placebo) is recommended. The three-arm, parallel-group design allows the establishment of non-inferiority/equivalence of the TCM as compared to the active control (WM) and the demonstration of the superiority of the TCM with respect to the placebo. One of the advantages of a crossover clinical trial is that a comparison within each individual can be made, although it will take a longer time to complete the study. Although a crossover design requires a smaller sample size as compared to a parallel-group design, there are some limitations for the use of crossover design. First, baselines prior to dosing may not be the same. Second, when a significant sequence effect is observed, we would not be able to isolate the effects of period effect, carryover effect, and subject-by-treatment effect, which are confounded to one another.

In many cases, factorial designs are used to evaluate the impact of specific components (with respect to the therapeutic effect) by fixing some of the components. For example, we may consider a parallel-group design comparing two treatment groups (one group is treated with the TCM with a specific component, and the other group is treated with the TCM without the specific component). The design of this kind may be useful to identify the most active component of the TCM with respect to the diseases under study. However, it does not address the possible drug-to-drug interactions among the components.

23.3.2 Validation of Quantitative Instrument

In TCM medical practice, a Chinese doctor usually collects information from the patient with a certain disease through the four subjective approaches as described in the previous section. The purpose of these subjective approaches is to collect information on various aspects of the disease under study such as signs, symptoms, patient's performance, and functional activities, so a quantitative instrument with a large number of questions/items is necessary and helpful. For a simple analysis and an easy interpretation, these questions are usually grouped to form subscales, composite scores (domains), or an overall score. The items (or subscales) in each subscale (or composite score) are correlated. As a result, the structure of responses to a quantitative instrument is multidimensional, complex, and correlated. As mentioned above, a standardized quantitative tool (instrument) is necessary to reduce variability from one Chinese doctor to another (prior to the conduct of a clinical trial).

Guilford (1954) discussed several methods such as Cronbach's α for measuring the reliability of internal consistency of a quantitative instrument. Guyatt et al. (1989) indicated that a quantitative instrument should be validated in terms of its validity, reproducibility, and responsiveness. Hollenberg et al. (1991) discussed several methods for validation of a quantitative instrument, such as consensual validation, construct validation, and criterion-related validation. There is, however, no gold standard as to how a quantitative instrument should be validated. In this paper, we will focus on the validation of a quantitative instrument in terms of validity, reliability (or reproducibility), and responsiveness (see, e.g., Chow and Ki, 1994, 1996). As indicated in Chow and Shao (2002), the validity of a quantitative instrument is the extent to which the instrument measures what it is designed to measure. It is a measure of biasedness of the instrument. The biasedness of a quantitative instrument reflects the accuracy of the instrument. The reliability of a quantitative instrument measures the variability of the instrument, which directly relates to the precision of the instrument. On the other hand, the responsiveness of a quantitative instrument is usually referred to as the ability of the instrument to detect a difference of clinical significance within a treatment.

Hsiao et al. (2009) considered a specific design for calibration/validation of the CDP. In the proposed study design, qualified subjects are randomly assigned to receive either a TCM or a WM. Each patient will be evaluated by a Chinese doctor and a Western clinician independently, regardless of which treatment group he/she is in. As a result, there are four groups of data, namely, (1) patients who receive TCM and are evaluated by a Chinese doctor, (2) patients who receive TCM but are evaluated by a Western clinician, (3) patients who receive WM but are evaluated by a Chinese doctor, and (4) patients who receive WM and are evaluated by a Western clinician. Groups (3) and (4) are used to establish a standard curve for calibration between the TCM and the WM. Groups (1) and (2) are then used to validate the CDP based on the established standard curve.

23.3.3 Clinical Endpoint

Unlike WMs, the primary study endpoints for the assessment of safety and effectiveness of a TCM are usually assessed subjectively by a quantitative instrument by experienced Chinese doctors. Although the quantitative instrument is developed by the community of Chinese doctors and is considered a gold standard for the assessment of safety and effectiveness of the TCM under investigation, it may not be accepted by the Western clinicians due to fundamental differences in medical theory, perception, and practice. In practice, it is very difficult for a Western clinician to conceptually understand the clinical meaning of the difference detected by the subjective Chinese quantitative instrument. Consequently, whether the subjective quantitative instrument can accurately and reliably assess the safety and effectiveness of the TCM is always a concern to Western clinicians.

As an example, for the assessment of safety and efficacy of a drug product for the treatment of ischemic stroke, a commonly considered primary clinical endpoint is the functional status assessed by the so-called Barthel index. The Barthel index is a weighted functional assessment scoring technique composed of 10 items with a minimum score of 0 (functional incompetence) and a maximum score of 100 (functional competence). The Barthel index is a weighted scale measuring performance in self-care and mobility, which is widely accepted in ischemic stroke clinical trials. A patient may be considered a responder if his/her Barthel index is greater than or equal to 60. On the other hand, Chinese doctors usually consider a quantitative instrument developed by the Chinese medical community as the standard diagnostic procedure for the assessment of ischemic stroke. The standard quantitative instrument is composed of six domains, which capture different information regarding a patient's performance, functional activities, and signs and symptoms and status of the disease.

In practice, it is of interest to both Western clinicians and Chinese doctors how an observed clinically meaningful difference by the Chinese quantitative instrument can be translated to that of the primary study endpoint assessed by the Barthel index. To reduce the fundamental differences in medical theory/perception and practice, it is suggested that the subjective Chinese quantitative instrument be calibrated and validated with respect to that of the clinical endpoint assessed by the Barthel index before it can be used in TCM ischemic stroke clinical trials.

23.3.4 Matching Placebo

In clinical development, double-blind, placebo-controlled randomized clinical trials are often conducted for the evaluation of the safety and effectiveness of a test treatment under investigation. To maintain blindness, a matching placebo should be identical to the active drug in all aspects of, size, color, coating, taste, texture, shape, and order except that it contains no active ingredient. In clinical trials, as an advanced technique available for formulation, a matching placebo is not difficult to make because most WMs contain a single active ingredient. Unlike WMs, TCMs usually consist of a number of components, which often have different taste. In TCM clinical trials, the TCM under investigation is often encapsulated. However, the test treatment will be easily unblinded if either the patient or the Chinese doctor breaks the capsule. As a result, the preparation of matching placebo in TCM clinical trials not only plays an important role for the success of the TCM clinical trials, but also posts a major-challenging to clinical scientists.

23.3.5 Sample Size Calculation

In clinical trials, sample size is usually selected to achieve a desired power for detecting a clinically meaningful difference in one of the primary study endpoints for the intended indication of the treatment under investigation (see, e.g., Chow et al., 2002b). As a result, sample size calculation depends upon the primary study endpoint and the clinically meaningful difference that one would like to detect. Different primary study endpoints may result in very different sample sizes.

For illustration purpose, consider the example concerning a TCM for the treatment of ischemic stroke, which was developed with more than 30 years of clinical experience with humans. Suppose a sponsor would like to conduct a clinical trial to scientifically evaluate the safety and efficacy of the TCM the Western way as compared to an active control (e.g., aspirin). Thus, the intended clinical trial is a double-blind, parallel-group, placebo-controlled, randomized trial. The primary clinical endpoint is the response rate (a patient is considered a responder if his/her Barthel index is greater than or equal to 60) based on the functional status assessed by the Barthel index. Sample size calculation is performed based on the response rate after 4 weeks of treatment under the hypotheses of testing for superiority. As a result, a sample size of 150 patients per treatment group is required for achieving an 80% power for the establishment of superiority of the TCM over the active control agent. Alternatively, we may consider the quantitative instrument developed by experienced Chinese doctors as the primary study endpoint for sample size calculation. Based on a pilot study, about 80% (79 out of 122) of ischemic stroke patients were diagnosed by one domain of the quantitative instrument. A patient is considered a responder if his/her domain score is greater than or equal to 7. Based on this primary study endpoint, a sample size of 90 per treatment group is required to achieve an 80% power for the establishment of superiority.

The difference in sample size leads to the question of whether the use of the primary endpoint of response rate based on one domain of the Chinese quantitative instrument could provide substantial evidence of safety and effectiveness of the TCM under investigation.

23.4 Controversial Issues

Although TCM has a long history of being used in humans, no scientifically valid documentations are available. As indicated by the United States Food and Drug Administration (FDA), substantial evidence regarding safety and effectiveness of the test treatment under investigation can only be obtained by conducting adequate and well-controlled clinical trials. However, before the test treatment under investigation can be used in humans, sufficient information regarding chemistry, manufacturing, and control (CMC), clinical pharmacology, and toxicology are necessary (see, e.g., Chow and Liu, 1995). Since most TCMs consist of multiple components with unknown pharmacological activities, valid information regarding CMC, clinical pharmacology, and toxicology are difficult to obtain. In what follows, these difficulties are briefly described.

23.4.1 Test for Consistency

As mentioned earlier, unlike most WMs, TCMs usually consist of a number of components. The pharmacological activities, interactions, and relative proportions of these components are usually unknown. In practice, TCM is usually prescribed subjectively by an experienced Chinese doctor. As a result, the actual dose received by each individual varies depending upon the signs and symptoms as perceived by the Chinese doctor. Although the purpose of this medical practice is to reduce the within-subject (or intra-subject) variability, it could also introduce nonnegligible variability such as variations from component to component and from rater to rater (a Chinese doctor to another). Consequently, reproducibility or consistency of clinical results is questionable. Thus, how to ensure the reproducibility or consistency of the observed clinical results has become a great concern to regulatory agencies in the review and approval process. It is also a great concern to the sponsor of the manufacturing process. To address the question of reproducibility or consistency, a valid statistical quality control (QC) process on the raw materials and final product is suggested.

Tse et al. (2006) proposed a statistical QC method to assess a proposed consistency index of raw materials obtained from different resources and/or final product, which may be manufactured at different sites. The consistency index is defined as the probability that the ratio of the characteristics (e.g., extract) of the most active component among the multiple components of a TCM from two different sites (locations) is within a limit of consistency. A consistency index close to 1 indicates that the components from the two sites or locations are almost identical. The idea for testing consistency is to construct a 95% confidence interval for the proposed consistency index under a sampling plan. If the constructed 95% confidence lower limit is greater than a prespecified QC lower limit, then we claim that the raw materials or final product has passed the QC and hence can be released for further process or use. Otherwise, the raw materials and/or final product should be rejected. More details regarding the statistical methods proposed by Tse et al. (2006) are given in the next section.

23.4.2 Animal Studies

The purpose of animal studies is not only to study possible toxicity in animals but also to suggest an appropriate dose for use in humans, assuming that the established animal model is predictive of the human model. For a newly developed drug product, animal studies are necessary. However, for some well-known TCMs, which have been used in humans for years and have a very mild toxicity profile, it is questionable whether animal studies are necessary. It is suggested that all components of TCMs as described in Chinese Pharmacopedia (CP) be classified into several categories depending upon their potential toxicities and/or safety profiles as a basis for regulatory requirements for animal studies. In other words, for some well-known TCM components such as Ginseng, animal studies for testing toxicity may be waived depending upon past experiences of human use, although health risks or side effects following the proper administration of designated therapeutic dosages were not recorded in human use. Note that the German regulatory authority's herbal watchdog agency, commonly called Commission E, has conducted an intensive assessment of the peer-reviewed literature on some 300 common botanicals with respect to the quality of the clinical evidence and the uses for which the herb can be reasonably considered effective (PDR, 1998).

23.4.3 Stability Analysis

Most regulatory agencies require that the expiration dating period (or shelf life) of a drug product must be indicated in the immediate container label before it can be released for use. To fulfill this requirement, stability studies are usually conducted in order to characterize the degradation of the drug product. For drug products with a single active ingredient, statistical methods for determination of drug shelf life are well established (e.g., FDA, 1987; ICH, 1993). However, regulatory requirements for estimation of drug shelf life for drug products with multiple components are not available.

Following the concept of estimating shelf life for drug products with a single active ingredient, two approaches are worth considering. First, we may (conservatively) consider the minimum of the shelf-lives obtained from each component of the drug product. This approach is conservative, and yet may not be feasible due to the fact that (1) not all of the components of a TCM can be accurately and reliably quantitated and (2) the resultant shelf life may be too short to be useful (see, e.g., Pong and Raghavarao, 2002).

Alternatively, we may consider a two-stage approach for determining drug shelf life. In the first stage, an attempt should be made to identify the most active component(s) whenever possible. A shelf life can then be obtained based on the method suggested in the FDA and International Conference on Harmonization (ICH) guidelines. In the second stage, the obtained shelf life is adjusted based on the relationship and/or interactions of the most active ingredient(s) and other components. As an alternative, Chow and Shao (2005) proposed a statistical method for determining the shelf life of a TCM following a similar idea suggested by the FDA, assuming that the components are linear combinations of some factors.

23.4.4 Regulatory Requirements

Although the use of TCMs in humans has a long history, there have been no regulatory requirements regarding the assessment of safety and effectiveness of the TCMs until recently. For example, both regulatory authorities of China and Taiwan have published guidelines/guidances for the clinical development of TCMs (see, e.g., MOPH, 2002; DOH, 2004a,b). In addition, the FDA has also published a guidance for botanical drug products (FDA, 2004). These regulatory requirements for TCM research and development, especially for clinical development, are very similar to well-established guidelines/guidances for pharmaceutical research and development for WMs. It is a concern whether these regulatory requirements and the corresponding statistical methods are feasible for research and development of TCM, based on the fact that there are so many fundamental differences in medical practice, drug administration, and diagnostic procedure. As a result, it is suggested that current regulatory requirements and the corresponding statistical methods should be modified in order to reflect these fundamental differences.

It is strongly recommended that regulatory requirements for the development, review, and approval process for Premarin (conjugated estrogens tablets, United States Pharmacopedia [USP]) be consulted because Premarin is a WM consisting of multiple components which are similar to a TCM (FDA, 1991; Liu and Chow, 1996). Premarin, which contains multiple components of estrone, equilin, 17α-dihydroequilin, 17α-estradiol, and 17β-dihydroequilin, is intended for treatment of moderate to severe vasomotor symptoms associated with menopause. The experience with Premarin is helpful in developing appropriate guidelines/guidances for TCM drug products with multiple components.

23.4.5 Indication and Label

As indicated earlier, it is very important to clarify the intention for the use of a TCM (by Chinese doctors alone, Western clinicians alone, or both Chinese doctors and Western clinicians) once it is approved by the regulatory agencies. If a TCM is intended for use by Chinese doctors alone, the clinical trials conducted for obtaining substantial evidence should reflect the medical theory of TCM and the medical practice of Chinese doctors. The label should provide sufficient information as to how to prescribe the TCM the Chinese way. On the other hand, if the TCM under investigation is intended for use by Western clinicians alone, patients under study should be evaluated based on clinical study endpoints for safety and efficacy the Western way. Consequently, the label should provide sufficient information for prescribing the TCM the Western way. If the TCM is intended for both Western clinicians and Chinese doctors, patients are necessarily evaluated by both Western clinical study endpoints and CDPs (e.g., some standardized quantitative instrument) provided that the CDP has been calibrated and validated against the well-established Western clinical endpoint. In this case, there is a clear understanding of how an observed difference by CDP can be translated to a clinical effect which is familiar to Western clinicians, and vice versa.

23.5 Recent Development

23.5.1 Statistical Quality Control Method for Assessing Consistency

Tse et al. (2006) proposed a statistical QC method to assess a proposed consistency index of raw materials obtained from different resources and/or final product, which may be manufactured at different sites. The idea is to construct a 95% confidence interval for a proposed consistency index under a sampling plan. If the constructed 95% confidence lower limit is greater than a prespecified QC lower limit, then we claim that the raw materials and/or final product have passed the QC and hence can be released for further process or use. Otherwise, the raw materials and/or final product should be rejected. For a given component (the most active component if possible), a sampling plan is derived to ensure that there is a desired probability for establishing consistency between sites when truly there is no difference in raw materials or final products between sites. The statistical QC method for the assessment of consistency proposed by Tse et al. (2006) is described below.

Let U and W be the characteristics of the most active component among the multiple components of a TCM from two different sites, where $X = \log U$ and $Y = \log W$ follows normal distributions with means μ_X, μ_Y and variances V_X, V_Y, respectively. Similar to the idea of using $P(X < Y)$ to assess reliability in statistical QC (Church and Harris, 1970; Enis and Geisser, 1971), Tse et al. (2006) propose the following probability as an index to assess the consistency of raw materials and/or final product from two different sites

$$p = P\left(1 - \delta < \frac{U}{W} < \frac{1}{1-\delta}\right), \tag{23.1}$$

where $0 < \delta < 1$ and is defined as a limit that allows for consistency. Tse et al. (2006) refer p as the consistency index. Thus, p tends to 1 as δ tends to 1. For a given δ, if p is close to 1, materials U and W are considered to be identical. It should be noted that a small δ implies the requirement of a high degree of consistency between material U and material W. In practice, it may be difficult to meet this narrow specification for consistency. Under the normality assumption of $X = \log U$ and $Y = \log W$, (23.1) can be rewritten as

$$p = P(\log(1-\delta) < \log U - \log W < -\log(1-\delta))$$

$$= \Phi\left(\frac{-\log(1-\delta) - (\mu_X - \mu_Y)}{\sqrt{V_X + V_Y}}\right) - \Phi\left(\frac{\log(1-\delta) - (\mu_X - \mu_Y)}{\sqrt{V_X + V_Y}}\right).$$

where $\Phi(z_0) = P(Z < z_0)$ with Z being a standard normal random variable. Therefore, the consistency index p is a function of the parameters $\theta = (\mu_X, \mu_Y, V_X, V_Y)$, i.e., $p = h(\theta)$. Suppose that observations $X_i = \log U_i, i = 1, \ldots, n_X$ and $Y_i = \log W_i, i = 1, \ldots, n_Y$ are collected in an assay study. Then, using the invariance principle, the maximum likelihood estimator (MLE) of p can be obtained as

$$\hat{p} = \Phi\left(\frac{-\log(1-\delta) - (\bar{X} - \bar{Y})}{\sqrt{\hat{V}_X + \hat{V}_Y}}\right) - \Phi\left(\frac{\log(1-\delta) - (\bar{X} - \bar{Y})}{\sqrt{\hat{V}_X + \hat{V}_Y}}\right), \tag{23.2}$$

where $\bar{X} = (1/n_X)\sum_{i=1}^{n_X} X_i$, $\bar{Y} = (1/n_Y)\sum_{i=1}^{n_Y} Y_i$, $\hat{V}_X = (1/n_X)\sum_{i=1}^{n_X}\left(X_i - \bar{X}\right)^2$, and $\hat{V}_Y = (1/n_Y)\sum_{i=1}^{n_Y}(Y_i - \bar{Y})^2$. In other words, $\hat{p} = h(\hat{\theta}) = h(\bar{X}, \bar{Y}, \hat{V}_X, \hat{V}_Y)$. Furthermore, it can be easily verified that the following asymptotic result holds.

Theorem 23.1

$\hat{p}$ as given in (23.2) is asymptotically normal with mean $E(\hat{p})$ and variance $\mathrm{Var}(\hat{p})$. In other words,

$$\frac{\hat{p} - E(\hat{p})}{\sqrt{\mathrm{var}(\hat{p})}} \rightarrow N(0,1), \tag{23.3}$$

where
$$E(\hat{p}) = p + B(p) + o(1/n),$$
$$\text{Var}(\hat{p}) = C(p) + o(1/n).$$

The detailed expressions of $B(p)$ and $C(p)$ are given in the proof below.

Proof

Based on the definitions of $\bar{X}$ and $\hat{V}_X$, it is easy to show that

$$E(\bar{X}) = \mu_X, \quad E(\hat{V}_X) = \frac{n_X - 1}{n_X} V_X,$$

$$\text{var}(\bar{X}) = \frac{V_X}{n_X} \quad \text{and} \quad \text{var}(\hat{V}_X) = \frac{2(n_X - 1)}{n_X^2} V_X^2.$$

Similarly,

$$E(\bar{Y}) = \mu_Y, \quad E(\hat{V}_Y) = \frac{n_Y - 1}{n_Y} V_Y,$$

$$\text{var}(\bar{Y}) = \frac{V_Y}{n_Y} \quad \text{and} \quad \text{var}(\hat{V}_Y) = \frac{2(n_Y - 1)}{n_Y^2} V_Y^2.$$

Applying expansion of $\hat{p}$ at p, we have

$$\hat{p} = p + \frac{\partial \hat{p}}{\partial \mu_X}(\bar{X} - \mu_X) + \frac{\partial \hat{p}}{\partial \mu_Y}(\bar{Y} - \mu_Y)$$

$$+ \frac{\partial \hat{p}}{\partial V_X}(\hat{V}_X - V_X) + \frac{\partial \hat{p}}{\partial V_Y}(\hat{V}_Y - V_Y)$$

$$+ \frac{1}{2}\left[\frac{\partial^2 \hat{p}}{\partial \mu_X^2}(\bar{X} - \mu_X)^2 + \frac{\partial^2 \hat{p}}{\partial \mu_Y^2}(\bar{Y} - \mu_Y)^2 + \frac{\partial^2 \hat{p}}{\partial V_X^2}(\hat{V}_X - V_X)^2 + \frac{\partial^2 \hat{p}}{\partial V_Y^2}(\hat{V}_Y - V_Y)^2 \right] + \cdots$$

The other second-order partial derivatives are not considered because they will lead to expected values of order $O(n^{-2})$ or higher. Taking expectation

$$E(\hat{p}) = p + \frac{1}{2}\left[\frac{\partial^2 \hat{p}}{\partial \mu_X^2}\frac{V_X}{n_X} + \frac{\partial^2 \hat{p}}{\partial \mu_Y^2}\frac{V_Y}{n_Y} + \frac{\partial^2 \hat{p}}{\partial V_X^2}\left(\frac{2V_X^2}{n_X}\right) + \frac{\partial^2 \hat{p}}{\partial V_Y^2}\left(\frac{2V_Y^2}{n_Y}\right) \right] + O(n^{-2})$$

and

$$\text{var}(\hat{p}) = \left[\left(\frac{\partial \hat{p}}{\partial \mu_X}\right)^2 \frac{V_X}{n_X} + \left(\frac{\partial \hat{p}}{\partial \mu_Y}\right)^2 \frac{V_Y}{n_Y} + \left(\frac{\partial \hat{p}}{\partial V_X}\right)^2 \left(\frac{2V_X^2}{n_X}\right) + \left(\frac{\partial \hat{p}}{\partial V_Y}\right)^2 \left(\frac{2V_Y^2}{n_Y}\right) \right] + O(n^{-2}).$$

Therefore,

$$B(p) = \frac{1}{2}\left[\frac{\partial^2 \hat{p}}{\partial \mu_X^2}\frac{V_X}{n_X} + \frac{\partial^2 \hat{p}}{\partial \mu_Y^2}\frac{V_Y}{n_Y} + \frac{\partial^2 \hat{p}}{\partial V_X^2}\left(\frac{2V_X^2}{n_X}\right) + \frac{\partial^2 \hat{p}}{\partial V_Y^2}\left(\frac{2V_Y^2}{n_Y}\right)\right],$$

and

$$C(p) = \left[\left(\frac{\partial \hat{p}}{\partial \mu_X}\right)^2\frac{V_X}{n_X} + \left(\frac{\partial \hat{p}}{\partial \mu_Y}\right)^2\frac{V_Y}{n_Y} + \left(\frac{\partial \hat{p}}{\partial V_X}\right)^2\left(\frac{2V_X^2}{n_X}\right) + \left(\frac{\partial \hat{p}}{\partial V_Y}\right)^2\left(\frac{2V_Y^2}{n_Y}\right)\right].$$

For the sake of simplicity, denote

$$z_1 = \frac{\log(1-\delta)-(\mu_X-\mu_Y)}{\sqrt{V_X+V_Y}}, \quad z_2 = \frac{-\log(1-\delta)-(\mu_X-\mu_Y)}{\sqrt{V_X+V_Y}}$$

and

$$\phi(z) = \frac{1}{\sqrt{2\pi}}\exp\left(-\frac{z^2}{2}\right).$$

Then after some algebra, the partial derivatives are given as

$$\frac{\partial \hat{p}}{\partial \mu_X} = -\frac{\partial \hat{p}}{\partial \mu_Y} = \left(\frac{-1}{\sqrt{V_X+V_Y}}\right)[\varphi(z_2)-\varphi(z_1)],$$

$$\frac{\partial \hat{p}}{\partial V_X} = \frac{\partial \hat{p}}{\partial V_Y} = \left(\frac{-1}{2\sqrt{V_X+V_Y}}\right)[z_2\varphi(z_2)-z_1\varphi(z_1)],$$

$$\frac{\partial^2 \hat{p}}{\partial \mu_X^2} = \frac{\partial^2 \hat{p}}{\partial \mu_Y^2} = \left(\frac{-1}{V_X+V_Y}\right)[z_2\phi(z_2)-z_1\phi(z_1)]$$

and

$$\frac{\partial^2 \hat{p}}{\partial V_X^2} = \frac{\partial^2 \hat{p}}{\partial V_Y^2} = \frac{1}{4(V_X+V_Y)^{3/2}}[(2z_2-z_2^3)\varphi(z_2)-(2z_1-z_1^3)z_1\varphi(z_1)].$$

This completes the proof.

Based on the result of Theorem 23.1, an approximate $(1 - \alpha)100\%$ confidence interval for p, i.e., $(LL(\hat{p}), UL(\hat{p}))$, can be obtained. In particular,

$$LL(\hat{p}) = \hat{p} - B(\hat{p}) - z_{\alpha/2}\sqrt{C(\hat{p})} \quad \text{and} \quad UL(\hat{p}) = \hat{p} - B(\hat{p}) + z_{\alpha/2}\sqrt{C(\hat{p})}, \qquad (23.4)$$

where z_α is the upper α percentile of a standard normal distribution.

For a valid statistical QC process, a testing procedure is necessarily performed according to some prespecified acceptance criteria under a sampling plan. In this section, we propose a statistical QC method for assessing the consistency of raw materials and/or final product of TCM. The idea is to construct a 95% confidence interval for a proposed consistency index described above under a sampling plan. If the constructed 95% confidence lower limit is greater than a prespecified QC lower limit, then we claim that the raw material or final product has passed the QC and hence can be released for further processing or use. Otherwise, the raw materials and/or final product should be rejected. For a given component (the most active component if possible), a sampling plan is derived to ensure that there is a desired probability for establishing consistency between sites when truly there is no difference in raw materials or final products between sites. In what follows, details regarding the choice of acceptance criteria, sampling plan and the corresponding testing procedure are briefly outlined.

23.5.1.1 Acceptance Criteria

In terms of consistency, we propose the following QC criterion. If the probability that the lower limit $LL(\hat{p})$ of the constructed $(1 - \alpha)100\%$ confidence interval of p is greater than or equal to a prespecified QC lower limit, say, QC_L, and exceeds a prespecified number β (say $\beta = 80\%$), then we claim that U and W are consistent or similar. In other words, U and W are consistent or similar if $P(QC_L \leq LL(\hat{p})) \geq \beta$, where β is a prespecified constant.

23.5.1.2 Sampling Plan

In practice, it is necessary to select a sample size to ensure that there is a high probability, say β, of consistency between U and W when in fact U and W are consistent. It is suggested that the sample size is chosen such that there is more than 80% chance that the lower confidence limit of p is greater than or equal to the QC lower limit, i.e., $\beta = 0.8$. In other words, the sample size is determined such that

$$P\{QC_L \leq LL(\hat{p})\} \geq \beta. \qquad (23.5)$$

Using (23.5), this leads to

$$P\left\{QC_L \leq \hat{p} - B(\hat{p}) - z_{\alpha/2}\sqrt{\text{Var}(\hat{p})}\right\} \geq \beta.$$

Thus,

$$P\left\{QC_L + z_{\alpha/2}\sqrt{\mathrm{Var}(\hat{p})} - p \le \hat{p} - p - B(p)\right\} \ge \beta.$$

This gives

$$P\left\{\frac{QC_L - p}{\sqrt{\mathrm{Var}(\hat{p})}} + z_{\alpha/2} \le \frac{\hat{p} - p - B(p)}{\sqrt{\mathrm{Var}(\hat{p})}}\right\} \ge \beta.$$

Therefore, the sample size required for achieving a probability higher than β can be obtained by solving the following equation:

$$\frac{QC_L - p}{\sqrt{\mathrm{Var}(\hat{p})}} + z_{\alpha/2} \le -z_{1-\beta}. \tag{23.6}$$

Assuming that $n_X = n_Y = n$, the common sample size is given by

$$n \ge \frac{(z_{1-\beta} + z_{\alpha/2})^2}{(p - QC_L)^2}\left\{\left(\frac{\partial\hat{p}}{\partial\mu_X}\right)^2 V_X + \left(\frac{\partial\hat{p}}{\partial\mu_Y}\right)^2 V_Y + \left(\frac{\partial\hat{p}}{\partial V_X}\right)^2 (2V_X^2) + \left(\frac{\partial\hat{p}}{\partial V_Y}\right)^2 (2V_Y^2)\right\}. \tag{23.7}$$

The above result suggests that the required sample size will depend on the choices of α, β, V_X, V_Y, $\mu_X - \mu_Y$, and $p - QC_L$. It is clear from the expression in (23.7) that a larger sample size is required for smaller α and larger β, i.e., the interval is expected to have high confidence level $(1 - \alpha)$ and high chance that the lower confidence limit is larger than QC_L. Furthermore, if we require the QC_L to be close to p, i.e., $p - QC_L$ is small, a relatively large sample size is required. The dependence of the sample size n on the other parameters V_X, V_Y, and $\mu_X - \mu_Y$ is relatively unclear because these parameters are linked to the corresponding partial derivatives. A numerical study is conducted to explore the pattern. Given the large number of parameters involved in equation (23.7), it is impractical to list the value of n for all the parameter combinations. However, for illustration purpose, we only consider a certain combination of parameter values in an attempt to explore the pattern of dependence of n on the parameters. For the sake of simplicity, define

$$S = \frac{1}{(p - QC_L)^2}\left\{\left(\frac{\partial\hat{p}}{\partial\mu_X}\right)^2 V_X + \left(\frac{\partial\hat{p}}{\partial\mu_Y}\right)^2 V_Y + \left(\frac{\partial\hat{p}}{\partial V_X}\right)^2 (2V_X^2) + \left(\frac{\partial\hat{p}}{\partial V_Y}\right)^2 (2V_Y^2)\right\}.$$

Then, for given choices of α and β, the required sample size n is equal to $(z_{1-\beta} + z_{\alpha/2})^2 S$. In particular, in our study, $\delta = 0.10, 0.15$, and 0.20; $\mu_X - \mu_Y = 0.5$, 1.0, and 1.5; $p - QC_L = 0.02, 0.05$, and 0.08. V_X is chosen to be 1 and $V_Y = 0.2$, 0.5, 1.0, 2.0, and 5.0. For each combination of these parameter values, the corresponding value of S is listed in Table 23.2. Given the number of parameters involved and the complexity of the mathematical expression of S, it is not easy to detect a general pattern. However, in general, the results suggest that S increases as $\mu_X - \mu_Y$ decreases, and as the variances V_X and V_Y differ more from each other. In other words, a smaller sample size is required if the difference between the population means is large or the variability of the two sites are of similar magnitude.

As an illustration, if for a study with $\delta = 0.2$, $V_X = 1$, $V_Y = 0.5$, $\mu_X - \mu_Y = 1.0$, an experiment $p - QC_L$ is expected to be not larger than 0.05, then the results in Table 23.2 suggest that $S = 3.024$. Suppose a probability higher than $\beta = 0.8$ at the $\alpha = 0.05$ level of significance is required, the corresponding required sample size is given by

$$n \geq (z_{1-0.8} + z_{0.05/2})^2 S = (0.842 + 1.96)^2 (3.024) = 23.74,$$

i.e., a sample of size of at least 24 is required.

23.5.1.3 Testing Procedure

Hypotheses testing of the consistency index p can also be conducted based on the asymptotic normality of $\hat{p}$. Consider the following hypotheses:

$$H_0 : p \leq p_0 \quad \text{versus} \quad H_a : p > p_0.$$

We would reject the null hypothesis in favor of the alternative hypothesis of consistency. Under H_0, we have

$$\frac{\hat{p} - p_0 - B(\hat{p})}{\sqrt{\operatorname{var}(\hat{p})}} \sim N(0,1). \tag{23.8}$$

Thus, we reject the null hypothesis H_0 at the α level of significance if

$$\frac{\hat{p} - p_0 - B(\hat{p})}{\sqrt{\operatorname{var}(\hat{p})}} > Z_\alpha.$$

This is equivalent to rejecting the null hypothesis H_0 when

$$\hat{p} > p_0 + B(\hat{p}) + Z_\alpha \sqrt{\operatorname{var}(\hat{p})}.$$

Again, for illustration purpose, Table 23.3 provides critical values of the proposed test for consistency index for various combinations of parameters.

TABLE 23.2

Values of $n/(z_{1-\beta} + z_{\alpha/2})^2$, Where n Is the Required Sample Size

		$\delta = 0.10$			$\delta = 0.15$			$\delta = 0.20$		
		$\Delta = 0.5$	$\Delta = 1.0$	$\Delta = 1.5$	$\Delta = 0.5$	$\Delta = 1.0$	$\Delta = 1.5$	$\Delta = 0.5$	$\Delta = 1.0$	$\Delta = 1.5$
$D = 0.02$	$V_Y = 0.2$	5.693	5.376	4.955	13.403	12.681	11.702	24.861	23.594	21.810
	0.5	4.518	4.289	4.196	10.655	10.134	9.921	19.820	18.901	18.520
	1.0	3.939	3.336	3.237	9.310	7.894	7.662	17.370	14.761	14.333
	2.0	4.231	2.962	2.226	10.020	7.021	5.280	18.756	13.163	9.906
	5.0	5.728	4.159	2.469	13.595	9.876	5.866	25.534	18.558	11.032
$D = 0.05$	0.2	0.911	0.860	0.793	2.144	2.029	1.872	3.978	3.775	3.490
	0.5	0.723	0.686	0.671	1.705	1.622	1.587	3.171	3.024	2.963
	1.0	0.630	0.534	0.518	1.490	1.263	1.226	2.779	2.362	2.293
	2.0	0.677	0.474	0.356	1.603	1.123	0.845	3.001	2.106	1.585
	5.0	0.916	0.666	0.395	2.175	1.580	0.939	4.085	2.969	1.765
$D = 0.08$	0.2	0.356	0.336	0.310	0.838	0.793	0.731	1.554	1.475	1.363
	0.5	0.282	0.268	0.262	0.666	0.633	0.620	1.239	1.181	1.158
	1.0	0.246	0.208	0.202	0.582	0.493	0.479	1.086	0.923	0.896
	2.0	0.264	0.185	0.139	0.626	0.439	0.330	1.172	0.823	0.619
	5.0	0.358	0.260	0.154	0.850	0.617	0.367	1.596	1.160	0.690

Notation: $\Delta = \mu_X - \mu_Y$, $D = p - QC_L$.

TABLE 23.3

Critical Values of the Proposed Test for Consistency Index p_0

p_0	δ	V_Y	$\Delta = 0.5$			$\Delta = 1.0$			$\Delta = 1.5$		
			$n = 15$	$n = 30$	$n = 50$	$n = 15$	$n = 30$	$n = 50$	$n = 15$	$n = 30$	$n = 50$
0.75	0.10	0.2	0.7695	0.7640	0.7609	0.7683	0.7632	0.7604	0.7680	0.7629	0.7601
		0.5	0.7673	0.7624	0.7597	0.7665	0.7619	0.7593	0.7665	0.7619	0.7593
		1.0	0.7662	0.7616	0.7590	0.7646	0.7605	0.7582	0.7645	0.7604	0.7581
		2.0	0.7668	0.7620	0.7594	0.7639	0.7600	0.7578	0.7620	0.7586	0.7567
		5.0	0.7697	0.7640	0.7609	0.7667	0.7619	0.7593	0.7628	0.7592	0.7572
	0.20	0.2	0.7907	0.7791	0.7727	0.7884	0.7777	0.7717	0.7878	0.7771	0.7712
		0.5	0.7863	0.7760	0.7703	0.7846	0.7749	0.7695	0.7847	0.7749	0.7695
		1.0	0.7839	0.7743	0.7689	0.7807	0.7721	0.7673	0.7805	0.7719	0.7671
		2.0	0.7853	0.7753	0.7697	0.7793	0.7710	0.7664	0.7754	0.7682	0.7642
		5.0	0.7915	0.7797	0.7731	0.7853	0.7752	0.7697	0.7771	0.7694	0.7651
0.85	0.10	0.2	0.8695	0.8640	0.8609	0.8683	0.8632	0.8604	0.8680	0.8629	0.8601
		0.5	0.8673	0.8624	0.8597	0.8665	0.8619	0.8593	0.8665	0.8619	0.8593
		1.0	0.8662	0.8616	0.8590	0.8646	0.8605	0.8582	0.8645	0.8604	0.8581
		2.0	0.8668	0.8620	0.8594	0.8639	0.8600	0.8578	0.8620	0.8586	0.8567
		5.0	0.8697	0.8640	0.8609	0.8667	0.8619	0.8593	0.8628	0.8592	0.8572
	0.20	0.2	0.8907	0.8791	0.8727	0.8884	0.8777	0.8717	0.8878	0.8771	0.8712
		0.5	0.8863	0.8760	0.8703	0.8846	0.8749	0.8695	0.8847	0.8749	0.8695
		1.0	0.8839	0.8743	0.8689	0.8807	0.8721	0.8673	0.8805	0.8719	0.8671
		2.0	0.8853	0.8753	0.8697	0.8793	0.8710	0.8664	0.8754	0.8682	0.8642
		5.0	0.8915	0.8797	0.8731	0.8853	0.8752	0.8697	0.8771	0.8694	0.8651

(continued)

TABLE 23.3 (continued)

Critical Values of the Proposed Test for Consistency Index p_0

			$\Delta = 0.5$			$\Delta = 1.0$			$\Delta = 1.5$		
p_0	δ	V_Y	$n = 15$	$n = 30$	$n = 50$	$n = 15$	$n = 30$	$n = 50$	$n = 15$	$n = 30$	$n = 50$
0.90	0.10	0.2	0.9195	0.9140	0.9109	0.9183	0.9132	0.9104	0.9180	0.9129	0.9101
		0.5	0.9173	0.9124	0.9097	0.9165	0.9119	0.9093	0.9165	0.9119	0.9093
		1.0	0.9162	0.9116	0.9090	0.9146	0.9105	0.9082	0.9145	0.9104	0.9081
		2.0	0.9168	0.9120	0.9094	0.9139	0.9100	0.9078	0.9120	0.9086	0.9067
		5.0	0.9197	0.9140	0.9109	0.9167	0.9119	0.9093	0.9128	0.9092	0.9072
	0.20	0.2	0.9407	0.9291	0.9227	0.9384	0.9277	0.9217	0.9378	0.9271	0.9212
		0.5	0.9363	0.9260	0.9203	0.9346	0.9249	0.9195	0.9347	0.9249	0.9195
		1.0	0.9339	0.9243	0.9189	0.9307	0.9221	0.9173	0.9305	0.9219	0.9171
		2.0	0.9353	0.9253	0.9197	0.9293	0.9210	0.9164	0.9254	0.9182	0.9142
		5.0	0.9415	0.9297	0.9231	0.9353	0.9252	0.9197	0.9271	0.9194	0.9151

Notation: $\Delta = \mu_X - \mu_Y$.

In particular, $\alpha = 0.1$, $p_0 = 0.75$, 0.85, and 0.9, $\delta = 0.10$ and 0.20; $\mu_X - \mu_Y = 0.5$, 1.0, and 1.5. V_X is chosen to be 1 and $V_Y = 0.2$, 0.5, 1.0, 2.0, and 5.0. Note that the critical value is closer to the corresponding p_0 either for larger sample size n, smaller δ, or smaller $\mu_X - \mu_Y$.

23.5.1.4 Strategy for Statistical Quality Control

In practice, raw materials, in-process materials, and/or final products at different sites are manufactured sequentially in batches or lots. As a result, it is important to perform statistical QC on batches. A typical approach is to randomly select samples from several (consecutive) batches for testing. In this case, observations from the study would be subject to batch-to-batch variability. For the sake of administrative convenience, it is common to have an equal number of observations from the batches. Consider the following model:

$$X_{ij} = \mu_X + A_i^X + \varepsilon_{ij}^X, \quad i = 1, \ldots, m_X; \quad j = 1, \ldots, n_X,$$

where

A_i^X accounts for the batch-to-batch variability for the observations collected in site 1 and is normally distributed with mean 0 and variance σ_{b1}^2,
m_X is the number of batches collected in the study at site 1,
ε_{ij}^X are normal random variables with mean 0 and variance σ_1^2.

Similarly,

$$Y_{ij} = \mu_Y + A_i^Y + \varepsilon_{ij}^Y, \quad i = 1, \ldots, m_Y; \quad j = 1, \ldots, n_Y,$$

where

A_i^Y accounts for the batch-to-batch variability of the observations collected in site 2 and is normally distributed with mean 0 and variance σ_{b2}^2,
m_Y is the number of batches collected in the study at site 2,
ε_{ij}^Y are normal random variables with mean 0 and variance σ_2^2.

Therefore, the total variability of the most active component at the two sites are given by $\operatorname{var} X = V_X = \sigma_{b1}^2 + \sigma_1^2$ and $\operatorname{var} Y = V_Y = \sigma_{b2}^2 + \sigma_2^2$, respectively. Furthermore, let

$$\bar{X}_{i.} = \frac{1}{n_X} \sum_{j=1}^{n_X} X_{ij} \quad \text{and} \quad \bar{X} = \frac{1}{m_X} \sum_{i=1}^{m_X} \bar{X}_{i..}$$

Then, the observed sums of squares are

$$\mathrm{SSA}_1 = n_X \sum_{i=1}^{m_X} (\bar{X}_{i.} - \bar{X})^2, \quad \mathrm{SSE}_1 = \sum_{i=1}^{m_X} \sum_{j=1}^{n_X} (X_{ij} - \bar{X}_{i.})^2$$

and

$$SST_1 = SSA_1 + SSE_1.$$

Following the results in Chow and Tse (1991), the MLE of σ_{b1}^2 and σ_1^2 are

$$\hat{\sigma}_{b1}^2 = \begin{cases} \dfrac{1}{n_X}\left(\dfrac{1}{m_X}SSA_1 - \dfrac{1}{m_X(n_X-1)}SSE_1\right) & \dfrac{1}{m_X}SSA_1 \geq \dfrac{1}{m_X(n_X-1)}SSE_1 \\[2em] & \text{if} \\[2em] 0 & \dfrac{1}{m_X}SSA_1 < \dfrac{1}{m_X(n_X-1)}SSE_1 \end{cases}$$

(23.9)

and

$$\hat{\sigma}_1^2 = \begin{cases} \dfrac{1}{m_X(n_X-1)}SSE_1 & \dfrac{1}{m_X}SSA_1 \geq \dfrac{1}{m_X(n_X-1)}SSE_1 \\[2em] \text{if} & \\[2em] \dfrac{1}{n_X m_X}SST_1 & \dfrac{1}{m_X}SSA_1 < \dfrac{1}{m_X(n_X-1)}SSE_1 \end{cases}$$

(23.10)

Furthermore, the MLE of the total variability V_X is given by $\hat{V}_X = 1/(n_X m_X)$ SST_1. The MLE of σ_{b2}^2, σ_2^2, and V_Y, denoted by $\hat{\sigma}_{b2}^2$, $\hat{\sigma}_2^2$, and $\hat{V}_Y$, respectively, can be obtained in a similar way by using observations Y_{ij}. Comparison of the estimates $\hat{\sigma}_{b2}^2$ and $\hat{\sigma}_{b1}^2$ would give an idea of the magnitude of the batch-to-batch variability at the two sites.

23.5.1.5 Remarks

Note that the method proposed by Tse et al. (2006) only focuses on a single (i.e., the most active) component assuming that the most active component can be quantitatively identified among multiple active components. Following a similar idea, Lu et al. (2007) extended their results to the case of two correlative components by considering p_1 and p_2, the consistency indices of the two most active components of a TCM from two different sites. Lu et al. (2007) proposed to define the consistency index of a TCM with two correlative components by $\min(p_1, p_2)$ and denote it by p; where

$$p_i = P\left(1 - \delta_i < \frac{U_i}{W_i} < \frac{1}{1 - \delta_i}\right), \quad 0 < \delta_i < 1, \quad i = 1, 2$$

and δ_i is a limit that allows for consistency. Therefore, the consistency index p is a function of the parameter $\theta = (\mu_{X_1}, \mu_{X_2}, \mu_{Y_1}, \mu_{Y_2}, V_{X_1}, V_{X_2}, V_{Y_1}, V_{Y_2})$, i.e., $p = h(\theta)$. By invariance principle, the MLE of p_1 and p_2 are given by

$$\hat{p}_i = \Phi\left(\frac{-\log(1-\delta_i)-(\bar{X}_i-\bar{Y}_i)}{\sqrt{\hat{V}_{X_i}+\hat{V}_{Y_i}}}\right) - \Phi\left(\frac{\log(1-\delta_i)-(\bar{X}_i-\bar{Y}_i)}{\sqrt{\hat{V}_{X_i}+\hat{V}_{Y_i}}}\right), \qquad (23.11)$$

where $\Phi(z_0) = P(Z < z_0)$ with Z being a standard normal random variable,

$$\bar{X}_i = \frac{1}{n}\sum_{j=1}^{n} X_{ij}, \quad \bar{Y}_i = \frac{1}{n}\sum_{j=1}^{n} Y_{ij},$$

and

$$\hat{V}_{X_i} = \frac{1}{n}\sum_{j=1}^{n}(X_{ij}-\bar{X}_i)^2, \quad \hat{V}_{Y_i} = \frac{1}{n}\sum_{j=1}^{n}(Y_{ij}-\bar{Y}_i)^2, \quad i = 1,2.$$

Thus, the MLE of the proposed consistency index p is given by $\hat{p} = \min(\hat{p}_1, \hat{p}_2)$. Furthermore, it can be verified that the following asymptotic result holds (see also Lu et al., 2007).

Theorem 23.2

$\log \hat{p}$ as given in (23.1) with mean $E(\log \hat{p})$ and variance $\mathrm{Var}(\log \hat{p})$, where $E(\log \hat{p}) = \log p + B(p) + o(n^{-1})$ and $\mathrm{Var}(\log \hat{p}) = C(p) + o(n^{-1})$. The detailed expressions of $B(p)$ and $C(p)$ are given in the Appendix. Furthermore,

$$\frac{\log \hat{p} - \log p - B(\hat{p})}{\sqrt{C(\hat{p})}} \to N(0,1),$$

where $B(\hat{p})$ and $C(\hat{p})$ are estimates of $B(p)$ and $C(p)$ with the unknown population parameter $\theta = (\mu_{X_1}, \mu_{X_2}, \mu_{Y_1}, \mu_{Y_2}, V_{X_1}, V_{X_2}, V_{Y_1}, V_{Y_2})$ estimated by their corresponding MLEs $\hat{\theta} = (\bar{X}_1, \bar{X}_2, \bar{Y}_1, \bar{Y}_2, \hat{V}_{X_1}, \hat{V}_{X_2}, \hat{V}_{Y_1}, \hat{V}_{Y_2})$.

Proof
The details of the derivation of $B(p)$ and $C(p)$ can be found in Tse et al. (2006). In particular,

$$B(p) = \frac{1}{np_k}\frac{\partial^2 \hat{p}_k}{\partial V_{X_k}^2}(V_{X_k}^2 + V_{Y_k}^2) - \frac{1}{2np_k^2}\left[\left(\frac{\partial \hat{p}_k}{\partial \mu_{X_k}}\right)^2(V_{X_k}+V_{Y_k})+2\left(\frac{\partial \hat{p}_k}{\partial V_{X_k}}\right)^2(V_{X_k}^2+V_{Y_k}^2)\right]$$

and

$$C(p) = \frac{1}{np_k^2}\left[\left(\frac{\partial \hat{p}_k}{\partial \mu_{X_k}}\right)^2 (V_{X_n} + V_{Y_n}) + 2\left(\frac{\partial \hat{p}_k}{\partial V_{X_k}}\right)^2 (V_{X_k}^2 + V_{Y_k}^2)\right];$$

where the subscript k is defined by $k = j$ if $\hat{p} = \hat{p}_j$, $j = 1$ or 2. Note that $B(p)$ converges to 0 as n tends to infinity. Thus, $\hat{p}$ is asymptotically unbiased. Since $\hat{\theta} = (\bar{X}_1, \bar{X}_2, \bar{Y}_1, \bar{Y}_2, \hat{V}_{X_1}, \hat{V}_{X_2}, \hat{V}_{Y_1}, \hat{V}_{Y_2})$ is asymptotically multivariate normally distributed and $\hat{p}$ is a function of $\hat{\theta}$, it follows from Serfling (1980) that $\log \hat{p} - E(\log \hat{p})/\sqrt{\mathrm{var}(\log \hat{p})} \to N(0,1)$. Using Slutsky's theorem, it can be shown that $\log \hat{p} - \log p - B(\hat{p})/\sqrt{C(\hat{p})}$ is asymptotically normal since $B(\hat{p})$ and $C(\hat{p})$ are consistent estimates of $B(p)$ and $C(p)$, respectively.

Based on the results given in Theorem 23.2, for a given level $0 < \alpha < 1$, an approximate $(1-\alpha)100\%$ confidence interval for $\log p$, denoted by $(LL(\log \hat{p})$, $UL(\log \hat{p}))$, can be obtained based on (23.4). In particular,

$$LL(\log \hat{p}) = \log \hat{p} - B(\hat{p}) - z_{\alpha/2}\sqrt{C(\hat{p})}, \tag{23.12}$$

and

$$UL(\log \hat{p}) = \log \hat{p} - B(\hat{p}) + z_{\alpha/2}\sqrt{C(\hat{p})}, \tag{23.13}$$

where $z_{\alpha/2}$ is the upper $\alpha/2$-percentile of the standard normal distribution. Consequently, an approximate $(1-\alpha)100\%$ confidence interval for p, denoted by $(LL(\hat{p}), UL(\hat{p}))$, is given as

$$(e^{LL(\log \hat{p})}, e^{UL(\log \hat{p})}). \tag{23.14}$$

23.5.2 Stability Analysis for TCM

In the pharmaceutical industry, stability analysis refers to a study conducted for determining the expiration dating period (shelf life) of a drug product under appropriate storage conditions. The shelf life of a drug is defined as the time interval in which the potency of the drug remains within the approved specification limit, e.g., the specification limit given in the United States Pharmacopedia (USP) and National Formulary (NF) (USP/NF, 2000). The FDA requires that the shelf life be indicated on the immediate container label for every drug product in the marketplace. While many drug products consist of a single active ingredient, there are drug products containing multiple active ingredients (see, e.g., Pong and Raghavarao, 2002). For example, as indicated by Chow and Shao (2007), Premarin (conjugated estrogens, USP) contains at least five active ingredients, estrone, equilin, 17α-dihydroequilin, 17α-estradiol, and 17β-dihydroequilin. Other examples include combinational drug products, such as the TCMs, which are known to contain multiple active components. For a drug product with multiple active ingredients, an ingredient-by-ingredient stability analysis may not be appropriate, since these active ingredients may have some unknown interactions. Chow and Shao (2007) proposed a statistical method for determining the shelf life of a drug product with multiple active components or ingredients following a similar idea as suggested by the FDA and assuming that these active components or ingredients are linear combinations of some factors. The method proposed by Chow and Shao (2007) is described below.

Let $y(t, k)$ be the potency of the kth component or ingredient at time t after the manufacture of a given drug product, $k = 1, \ldots, p$. For ingredient k, its shelf life is the time interval in which $E[y(t, k)]$ (the expectation of $y(t, k)$) remains within a specified limit, whereas the shelf life for the drug product may be the time interval at which $E[f(y(t, 1), \ldots, y(t, p))]$ remains within the specified limits, where f is a function (such as a linear combination of $y(t, 1), \ldots, y(t, p)$) that characterizes the impact of all active components or ingredients. In general, f is a vector-valued function with dimension $q \leq p$.

If data are observed from $y(t, 1), \ldots, y(t, p)$ and the function f is a known function, then the stability analysis can be made by using the transformed data $z(t) = f(y(t, 1), \ldots, y(t, p))$. If the dimension of f is 1, then $z(t)$ can be treated as a single component or ingredient. If the dimension of f is $q > 1$, then one may define the shelf life to be the minimum of the shelf-lives $\tau_1, \ldots, \tau_q$, where τ_h is the shelf life when the hth component or ingredient of $z(t)$ is treated as a single component or ingredient. One special case is where f is the identity function, so that the shelf life is the minimum of all shelf-lives corresponding to different components or ingredients $y(t, k), k = 1, \ldots, p$.

In practice, however, f is typically unknown. Although the best way to estimate f is to fit a model between the y and z variables, it requires data observed from both y and z, which is not a common practice in pharmaceutical industry, because the variable z in many problems, such as the TCMs, is not clearly defined. In this chapter, we assume that the components of z are linear combinations of the components of y and propose a method to establish the shelf life. Note that the approach proposed by Chow and Shao (2007) is basically an application of the factor model in multivariate analysis.

23.5.2.1 Models and Assumptions

Let $y(t)$ denote the p-dimensional vector whose kth component is the potency of the kth component or ingredient at time t after the manufacture of a given drug product, $k = 1, \ldots, p$. We assume that the drug potency is expected to decrease with time t. If $p = 1$, i.e., $y(t)$ is univariate, the current established procedure for determination of a shelf life is to use the time at which a 95% lower confidence bound for the mean degradation curve $E[y(t)]$ intersects the acceptable lower product specification limit as specified in USP/NF (2000) (see, also, FDA, 1987; ICH, 1993). Let η be the vector whose kth component or ingredient is the lower product specification limit as specified in the USP/NF for the kth component or ingredient of $y(t)$. Assume that for any t

$$y(t) - E[y(t)] = LF_t + \varepsilon_t, \tag{23.15}$$

where
　　L is a $p \times q$ nonrandom unknown matrix of full rank,
　　F_t and ε_t are unobserved independent random vectors of dimensions q and p,
　　　respectively.

$E(F_t) = 0$, $\text{Var}(F_t) = I_q$ (the identity matrix of order q), $E(\varepsilon_t) = 0$, $\text{Var}(\varepsilon_t) = \Psi$, and Ψ is an unknown diagonal matrix of order p. Note that model (23.15) with the assumptions on F_t and ε_t is the so-called orthogonal factor model. If ε_t is treated as a random error, then model (23.15) assumes that the p-dimensional component or ingredient vector $y(t)$ is governed by a q-dimensional unobserved vector F_t. Normally q is much smaller than p. Let $z(t) = (L'L)^{-1} L'[y(t) - \eta]$. It follows from (23.15) that

$$z(t) - E[z(t)] = F_t + (L'L)^{-1} L' \varepsilon_t. \tag{23.16}$$

If L is known, then (23.16) suggests performing a stability analysis based on the transformed data observed from $z(t)$. In practice, since L is unknown, if we can estimate L based on model (23.15) and the observed data from $y(t)$, then we can carry out a stability analysis using the transformed $z(t)$ with L replaced by its estimate.

Let $x(t)$ be an s-dimensional covariate vector associated with $y(t)$ at time t. For example, $x(t) = (1, t)'$ $(s = 2)$ or $x(t) = (1, t, t^2)'$ $(s = 3)$. We assume the following model at any time t:

$$E[y(t) - \eta] = Bx(t), \quad \text{Var}[y(t)] = \Sigma, \quad i = 1, \ldots, m, \quad j = 1, \ldots, n, \tag{23.17}$$

where
　　B is a $p \times s$ matrix of unknown parameters,
　　$\Sigma > 0$ is an unknown $p \times p$ positive definite covariance matrix.

Since $z(t) = (L'L)^{-1} L'[y(t) - \eta]$, it follows from (23.17) that

$$E[z(t)] = \gamma' x(t), \quad i = 1, \ldots, m, \quad j = 1, \ldots, n, \tag{23.18}$$

where $\gamma = B'L(L'L)^{-1}$.

23.5.2.2 Shelf-Life Determination

Suppose that we independently observe data y_{ij}, $i = 1, \ldots, m$, $j = 1, \ldots, n$, where y_{ij} is the jth replicate of $y(t_i)$ and $t_1, \ldots, t_m$ are designed time points for the stability analysis. Define

$$x_i = x(t_i), \quad z_{ij} = (L'L)^{-1}L'(y_{ij} - \eta), \quad i = 1,\ldots,m, \quad j = 1,\ldots,n. \tag{23.19}$$

Consider first the case of $q = 1$, i.e., z_{ij} in (23.19) is univariate. If z_{ij}'s are observed, then an approximate 95% lower confidence bound for $E[z(t)] = \gamma'x(t)$ is

$$l(t) = \hat{\gamma}x(t) - t_{0.95,mn-s}\hat{\sigma}\sqrt{D(t)}, \tag{23.20}$$

where
  $\hat{\gamma}$ is the least squares estimator of γ in model (23.20) based on data z_{ij}'s and
    x_i's
  $\hat{\sigma}^2$ is the usual sum of squared residuals divided by its degrees of freedom
    $mn - s$
  $t_{0.95,mn-s}$ is the 95th percentile of the t distribution with degrees of freedom
    $mn - s$

$$D(t) = \left[n \sum_{i=1}^{m} x(t)'x_i x_i'x(t) \right]^{-1}.$$

Hence, if z_{ij}'s are observed, a shelf life according to the 1987 FDA guideline for stability (FDA, 1987) is

$$\tau = \inf\{t : l(t) \le 0\}. \tag{23.21}$$

For TCM, y_{ij}'s, not z_{ij}'s are observed. Hence, the lower confidence bound $l(t)$ in (23.20) needs to be modified. Since $\gamma' = (L'L)^{-1} L'B$, we can obtain an estimator of γ in two steps. In the first step, we use model (23.17), observed data y_{ij}'s and x_i's, and the multivariate linear regression to obtain a least squares estimator $\hat{B}$ of B. In the second step, we consider the orthogonal factor model (23.15) and apply the method of principal components. To obtain an estimator $\hat{L}$ of L, using data $y_{ij} - \eta - \hat{B}x_i$, $i = 1, \ldots, m$, $j = 1, \ldots, n$. More precisely, $\hat{L}$ is the normalized eigenvector corresponding to the largest eigenvalue of the sample covariance matrix based on data $y_{ij} - \hat{B}x_i$, $i = 1, \ldots, m$, $j = 1, \ldots, n$. Let $\hat{\gamma} = \hat{B}'\hat{L}(\hat{L}'\hat{L})^{-1}$.

The lower confidence bound in (23.20) is modified to

$$l(t) = \hat{\gamma}'x(t) - t_{0.95,mn-s}\sqrt{x(t)'Vx(t)}, \tag{23.22}$$

where V is the jackknife variance estimator of $\hat{\gamma}$ (see, e.g., Shao and Tu, 1995), i.e.,

$$V = \frac{mn-1}{mn} \sum_{i=1}^{m} \sum_{j=1}^{n} (\hat{\gamma}_{i,j} - \hat{\gamma})(\hat{\gamma}_{i,j} - \hat{\gamma})',$$

where $\hat{\gamma}_{i,j}$ is the estimator of γ calculated using the same method as in the calculation of $\hat{\gamma}$ but with the (i, j)th data point deleted. The result for $q = 1$ is sufficient for applications with a small or moderate p. When p is large, Chow and Shao (2007) proposed the following procedure with $1 < q < p$. Let $\hat{B}$ be defined as before, λ_k be the kth largest eigenvalue of the sample covariance matrix based on $y_{ij} - \eta - \hat{B}x_i$, $i = 1, \ldots, m$, $j = 1, \ldots, n$, and e_k be the normalized eigenvector corresponding to λ_k. Then, the estimator $\hat{L}$ of L is the $p \times q$ matrix whose kth column is $\lambda_k e_k$, $k = 1, \ldots, q$. The estimator of γ is still $\hat{\gamma} = \hat{B}'\hat{L}(\hat{L}'\hat{L})^{-1}$, which is an $s \times q$ matrix. Let $\hat{\gamma}_k$ be the kth column of $\hat{\gamma}$, $k = 1, \ldots, q$

$$l_k(t) = \hat{\gamma}_k' x(t) - t_{1-0.05/q, mn-s}\sqrt{x(t)'V_k x(t)} \tag{23.23}$$

and

$$V_k = \frac{mn-1}{mn} \sum_{i=1}^{m} \sum_{j=1}^{n} (\hat{\gamma}_{k,i,j} - \hat{\gamma}_k)(\hat{\gamma}_{k,i,j} - \hat{\gamma}_k)',$$

where $\hat{\gamma}_{k,i,j}$ is the same as $\hat{\gamma}_k$ but calculated with the (i, j)th data point deleted. Then, $l_k(t)$, $k = 1, \ldots, q$, are approximate 95% simultaneous lower confidence bounds for $\zeta_k(t)$, $k = 1, \ldots, q$, where $\zeta_k(t)$ is the kth component of $E[z(t)] = \gamma'x(t)$. An approximate level 95% shelf life for the drug product (when the sample size mn is large) is

$$\tau = \min_{k=1, \ldots, q} \tau_k,$$

where each τ_k is defined by the right-hand side of (23.21) with $l(t)$ replaced by $l_k(t)$ and is in fact a shelf life for the kth component of z with confidence level $(1 - 0.05/q)\%$.

23.5.2.3 An Example

To illustrate the proposed method for determining the shelf life of a drug product with multiple active ingredients, consider a stability study conducted for a traditional Chinese herbal medicine, which is newly developed for the treatment of patients with rheumatoid arthritis. This medicine contains

TABLE 23.4

Components of a TCM

Component	Formulation (mg)
Herba Epimedii	60
B	25
C	25
Excipient	90
Total	200

three active botanical components, namely, Herba Epimedii, B extract, and C extract. Each of the three components has been used as herbal remedies since ancient China and is well documented in the CP. The proportions of each component are summarized in Table 23.4.

To establish a shelf life for this product, a stability study was conducted for a period of 18 months under a testing condition of 25°C/60% relative humidity. The lower product specification limit for each component is 90%. Stability data (percent of label claim) at each sampling time point for the three components are given in Table 23.5.

Since $p = 3$, we consider $q = 1$. Using the proposed procedure described in the previous sections, $l(t)$ in (23.22) for various t (month) as given in Table 23.6.

Hence, the estimated shelf life for this product is 27 months.

TABLE 23.5

Stability Data of a TCM

Component	Sampling Time Point (Months)					
	0	3	6	9	12	18
Herba Epimedii	99.6	97.5	96.8	96.2	94.8	95.3
	99.7	98.3	97.0	96.0	95.1	94.8
	100.2	99.0	98.2	97.1	95.3	94.6
B	99.5	98.4	96.3	95.4	93.2	91.0
	100.5	98.5	97.4	94.9	94.5	92.1
	99.3	99.0	97.3	95.0	93.1	91.5
C	100.0	99.5	98.9	98.2	97.9	97.5
	99.8	99.4	99.0	98.5	98.0	97.9
	101.2	99.9	100.3	99.5	98.9	98.0

TABLE 23.6

$l(t)$ Values with Various t

t	19	20	21	22	23	24	25	26	27	28
$l(t)$	4.97	4.36	3.75	3.14	2.52	1.90	1.28	0.66	0.03	−0.60

23.5.2.4 Discussion

The statistical method for determining the shelf life of a drug product with p active ingredients proposed by Chow and Shao (2007) assumed that these active ingredients are linear combinations of q factors. Since we propose to choose these factors using principal components, the first factor can be viewed as the primary active factor and the second factor can be viewed as the secondary active factor. We assume that active ingredients decrease with time. If one or more ingredients increase with time, then a transformation such as $g(y) = -y$ or $g(y) = 1/y$ may be applied. If p is small or moderate, then $q = 1$ is recommended. If p is large, then adding a few more factors may be considered. Since the principal components are orthogonal, adding more factors will not affect the previous selected factors (except that $t_{0.95,mn-s}$ changed to $t_{1-0.05/q,mn-s}$) so that one can compare the results in a sensitivity analysis. Finally, adding more factors always results in a more conservative procedure.

Note that in our proposed approach, we assume that there is no significant toxic degradant in the test drug product with multiple components. This is a reasonable assumption for most TCM since multiple ingredients are used to reduce toxicities when used in conjunction with primary therapy. However, in the case where toxic degradation products are detected, special attention should be paid to (1) identity (chemical structure), (2) cross reference to information about biological effects and significance of concentration likely to be encountered, and (3) indications of pharmacological action or inaction as indicated in the FDA guideline for stability analysis. (FDA, 1987).

The approach proposed by Chow and Shao (2007) is useful when different ingredients degrade not independently of each other, which is the case for most TCM. If multiple ingredients degrade independently, then an ingredient-by-ingredient analysis may be appropriate. If our approach is applied, then we will select $q = 1$ or $q = 2$ factors that are ingredients having the most variability.

23.5.3 Calibration of Study Endpoints

When planning a clinical trial, it is suggested that the study objectives should be clearly stated in the study protocol. Once the study objectives are confirmed, a valid study design can be chosen and the primary clinical endpoints can be determined accordingly. For the evaluation of treatment effect of a TCM, however, the commonly used clinical endpoint is usually not applicable due to the nature of the CDPs as described earlier. The CDP is in fact an instrument (or questionnaire) which consists of a number of questions to capture the information regarding the patient's activity, function, disease status and severity. As required by most regulatory agencies, such a subjective instrument must be validated before it can be used for assessment of treatment effect in clinical trials. However, without a reference marker, not only can the CDP not be validated but we also do not know whether the

TCM has achieved clinically significant effect at the end of the clinical trial. In this section, we will study the calibration and validation of the CDP for the evaluation of a TCM with respect to a well-established clinical endpoint for the evaluation of a WM.

To address these issues described above, Hsiao et al. (2009) proposed a study design, which allows calibration and validation of a CDP with respect to a well-established clinical endpoint for WM (as a reference marker). Subjects will be screened based on criteria for Western indication. Qualified subjects will be diagnosed by the CDP to establish a baseline. Qualified subjects will then be randomized to receive either the test TCM or an active control (a well-established WM). Participating physicians including Chinese doctors and Western clinicians will also be randomly assigned to either the TCM arm or the WM arm. As a result, this study design will be divided into three groups:

Group 1: Subjects who receive WM but are evaluated by both a Chinese doctor and a Western clinician.

Group 2: Subjects who receive TCM and are evaluated by a Chinese doctor A.

Group 3: Subjects who receive TCM and are evaluated by a Chinese doctor B.

Group 1 can be used to calibrate the CDP against the well-established clinical endpoint, while groups 2 and 3 can be used to validate the CDP based on the established standard curve for calibration.

23.5.3.1 Chinese Diagnostic Procedure

As indicated earlier, the diagnostic procedure for TCM consists of four major techniques, namely, inspection, auscultation and olfaction, interrogation, and pulse taking and palpation. All these diagnostic techniques aim mainly at providing an objective basis for the differentiation of syndromes by collecting symptoms and signs from the patient. Inspection involves observing the patient's general appearance (strong or week, fat or thin), mind, complexion (skin color), five sense organs (eye, ear, nose, lip, and tongue), secretions, and excretions. Auscultation involves listening to the voice, expression, respiration, vomit, and cough. Olfaction involves smelling the breath and body odor. Interrogation involves asking questions about specific symptoms and the general condition including history of the present disease, past history, personal life history, and family history. Pulse taking and palpation can help to judge the location and nature of a disease according to the changes in the pulse. The smallest detail can have a strong impact on the treatment scheme as well as on the prognosis. While the pulse diagnosis and examination of the tongue receive much attention due to their frequent mention, the other aspects of diagnosis cannot be ignored.

As indicated earlier, after these four diagnostic techniques have been performed, the TCM doctor has to configure a syndrome diagnosis describing the fundamental substances of the body and how they function in the body based on the eight principles, five-element theory, five Zang and six Fu, and information regarding channels and collaterals.

In addition to providing diagnostic information, these elements of TCM can also help to describe the etiology of disease including the six exogenous factors (i.e., wind, cold, summer, dampness, dryness, and fire), seven emotional factors (i.e., anger, joy, worry, grief, anxiety, fear, and fright), and other pathogenic factors. Once all this information is collected and processed into a logical and workable diagnosis, the traditional Chinese medical doctor can determine the treatment approach.

23.5.3.2 Calibration

Let N be the number of patients collected in group 1. For the data from group 1, let x_j be the measurement of the well-established clinical endpoint of the jth patient. For simplicity, we assume that the measurement of well-established clinical endpoints is continuous. Suppose that the TCM diagnostic procedure consists of K items. Let z_{ij} denote the TCM diagnostic score of jth patient from the ith item, $i = 1, \ldots, K, j = 1, \ldots, N$. Let y_j represent the scale (or score) of the jth patient summarized from the K TCM diagnostic items. For simplicity, we assume that

$$y_j = \sum_{i=1}^{K} \sum_{j}^{N} z_{ij}.$$

Similar to calibration of an analytical method (cf. Chow and Liu, 1995), we will consider the following five candidate models:

Model 1: $y_j = \alpha + \beta x_j + e_j,$

Model 2: $y_j = \beta x_j + e_j,$

Model 3: $y_j = \alpha + \beta_1 x_j + \beta_2 x_j^2 + e_j,$

Model 4: $y_j = \alpha x_j^\beta e_j,$

Model 5: $y_j = \alpha e^{\beta x_j} e_j,$

where α, β, β_1, and β_2 are unknown parameters and e's are independent random errors with $E(e_j) = 0$ and finite $\mathrm{Var}(e_j)$ in models 1–3 and $E(\log(e_j)) = 0$ and finite $\mathrm{Var}(\log(e_j))$ in models 4 and 5.

Model 1 is a simple linear regression model which is probably the most commonly used statistical model for the establishment of standard curves for calibration. When the standard curve passes through the origin, model 1 reduces to model 2. Model 3 indicates that the relationship between y and x is quadratic. When there is a nonlinear relationship between y and x, models 4 and 5 are useful. Note that both models 4 and 5 are equivalent to a simple linear regression model after logarithm transformation. If all the above models cannot fit the data, generalized linear models can be used.

By fitting an appropriate statistical model between these standards (well-established clinical endpoints) and their corresponding responses (TCM scores), an estimated calibration curve can be obtained. The estimated calibration curve is also known as the standard curve. For a given patient, his/her unknown measurement of well-established clinical endpoint can be determined based on the standard curve by replacing the dependent variable with its TCM score.

23.5.3.3 Validity

The validity itself is a measure of biasedness of the TCM instrument. Since a TCM instrument usually contains four categories or domains, which in turn consist of a number of questions agreed by the community of the Chinese doctors, it is a great concern that the questions may not be the right ones to capture the information regarding the patient's activity/function, disease status, and disease severity. We will use group 2 to validate the CDP based on the previously established standard curve for calibration. Let X be the unobservable measurement of the well-established clinical endpoint which can be quantified by the TCM items, Z_i, $i = 1, \ldots, K$ based on the estimated standard curve discussed in the previous section. For convention, we assume that

$$X = \frac{(Y - \alpha)}{\beta},$$

where $Y = \sum_{i=1}^{K} Z_i$. That is, model 1 was used for calibration. Suppose that X is distributed as a normal distribution with mean θ and variance τ^2. Let $Z = (Z_1, \ldots, Z_K)'$. Again suppose Z follows a distribution with mean $\mu = (\mu_1, \ldots, \mu_K)'$ and variance Σ. To assess the validity, it is desired to see whether the mean of Z_i, $i = 1, \ldots, K$ is close to $(\alpha + \beta\theta)/K$. Let $\bar{\mu} = 1/K \sum_{i=1}^{K} \mu_i$. Then $\theta = (\bar{\mu} - \alpha)/\beta$. Consequently, we can claim that the instrument is validated in terms of its validity if

$$|\mu_i - \bar{\mu}| < \delta, \quad \forall i = 1, \ldots, K, \tag{23.24}$$

for some small prespecified δ. To verify (23.24), we can consider constructing a simultaneous confidence interval for $\mu_i - \bar{\mu}$. Assume that the TCM instrument is administered to N patients from group 2. Let $\hat{\mathbf{m}} = 1/N \sum_{j=1}^{N} \mathbf{Z}_j = \bar{\mathbf{Z}}$. Then the $(1 - \alpha)100\%$ simultaneous confidence interval for $\mu_i - \bar{\mu}$ are given by

$$\mathbf{a}_i'\hat{\mathbf{m}} - \sqrt{\frac{1}{N} \mathbf{a}_i' S \mathbf{a}_i} T(\alpha, K, N - K) \le \mu_i - \bar{\mu} \le \mathbf{a}_i'\hat{\mathbf{m}}$$

$$+ \sqrt{\frac{1}{N} \mathbf{a}_i' S \mathbf{a}_i} T(\alpha, K, N - K), \quad i = 1, \ldots, K,$$

where

$$\mathbf{a}_i' = \begin{pmatrix} -\dfrac{1}{K} \mathbf{1}_{i-1} \\ 1 - \dfrac{1}{K} \\ -\dfrac{1}{K} \mathbf{1}_{k-1} \end{pmatrix}$$

$$S = \frac{1}{N-1} \sum_{j=1}^{N} (\mathbf{Z}_j - \bar{\mathbf{Z}})(\mathbf{Z}_j - \bar{\mathbf{Z}})',$$

$$T^2(\alpha, K, N - K) = \frac{(N-1)K}{N-K} F(\alpha, K, N - K),$$

and

$$P(T^2(K, N - K) \le T^2(\alpha, K, N - K)) = 1 - \alpha.$$

The Bonferroni adjustment of an overall α level might be conducted as follows:

$$\mathbf{a}_i'\hat{\mathbf{m}} - \sqrt{\frac{1}{N} \mathbf{a}_i' S \mathbf{a}_i} T\left(\frac{\alpha}{2K}, N - 1\right) \le \mu_i - \bar{\mu} \le \mathbf{a}_i'\hat{\mathbf{m}} + \sqrt{\frac{1}{N} \mathbf{a}_i' S \mathbf{a}_i} T\left(\frac{\alpha}{2K}, N - 1\right).$$

We can reject the null hypothesis that

$$H_0 : |\mu_i - \bar{\mu}| \ge \delta, \quad \forall i = 1, \ldots, K, \tag{23.25}$$

if any confidence interval falls completely within $(-\delta, \delta)$.

23.5.3.4 Reliability

The calibrated well-established clinical endpoints derived from the estimated standard curve are considered reliable if the variance of X is small. In this regard, we can test the hypothesis

$$H_0 : \tau^2 >= \Delta \text{ for some fixed } \Delta \tag{23.26}$$

to verify the reliability of estimating θ by X. We will use group 2 to verify the reliability based on the previously established standard curve for calibration. Based on the estimated standard curve, we can derive that

$$\tau^2 = \frac{1}{\beta^2} \text{Var}\left(\sum_{i=1}^{K} Z_i \right) = \frac{1}{\beta^2} \mathbf{1}' \Sigma \mathbf{1}.$$

Note that the sample distribution of

$$\sum_{j=1}^{N} \frac{(X_j - \bar{X})^2}{\tau^2}$$

has a chi-square distribution with $N - 1$ degrees of freedom. Thus, we can construct a $(1 - \alpha)100\%$ one-sided confidence interval for τ^2 as follows:

$$\tau^2 \geq \frac{\sum_{j=1}^{N} (X_i - \bar{X})^2}{\chi^2(\alpha, N - 1)} = \xi.$$

We can reject the null hypothesis of (23.26) and conclude that the items are not reliable in estimation of θ if $\xi > \Delta$.

23.5.3.5 Ruggedness

In addition to validity and reliability, an acceptable TCM diagnostic instrument should produce similar results on different raters. In other words, it is desirable to quantify the variation due to rater and the proportion of rater-to-rater variation to the total variation. We will use the one-way nested random model to evaluate instrument ruggedness (Chow and Liu, 1995). The one-way nested random model can be expressed as

$$X_{ij} = \mu + A_i + e_{j(i)}, \quad i = 1 \text{ (group 2)}, 2 \text{ (group 3)}; \quad j = 1, \dots, N,$$

where

X_{ij} is the calibrated scale of the jth patient obtained from the ith rater
μ is the overall mean
A_i is the random effect due to the ith rater
$e_{j(i)}$ is the random error of the jth patient's scale nested within the ith rater.

For the one-way nested random model, we need the following assumptions: A_i are independent and identically distributed (i.i.d.) normal with mean 0 and variance σ_A^2; $e_{j(i)}$ are i.i.d. normal with mean 0 and variance σ^2; A_i and $e_{j(i)}$ are mutually independent for all i and j.

Let $\ \bar{X}_{i\cdot} = (1/J)\sum_{j=1}^{N} X_{ij}\ $ and $\ \bar{X}.. = (1/2N)\sum_{i=1}^{2}\sum_{j=1}^{N} X_{ij} = (1/2)\sum_{i=1}^{2} \bar{X}_{i\cdot}.$

Also, let SSA and SSE denote the sum of squares of factor A and the sum of squares of errors, respectively. In other words,

$$SSA = N\sum_{i=1}^{2}\left(\bar{X}_{i\cdot} - \bar{X}..\right)^2$$

and

$$SSE = \sum_{i=1}^{2}\sum_{j=1}^{N}\left(X_{ij} - \bar{X}_{i\cdot}\right)^2.$$

Also let MSA and MSE denote mean squares for factor A and mean square error. Then MSA = SSA and MSE = SSE/[2(N − 1)]. As a result, the analysis of variance estimators of σ_A^2 and σ^2 can be obtained as follows:

$$\hat{\sigma}^2 = MSE$$

and

$$\hat{\sigma}_A^2 = \frac{MSA - MSE}{N}.$$

Note that $\hat{\sigma}_A^2$ is obtained from the difference between MSA and MSE, and thus it is possible to obtain a negative estimate for σ_A^2.

Three criteria can be used to evaluate instrument ruggedness. The first criterion is to compute the probability for obtaining a negative estimate of σ_A^2 given by

$$P(\hat{\sigma}_A^2 < 0) = P(F[1, 2(N-1)] < (F)^{-1}),$$

where $F[1, 2(N-1)]$ is a central F distribution with 1 and $2(N-1)$ degrees of freedom and

$$F = \frac{\sigma^2 + N\sigma_A^2}{\sigma^2}.$$

If $P(\hat{\sigma}_A^2 < 0)$ is large, it may suggest that $\sigma_A^2 = 0$. The second criterion is to test whether the variation due to factor A is significantly larger than zero:

$$H_0 : \sigma_A^2 = 0 \quad \text{versus} \quad H_1 : \sigma_A^2 > 0. \tag{23.27}$$

The null hypothesis (23.27) is rejected at the α level of significance if

$$F_A > F_C = F(\alpha, 1, 2(N-1)),$$

where $F_A = \text{MSA}/\text{MSE}$. The third criterion is to evaluate the proportion of the variation due to factor A, which is defined as follows:

$$\rho_A = \frac{\sigma_A^2}{\sigma^2 + \sigma_A^2}.$$

According to Searle et al. (1992), the estimator and the $(1-\alpha)100\%$ confidence interval for σ_A^2 are given by

$$\hat{\rho}_A = \frac{\text{MSA} - \text{MSE}}{\text{MSA} + (N-1)\text{MSE}},$$

$$L_\rho = \frac{F_A/F_U - 1}{N + (F_A/F_U - 1)},$$

$$U_\rho = \frac{F_A/F_L - 1}{N + (F_A/F_L - 1)},$$

where $F_L = F(1 - 0.5\alpha, 1, 2(N-1))$ and $F_U = F(0.5\alpha, 1, 2(N-1))$.

It may also be desired to test whether or not the rater-to-rater variability is within an acceptable limit ω. In this case, Hsiao et al. (2007) have considered testing the following hypothesis:

$$H_0 : \sigma_A^2 \quad \text{versus} \quad H_1 : \sigma_A^2 < \omega. \tag{23.28}$$

Since there exists no exact $(1 - \alpha)100\%$ confidence interval for σ_A^2, we can derive the Williams–Tukey interval with a confidence level between $(1 - 2\alpha)100\%$ and $(1 - \alpha)100\%$ which is given by (L_A, U_A), where

$$L_A = \frac{\text{SSA}(1 - F_U/F_A)}{N\chi_{UA}^2}, \quad U_A = \frac{\text{SSA}(1 - F_L/F_A)}{N\chi_{LA}^2},$$

where $F_L = F(1 - 0.5\alpha, 1, 2(N - 1))$ and $F_U = F(0.5\alpha, 1, 2(N - 1))$ represent the $(1 - 0.5\alpha)$th and (0.5α)th upper quantiles of a central F distribution with 1 and $2(N - 1)$ degrees of freedom, $\chi_{LA}^2 = \chi^2(1 - 0.5\alpha, 1)$ and $\chi_{UA}^2 = \chi^2(0.5\alpha, 1)$ are the $(1 - 0.5\alpha)$th and (0.5α)th upper quantiles of a central chi-square distribution with 1 degree of freedom, and $F_A = \text{MSA}/\text{MSE}$. The null hypothesis (23.28) is rejected at α level of significance if $U_A < \omega$.

23.6 Concluding Remarks

As indicated earlier, a TCM is defined as a Chinese herbal medicine developed for treating patients with certain diseases as diagnosed by the four major techniques of inspection, auscultation and olfaction, interrogation, and pulse taking and palpation based on the traditional Chinese medical theory of global balance among the functions/activities of all the organs of the body. When conducting a TCM clinical trial, it is suggested that the fundamental differences between a WM and a TCM, as described in Section 23.2, should be evaluated carefully for a valid and unbiased assessment of the safety and effectiveness of the TCM under investigation.

One of the key issues in TCM research and development is to clarify the difference between Westernization of TCM and modernization of TCM. For Westernization of TCM, we follow regulatory requirements at critical stages of the process for pharmaceutical development including drug discovery, formulation, laboratory development, animal studies, clinical development, manufacturing process validation and QC, regulatory submission, review, and process, despite the fundamental differences between WM and TCM. For modernization of TCM, it is suggested that regulatory requirements should be modified in order to account for the fundamental differences between WM and TCM. In other words, we still ought to be able to see if TCM is really working with modified regulatory requirements using Western clinical trials as a standard for comparison.

In practice, it is recognized that WMs tend to achieve the therapeutic effect sooner than TCMs for critical and/or life-threatening diseases. TCMs are found to be useful for patients with chronic diseases or non-life-threatening diseases. In many cases, TCMs have shown to be effective in reducing toxicities or improving the safety profile for patients with critical and/or life-threatening diseases. As a strategy for TCM research and development, it is suggested that (1) TCM be used in conjunction with a well-established WM as a supplement to improve its safety profile and/or enhance therapeutic effect whenever possible and (2) TCM should be considered as the second-line or third-line treatment for patients who fail to respond to the available treatments. However, some sponsors are interested in focusing on the development of TCM as a dietary supplement due to (1) the lack or ambiguity of regulatory requirements, (2) the lack of understanding of the medical theory/mechanism of TCM, (3) the confidentiality of nondisclosure of the multiple components, and (4) the lack of understanding of pharmacological activities of the multiple components of TCM.

Since TCM consists of multiple components which may be manufactured from different sites or locations, the post-approval consistency in the quality of the final product is both a challenge to the sponsor and a concern to the regulatory authority. As a result, some post-approval tests, such as tests for content uniformity, weight variation, and/or dissolution and (manufacturing) process validation, must be performed for quality assurance before the approved TCM can be released for use.

24

The Assessment of Follow-On Biologic Products

24.1 Introduction

When an innovative (brand-name) drug product is going off patent, pharmaceutical and generic companies may file an abbreviated new drug application (ANDA) for approval of generic copies of the innovative drug product. In 1984, the United States Food and Drug Administration (FDA) was authorized to approve generic drug products under the Drug Price Competition and Patent Term Restoration Act (which is also known as the Hatch and Waxman Act). For the approval of generic drug products, the FDA requires that evidence in average bioavailability (in terms of rate and extent of drug absorption) be provided through the conduct of bioavailability and bioequivalence studies. As indicated by Chow and Liu (2008), the assessment of bioequivalence as a surrogate for the evaluation of drug safety and efficacy is based on the Fundamental Bioequivalence Assumption that if two drug products are shown to be bioequivalent in average bioavailability, it is assumed that they will reach the same therapeutic effect or they are therapeutically equivalent and can be used interchangeably. Under the Fundamental Bioequivalence Assumption, regulatory requirements, study design (e.g., a two-sequence, two-period crossover design or a replicated crossover design), criteria (e.g., 80/125 rule based on log-transformed data), and statistical methods (e.g., Shuirmann's two one-sided tests or confidence interval approach) for assessment of bioequivalence have been well established (see, e.g., Schuirmann, 1987; FDA 2001, 2003b; Chow and Liu, 2008).

Unlike drug products, the concept for development of "generic" versions of biologic products, which are usually referred to as follow-on biologics (FOB) by the U.S. FDA or biosimilars by the European Medicines Agency (EMEA) of the European Union (EU), or subsequent entered biologics (SEB) by Health Canada, is different. Webber (2007) defines follow-on (protein) biologics as products that are intended to be sufficiently similar to an approved product to permit the applicant to rely on certain existing scientific knowledge about safety and efficacy of an approved reference product. Under this definition, follow-on biologic products are not only intended to be similar to the reference product, but also intended to be interchangeable with the reference product. As a number of biologic products are due to expire in the next few years, the subsequent follow-on biologic products have generated considerable interest within the pharmaceutical/biotechnological industry as biosimilar manufacturers strive to obtain part of an already large and rapidly growing market. The potential opportunity for price reductions versus the originator biologic products remains to be determined, as the advantage of a cheaper price may be outweighed by the hypothetical increased risk of side effects from biosimilar molecules that are not exact copies of their originators. In this chapter, we will focus on the issues surrounding biosimilars, including manufacturing, quality control (QC), clinical efficacy, side effects (safety), and immunogenicity. In addition, we will also attempt to address the challenges regarding how regulatory agencies and industry regulations are evolved in dealing with these issues.

Biosimilars are fundamentally different from generic chemical drugs. Important differences include the size and complexity of the active substance and the nature of the manufacturing process. Because biosimilars are not the exact copy of their originator products, different criteria for regulatory approval may be required although the principles of evaluating bioequivalence are the same. This is partly a reflection of the complexities of manufacturing and safety and efficacy controls of biosimilars when compared to their small-molecule generic counterparts (see, e.g., Chirino and Mire-Sluis, 2004; Schellekens, 2004; Crommelin et al., 2005; Roger, 2006; Roger and Mikhail, 2007; Keith, 2007; Webber, 2007). Since biologic products are usually recombinant protein molecules manufactured in living cells (Kuhlmann and Covic, 2006), manufacturing processes for biologic products are highly complex and require hundreds of specific isolation and purification steps. In practice, it is impossible to produce an identical copy of a biologic product, as changes to the structure of the molecule can occur with changes in the production process. Since a protein can be modified (e.g., side chain may be added and structure may have changed due to protein misfolding) during the process, different manufacturing processes may invariably lead to structural differences in the final product, which may result in differences in efficacy and may have a negative impact on patient immune responses. It should be noted that the above issues also occur for the post-approval changes for the innovator biological products.

In the next section, regulatory requirements for approval of biosimilars by the EMEA of EU are briefly outlined. Also included in the section is the current position of the FDA. Section 24.3 reviews various criteria for the assessment of bioequivalence, similarity, and consistency of chemical generics that appeared in either regulatory guidances and/or literature. Some scientific issues for the assessment of biosimilars are discussed in Section 24.4. In Section 24.5, an approach to assessment of similarity using genomic data is proposed. Some concluding remarks are given in the last section.

24.2 Regulatory Requirements

For the approval of biosimilars in the EU community, the EMEA has issued a new guideline describing general principles for the approval of similar biological medicinal products, or biosimilars. The guideline is accompanied by six concept papers that outline areas in which the agency intends to provide more targeted guidance (EMEA 2003a,b, 2005a–g). Specifically, the concept papers discuss approval requirements for several classes of human recombinant products containing erythropoietin, human growth hormone, granulocyte colony-stimulating factor, and insulin. The guideline consists of a checklist of documents published to date relevant to data requirements for biological pharmaceuticals. It is not clear what specific scientific requirements will be applied to biosimilar applications. In addition, it is not clear how the agency will treat innovator data contained in the reference product dossiers. The guideline provides a useful summary of the biosimilar legislation and previous EU publications, and it also provides a few answers to the issues.

On the other hand, for the approval of follow-on biologics in the United States, it depends on whether the biologic product is approved under the U.S. Food, Drug, and Cosmetic Act (FD&C) or whether it is licensed under the U.S. Public Health Service Act (PHS) (Kozlowski, 2007; Liang, 2007). As indicated, some proteins are licensed under the PHS Act, while some are approved under the FD&C Act. For products approved under an NDA (FD&C Act), a generic version of the products can be approved under an ANDA, e.g., under Section 505(b)(2) of the FD&C Act. For products that are licensed under a Biologics License Application (BLA) (PHS Act), there exists no abbreviated BLA. As pointed out by Woodcock et al. (2007), for the assessment of similarity of follow-on biologics, the FDA would consider the following factors regarding (1) the robustness of the manufacturing process, (2) the degree to which structural similarity could be assessed, (3) the extent to which the mechanism of action was understood, (4) the existence of valid, mechanistically related pharmacodynamic (PD) assays, (5) the comparative pharmacokinetics (PK), (6) the comparative immunogenicity, (7) the amount of clinical data available, and (7) the extent of experience with the original product (ICH, 1996c, 1999, 2005b) A typical example would be a recent regulatory approval of Omnitrope® (Somatropin), which was approved in 2006 under Section 505(b)(2) of the FD&C Act. Omnitrope was approved based on the following evaluations: (1) physicochemical testing that established highly similar structure to Genotropin, (2) new nonclinical pharmacology and toxicology data specific to Omnitrope, (3) PK, PD, and comparative bioavailability data, (4) clinical efficacy and safety data from comparative controlled trials and from long-term trials with Omnitrope, (5) vast clinical experience and a wealth of published literature concerning the clinical effects (safety and effectiveness) of human growth hormone. The approval of Omnitrope is based on an ad hoc case-by-case review of individual biosimilar application. In practice, there is a stronger industrial interest and desire for the regulatory agencies to develop review standards and an approval process for biosimilars than an ad hoc case-by-case review of individual biosimilar applications. As more biologic products are going off patents in the next few years, the FDA hosted a *Public Hearing on Approval Pathway for Biosimilar and Interchangeable Biological Products* between November 2 and 3, 2010, at Silver Spring, Maryland, to address some scientific factors regarding the assessment of biosimilarity (e.g., criteria, design, and statistical methods), drug interchangeability (e.g., the issues of alternating and switching), and quality (e.g., test for comparability in manufacturing process) of follow-on biologics. As a result, the FDA indicated that the following guidances are currently under development: (1) a guidance for the industry on scientific considerations demonstrating the safety and effectiveness of follow-on protein products and (2) a guidance for the industry on CMC issues for follow-on protein products.

24.3 Criteria for Biosimilarity

For comparison between drug products, some criteria for the assessment of bioequivalence, similarity (e.g., dissolution profiles comparison), and consistency (e.g., comparison between manufacturing processes) are available in either regulatory guidelines/guidances or the literature. These criteria, however, can be classified into (1) absolute change versus relative change, (2) aggregated versus disaggregated, or (3) moment based versus probability based. In this section, different categories of criteria are briefly reviewed.

24.3.1 Absolute Change versus Relative Change

In clinical research and development, for a given study endpoint, posttreatment absolute change from baseline or posttreatment relative change from baseline is usually considered for comparison between treatment groups. A typical example would be the study of weight reduction in an obese patient population. In practice, it is not clear whether a clinically meaningful difference in terms of absolute change from baseline can be translated to a clinically meaningful difference in terms of relative change from baseline. Sample size calculation based on power analysis in terms of absolute change from baseline or relative change from baseline could lead to a very different result.

Current regulations for the assessment of bioequivalence between drug products in terms of average bioavailability are based on relative change. In other words, we conclude bioequivalence between a test product and a reference product if the 90% confidence interval for the ratio of means of the primary PK response such as area under the blood or plasma-concentration time curve (AUC) between the two drug products is totally within 80% and 125%.

Note that regulatory agencies suggest that a log-transformation be performed before data analysis for the assessment of bioequivalence.

24.3.2 Aggregated versus Disaggregated

As indicated by Chow and Liu (2008), bioequivalence can be assessed by evaluating differences in averages, intra-subject variabilities, and variance due to subject-by-formulation interaction between drug products separately. Individual criteria for the assessment of differences in averages, intra-subject variabilities, and variance due to subject-by-formulation interaction between drug products are referred to as disaggregated criteria. If the criterion is a single summary measure composed of these individual criteria, it is called an aggregated criterion.

For the assessment of average bioequivalence (ABE), most regulatory agencies including the FDA recommend the use of a disaggregate criterion based on average bioavailability. In other words, bioequivalence is concluded if the average bioavailability of the test formulation is within (80%, 125%) that of the reference formulation, with a certain assurance. Note that the EMEA (2001) and the World Health Organization (WHO) (2005) use the same equivalence criterion of 80%–125% for the log-transformed PK responses such as AUC. However, for C_{max}, in certain cases, the EMEA and WHO allow a wider interval of 75%–133% for the ratio of average bioavailability to address any safety and efficacy concerns for patients switched between formulations. If a wider interval is used, it must be prespecified in the protocol. More details can be found in Chow and Liu (2008).

On the other hand, for the assessment of population bioequivalence (PBE) and individual bioequivalence (IBE), the following aggregated criteria are often considered. For the assessment of IBE, a criterion proposed in the FDA guidance (FDA, 2001) can be expressed as

$$\theta_I = \frac{(\delta^2 + \sigma_D^2 + \sigma_{WT}^2 - \sigma_{WR}^2)}{\max\{\sigma_{W0}^2, \sigma_{WR}^2\}},$$ (24.1)

where
$\delta = \mu_T - \mu_R$ is the true difference in means
$\sigma_{WT}^2, \sigma_{WR}^2, \sigma_D^2$ are intra-subject variabilities of the test product and the reference product, and variance due to subject-by-formulation interaction between drug products, respectively
σ_{W0}^2 is the scale parameter specified by the user

Similarly, the criterion for the assessment of PBE suggested in the FDA guidance (FDA, 2001) is given by

$$\theta_P = \frac{(\delta^2 + \sigma_{TT}^2 - \sigma_{TR}^2)}{\max\{\sigma_{T0}^2, \sigma_{TR}^2\}},$$ (24.2)

where
   σ_{TT}^2 and σ_{TR}^2 are the total variances for the test product and the reference
      product, respectively
   σ_{T0}^2 is the scale parameter specified by the user

A typical approach is to construct a one-sided 95% confidence interval for
$\theta_I(\theta_P)$ for the assessment of individual (population) bioequivalence. If the
one-sided 95% upper confidence limit is less than the bioequivalence limit
of $\theta_I(\theta_P)$, we then conclude that the test product is bioequivalent to that of the
reference product in terms of individual (population) bioequivalence. More
details regarding IBE and PBE can be found in Chow and Liu (2008).

24.3.3 Moment-Based Criteria versus Probability-Based Criteria

Schall and Luus (1993) proposed the moment-based and probability-based
measures for the expected discrepancy in PK responses between drug products. The moment-based measure suggested by Schall and Luus (1993) is
based on the following expected mean-squared differences:

$$d(Y_j; Y_{j'}) = \begin{cases} E(Y_T - Y_R)^2 & \text{if } j = T \text{ and } j' = R, \\ E(Y_R - Y_R')^2 & \text{if } j = R \text{ and } j' = R. \end{cases} \tag{24.3}$$

For some prespecified positive number r, one of the probability-based
measures for the expected discrepancy is given as (Schall and Luus, 1993)

$$d(Y_j; Y_{j'}) = \begin{cases} P\{|Y_T - Y_R| < r\} & \text{if } j = T \text{ and } j' = R, \\ P\{|Y_R - Y_R'| < r\} & \text{if } j = R \text{ and } j' = R. \end{cases} \tag{24.4}$$

where
   $d(Y_T; Y_R)$ measures the expected discrepancy for some PK metric between
      test and reference formulations
   $d(Y_R; Y_R')$ provides the expected discrepancy between the repeated admin-
      istrations of the reference formulation

The role of $d(Y_R; Y_R')$ in the formulation of bioequivalence criteria is to serve
as a control. The rationale is that the reference formulation should be bioequivalent to itself. Therefore, for the moment-based measures, if the test formulation is indeed bioequivalent to the reference formulation, then $d(Y_T; Y_R)$
should be very close to $d(Y_R; Y_R')$. It follows that if the criteria are functions
of the difference (or ratio) between $d(Y_T; Y_R)$ and $d(Y_R; Y_R')$, bioequivalence is
concluded if they are smaller than some prespecified limit. On the other hand,
for probability-based measures, if the test formulation is indeed bioequivalent

to the reference formulation, as compared with $d(Y_R; Y_R')$, $d(Y_T; Y_R)$ should be relatively large. As a result, bioequivalence is concluded if the criteria based on the probability-based measure is greater than some prespecified limit.

24.3.4 Similarity Factor for Dissolution Profile Comparison

In vivo bioequivalence studies are surrogate trials for assessing equivalence between test and reference formulations based on the rate and extent of drug absorption in humans to establish similar effectiveness and safety under the fundamental bioequivalence assumption. However, drug absorption depends on the dissolved state of drug product and dissolution testing provides a rapid *in vitro* assessment of the rate and extent of drug release. Leeson (1995), therefore, suggested that *in vitro* dissolution testing be used as a surrogate for *in vivo* bioequivalence studies to assess equivalence between the test and reference formulations for post-approval changes. For comparison of dissolution profiles, the FDA guidance suggests considering the assessment of (1) the overall profile similarity and (2) similarity at each sampling time point (FDA, 1997). In order to achieve these two objectives, based on Moore and Flanner (1996), both the FDA SUPAC guidance (SUPAC-IR, 1995) and guidance on dissolution testing (FDA, 1997) suggest the similarity and difference factor for the assessment of similarity. The similarity factor is then defined as the logarithmic reciprocal square root transformation of 1 plus the mean-squared (the average sum of squares) difference in mean cumulative percentage dissolved between the test and reference formulations over all sampling time points. That is,

$$f_2 = 50 \log \left\{ \left[1 + \frac{Q}{n} \right]^{-0.5} 100 \right\},$$ (24.5)

where

$$Q = \sum_{t=1}^{n} \left(\mu_{Rt} - \mu_{Tt} \right)^2,$$

where log denotes the logarithm based on 10.

On the other hand, the difference factor is the sum of the absolute difference in mean cumulative percentage dissolved between the test and reference formulations divided by the sum of the mean cumulative dissolved of the reference formulation.

$$f_1 = \frac{\sum_{t=1}^{n} \left| \mu_{Ri} - \mu_{Ti} \right|}{\sum_{t=1}^{n} \mu_{Ri}}.$$ (24.6)

It should be noted that it is not clear whether the definitions of f_1 and f_2 provided by Moore and Flanner (1996) and in the SUPAC and guidance on dissolution testing are defined based on the population means or the sample averages. However, following the traditional statistical inference with ability for the evaluation of error probability, we define both f_1 and f_2 based on the population mean dissolution rates. It follows that f_1 and f_2 are population parameters for the assessment of similarity of dissolution profiles between the test and reference formulations.

24.3.5 Consistency in Manufacturing Process/Quality Control

Tse et al. (2006) proposed a statistical QC method to assess a proposed index to test consistency between raw materials (which are from different resources) and/or between final products manufactured by different manufacturing processes. The consistency index is defined as the probability that the ratio of the characteristics (e.g., potency) of the drug products produced by two different manufacturing processes is within a prespecified limit of consistency. A consistency index close to 1 indicates that the characteristics of the drug products from the two manufacturing processes are almost identical. The idea for testing consistency is to construct a 95% confidence interval for the proposed consistency index under a sampling plan. If the constructed 95% confidence lower limit is greater than a prespecified QC lower limit, then we claim that the final products produced by the two manufacturing processes are consistent.

Let U and W be the characteristics of the drug products from two different manufacturing processes, where $X = \log U$ and $Y = \log W$ follows normal distributions with means μ_X, μ_Y and variances V_X, V_Y, respectively. Similar to the idea of using $P(X < Y)$ to assess reliability in statistical QC (Church and Harris, 1970; Enis and Geisser, 1971), Tse et al. (2006) proposed the following probability as an index to assess the consistency between the two different manufacturing processes:

$$p = P\left(1 - \delta < \frac{U}{W} < \frac{1}{1-\delta}\right),$$

where $0 < \delta < 1$ and is defined as a limit that allows for consistency. Tse et al. (2006) refer to p as the consistency index. Thus p tends to 1 as δ tends to 1. For a given δ, if p is close to 1, the materials U and W are considered to be identical. It should be noted that a small δ implies the requirement of a high degree of consistency between material U and material W. In practice, it may be difficult to meet this narrow specification for consistency. Tse et al. (2006) proposed the following QC criterion. If the probability that the lower limit $LL(\hat{p})$ of the constructed $(1 - \alpha)100\%$ confidence interval of p is greater than or equal to a prespecified QC lower limit, say, QC_L, and exceeds a prespecified number β (say $\beta = 80\%$), then we claim that U and W are consistent or similar. In other words, U and W are consistent or similar if $P(QC_L \leq LL(\hat{p})) \geq \beta$, where β is a prespecified constant.

24.4 Scientific Issues

24.4.1 Biosimilarity in Biological Activity

Pharmacological or biological activity is an expression describing the beneficial or adverse effects of a drug on living matter. When the drug is a complex chemical mixture, this activity is exerted by the substance's active ingredient or pharmacophore but can be modified by other constituents. A crucial component of biological activity is a substance's toxicity. Activity is generally dosage-dependent and it is not uncommon to have effects ranging from beneficial to adverse for one substance when going from low to high doses. Activity depends critically on fulfillment of the ADME criteria.

Note that the new EU Pharmaceutical Review legislation published on April 30, 2004 amended the EU community code on medicinal products to provide for the approval of biosimilars based on fewer preclinical and clinical data than had been required for the original reference product. The complexity of the protein and the knowledge of its structure–function relationships determine the types of information needed to establish similarity.

24.4.2 Similarity in Size and Structure

In practice, various *in vitro* tests such as the assessments of the primary amino acid sequence, charge, and hydrophobic properties are performed to compare the structural aspects of biosimilars with their originator molecules. However, it is of concern whether *in vitro* tests can be predictive of biological activity *in vivo* due to the fact that there are significant differences in biological activity despite similarities in size and structure. Besides, it is difficult to assess biological activity adequately as few animal models are able to provide data that can be extrapolated for an accurate and reliable prediction of biological activity in humans. Thus, controlled clinical trials remain the most reliable means of demonstrating similarity between a biosimilar molecule and the originator product in the clinic.

24.4.3 The Problem of Immunogenicity

Since all biologic products are biologically active molecules derived from living cells and have the potential to evoke an immune response, the immunogenicity is probably the most critical safety concern for the assessment of biosimilarity of follow-on biologics. The commonly seen possible causes of immunogenicity include, but are not limited to, (1) sequence differences between therapeutic protein and endogenous protein, (2) nonhuman sequences or epitopes, (3) structural alterations, (4) storage conditions, (5) purification during the manufacturing process, (6) formulation, (7) route, dose, and frequency of administration, and (8) patient status such as genetic background. Thus, the following questions are necessarily asked when assessing biosimilarity between biological products: (1) What is the immunogenic potential of the therapeutic protein? (2) What is the impact of the generating antibodies to the self-protein? (3) What is the impact of immunogenicity in preclinical toxicity (e.g., PK levels and dose-limiting toxicity)? (4) What is the impact of immunogenicity of the therapeutic protein on safety? (5) What are the risk evaluation and mitigation strategy processes required by the regulatory agency such as the FDA?

The immune responses to biologic products can lead to (1) anaphylaxis, (2) injection site reactions, (3) flu-like syndromes, and (4) allergic responses. Note that one of the most serious adverse events occurs when neutralizing antibodies to product cross-react with endogenous proteins that have a unique physiological role. The risk of immunogenicity can be reduced through stringent testing of the products during its development. It, however, should be noted that not only immunogenicity in animal does not predict immunogenicity in clinical trials, but analytical techniques also may not detect changes that may impact immunogenicity. Therefore, the immunogenicity of a biological product depends heavily upon the product quality attributes such as the physical, structural, and functional properties of the active pharmaceutical ingredients; and excipients, container closure, and deliver system. It turns out that similarity of the acceptable ranges of these quality attributes is crucial to the evaluation of equivalence between biosimilar and innovator products.

24.4.4 Manufacturing Process

Unlike small-molecule drug products, biological products are made of living cells. Thus, manufacturing of biologic products is a very complicated process, which involves the steps of (1) cell expansion, (2) cell production (in bioreactors), (3) recovery (through filtration or centrifugation), (4) purification (through chromatography), and (5) formulation. A small discrepancy at each step (e.g., purification) could lead to a significant difference in the final product, which might cause drastic changes in clinical outcomes. Thus, process control and validation play an important role for the success of the manufacturing of biological products. In addition, since at each step (e.g., purification), different methods may be used for different biological manufacturing processes (within the same company or at different biotech companies), a test for consistency is necessarily performed. Note that at the step of purification, the following chromatography media or resins are commonly considered: (1) gel filtration, (2) ion exchange, (3) hydrophobic interaction, (4) reversed-phase normal phase, and (5) affinity. Thus, at each step of the manufacturing process, primary performance characteristics should be identified, controlled, and tested for consistency for process control and validation.

24.4.5 Statistical Considerations

24.4.5.1 Fundamental Biosimilarity Assumption

Because of the complexity of biosimilar products, unlike chemical generics, although EMEA approved Omnitrope and Valtropin® in 2006 and the FDA approved Omnitrope also in 2006, both regulatory agencies required clinical equivalence trials. For example, the application documents of Valtropin submitted to the EMEA include the following clinical evaluations:

One pivotal bioequivalence study in healthy volunteers ($n = 24$)

One phase III pivotal clinical equivalence trial for efficacy and safety in children with growth hormone deficiency (GHD) ($n = 147$)

One phase III trial for efficacy and safety in children with Turner's syndrome (TS) ($n = 30$)

One phase III trial for efficacy and safety in children with TS conducted in Korea ($n = 60$)

One phase III trial for efficacy and safety in adults with GHD ($n = 92$)

Therefore, substantial clinical information was required by both regulatory agencies for the approval of Valtropin as compared to one pivotal bioequivalence study and possibly one food interaction study required by chemical generics. In fact, the amount of clinical information requested by the regulatory agencies is almost equivalent to an NDA or BLA of the innovator biological products. Consequently, reduction of the price of biosimilar products is severely limited and affordability and accessibility of the biological products including biosimilar products to the patients in need is seriously hampered.

However, similar to the chemical generic drug products, approval for biosimilar drug products can be treated as the evaluation of post-approval changes (ICH, 2005b) and this post-approval change is the change of drug manufacturers. Therefore, biosimilar drug products and the corresponding innovative biological products appear highly similar. In addition, based on the accumulated experience of relevant information and data, minute differences observed in the product characteristics are expected to have no clinically meaningful adverse effect of safety and efficacy profiles. Under this circumstance, biosimilar drug products and innovator products can be considered similar. Therefore, except for the traditional pivotal bioequivalence study, no further data from pivotal phase III trials should be requested. However, the above statement is based on a crucial assumption that at least one of the product characteristics are validated and reliable predictors of the safety and efficacy profiles of the biological products. As a result, the Fundamental Biosimilarity Assumption is met when a biosimilar product is claimed to be biosimilar to an innovator product based on some well-defined product characteristics and is therapeutically equivalent provided that the well-defined product characteristics are validated and reliable predictors of safety and efficacy of the products.

For the chemical generic products, the well-defined product characteristics are the exposure measures for early, peak, and total portions of the concentration–time curve. The fundamental bioequivalence assumption assumes that the equivalence in the exposure measures implies that they are therapeutically equivalent. However, due to the complexity of biosimilar drug products, one has to verify that some validated product characteristics are indeed reliable predictors of the safety and efficacy. It follows that the design and analysis for the evaluation of equivalence between a biosimilar drug product and innovator products are substantially different from those of chemical generic products.

24.4.5.2 Study Design

Since some of the biological products such as therapeutic antibody or pegylated proteins have a long half-life and equivalence in terms of absorption/bioavailability may not be sufficient. Demonstration of equivalence on clearance and half-life may be required to assess the risk of difference in elimination rate. As a result, the traditional crossover designs may not be optimal for the evaluation of equivalence between follow-on or biosimilar and innovator biological products. On the other hand, if the well-defined and validated product characteristics are PK/PD responses, it is then very important to investigate the extrapolation ability of equivalence in PK responses to equivalence in PD and to equivalence in efficacy responses. In order to ensure the internal validity of treatment comparisons, PK, PD, and efficacy responses should be evaluated simultaneously in the same trial. We consider a design (a) proposed in Figure 24.1.

Design (a) is a two-group parallel design in patients for the PK/PD/ efficacy bridging study with the disease which the innovator biological

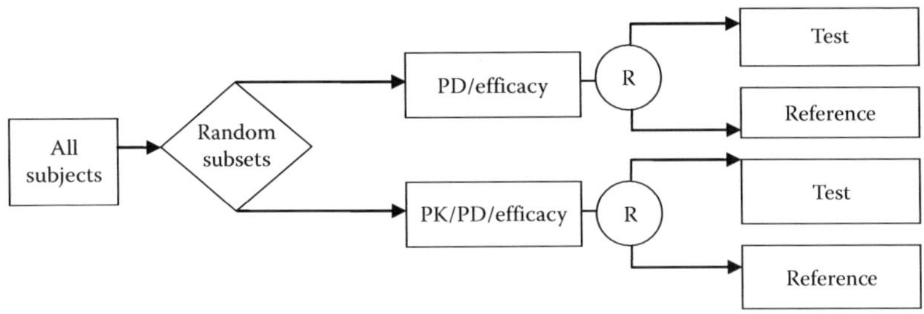

FIGURE 24.1
Diagram of the PK/PD/efficacy bridging study.

product indicates. After meeting the inclusion and exclusion criteria, patients are randomly divided into two groups. PD/efficacy/safety will be evaluated for the first group of patients (validation set). Additional PK responses will be assessed for the second group of patients (training set). A randomization in a 1:1 ratio will be performed separately for each group. The sample size of the second group will be large enough to provide sufficient power for the evaluation of bioequivalence based on PK responses. Calibration models will be built based on the PK/PD/efficacy response obtained from the patients in the training set. The PD/efficacy data from the validation set will be used to provide independent assessment of extrapolation ability of equivalence in PK responses to equivalence in PD/efficacy responses.

Design (a) may require quite a large sample size because of simultaneous evaluation of extrapolation ability of equivalence in PK to equivalence in PD and to equivalence in efficacy. One way to resolve this issue is to adopt the design for dose–response trials for evaluation of extrapolation ability of equivalence in some well-defined product characteristics to equivalence in efficacy. This design is referred to as design (b). Design (b) in fact consists of two dose–response trials: one for the biosimilar product and one for the innovator biological product, each with at least three dose levels with a placebo group. Eligible patients are first randomized into biosimilar or innovator groups. Within each group, patients are randomized again to receive one of the doses for the respective products. Well-defined product characteristics and primary efficacy endpoints are evaluated for all patients at their respective doses. Suppose that a statistically significant relationship represented by a simple linear regression equation can be established between the well-defined product characteristics and the primary efficacy endpoint through dose levels for the innovator product, after a suitable transformation. If a similar linear relationship can be also obtained for the biosimilar product and its corresponding linear regression equation is very close to the one for the innovator product, then equivalence in efficacy based on the primary efficacy endpoint may be claimed. Because the innovator product has been approved by the regulatory agencies due to its confirmed efficacy, the objective of design (b) is not to establish the efficacy of either biological products but to establish the similar patterns of the relationship between the well-defined product characteristics and primary efficacy endpoint for the two products. As a result, the sample size of design (b) can be reduced significantly (Figure 24.2).

24.4.5.3 Alternative Criteria for Biosimilarity

Due to the complexity, heterogeneity, and complication mechanisms of biological drug products, the difference in variability between biosimilar and innovator biological products in PK, PD, and clinical responses will be much larger than the difference observed between the conventional generic and

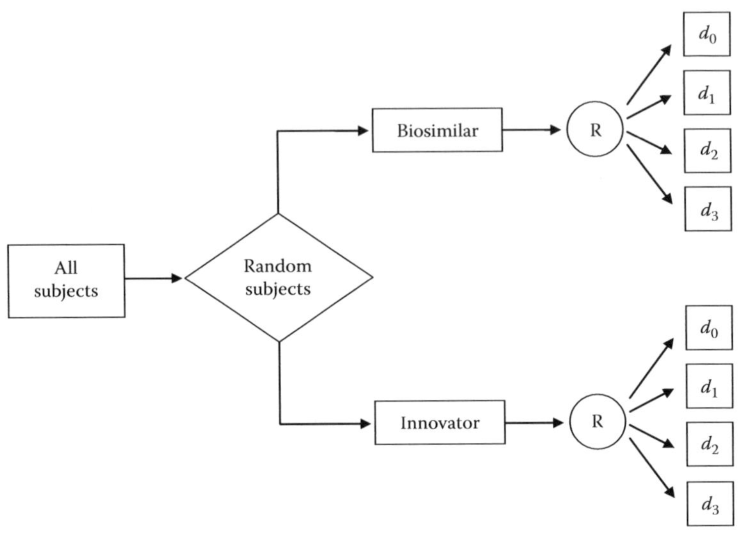

FIGURE 24.2
Design (b) for the evaluation of extrapolation ability.

innovator chemical drug product. Therefore ABE alone may not be sufficient to establish equivalence between the follow-on and innovator biological products. On the other hand, because of the masking effect, the aggregate metrics for population and IBE fail to address the closeness of the distributions of the responses between the follow-on and innovator biological products (Liu, 1998; Carrasco and Jover, 2003). Disaggregate metrics can address the masking effect suffered by the aggregate metrics and find the sources of in-equivalence. However, determination of individual equivalence margins with different interpretations is not an easy task. In addition, because of the involvement of multiparameters, any procedures based on a disaggregate metric for the evaluation of equivalence between follow-on and innovator biological products will tend to be conservative, especially in small samples. Furthermore, all current methods derived from the probability-based, moment-based, aggregate or disaggregate criteria are based on the normality assumption which is either extremely difficult to verify or simply not true. To resolve the above-mentioned dilemmas for evaluation of equivalence, the following concept of stochastic equivalence or stochastic non-inferiority is proposed.

Let $F(x)$ and $G(y)$ be the cumulative distribution functions of the responses for biosimilar and innovator biological products, respectively. Assuming that a large response value indicates a better efficacy, the follow-on and innovator biological products are said to be stochastically equivalent (two-sided) if the absolute difference between $F(x)$ and $G(x)$ is within some prespecified margins for all x. In other words, metric $\theta = \sup|F(x) - G(x)|$ and hypothesis for equivalence becomes

$$H_0 : \sup|F(x) - G(x)| \geq \eta \quad \text{for some } x$$

$$\text{versus} \tag{24.7}$$

$$H_a : \sup|F(x) - G(x)| < \eta \quad \text{for all } x.$$

Similarly, the biosimilar product is said to be stochastically non-inferior to the innovator counterpart if the difference between $F(x)$ and $G(x)$ is greater than $-\eta$. The corresponding hypothesis is given as

$$H_0 : \sup|F(x) - G(x)| \leq -\eta \quad \text{for some } x$$

$$\text{versus} \tag{24.8}$$

$$H_a : \sup|F(x) - G(x)| > -\eta \quad \text{for all } x.$$

However, the hypotheses in (24.7) and (24.8) can only be used for the evaluation of equivalence with respect to one study endpoint such as AUC or some primary efficacy endpoint. They cannot be utilized to assess whether the equivalence in product characteristic such as AUC can be extrapolated to the equivalence in primary efficacy endpoint. Both well-defined product characteristics and primary efficacy endpoint are measured for each patient. Therefore, group means of a well-defined product characteristic can be computed for each dose level for biosimilar and innovator product. Using the group means of the well-defined characteristic as the independent variable, a simple linear regression equation can be fit to the primary efficacy endpoint (dependent variable) for biosimilar and innovator biological products. It follows that the concept of the relative potency in the parallel-line bioassay can be then employed to investigate the extrapolation ability of equivalence in product characteristic to equivalence in efficacy (Finney, 1979). In other words, if the relative potency between the biosimilar and innovator biological products is within some predefined margins, then it can be concluded that equivalence in the product characteristic can be extrapolated to equivalence in efficacy. Figure 24.3 provides a graphical depiction of the application of the parallel-line assay to the evaluation of the extrapolation ability of equivalence in product characteristic to equivalence in efficacy. Let ρ be the relative potency of the biosimilar product to the innovator biological product. The hypothesis of extrapolation ability is given below:

$$H_0 : \rho \leq \rho_L \text{ or } \rho \geq \rho_U \quad \text{versus} \quad H_a : \rho_L < \rho < \rho_U, \tag{24.9}$$

where $0 < \rho_L < 1 < \rho_U$.

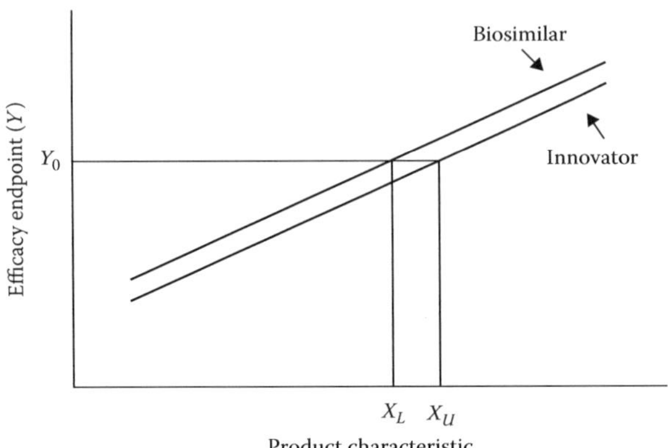

FIGURE 24.3
Parallel-line assays for the evaluation of extrapolation ability.

24.4.5.4 Statistical Methods

Although the methods based on the Kolmogorov–Simirnov type of statistics have been extensively investigated (Serfling, 1980), relatively few literature exists on the statistical tests for stochastic equivalence or non-inferiority. One method is to employ the naive asymptotic confidence band for $\theta = \sup|F(x) - G(x)|$ as the test statistics for hypotheses (24.7) and (24.8). If the $(1 - 2\alpha)100\%$ confidence band is totally contained with the band formed by the equivalence margins $(-\eta, \eta)$, then equivalence between the biosimilar and biological products can be concluded at the α significance level. Similarly, if the $(1 - \alpha)100\%$ lower confidence band is above the lower band formed by the lower margin $-\eta$, then the non-inferiority of the biosimilar product to the innovator biological product can be established at the α significance level. Derivation of the test statistics for hypotheses (24.7) and (24.8) at the boundary margins of the null hypothesis and the corresponding distribution and confidence interval requires further research. However, permutation and bootstrap technique can also be used to find the distribution of the test statistics and the corresponding confidence intervals empirically.

Because of the nature of design (a) proposed above, it is in fact an active control trial without a placebo-controlled arm where the follow-on biological product is the test treatment and its innovator counterpart is the active control treatment. Assuming that the innovator biological product was approved due to its superior efficacy over placebo, the equivalence in PD/efficacy is in fact the equivalence in relative efficacy as compared to the (putative) placebo of the follow-on biological product with the innovator counterpart. On the other hand, the prior information of the comparison of the innovator biological product to the placebo can be incorporated into the determination of equivalence margins and the evaluation of equivalence between the follow-on and innovator biological products. The Bayesian design proposed by Simon (1999) may be applied to derive the procedures for the assessment of equivalence based on PD/efficacy endpoints.

Calibration models have been used to correlate the surrogate responses with the true endpoints (Sargent, 2005). Because PK, PD, and efficacy responses are random variables, a mixed-models approach was suggested to assess surrogates as trial endpoints (Korn et al., 2005). The measurement error models can be used to establish calibration models (Cheng and Van Ness, 1999). With the established calibration model obtained from the data of the training set of design (a),

$$P_{efpk} = P(\text{equivalence in efficacy responses}|\text{equivalence in PK responses}),$$

and

$$P_{pdpk} = P(\text{equivalence in PD responses}|\text{equivalence in PK responses})$$

can be estimated for the evaluation of the extrapolation ability of the PK responses using the data from the validation set of design (a). If P_{efpk} is sufficiently high, then the equivalence in PK responses can be extrapolated to the equivalence in efficacy, and no further phase III clinical evaluation of biosimilar product based on efficacy responses may be required. On the other hand, if P_{efpk} is low, then equivalence in PK responses cannot predict the equivalence in efficacy responses, and phase III clinical trials for the evaluation of follow-on biological products are required. Under design (a), the sample size required for the bioequivalence evaluation based on PK responses of the training set may be different from the validation set for the assessment of extrapolation. It is of interest to compare the sample size required by design (a) with the total sample size required for the full clinical evaluation of the biosimilar product.

For design (b), the standard statistical method for the analysis of parallel-line assays can be used to construct the $(1 - 2\alpha)100\%$ confidence interval for the relative potency in the following steps:

Step 1: Fit a linear regression equation to the primary efficacy endpoint with the group mean of the product characteristic at each dose level as the independent variable separately for the biosimilar and innovator biological products. This may be done after a suitable transformation.

Step 2: If the estimate of the slope of any one product is not significant at the predefined level, then conclude that no simple relationship can be established between the product characteristic and primary endpoint, and hence, a full clinical evaluation of the biosimilar product is required. Otherwise, go to step 3.

Step 3: Test whether the two estimated simple linear regressions are parallel at the predefined significance level. If the two estimated linear regressions are not parallel, then further clinical evaluation of the biosimilar product is warranted. Otherwise, proceed to step 4.

Step 4: Compute the estimated relative potency and its corresponding $(1 - 2\alpha)$ 100% confidence interval. If the $(1 - 2\alpha)$100% confidence interval for the relative potency is within the predefined margins (ρ_L, ρ_U), then equivalence in the product characteristic can be extrapolated to equivalence in the primary efficacy endpoint at the α significance level.

Otherwise, further clinical investigation of the biosimilar product is needed. The objective of application of design (b) is not to establish the efficacy of the biosimilar product but rather to test whether the relative potency is within some predefined limits. Therefore, the required sample size for design (b) is determined upon the test for positive slope and equivalence margins. Therefore, the sample size required by design (b) may be smaller than that for design (a) or for the full clinical evaluation of a biosimilar product. However, further theoretical work or simulation studies are needed to address this issue.

24.5 Assessing Similarity Using Genomic Data

Although factors such as age, gender, education or social-economic status, smoking habit, weight, sexual orientation, and underlying disease characteristics at the baseline may contribute to the variation among patients, one of the most important reasons is the genetic or genomic variations among trial participants. As a result, due to genetic variations and genetic-by-environmental interaction, patients respond differently to the same treatment or therapeutic regimen. After the completion of the Human Genome Project, the disease targets at the molecular level can be identified and hence biochip products based on heritable DNA markers, mutations, and expression patterns for the detection of diseases using microarrays technology are possible. Genomic technologies such as DNA sequencing, mRNA transcript profiling, and comparative genomic hybridization have increased the possibility of identifying those patients who are most likely to benefit from a molecularly targeted drug, and this also indicates an increasing importance of diagnostic tests for the identification of molecular targets, and an increasing demand of targeted clinical trials conducted for the individualized treatment of patients (Liu and Chow, 2008; Liu and Lin, 2008; Liu, et al., 2009). A new generation of molecularly targeted agents has been developed and approved by the regulatory agencies such as the FDA and EMEA. Many of these drugs benefit only a subset of treated patients and may be overlooked by the traditional, broad-eligibility approach to randomized clinical trials.

A major portion of targeted drugs are biological products. It follows that the evaluation of biosimilar products should take into account genomic information. Chow, Shao, and Li (2004) proposed methods for the evaluation of bioequivalence using genomic data. Although their methods were derived for chemical generics, the same principles can be applied to the evaluation of biosimilar products. Similarly, the genomic information can be incorporated into the linear regression equations as possible covariates to reduce the variability for estimation and inference about the relative potency.

Because biological products are peptides or protein products with primary, secondary, tertiary, and quaternary structures, immunogenicity is an extremely important safety issue. On the other hand, unlike chemical generic products, no two biological products are exactly the same, even for the different batches of any innovator product. It follows that the evaluation of immunogenicity of biosimilar products is more important and difficult than that of chemical generic products and usually requires large clinical trials. However, identification of potential immunodominant positions and prediction of antigenic variants may provide a way to evaluate the immunogenicity of biosimilar products without extensive clinical immunogenicity trials (Lee et al., 2007; Liao et al., 2008). Because the linear sequence of the primary structure usually determines the tertiary structure, one can conduct pairwise comparisons between the amino acid sequences of the biosimilar product with those of the innovator biological product and use the antigenic distance as a measure for the evaluation of the similarity of the amino acid sequences between the biosimilar and innovator products. If the potential antigenic sites of the amino acid sequences are known predictors of antigenic variants, this information along with other data can be used for equivalence evaluation of the amino acid sequences between the biosimilar and innovator biological products. The cut-off margin based on the antigenic distance is 4 as recommended by Lee and Chen (2004). However, a more stringent margin could be applied if the potential antigenic sites are known to induce serious immunogenicity reactions.

24.6 Concluding Remarks

Because of the size and complexity of the active ingredients and the nature of the manufacturing process, biological products are different from traditional chemical drugs with small molecular weights. Many pointed out that no two biological products are the same (Schellekens, 2004; Roger and Mikhail, 2007). It is also important to know that traditional chemical generics are not exactly the same as their corresponding innovator product either. They are different in the excipients and their compositions and manufacturing methods and processes. In addition, no two batches of the same innovator biological products are the same. This is the reason why ICH issued Q5E Guideline on Comparability of Biotechnological/Biological Products Subject to Changes in Their Manufacturing Process in 2005 to address the issues of post-approval changes. The same principles to evaluate post-approval changes for the innovator biological products can be and should be also applied to assess the similarity between the biosimilar and innovator biological products because the change is in the manufacture.

Under the authorization of the Drug Price Competition and Patent Restoration Act (Hatch-Waxman Act), the FDA started in 1984 to approve traditional chemical generic drug products through the ANDA process. Although at the beginning of the implementation of the Hatch-Waxman Act there were also serious concerns about whether the chemical generic products can deliver equivalent efficacy and safety, experience accumulated over the past 25 years indicates that the fundamental bioequivalence assumption is a sound basis for the approval of chemical generic products without conducting costly and time-consuming clinical trials. However, for the biological products which are much more expensive than the traditional chemical drug products, there are no similar legislatures passed by the U.S. Congress.

One of the key scientific issues is to search product characteristics which are strongly correlated with efficacy, immunogenicity, and safety so that the fundamental biosimilar assumption can be verified. The crucial statistical methodology that should be rapidly developed is to evaluate the extrapolation ability from equivalence in product characteristics to equivalence in efficacy or immunogenicity. Until the above issues are satisfactorily and adequately addressed, biosimilar products will be still approved on individual basis with requirement of clinical data without any possibility of reducing price and increasing patient accessibility.

Note that as more innovative biologic products are going off patents, there is a need for the establishment of a pathway for regulatory approval of follow-on biologics. As a result, the U.S. FDA hosted a *Public Hearing on Approval Pathway for Biosimilar and Interchangeable Biological Products* between November 2 and 3, 2010, at Silver Spring, Maryland. At the public hearing, some scientific factors regarding the assessment of biosimilarity (e.g., criteria, design, and statistical methods), drug interchangeability (e.g., the issues of alternating and switching), and quality (e.g., test for comparability in manufacturing process) of follow-on biologics were discussed. However, these scientific issues remain unanswered and require further research and investigation.

25

Generalizability/Reproducibility Probability

25.1 Introduction

For marketing approval of a new drug product, the United States Food and Drug Administration (FDA) requires that at least two adequate and well-controlled clinical trials be conducted to provide substantial evidence regarding the effectiveness of the drug product under investigation (FDA, 1988). The purpose of conducting the second trial is to study whether the observed clinical result from the first trial is reproducible on the same target patient population. Let H_0 be the null hypothesis that the mean response of the drug product is the same as the mean response of a control (e.g., placebo) and H_a be the alternative hypothesis. An observed result from a clinical trial is said to be significant if it leads to the rejection of H_0. It is often of interest to determine whether clinical trials that produced significant clinical results provide substantial evidence to assure that the results will be reproducible in a future clinical trial with the same study protocol. Under certain circumstance, the FDA Modernization Act (FDAMA) of 1997 includes a provision (Section 115 of FDAMA) to allow data from one adequate and well-controlled clinical trial investigation and confirmatory evidence to establish the effectiveness for the risk–benefit assessment of drug and biological candidates for approval. Suppose that the null hypothesis H_0 is rejected if and only if $|T| > c$, where c is a positive known constant and T is a test statistic. This is usually related to a two-sided alternative hypothesis. The discussion for one-sided alternative hypotheses is similar. In statistical theory, the probability of observing a significant clinical result when H_a is indeed true is referred to as the power of the test procedure. If the statistical model under H_a is a parametric model, then the power is

$$P(|T| > c \mid H_a) = P(|T| > c \mid \theta), \tag{25.1}$$

where θ is an unknown parameter or vector of parameters. Suppose now that one clinical trial has been conducted and the result is significant. What is the probability that the second trial will produce a significant result, that is, the significant result from the first trial is reproducible? Mathematically, if the two trials are independent, the probability of observing a significant result from the second trial when H_a is true is still given by (25.1), regardless of whether the result from the first trial is significant or not. However, information from the first clinical trial should be useful in the evaluation of the probability of observing a significant result in the second trial. This leads to the concept of reproducibility probability, which is different from the power defined by (25.1).

In general, the reproducibility probability is a person's subjective probability of observing a significant clinical result from a future trial, when he/she observes significant results from one or several previous trials. For example, Goodman (1992) considered the reproducibility probability as the probability in (25.1) with θ replaced by its estimate based on the data from the previous trial(s). In other words, the reproducibility probability can be defined as an estimated power of the future trial using the data from the previous trial(s). In Section 25.2, we will study how to evaluate the reproducibility probability using this approach, under several study designs. When the reproducibility probability is used to provide an evidence of the effectiveness of a drug product, the estimated power approach may produce a rather optimistic result. A more conservative approach is to define the reproducibility probability as a lower confidence bound of the power of the second trial. This will be studied in Section 25.3. Perhaps a more sensible definition of reproducibility probability can be obtained by using the Bayesian approach. Under the Bayesian approach, the unknown parameter θ is a random vector with a prior distribution $\pi(\theta)$ assumed to be known. Thus, the reproducibility probability can be defined as the conditional probability of $|T| > c$ in the future trial, given the data set x observed from the previous trial(s), that is,

$$P(|T| > c \mid x) = \int P(|T| > c \mid \theta)\pi(\theta \mid x)d\theta, \tag{25.2}$$

where
$T = T(y)$ is based on the data set y from the future trial,
$\pi(\theta|x)$ is the posterior density of θ, given x.

More discussion about this Bayesian approach will be given in Section 25.4. In applications, the reproducibility probability is useful when the clinical trials are conducted sequentially. It provides important information for regulatory agencies in deciding whether it is necessary to require a second clinical trial when the result from the first clinical trial is strongly significant (Section 25.5). On the other hand, if the second trial is necessary, the reproducibility probability can be used for sample size adjustment of the second trial.

In the pharmaceutical industry, the sponsors are often interested in evaluating their drug products from one patient population (e.g., adult patient population) to another patient population (e.g., elderly patient population) to increase the exclusivity of the drug products in the marketplace. However, it is a concern whether the clinical results can be generalized from the target patient population to a similar but different patient population due to differences in demographic or ethnic factors. In Section 25.5 we show how to use the reproducibility probability to study the generalizability of the clinical results from one patient population to a similar but different patient population. Some concluding remarks are given in the last section.

25.2 The Estimated Power Approach

To study the reproducibility probability, we need to specify the test procedure, that is, the form of the test statistic T. We will consider several different study designs.

25.2.1 Two Samples with Equal Variances

Suppose that a total of $n = n_1 + n_2$ patients are randomly assigned to two groups, a treatment group and a control group. In the treatment group, n_1 patients receive the treatment (or a test drug) and produce responses $x_{11}, \ldots, x_{1n_1}$. In the control group, n_2 patients receive the placebo (or a reference drug) and produce responses $x_{21}, \ldots, x_{2n_2}$. This design is a typical two-group parallel design in clinical trials. We assume that x_{ij}'s are independent and normally distributed with means μ_i, $i = 1, 2$, and a common variance σ^2. Suppose that the hypotheses of interest are

$$H_0 : \mu_1 - \mu_2 = 0 \quad \text{versus} \quad H_a : \mu_1 - \mu_2 \neq 0. \tag{25.3}$$

The discussion for a one-sided H_a is similar.

Consider the commonly used two-sample t-test which rejects H_0 if and only if $|T| > t_{0.975, n-2}$, where $t_{0.975, n-2}$ is the 97.5th percentile of the t distribution with $n-2$ degrees of freedom

$$T = \frac{\bar{x}_1 - \bar{x}_2}{\sqrt{\left(((n_1 - 1)s_1^2 + (n_2 - 1)s_2^2)/(n - 2)\right)}\sqrt{(1/n_1 + 1/n_2)}} \tag{25.4}$$

and $\bar{x}_i$ and s_i^2 are, respectively, the sample mean and variance based on the data from the ith treatment group. The power of T for the second trial is then given by

$$p(\theta) = P(|T(y)| > t_{0.975, n-2}) = 1 - \Im_{n-2}(t_{0.975, n-2} \mid \theta) + \Im_{n-2}(-t_{0.975, n-2} \mid \theta) \tag{25.5}$$

where

$$\theta = \frac{\mu_1 - \mu_2}{\sigma\sqrt{(1/n_1 + 1/n_2)}} \tag{25.6}$$

and $\Im_{n-2}(\bullet \mid \theta)$ denotes the distribution function of the noncentral t distribution with $n - 2$ degrees of freedom and the noncentrality parameter θ. Note that $p(\theta) = p(|\theta|)$.

TABLE 25.1

Values of the Power Function $p(\theta)$ in (25.5)

	Total Sample Size							
$\|\theta\|$	10	20	30	40	50	60	100	∞
1.96	0.407	0.458	0.473	0.480	0.484	0.487	0.492	0.500
2.02	0.429	0.481	0.496	0.504	0.508	0.511	0.516	0.524
2.08	0.448	0.503	0.519	0.527	0.531	0.534	0.540	0.548
2.14	0.469	0.526	0.542	0.550	0.555	0.557	0.563	0.571
2.20	0.490	0.549	0.565	0.573	0.578	0.581	0.586	0.594
2.26	0.511	0.571	0.588	0.596	0.601	0.604	0.609	0.618
2.32	0.532	0.593	0.610	0.618	0.623	0.626	0.632	0.640
2.38	0.552	0.615	0.632	0.640	0.645	0.648	0.654	0.662
2.44	0.573	0.636	0.654	0.662	0.667	0.670	0.676	0.684
2.50	0.593	0.657	0.675	0.683	0.688	0.691	0.697	0.705
2.56	0.613	0.678	0.695	0.704	0.708	0.711	0.717	0.725
2.62	0.632	0.698	0.715	0.724	0.728	0.731	0.737	0.745
2.68	0.652	0.717	0.735	0.743	0.747	0.750	0.756	0.764
2.74	0.671	0.736	0.753	0.761	0.766	0.769	0.774	0.782
2.80	0.690	0.754	0.771	0.779	0.783	0.786	0.792	0.799
2.86	0.708	0.772	0.788	0.796	0.800	0.803	0.808	0.815
2.92	0.725	0.789	0.805	0.812	0.816	0.819	0.824	0.830
2.98	0.742	0.805	0.820	0.827	0.831	0.834	0.839	0.845
3.04	0.759	0.820	0.835	0.842	0.846	0.848	0.853	0.860
3.10	0.775	0.834	0.849	0.856	0.859	0.862	0.866	0.872
3.16	0.790	0.848	0.862	0.868	0.872	0.874	0.879	0.884
3.22	0.805	0.861	0.874	0.881	0.884	0.886	0.890	0.895
3.28	0.819	0.873	0.886	0.892	0.895	0.897	0.901	0.906
3.34	0.832	0.884	0.897	0.902	0.905	0.907	0.911	0.916
3.40	0.844	0.895	0.907	0.912	0.915	0.917	0.920	0.925
3.46	0.856	0.905	0.916	0.921	0.924	0.925	0.929	0.932
3.52	0.868	0.914	0.925	0.929	0.932	0.933	0.936	0.940
3.58	0.879	0.923	0.933	0.937	0.939	0.941	0.943	0.947
3.64	0.889	0.931	0.940	0.944	0.946	0.947	0.950	0.953
3.70	0.898	0.938	0.946	0.950	0.952	0.953	0.956	0.959
3.76	0.907	0.944	0.952	0.956	0.958	0.959	0.961	0.965
3.82	0.915	0.950	0.958	0.961	0.963	0.964	0.966	0.969
3.88	0.923	0.956	0.963	0.966	0.967	0.968	0.970	0.973
3.94	0.930	0.961	0.967	0.970	0.971	0.972	0.974	0.977

Source: Shao, J. and Chow, S.C., *Stat. Med.*, 21, 1727, 2002. With permission.

Values of $p(\theta)$ as a function of $|\theta|$ are provided in Table 25.1. Using the idea of replacing θ by its estimate $T(x)$ (Goodman, 1992), where T is defined by (25.4), we obtain the following reproducibility probability:

$$\hat{P} = 1 - \Im_{n-2}(t_{0.975,n-2} \mid T(x)) + \Im_{n-2}(-t_{0.975,n-2} \mid T(x)), \qquad (25.7)$$

which is a function of $|T(x)|$. When $|T(x)| > t_{0.975,n-2}$,

$$\hat{P} \approx \begin{cases} 1 - \Im_{n-2}(t_{0.975,n-2} \mid T(x)) & \text{if } T(x) > 0, \\ \Im_{n-2}(-t_{0.975,n-2} \mid T(x)) & \text{if } T(x) < 0. \end{cases} \tag{25.8}$$

If $\Im_{n-2}$ is replaced by the normal distribution and $t_{0.975,n-2}$ is replaced by the normal percentile, then formula (25.8) is the same as that given by Goodman (1992), who studied the case where the variance σ^2 is known. Table 25.1 can be used to find the reproducibility probability $\hat{P}$ in (25.7) with a fixed sample size n. For example, if $T(x) = 2.9$ was observed in a clinical trial with $n = n_1 + n_2 = 40$, then the reproducibility probability is 0.807. If $T(x) = 2.9$ was observed in a clinical trial with $n = 36$, then an extrapolation of the results in Table 25.1 (for $n = 30$ and 40) leads to a reproducibility probability of 0.803.

25.2.2 Two Samples with Unequal Variances

Consider the problem of testing hypotheses (25.3) under the two-group parallel design without the assumption of equal variances. That is, x_{ij}'s are independently distributed as $N(\mu_i, \sigma_i^2)$, $i = 1, 2$. When $\sigma_1^2 \neq \sigma_2^2$, there exists no exact testing procedure for the hypotheses in (25.3). When both n_1 and n_2 are large, an approximate 5% level test rejects H_0 when $|T| > z_{0.975}$, where

$$T = \frac{\bar{x}_1 - \bar{x}_2}{\sqrt{(s_1^2/n_1 + s_2^2/n_2)}}. \tag{25.9}$$

Since T is approximately distributed as $N(\theta, 1)$ with

$$\theta = \frac{\mu_1 - \mu_2}{\sqrt{(\sigma_1^2/n_1 + \sigma_2^2/n_2)}} \tag{25.10}$$

the reproducibility probability obtained by using the estimated power approach is given by

$$\hat{P} = \Phi(T(x) - z_{0.975}) + \Phi(-T(x) - z_{0.975}). \tag{25.11}$$

When the variances under different treatments are different and the sample sizes are not large, a different study design, such as a matched-pair parallel design or a 2×2 crossover design, is recommended. A matched-pair parallel design involves m pairs of matched patients. One patient in each pair is assigned to the treatment group and the other is assigned to the control group. Let x_{ij} be the observation from the jth pair and the ith group. It is assumed that the differences $x_{1j} - x_{2j}$, $j = 1, \ldots, m$, are independent and identically distributed as $N(\mu_1 - \mu_2, \sigma_D^2)$. Then the null hypothesis H_0 is rejected at the 5% level of significance if $|T| > t_{0.975, m-1}$, where

$$T = \frac{\sqrt{m}(\bar{x}_1 - \bar{x}_2)}{\hat{\sigma}_D^2} \qquad (25.12)$$

and $\hat{\sigma}_D^2$ is the sample variance based on the differences $x_{1j} - x_{2j}$, $j = 1, \ldots, m$. Note that T has the noncentral t distribution with $m - 1$ degrees of freedom and the noncentrality parameter

$$\theta = \frac{\sqrt{m}(\mu_1 - \mu_2)}{\sigma_D^2}. \qquad (25.13)$$

Consequently, the reproducibility probability obtained by using the estimated power approach is given by (25.7) with T defined by (25.12) and $n - 2$ replaced by $m - 1$.

Suppose that the study design is a 2×2 crossover design in which n_1 patients receive the treatment in the first period and the placebo in the second period and n_2 patients receive the placebo in the first period and the treatment in the second period. Let x_{lij} be the normally distributed observation from the jth patient at the ith period and lth sequence. Then the treatment effect μ_D can be unbiasedly estimated by

$$\hat{\mu}_D = \frac{\bar{x}_{11} - \bar{x}_{12} - \bar{x}_{21} + \bar{x}_{22}}{2} \sim N\left(\mu_D, \frac{\sigma_D^2}{4}\left(\frac{1}{n_1} + \frac{1}{n_2}\right)\right),$$

where $\bar{x}_{li}$ is the sample mean based on x_{lij}, $j = 1, \ldots, n_l$ and $\sigma_D^2 = \mathrm{var}(x_{l1j} - x_{l2j})$. An unbiased estimator of σ_D^2 is

$$\hat{\sigma}_D^2 = \frac{1}{n_1 + n_2 - 2}\sum_{l=1}^{2}\sum_{j=1}^{m}(x_{l1j} - x_{l2j} - \bar{x}_{l1} + \bar{x}_{l2})^2,$$

which is independent of $\hat{\mu}_D$ and distributed as $\sigma_D^2/(n_1 + n_2 - 2)$ times the chi-square distribution with $n_1 + n_2 - 2$ degrees of freedom. Thus, the null hypothesis $H_0 : \mu_D = 0$ is rejected at the 5% level of significance if $|T| > t_{0.975, n-2}$, where $n = n_1 + n_2$ and

$$T = \frac{\hat{\mu}_D}{(\hat{\sigma}_D/2)\sqrt{(1/n_1 + 1/n_2)}}. \qquad (25.14)$$

Note that T has the noncentral t distribution with $n - 2$ degrees of freedom and the noncentrality parameter

$$\theta = \frac{\mu_D}{(\sigma_D/2)\sqrt{(1/n_1 + 1/n_2)}}. \tag{25.15}$$

Consequently, the reproducibility probability obtained by using the estimated power approach is given by (25.7) with T defined by (25.14).

25.2.3 Parallel-Group Designs

Parallel-group designs are often adopted in clinical trials to compare more than one treatment with a placebo control or to compare one treatment, one placebo control, and one active control. Let $a \geq 3$ be the number of groups and x_{ij} be the observation from the jth patient in the ith group, $j = 1,\ldots,n_i$, $i = 1,\ldots,a$. Assume that x_{ij}'s are independently distributed as $N(\mu_i, \sigma^2)$. The null hypothesis H_0 is then $H_0 : \mu_1 = \mu_2 = \cdots = \mu_a$, which is rejected at the 5% level of significance if $T > F_{0.95;a-1,n-a}$, where $F_{0.95;a-1,n-a}$ is the 95th percentile of the F distribution with $a - 1$ and $n - a$ degrees of freedom, $n = n_1 + n_2 + \cdots + n_a$:

$$T = \frac{\text{SST}/(a-1)}{\text{SSE}/(n-a)}, \tag{25.16}$$

and

$$\text{SST} = \sum_{i=1}^{a} n_i(\bar{x}_i - \bar{x})^2, \quad \text{SSE} = \sum_{i=1}^{a}\sum_{j=1}^{n_i}(x_{ij} - \bar{x}_i)^2,$$

where
$\bar{x}_i$ is the sample mean based on the data in the ith group,
$\bar{x}$ is the overall sample mean.

Note that T has the noncentral F distribution with $a - 1$ and $n - a$ degrees of freedom and the noncentrality parameter

$$\theta = \sum_{i=1}^{a} \frac{n_i(\mu_i - \bar{\mu})^2}{\sigma^2},$$

where $\bar{\mu} = \sum_{i=1}^{a} n_i\mu_i/n$. Let $\Im_{a-1,v-\alpha}(\bullet|\theta)$ be the distribution function of T. Then, the power of the second clinical trial is

$$P(T(y) > F_{0.95;a-1,n-a}) = 1 - \Im_{a-1,n-a}(F_{0.95;a-1,n-a} \mid \theta).$$

Thus, the reproducibility probability obtained by using the estimated power approach is

$$\hat{P} = 1 - \Im_{a-1,n-a}(F_{0.95;a-1,n-a} \mid T(x)), \tag{25.17}$$

where $T(x)$ is the observed T based on the data x from the first clinical trial.

25.3 The Confidence Bound Approach

Since $\hat{P}$ in (25.7) or (25.11) is an estimated power, it provides a rather optimistic result. Alternatively, we may consider a more conservative approach, which considers a 95% lower confidence bound of the power as the reproducibility probability. Consider first the case of the two-group parallel design with a common unknown variance σ^2. Note that $T(x)$ defined by (25.4) has the noncentral t distribution with $n - 2$ degrees of freedom and the noncentrality parameter θ given by (25.6). Let $\Im_{n-2}(\bullet \mid \theta)$ be the distribution function of $T(x)$ for any given θ. It can be shown that $\Im_{n-2}(t \mid \theta)$ is a strictly decreasing function of θ for any fixed t. Consequently, a 95% confidence interval for θ is given by $(\hat{\theta}_-, \hat{\theta}_+)$, where $\hat{\theta}_-$ is the unique solution of $\Im_{n-2}(T(x) \mid \theta) = 0.975$ and $\hat{\theta}_+$ is the unique solution of $\Im_{n-2}(T(x) \mid \theta) = 0.025$ (see, e.g., Theorem 7.1 in Shao, 1999). Then, a 95% lower confidence bound for $|\theta|$ is

$$|\hat{\theta}|_- = \begin{cases} \hat{\theta}_- & \text{if } \hat{\theta}_- > 0 \\ -\hat{\theta}_+ & \text{if } \hat{\theta}_+ > 0 \\ 0 & \text{if } \hat{\theta}_- \leq 0 \leq \hat{\theta}_+ \end{cases} \tag{25.18}$$

and a 95% lower confidence bound for the power $p(\theta)$ in (25.5) is

$$\hat{P}_- = 1 - \Im_{n-2}(t_{0.975,n-2} \mid\mid \hat{\theta}\mid_-) + \Im_{n-2}(-t_{0.975,n-2} \mid\mid \hat{\theta}\mid_-) \tag{25.19}$$

if $|\hat{\theta}|_- > 0$ and $\hat{P}_- = 0$ if $|\hat{\theta}|_- = 0$. The lower confidence bound in (25.19) is useful when the clinical result from the first trial is highly significant.

Table 25.2 contains values of the lower confidence bound $|\hat{\theta}|_-$ corresponding to $|T(x)|$ values ranging from 4.5 to 6.5. If $4.5 \leq |T(x)| \leq 6.5$ and the value of $|\hat{\theta}|_-$ is found from Table 25.2, the reproducibility probability $\hat{P}_-$ in (25.19) can be obtained from Table 25.1. For example, suppose that $|T(x)| = 5$ was observed from a clinical trial with $n = 30$. From Table 25.2, $|\hat{\theta}|_- = 2.6$. Then, by Table 25.1, $\hat{P}_- = 0.709$.

TABLE 25.2

95% Lower Confidence Bound $|\hat{\theta}|_-$

$|T(x)|$	Total Sample Size							
	10	20	30	40	50	60	100	∞
4.5	1.51	2.01	2.18	2.26	2.32	2.35	2.42	2.54
4.6	1.57	2.09	2.26	2.35	2.41	2.44	2.52	2.64
4.7	1.64	2.17	2.35	2.44	2.50	2.54	2.61	2.74
4.8	1.70	2.25	2.43	2.53	2.59	2.63	2.71	2.84
4.9	1.76	2.33	2.52	2.62	2.68	2.72	2.80	2.94
5.0	1.83	2.41	2.60	2.71	2.77	2.81	2.90	3.04
5.1	1.89	2.48	2.69	2.80	2.86	2.91	2.99	3.14
5.2	1.95	2.56	2.77	2.88	2.95	3.00	3.09	3.24
5.3	2.02	2.64	2.86	2.97	3.04	3.09	3.18	3.34
5.4	2.08	2.72	2.95	3.06	3.13	3.18	3.28	3.44
5.5	2.14	2.80	3.03	3.15	3.22	3.27	3.37	3.54
5.6	2.20	2.88	3.11	3.24	3.31	3.36	3.47	3.64
5.7	2.26	2.95	3.20	3.32	3.40	3.45	3.56	3.74
5.8	2.32	3.03	3.28	3.41	3.49	3.55	3.66	3.84
5.9	2.39	3.11	3.37	3.50	3.58	3.64	3.75	3.94
6.0	2.45	3.19	3.45	3.59	3.67	3.73	3.85	4.04
6.1	2.51	3.26	3.53	3.67	3.76	3.82	3.94	4.14
6.2	2.57	3.34	3.62	3.76	3.85	3.91	4.03	4.24
6.3	2.63	3.42	3.70	3.85	3.94	4.00	4.13	4.34
6.4	2.69	3.49	3.78	3.93	4.03	4.09	4.22	4.44
6.5	2.75	3.57	3.86	4.02	4.12	4.18	4.32	4.54

Source: Shao, J. and Chow, S.C., *Stat. Med.*, 21, 1727, 2002. With permission.

Consider next the two-group parallel design with unequal variances σ_1^2 and σ_2^2. When both n_1 and n_2 are large, T given by (25.9) is approximately distributed as $N(\theta, 1)$ with θ given by (25.10). Hence, the reproducibility probability obtained by using the lower confidence bound approach is given by

$$\hat{P}_- = \Phi\big(|T(x)| - 2z_{0.975}\big)$$

with T defined by (25.9).

For the matched-pair parallel design described in Section 25.2, T given by (25.12) has the noncentral t distribution with $m - 1$ degrees of freedom and the noncentrality parameter θ given by (25.13). Hence, the reproducibility probability obtained by using the lower confidence bound approach is given by (25.19) with T defined by (25.12) and $n - 2$ replaced by $m - 1$. Suppose now that the study design is the 2×2 crossover design described in Section 25.2. Since T defined by (25.14) has the noncentral t distribution with $n - 2$ degrees of freedom and the noncentrality parameter θ given by (25.15), the reproducibility probability obtained by using the lower confidence bound approach is given by (25.19) with T defined by (25.14).

Finally, consider the parallel-group design described in Section 25.2.3. Since T in (25.16) has the noncentral F distribution with $a-1$ and $n-a$ degrees of freedom and the noncentrality parameter θ given by (25.17) and $\mathfrak{I}_{a-1,n-a}(t|\theta)$ is a strictly decreasing function of θ, the reproducibility probability obtained by using the lower confidence bound approach is

$$\hat{P}_- = 1 - \mathfrak{I}_{a-1,n-a}(F_{0.95;a-1,n-a}\,|\,\hat{\theta}_-),$$

where $\hat{\theta}_-$ is the solution of $\mathfrak{I}_{a-1,n-a}(T(x)|\theta) = 0.95$.

25.4 The Bayesian Approach

As discussed in Section 25.1, the reproducibility probability can be viewed as the posterior mean (see, e.g., Berger, 1985) of the power function $p(\theta) = P(|T| > c|\theta)$ for the future trial. Thus, under the Bayesian approach, it is essential to construct the posterior density $\pi(\theta|x)$ in formula (25.2), given the data set x observed from the previous trial(s).

Consider first the two-group parallel design described in Section 25.2.1 with equal variances, that is, x_{ij}'s are independent and normally distributed with means μ_1 and μ_2 and a common variance σ^2. If σ^2 is known, then the power for testing the hypotheses in (25.3) is $\Phi(\theta - z_{0.975}) + \Phi(-\theta - z_{0.975})$ with θ defined by (25.6). A commonly used prior for (μ_1, μ_2) is the noninformative prior $\pi(\mu_1, \mu_2) \equiv 1$. Consequently, the posterior density for θ is $N(T(x), 1)$, where

$$T = \frac{\bar{x}_1 - \bar{x}_2}{\sigma\sqrt{(1/n_1 + 1/n_2)}}$$

and the posterior mean given by (25.2) is

$$\int \left[\Phi(\theta - z_{0.975}) + \Phi(-\theta - z_{0.975})\right]\pi(\theta\,|\,x)d\theta = \Phi\left(\frac{T(x) - z_{0.975}}{\sqrt{2}}\right) + \Phi\left(\frac{-T(x) - z_{0.975}}{\sqrt{2}}\right).$$

When $T(x) > z_{0.975}$, this probability is nearly the same as

$$\Phi\left(\frac{|T(x)| - z_{0.975}}{\sqrt{2}}\right),$$

which is exactly the same as that in formula (1) in Goodman (1992).

For the more realistic situation where σ^2 is unknown, we need a prior for σ^2. A commonly used non-informative prior for σ^2 is the Lebesgue (improper) density $\pi(\sigma^2) = \sigma^{-2}$. Assume that the priors for μ_1, μ_2, and σ^2 are independent. The posterior density for (δ, u^2) is $\pi(\delta \,|\, u^2, x)\pi(u^2 \,|\, x)$, where

$$\delta = \frac{\mu_1 - \mu_2}{\sqrt{\{((n_1 - 1)s_1^2 + (n_2 - 1)s_2^2)/(n-2)\}}\sqrt{(1/n_1 + 1/n_2)}},$$

$$u^2 = \frac{(n-2)\sigma^2}{(n_1 - 1)s_1^2 + (n_2 - 1)s_2^2},$$

$$\pi(\delta \,|\, u^2, x) = \frac{1}{u}\phi\left(\frac{\delta - T(x)}{u}\right),$$

where
  ϕ is the density function of the standard normal distribution
  T is given by (25.4), and $\pi(u^2 \,|\, x) = f(u)$ with

$$f(u) = \left[\Gamma\left(\frac{n-2}{2}\right)\right]^{-1}\left(\frac{n-2}{2}\right)^{(n-2)/2} u^{-n}e^{-(n-2)/(2u^2)}.$$

Since θ in (25.6) is equal to δ/u, the posterior mean of $p(\theta)$ in (25.5) is

$$\hat{P} = \int_0^\infty \left[\int_{-\infty}^\infty p\left(\frac{\delta}{u}\right)\phi\left(\frac{\delta - T(x)}{u}\right)d\delta\right] 2f(u)du, \tag{25.20}$$

which is the reproducibility probability under the Bayesian approach. It is clear that $\hat{P}$ depends on the data x through the function $T(x)$.

The probability $\hat{P}$ in (25.20) can be evaluated numerically. A Monte Carlo method can be applied as follows. First, generate a random variate γ_j from the gamma distribution with the shape parameter $(n-2)/2$ and the scale parameter $2/(n-2)$, and generate a random variate δ_j from $N(T(x), u_j^2)$, where $u_j^2 = \gamma_j^{-1}$. Repeat this process independently N times to obtain (δ_j, u_j^2), $j = 1, \ldots, N$. Then $\hat{P}$ in (25.20) can be approximated by

$$\hat{P}_N = 1 - \frac{1}{N}\sum_{j=1}^N \left[\Im_{n-2}\left(t_{0.975,n-2}\left|\frac{\delta_j}{u_j}\right|\right) - \Im_{n-2}\left(-t_{0.975,n-2}\left|\frac{\delta_j}{u_j}\right|\right)\right]. \tag{25.21}$$

Values of $\hat{P}_N$ for $N = 10,000$ and some selected values of $T(x)$ and n are given by Table 25.3. It can be seen that in assessing reproducibility, the Bayesian

TABLE 25.3

Reproducibility Probability under the Bayesian Approach Approximated by Monte Carlo with Size 10,000

$|T(x)|$	Total Sample Size							
	10	**20**	**30**	**40**	**50**	**60**	**100**	**∞**
2.02	0.435	0.482	0.495	0.501	0.504	0.508	0.517	0.519
2.08	0.447	0.496	0.512	0.515	0.519	0.523	0.532	0.536
2.14	0.466	0.509	0.528	0.530	0.535	0.543	0.549	0.553
2.20	0.478	0.529	0.540	0.547	0.553	0.556	0.565	0.569
2.26	0.487	0.547	0.560	0.564	0.567	0.571	0.577	0.585
2.32	0.505	0.558	0.577	0.580	0.581	0.587	0.590	0.602
2.38	0.519	0.576	0.590	0.597	0.603	0.604	0.610	0.618
2.44	0.530	0.585	0.610	0.611	0.613	0.617	0.627	0.634
2.50	0.546	0.609	0.624	0.631	0.634	0.636	0.640	0.650
2.56	0.556	0.618	0.638	0.647	0.648	0.650	0.658	0.665
2.62	0.575	0.632	0.654	0.655	0.657	0.664	0.675	0.680
2.68	0.591	0.647	0.665	0.674	0.675	0.677	0.687	0.695
2.74	0.600	0.660	0.679	0.685	0.686	0.694	0.703	0.710
2.80	0.608	0.675	0.690	0.702	0.705	0.712	0.714	0.724
2.86	0.629	0.691	0.706	0.716	0.722	0.723	0.729	0.738
2.92	0.636	0.702	0.718	0.730	0.733	0.738	0.742	0.752
2.98	0.649	0.716	0.735	0.742	0.744	0.748	0.756	0.765
3.04	0.663	0.726	0.745	0.753	0.756	0.759	0.765	0.778
3.10	0.679	0.738	0.754	0.766	0.771	0.776	0.779	0.790
3.16	0.690	0.754	0.767	0.776	0.781	0.786	0.792	0.802
3.22	0.701	0.762	0.777	0.790	0.792	0.794	0.804	0.814
3.28	0.708	0.773	0.793	0.804	0.806	0.809	0.820	0.825
3.34	0.715	0.784	0.803	0.809	0.812	0.818	0.828	0.836
3.40	0.729	0.793	0.815	0.819	0.829	0.830	0.838	0.846
3.46	0.736	0.806	0.826	0.832	0.837	0.839	0.847	0.856
3.52	0.745	0.816	0.834	0.843	0.845	0.846	0.855	0.865
3.58	0.755	0.828	0.841	0.849	0.857	0.859	0.867	0.874
3.64	0.771	0.833	0.854	0.859	0.863	0.865	0.872	0.883
3.70	0.778	0.839	0.861	0.867	0.870	0.874	0.884	0.891
3.76	0.785	0.847	0.867	0.874	0.882	0.883	0.890	0.898
3.82	0.795	0.857	0.878	0.883	0.889	0.891	0.898	0.906
3.88	0.800	0.869	0.881	0.891	0.896	0.899	0.904	0.913
3.94	0.806	0.873	0.890	0.897	0.904	0.907	0.910	0.919

Source: Shao, J. and Chow, S.C., *Stat. Med.*, 21, 1727, 2002. With permission.
Note: Prior for $(\mu_1, \mu_2, \sigma^2) = \sigma^{-2}$ with respect to the Lebesgue measure.

approach is more conservative than the estimated power approach, but less conservative than the confidence bound approach.

Consider next the two-group parallel design with unequal variance and large n_i's. The approximate power for the second trial is

$$p(\theta) = \Phi(\theta - z_{0.975}) + \Phi(-\theta - z_{0.975}),$$

$$\theta = \frac{\mu_1 - \mu_2}{\sqrt{(\sigma_1^2/n_1 + \sigma_2^2/n_2)}}.$$

Suppose that we use the non-informative prior density $\pi(\mu_1, \mu_2, \sigma_1^2, \sigma_2^2) = \sigma_1^{-2}\sigma_2^{-2}, \sigma_1^2 > 0, \sigma_2^2 > 0$. Let $\tau_i^2 = \sigma_i^{-2}, i = 1, 2$ and $\varsigma^2 = (n_1\tau_1^2)^{-1} + (n_2\tau_2^2)^{-1}$. Then, the posterior density $\pi(\mu_1 - \mu_2 \mid \tau_1^2, \tau_2^2, x)$ is the normal density with mean $\bar{x}_1 - \bar{x}_2$ and variance ς^2 and the posterior density $\pi(\tau_1^2, \tau_2^2 \mid x) = \pi(\tau_1^2 \mid x)\pi(\tau_2^2 \mid x)$, where $\pi(\tau_i^2 \mid x)$ is the gamma density with the shape parameter $(n_i - 1)/2$ and the scale parameter $2/[(n_i - 1)s_i^2], i = 1, 2$. Consequently, the reproducibility probability is the posterior mean of $p(\theta)$ given by

$$\hat{P} = \int \left[\Phi\left(\frac{\bar{x}_1 - \bar{x}_2}{\sqrt{2}\varsigma} - \frac{z_{0.975}}{\sqrt{2}} \right) + \Phi\left(-\frac{\bar{x}_1 - \bar{x}_2}{\sqrt{2}\varsigma} - \frac{z_{0.975}}{\sqrt{2}} \right) \right] \pi(\varsigma \mid x)d\varsigma,$$

where $\pi(\varsigma \mid x)$ is the posterior density of ς constructed using $\pi(\tau_i^2 \mid x), i = 1, 2$. The Monte Carlo method previously discussed can be applied to approximate $\hat{P}$. Reproducibility probabilities under the Bayesian approach can be similarly obtained for the matched-pairs parallel design and the 2 × 2 crossover design described in Section 25.2.

Finally, consider the parallel-group design with a groups, where the power is given by

$$p(\theta) = 1 - \Im_{a-1, n-a}(F_{0.95; a-1, n-a} \mid \theta)$$

with θ given by (25.17). Under the non-informative prior

$$\pi(\mu_1, \ldots, \mu_a, \sigma^2) = \sigma^{-2}, \quad \sigma^2 > 0$$

the posterior density $\pi(\theta \mid \tau^2, x)$, where $\tau^2 = \mathrm{SSE}/[(n - a)\sigma^2]$, is the density of the noncentral chi-square distribution with $a - 1$ degrees of freedom and the noncentrality parameter $\tau^2(a - 1)T(x)$. The posterior density $\pi(\tau^2 \mid x)$ is the gamma distribution with the shape parameter $(n - a)/2$ and the scale parameter $2/(n - a)$. Consequently, the reproducibility probability under the Bayesian approach is

$$\hat{P} = \int_0^\infty \left[\int_0^\infty p(\theta)\pi(\theta \mid \tau^2, x)\, d\theta \right] \pi(\tau^2 \mid x)\, d\tau^2.$$

The reproducibility probability based on the Bayesian approach depends on the choice of the prior distributions. The non-informative prior we chose produces a more conservative reproducibility probability than that obtained using the estimated power approach, but is less conservative than that under the confidence bound approach. If a different prior such as an informative prior is used, a sensitivity analysis may be performed to evaluate the effects of different priors on the reproducibility probability.

25.5 Applications

In this section we discuss some applications of the results obtained in Sections 25.2 through 25.4.

25.5.1 Substantial Evidence with a Single Trial

An important application of the results derived in the previous sections is to address the following question: is it necessary to conduct a second clinical trial when the first trial produces a relatively strong significant clinical result (e.g., a relatively small p-value is observed), assuming that other factors (such as consistent results between centers, discrepancies related to gender, race, and other factors, and safety issues) have been carefully considered? As mentioned earlier, the FDA Modernization Act of 1997 includes a provision (Section 115 of FDAMA) to allow data from one adequate and well-controlled clinical trial investigation and confirmatory evidence to establish effectiveness for risk–benefit assessment of drug and biological candidates for approval. This provision essentially codified an FDA policy that had existed for several years but whose application had been limited to some biological products approved by the Center for Biologic Evaluation and Research of the FDA and a few pharmaceuticals, especially orphan drugs such as zidovudine and lamotrigine. A relatively strong significant result observed from a single clinical trial (say, p-value is less than 0.001) would have about 90% chance of reproducing the result in future clinical trials. Consequently, a single clinical trial is sufficient to provide substantial evidence for demonstration of efficacy and safety of the medication under study. In 1998, the FDA published a guidance which shed the light on this approach despite the fact that the FDA has recognized that advances in sciences and practice of drug development may permit an expanded role for the single controlled trial in contemporary clinical development (FDA, 1988). Suppose it is agreed that the second trial is not needed if the probability for reproducing a significant clinical result in the second trial is equal to or higher than 90%. If a significant clinical result is observed in the first trial and the confidence bound $\hat{P}_-$ derived in Section 25.3 is equal to or higher than 90%, then we have 95% statistical assurance that, with a probability of at least 90%, the significant result is reproducible in the second trial. For example, under the two-group parallel design with a common unknown variance and $n = 40$, the 95% lower confidence bound $\hat{P}_-$ given by (25.19) is equal to or higher than 90% if and only if $|T(x)| \geq 5.7$, that is, the clinical result in the first trial is highly significant. Alternatively, if the Bayesian approach is applied to the same situation, the reproducibility probability in (25.20) is equal to or higher than 90% if and only if $|T(x)| \geq 3.96$.

25.5.2 Sample Size Adjustments

When the reproducibility probability based on the result from the first trial is not higher than a desired level, the second trial must be conducted. The results on the reproducibility probability derived in Sections 25.2 through 25.4 can be used to adjust the sample size for the second trial. If the sample size for the first trial was determined based on a power analysis with some initial guessing values of the unknown parameters, then it is reasonable to make a sample size adjustment for the second trial based on the results from the first trial. If the reproducibility probability is lower than a desired power level of the second trial, then the sample size should be increased. On the other hand, if the reproducibility probability is higher than the desired power level of the second trial, then the sample size may be decreased to reduce costs. In the following we illustrate the idea using the two-group parallel design with a common unknown variance.

Suppose that $\hat{P}$ in (25.7) is used as the reproducibility probability when $T(x)$ given by (25.4) is observed from the first trial. Let $\hat{\sigma}^2 = [(n_1 - 1)s_1^2 + (n_2 - 1)s_2^2]/(n - 2)$. For simplicity, consider the case where the same sample size $n^*/2$ is used for two treatment groups in the second trial, where n^* is the total sample size in the second trial. With fixed $\bar{x}_i$ and $\hat{\sigma}^2$ but a new sample size n^*, the T-statistic becomes

$$T^* = \frac{\sqrt{n^*}\,(\bar{x}_1 - \bar{x}_2)}{2\hat{\sigma}}$$

and the reproducibility probability is $\hat{P}$ with T replaced by T^*. By letting T^* be the value to achieve a desired power, the new sample size n^* should be

$$n^* = \frac{(T^*/T)^2}{(1/4n_1 + 1/4n_2)}. \tag{25.22}$$

For example, if the desired reproducibility probability is 80%, then T^* needs to be 2.91 (Table 25.1). If $T = 2.58$ is observed in the first trial with $n = 30$ ($n_1 = n_2 = 15$), then $n^* \approx 1.27n \approx 38$ according to (25.22), that is, the sample size should be increased by about 27%. On the other hand, if $T = 3.30$ is observed in the first trial with $n = 30$ ($n_1 = n_2 = 15$), then $n^* \approx 0.78n \approx 24$, that is, the sample size can be reduced by about 22%.

25.5.3 Generalizability between Patient Populations

In clinical development, after the investigational drug product has been shown to be effective and safe with respect to a target patient population (e.g., adults), it is often of interest to study a similar but different patient population (e.g., elderly patients with the same disease under study or a patient population with different ethnic factors) to see how likely the clinical result is reproducible in the different population. This information is useful in regulatory submission for supplement new drug application (e.g., when generalizing the clinical results from adults to elderly patients) and regulatory evaluation for bridging studies (e.g., when generalizing clinical results from Gaussian to Asian patient population). Detailed information regarding bridging studies can be found in ICH (1997). For this purpose, we propose to consider the generalizability probability, which is the reproducibility probability with the population of a future trial slightly deviated from the population of the previous trial(s).

We consider a parallel-group design for two treatments with population means μ_1 and μ_2 and an equal variance σ^2. Other designs can be similarly treated. Suppose that in the future trial, the population mean difference is changed to $\mu_1 - \mu_2 + \varepsilon$ and the population variance is changed to $C^2\sigma^2$, where $C > 0$. The signal-to-noise ratio for the population difference in the previous trial is $|\mu_1 - \mu_2|/\sigma$, whereas the signal-to-noise ratio for the population difference in the future trial is

$$\frac{|\mu_1 - \mu_2 + \varepsilon|}{C\sigma} = \frac{|\Delta(\mu_1 - \mu_2)|}{\sigma},$$

where

$$\Delta = \frac{1 + \varepsilon/(\mu_1 - \mu_2)}{C} \tag{25.23}$$

is a measure of change in the signal-to-noise ratio for the population difference. For most practical problems, $|\varepsilon| < |\mu_1 - \mu_2|$ and, thus, $\Delta > 0$. Table 25.4 gives an example on the effects of changes of ε and C on Δ.

TABLE 25.4

Effects of Changes in Mean and Standard
Deviation (ε and C) on Δ in (25.23)

$\left|e/(m_1 - m_2)\right|$	C	Range of Δ
<5%	0.8	1.188–1.313
	0.9	1.056–1.167
	1.0	0.950–1.050
	1.1	0.864–0.955
	1.2	0.792–0.875
	1.3	0.731–0.808
	1.4	0.679–0.750
	1.5	0.633–0.700
≥5% but <10%	0.8	1.125–1.375
	0.9	1.000–1.222
	1.0	0.900–1.100
	1.1	0.818–1.000
	1.2	0.750–0.917
	1.3	0.692–0.846
	1.4	0.643–0.786
	1.5	0.600–0.733
≥10% but <20%	0.8	1.000–1.500
	0.9	0.889–1.333
	1.0	0.800–1.200
	1.1	0.727–1.091
	1.2	0.667–1.000
	1.3	0.615–0.923
	1.4	0.571–0.857
	1.5	0.533–0.800

Source: Shao, J. and Chow, S.C., *Stat. Med.*, 21,
1727, 2002. With permission.

If the power for the previous trial is $p(\theta)$, then the power for the future trial is $p(\Delta\theta)$. Suppose that Δ is known. Under the frequentist approach, the generalizability probability is $\hat{P}_\Delta$, which is $\hat{P}$ given by (25.7) with $T(x)$ replaced by $\Delta T(x)$, or $\hat{P}_{\Delta-}$, which is $\hat{P}_-$ given by (25.19) with $|\theta|_-$ replaced by $\Delta|\theta|_-$. Under the Bayesian approach, the generalizability probability is $\hat{P}_\Delta$, which is $\hat{P}$ given by (25.20) with $p(\delta|u)$ replaced by $p(\Delta\delta|u)$. When the value of Δ is unknown, we may consider a set of Δ-values to carry out a sensitivity analysis. An example is given as follows.

A double-blind randomized trial was conducted in patients with schizophrenia for comparing the efficacy of a test drug with a standard therapy. A total of 104 chronic schizophrenic patients participated in this study. Patients were randomly assigned to receive the treatment of the test drug or the standard therapy for at least 1 year, where the test drug group has 56 patients and the standard therapy group has 48 patients. The primary clinical endpoint of this trial was the total score of Positive and Negative Symptom Scales (PANSS). No significant differences in demographics and baseline characteristics were observed for baseline comparability. Mean changes from baseline in total PANSS for the test drug and the standard therapy are $\bar{x}_1 = -3.51$ and $\bar{x}_2 = 1.41$, respectively, with $s_1^2 = 76.1$ and $s_1^2 = 74.86$. The difference $\mu_1 - \mu_2$ is estimated by $\bar{x}_1 - \bar{x}_2 = -4.92$ and is considered to be statistically significant with $T = -2.88$, a p-value of 0.004, and a reproducibility probability of 0.814 under the estimated power approach or 0.742 under the Bayesian approach.

The sponsor of this trial would like to evaluate the probability for reproducing the clinical result for an elderly patient population where Δ, the change in the signal-to-noise ratio, ranges from 0.75 to 1.2. The generalizability probabilities are given in Table 25.5 (since $|T| < 4$, the confidence bound approach is not considered). In this example, $|T|$ is not very large and, thus, a clinical trial is necessary. The generalizability probability can be used to determine the sample size n^* for such a clinical trial. The results are given in Table 25.5. For example, if $\Delta = 0.9$ and the desired power (reproducibility probability) is 80%, then $n^* = 118$ under the estimated power approach and 140 under the Bayesian approach; if the desired power (reproducibility probability) is 70%, then $n^* = 92$ under the estimated power approach and 104 under the Bayesian approach. A sample size smaller than that of the original trial is allowed if $\Delta \geq 1$, that is, the new population is less variable.

TABLE 25.5

Generalizability Probabilities and Sample Sizes Required for Bridging Studies (under a Two-Group Parallel Design with $n_1 = 56$, $n_2 = 48$, and $T = -2.88$)

	Estimated Power Approach			Bayesian Approach		
		New Sample	Size n^*		New Sample	Size n^*
Δ	$\hat{P}_\Delta$	70% Power	80% Power	$\hat{P}_\Delta$	70% Power	80% Power
1.20	0.929	52	66	0.821	64	90
1.10	0.879	62	80	0.792	74	102
1.00	0.814	74	96	0.742	86	118
0.95	0.774	84	106	0.711	98	128
0.90	0.728	92	118	0.680	104	140
0.85	0.680	104	132	0.645	114	154
0.80	0.625	116	150	0.610	128	170
0.75	0.571	132	170	0.562	144	190

Source: Shao, J. and Chow, S.C., *Stat. Med.*, 21, 1727, 2002. With permission.

The sample sizes n^* in Table 25.5 are obtained as follows. Under the estimated power approach

$$n^* = \frac{(T^*/\Delta T)^2}{(1/4n_1 + 1/4n_2)},$$

where T^* is the value obtained from Table 25.1 for which the reproducibility probability has the desired level (e.g., 70% or 80%). Under the Bayesian approach, for each given Δ we first compute the value T_Δ^* at which the reproducibility probability has the desired level and then use it.

25.6 Concluding Remarks

In pharmaceutical/clinical development, as indicated earlier in this chapter, the U.S. FDA requires at least two adequate and well-controlled clinical trials be conducted to provide substantial evidence regarding the effectiveness of the test treatment under investigation. Thus, one of the most commonly asked controversial issues is that whether one large single trial (by combining the two trials) can fulfill with the FDA's requirement for at least two adequate and well-controlled clinical trials. This chapter provides scientific/statistical justification to address this issue in terms of the evaluation of reproducibility probability. The other controversial issue of particular interest to the investigator is that whether the observed clinical results can be generalized to a similar but different (e.g., due to differences in demographics or ethnic factors) target patient population. This chapter also provides some answers to the question by studying the impact of the signal-to-noise ratio in generalizability probability.

26

Good Review Practices

26.1 Introduction

The research, development, and approval of a drug product is a lengthy process involving drug discovery, laboratory development, animal studies, clinical trials, and regulatory registration. This lengthy process is necessary to assure the efficacy and safety of the drug product. In the United States, however, no regulations were put forth until the Pure Food and Drug Act was passed by the Congress in 1906. The purpose of this Act is to prevent misbranding and adulteration of food and drugs, yet it does not give the government any authority to inspect food and drugs. The Act was amended in 1912 (the Sherley Amendment) to prohibit labeling medicines with false and fraudulent claims. In 1931, the United States Food and Drug Administration (FDA) was formed. The provisions of the FDA are intended to ensure that (1) food is safe and wholesome, (2) drugs, biological products, and medical devices are safe and effective, (3) cosmetics are unadulterated, (4) the use of radiological products does not result in unnecessary exposure to radiation, and (5) all of these products are honestly and informatively labeled (Fairweather, 1994). The concept of testing marketed drugs in human subjects did not become a public issue until the late 1930s when the Elixir Sulfanilamide disaster occurred. The disaster was a safety concern of a liquid formulation of a sulfa drug which caused more than 100 deaths. This drug had never been tested in humans before its marketing. This safety concern led to the pass of the Federal Food, Drug, and Cosmetic Act (FD&C Act) in 1938. The FD&C Act extended its coverage to cosmetics and therapeutic devices. More importantly, the FD&C Act requires the pharmaceutical companies to submit full reports of investigations regarding the safety of new drugs. In 1962, a significant Kefauver-Harris Drug Amendments to the FD&C Act was passed, which not only strengthened the safety requirements for new drugs but also established an efficacy requirement for new drugs for the first time. In 1984, the Congress passed the Price Competition and Patent Term Restoration Act to provide for increased patent protection to compensate for patent life lost during the approval process. Based on this Act, the FDA was authorized to approve generic drugs through the evaluation of bioequivalency on healthy male subjects. In addition, the FDA also has the authority for designation of prescription drugs or over-the-counter (OTC) drugs.

Good regulatory (or review) practices can be defined as a quality system to ensure that the users of medicinal products, the applicants, and the regulators are satisfied with the scientific advice, opinions, establishment of maximum residue levels, inspection and assessment reports and related documents, taking into consideration legal requirements and guidance in order to protect and promote human and animal health (Korteweg, 2002). Thus, good regulatory (or review) practices consist of good laboratory practice (GLP), good manufacturing practice (GMP), good clinical practice (GCP), good statistics practice (GSP) which include good programming practice (GPP) and good data management practice (GDMP), and good review (regulatory) practice (GRP). In this chapter, our attention will be directed to GRP related to GSP and GCP in regulatory review and approval process for a pharmaceutical compound under investigation. Brief concluding remarks are given in the last section.

In the next section, regulatory process and requirement for pharmaceutical compounds in the United States are briefly described. GRPs developed by the Center for Drug Evaluation and Research (CDER) of the FDA are given and discussed in Section 26.3. Section 26.4 discusses some commonly seen controversial issues regarding obstacles and challenges in regulatory process. Brief concluding remarks are given in the last section.

26.2 Regulatory Process and Requirements

For approval of drug products, each country and/or region such as the European Community (EC), Japan, and the United States has similar but slightly different regulatory process and requirements for the conduct of clinical trials and the submission, review, and approval of clinical results. In this section, for illustration purposes, we will focus on the regulatory process and requirements adopted in the United States. For evaluation and approval of drug products, the sponsors are required to submit *substantial* evidence of effectiveness and safety accumulated from adequate and well-controlled clinical trials to the FDA. The current regulations for conducting clinical trials and the submission, review, and approval of clinical results for pharmaceutical compounds in the United States can be found in the Code of Federal Regulations (CFR) (see, e.g., 21 CFR Parts 50, 56, 312, and 314). These regulations are developed based on the FD&C Act. These regulations cover not only pharmaceutical entities such as drugs, biological products, and medical devices under investigation but also the welfare of participating subjects, labeling, and advertising of pharmaceutical products. The FDA has jurisdiction of administration of regulation and approval of drug products. These regulations include Investigational

New Drug Application (IND) and New Drug Application (NDA) for new drugs, orphan drugs, and OTC human drugs, Abbreviated New Drug Application (ANDA) for generic drugs, Establishment License Application (ELA) or Product License Application (PLA) for biological products, Investigational Device Exemptions (IDE), and Premarket Approval of Medical Devices (PMA) for medical devices and other means.

A treatment consisting of a combination of drugs, biological products, and/or medical devices is usually referred to as a combined therapy. If a treatment consists of a combination of drugs, biologics, and/or devices such as drug with device, biologic with device, drug with biologic, drug with biologic in conjunction with device, then it is defined as a combined product. For a combined product consisting of different pharmaceutical entities, the FDA requires that each of the entities should be reviewed separately by appropriate centers at the FDA. In order to avoid confusion of jurisdiction over a combination product and to improve efficiency of approval process, the principle of primary mode of action of a combination product was established in the Safe Medical Devices Act (SMDA) in 1990 (21 U.S.C. 353). In 1992, based on this principle, three inter-center agreements were signed between CDER and the Center for Biologics Evaluation and Research (CBER), between CDER and the Center for Devices and Radiological Health (CDRH), and between CBER and CDRH to establish the ground rules for assignment of a combined product and inter-center consultation (Margolies, 1994). Different regulations exist for different products, e.g., IND and NDA for drug products; ELA and PLA for biological products; IDE and PMA for medical devices. However, the spirit and principles for the conduct, submission, review, and approval of clinical trials are the same. Therefore, for the purpose of illustration, we will only give a detailed discussion on IND and NDA for drug products.

26.2.1 Investigational New Drug Application

Before a drug can be studied in humans, its sponsor must submit an IND to the FDA. Unless noticed otherwise, the sponsor may begin to investigate the drug 30 days after the FDA has received the application. The IND requirements extend throughout the period during which a drug is under study. As mentioned in Sections 312.1 and 312.3 of 21 CFR, an IND is synonymous with the Notice of Claimed Investigational Exemption for a New Drug. Therefore, an IND is, legally speaking, an exemption to the law that prevents the shipment of a new drug for interstate commerce. Consequently, the drug companies which file an IND have flexibility of conducting clinical investigations of products across the United States. Kessler (1989) indicated that there are two types of INDs, namely, the commercial IND and the noncommercial IND. A commercial IND permits the sponsor to gather the data on clinical safety and effectiveness that are needed for an NDA. If the drug is approved by the FDA, the sponsor is allowed to market the drug for specific uses.

On the other hand, a noncommercial IND allows the sponsor to use the drug in research or early clinical investigation to obtain advanced scientific knowledge of the drug. Note that the FDA itself does not investigate new drugs or conduct clinical trials. Pharmaceutical manufacturers, physicians, and other research organizations such as the NIH may sponsor INDs. If a commercial IND proves successful, the sponsor ordinarily submits an NDA. During this period, the sponsor and the FDA usually negotiate over the adequacy of the clinical data and the wording proposed for the label accompanying the drug which sets out description, clinical pharmacology, indications, and usage, contraindications, warnings, precautions, adverse reactions, and dosage and administration.

By the time an IND is filed, the sponsor should have enough information about chemistry, manufacturing, and controls (CMCs) of the drug substance and drug product to assure the identity, strength, quality, and purity of the investigational drug covered by the IND. In addition, the sponsor should provide adequate information about pharmacological studies for the absorption, distribution, metabolism, and excretion (ADME), and acute, subacute, and chronic toxicological studies and reproductive tests in various animal species to support the fact that the investigational drug is reasonably safe to be evaluated in clinical trials of various durations in human. The central focus of the initial IND submission should be on the general investigational plan and protocols for specific human studies. Therefore, a copy of protocol(s) which include study objectives, investigators, criteria for inclusion and exclusion, study design, dosing schedule, endpoint measurements, and clinical procedure should be submitted along with the investigational plan and other information such as CMCs, pharmacology and toxicology, previous human experiences with the investigational drug, and any additional and relevant information related to the investigational drug. Note that the FDA requires that all sponsors should submit an original and two copies of all submissions to the IND, i.e., including the original submission and all amendments and reports.

26.2.2 New Drug Application

For approval of a new drug, the FDA requires at least two adequate well-controlled clinical studies be conducted in humans to demonstrate substantial evidence of the effectiveness and safety of the drug. The substantial evidence as required in the Kefaurer-Harris amendments to the FD&C Act in 1962 is defined as the evidence consisting of adequate and well-controlled investigations, including clinical investigations, by experts qualified by scientific training and experience to evaluate the effectiveness of the drug involved, on the basis of which it could fairly and responsibly be concluded by such experts that the drug will have the effect it purports to is represented to have under the conditions of use prescribed, recommended, or suggested, in the labeling or proposed labeling thereof. Based on this amendment, the FDA requests that reports of adequate and well-controlled investigations provide the primary basis for determining whether there is substantial evidence to support the claims of new drugs and antibiotics. Section 314.122 of 21 CFR provides the definition of an adequate and well-controlled study.

An adequate and well-controlled study is judged by eight criteria specified in the CFR. These criteria are objectives, method of analysis, design of studies, selection of subjects, assignment of subjects, participants of studies, assessment of responses and effect. First, each study should have a very clear statement of objectives for clinical investigation which can be reformulated into statistical hypotheses and estimation procedures. In addition, proposed methods of analyses should be described in the protocol and actual statistical methods used for analyses of data should be described in detail in the report. Second, each clinical study should employ a design which allows a valid comparison with a control for an unbiased assessment of drug effect. Therefore, selection of a suitable control is one of the keys to integrity and quality of an adequate and well-controlled study. The CFR recognizes the following controls: placebo concurrent control, dose-comparison concurrent control, no treatment control, active concurrent control, and historical control. Next, the subjects in the study should have the disease or condition under study. Furthermore, subjects should be randomly assigned to different groups in the study to minimize potential bias and assure comparability of the groups with respect to pertinent variables such as age, gender, race, and other important prognostic factors. All statistical inferences are based on such randomization and possibly stratification to achieve these goals. However, bias will still occur if no adequate measures are taken on the part of subjects, investigator, and analysts of the study. Therefore, blinding is extremely crucial to eliminate the potential bias from this source. Usually, an adequate and well-controlled study should be at least double-blinded for which investigators and subjects are blinded to the treatments during the study. Currently, a triple-blinded study in which the sponsor (i.e., clinical monitor) of the study is also blinded to the treatment is not uncommon. Another critical criterion is the validity and reliability of assessment of responses. For example, the methods for measurements of response such as symptom scores for benign prostate hyperplasia should be validated before their usage in the study (Barry et al., 1992). Finally, appropriate statistical methods should be used for the assessment of comparability among treatment groups with respect to pertinent variables mentioned above and for unbiased evaluation of drug effects.

Section 314.50 of 21 CFR specifies the format and content of an NDA. The FDA requests that the applicant should submit a complete copy of the NDA form (A) to (F) with a cover letter. In addition, the sponsor needs to submit a review copy for each of the six technical sections with the cover letter and application form (356H). The reviewing disciplines include chemistry reviewers for the CMCs; pharmacology reviewers for nonclinical pharmacology and toxicology; medical reviewers for clinical data section; and statisticians for statistical technical section. The outline of review copies for clinical reviewing divisions include (1) cover letter, (2) application form (356H), (3) index,

(4) summary, and (5) clinical section. The outline of review copies for statistical reviewing division consists of (1) cover letter, (2) application form (356H), (3) index, (4) summary, and (5) statistical section. The information required by the FDA and ECC for marketing approval of a drug is essentially the same. However, no statistical technical section is required in ECC registration. In October 1988, to assist an applicant in presenting the clinical and statistical data required as part of an NDA submission, the CDER of the FDA issued the Guideline for the Format and Content of the Clinical and Statistical Sections of an Application under 21 CFR 314.50. The guideline indicates the preference of having an integrated clinical and statistical report rather than two separate reports. A complete submission should include clinical section [21 CFR 314.50(d)(5)], statistical section [21 CFR 314.50(d)(6)], and case report forms and tabulations [21 CFR 314.50(f)]. The same guideline also provides the content and format of the fully integrated clinical and statistical report of a controlled clinical study in an NDA. Based on the content and format of the fully integrated and statistical report of a controlled study required by the FDA, the Structure and Content of Clinical Study Reports was also issued by the EC in May 1993. In addition, EC also published a guideline entitled "Biostatistical Methodology in Clinical Trials in Applications for Marketing Authorizations for Medicinal Products" in March 1993. Detailed discussion for the preparation of clinical and statistical reports and an integrated summary of effectiveness and safety data for registration of a new drug can be found in Chow and Liu (1998b).

26.3 Good Review Practices

As indicated by the FDA, a GRP is a documented best practice within the CDER that discusses any aspect related to the process, format, content, and/or management of a product review. GRPs are developed over time as superior practices based on the CDER's collective experience to provide consistency to the overall review process of new products. GRPs are developed to improve the quality of reviews and review management. GRPs improve efficiency, clarity, and transparency of the review process and review management. GRPs are expected to be adopted by review staff as standard processes through supervisor mentoring, implementation teams, and formal training when necessary.

26.3.1 Fundamental Values

As described in the CDER/FDA GRPs, fundamental values for all GRPs include quality, efficiency, clarity, transparency, and consistency, which are briefly summarized below.

For quality, it is believed that consistent implementation of GRPs by review staff will enhance the quality of reviews, the review process, and the resultant regulatory action. GRPs will improve the efficiency of the review process through standardization. For clarity, GRPs support clarity throughout the review process, including critical review and decision activities that must be completed before a regulatory decision is made. Developing and documenting GRPs ensures that our review processes are readily available in one location via the Internet (through CDER's Web site) to sponsors and the public. For consistency, by offering a consistent approach and only deviating from it when appropriate (after supervisory concurrence), GRPs help reviewers achieve consistency with their reviews and provide standard review processes across divisions and offices.

26.3.2 Implementation of GRP

For the implementation of GRP within the FDA, review staff is expected to become thoroughly familiar with pertinent GRPs and to adhere to these GRPs when conducting their reviews unless a particular part of a GRP is not applicable to a particular review or the review staff receives supervisory instruction to do otherwise. The approving supervisor should separately document his or her reason for such a deviation in the electronic or paper document archive associated with that application. In addition, team leaders and supervisors are responsible for ensuring that GRPs are followed, and will provide specific instructions to deviate from the GRPs only when appropriate. Team leaders and supervisors will also mentor reviewers and provide appropriate instruction to review staff regarding content and policy within GRPs.

Most importantly, CDER/FDA will provide appropriate training courses and implementation teams when needed to inform review staff, team leaders, and supervisors of the content and policies contained in GRPs. Furthermore, similar to GSP and GCP, GRPs will be updated regularly with input from the appropriate CDER staff as needed. The updated GRPs will be posted to the GRP Web site after clearance.

26.3.3 Remarks

As indicated in the CDER/FDA GRP, although guidance documents do not legally bind the FDA, review staff may depart from guidance documents only with appropriate justification and supervisory concurrence. However, team leaders and supervisors will ensure that review staff follow GRPs. Documents that contain both GRPs and guidance for industry will continue to be issued as guidances for review staff and industry, but will be subtitled GRP. Note that the CDER/FDA GRP is available at http://www.fda.gov/cder/other/GRP.htm.

26.4 Obstacles and Challenges

In the past several decades, the current regulatory process for review and approval of a pharmaceutical compound under investigation has been criticized. The most commonly seen criticisms (or controversial issues) include, but are not limited to, the following: (1) there exist no gold standards for the evaluation of clinical data, (2) one-fits-all criterion for bioequivalence trials, and (3) the use of Bayesian statistics in drug evaluation. These criticisms have led to obstacles (or controversial issues) in the regulatory review/ approval process. In addition, most recently the FDA has been criticized to respond slowly to new concepts for clinical evaluation of efficacy and safety in pharmaceutical research and development. These new concepts refer to those listed in the *Opportunity List* of the *Critical Path Initiatives* such as the use of adaptive design methods in clinical trials, which has posted a great challenge to the review staff in the regulatory review/approval process. These obstacles (controversial issues) and challenges, which are related to quality, consistency, and validity of the review process, are described in the subsequent sections.

26.4.1 No Gold Standards for Evaluation of Clinical Data

One of the major criticisms in the regulatory review and approval process is probably the concern of a "gold standard" for the evaluation of the efficacy and safety of a test treatment under investigation although a number of drug-specific guidelines/guidances have been published by the regulatory agencies such as the FDA and EU EMEA. For a given regulatory submission, the FDA may accept the submission while the EU EMEA may not. Within a given regulatory agency, the review results may vary depending upon (1) the changing environment due to internal turnover and/or reorganization, (2) the interpretation of the related guidelines/guidances, and (3) the assigned review staff.

For the change in environment, e.g., a new review staff is assigned to an old submission, it will not only take some time for the newly assigned reviewer to get familiar with the submission and the history of the review (including all correspondences between the sponsor and the previous reviewer), but the newly assigned reviewer may have a different opinion and/or preference on certain aspects of the submission, which may result in an inconsistency of the review with the previous reviewer. In the case where this inconsistency occurs, the sponsor will make an attempt to convince the reviewer with documented correspondences with the previous reviewer. However, the success rate is low. In some cases, the sponsor receives a frustrating response from the reviewer that "That was then, this is now." This inconsistency has caused a waste of resources and time for the development of the test treatment under investigation and it is definitely not a GRP.

Regarding guidelines/guidances, although, as indicated in the CDER GRP, (1) team leaders and supervisors will ensure that review staff follows GRPs and (2) guidance documents do not legally bind FDA—the review staff may depart from guidance documents only with appropriate justification and supervisory concurrence. Based on the fact that different reviewers may interpret the guidelines/guidances differently, different reviewers may have set up his/her own rule or requirement for the evaluation of the test treatment under investigation and consequently provide different recommendations for the sponsors to follow. Thus, similar submissions may be accepted one over the other depending upon the assigned reviewer and his/her interpretation of the guidelines/guidances. This has resulted in the obstacle that "Guideline/guidance may not be the guide" in the regulatory review/approval process, which is very destructive to the trust and confidence of future regulatory submissions. As a result, in addition to following the guidelines and/or guidances issued by the regulatory agency, the sponsor always makes additional effort seeking for advice from the medical/statistical reviewers.

The quality of the review of a regulatory submission depends upon the knowledge and experience of the assigned review staff. If the assigned review staff is fresh out of school, the review may be performed by book without taking into consideration the practical experience of clinical research and development. One of the controversial issues regarding the review of clinical results submitted from the sponsors is how the regulatory agency can be the judge of the clinical results based on the fact that the regulatory agent does not conduct clinical studies. As a result, the quality of the review from a review staff that has little experience in clinical trials could be a concern.

As an example of no gold standards in regulatory review, consider endpoints supporting cancer drug approvals. As indicated by Williams et al. (2004), the FDA summarized the endpoints supporting 71 new cancer drug approvals over a 13-year period from 1990 to 2002 (Johnson et al., 2003). Fourteen of these applications received accelerated approval, while 54 applications were given regular approval based on the evidence of effects on clinical benefit endpoints or effects on accepted surrogate for clinical benefit. Table 26.1 provides a summary of endpoints used to support cancer drug approvals in 57 applications from January 1, 1990 to November 1, 2002. As can be seen from Table 26.1, the commonly used endpoint for regulatory approval is survival (18 out of 57) and response rate and/or time-to-disease progression (TTP) alone (18 out of 57). One of the controversial issues is whether the sponsor can switch from one endpoint (e.g., survival which is the primary endpoint as specified in the study protocol) to another (e.g., response rate and/or TTP which is considered a secondary endpoint in the study protocol) if the primary endpoint fails to demonstrate expected clinical benefit but other endpoints show promising clinical benefit to the patients.

TABLE 26.1

Endpoints Supporting Regulatory Approval of Oncology Drug
Marketing Applications, January 1, 1990 to November 1, 2002

Total	57
Survival	18
Response rate and/or TTP alone (predominantly hormone treatment of breast cancer or hematologic malignancies	18
Tumor-related signs and symptoms	13
Response rate + tumor-related signs and symptoms	(9)
Tumor-related signs and symptoms alone	(4)
Disease-free survival (adjuvant setting)	2
Recurrence of malignant pleural effusion	2
Decreased incidence of new breast cancer occurrence	2
Decreased impairment creatinine clearance	1
Decreased xerostomia	1

26.4.2 One-Fits-All Criterion for Bioequivalence Trials

As discussed above, those commonly encountered obstacles in regulatory
review and approval process described above may be due to the fact that
there exist no gold standards for the evaluation of clinical results obtained
from clinical trials. For review and approval of generic drug products, however, the FDA does employ "one-fits-all" criterion, which has been criticized
as well.

For the approval of generic drug products, the FDA requires that evidence
on bioequivalence in average bioavailability (in terms of the rate and extent of
drug absorption in the bloodstream) must be provided through the conduct
of bioequivalence studies. Two drug products are considered bioequivalent if
their rate and extent of drug absorption are similar. Average bioequivalence
can be demonstrated if the 90% confidence interval of the ratio of means
of a given study endpoint such as area under the blood concentration-time
curve (AUC) or maximum concentration (C_{max}) between a test treatment (e.g.,
a generic copy of the brand name drug) and the reference treatment (e.g., the
brand name drug) *totally* falls within the bioequivalence limits of 80% and
125%. The bioequivalence limits range of 80%–125% is a one-fits-all criterion
which is applicable to all drug products across all therapeutic areas even if
different drug products from different therapeutic areas may have different
therapeutic indices.

In the early 1990s, it was suggested that a more flexible criterion be
considered depending upon the therapeutic index and intra-subject variability of the drug product. Some drug products may be robust to efficacy,
while some drug products may be sensitive to safety. It is then suggested
that the lower limit and/or the upper limit of the bioequivalence limits be
adjusted in order to reflect the nature of the drug products. However, the

one-fits-all criterion is still a requirement for the approval of generic drug products in the United States.

Unlike clinical trials, the United States enforces the one-fits-all criterion for the evaluation of bioequivalence studies. In the past decade, many submissions were rejected by the FDA because either the lower limit or the upper limit of the constructed 90% confidence of the ratio of means is slightly off the lower bioequivalence limit of 80% or the upper bioequivalence limit of 125%. The sponsors may make an attempt to perform a test for outlier and reanalyze the data by excluding the identified outliers. However, there is little success. The FDA's response that "Rule is rule" has led to another controversial issue in clinical trials. That is, why the one-fits-all criterion for bioequivalence trials cannot be applied to clinical trials.

26.4.3 Bayesian Statistics in Drug Evaluation

For the evaluation of clinical efficacy and safety of a pharmaceutical entity (e.g., drug product, biological product, and medical device), unless it is specified in the regulatory guidelines/guidances, the assigned review staff's preference on statistical methods to be used will have an impact on policy and/or procedure of the review process. For example, for bioequivalence trials, the use of nonparametric methods for the assessment of bioequivalence in AUC and/or C_{max} is not encouraged (and may not be accepted by the FDA) in the United States without any scientific/statistical justification. On the other hand, the use of nonparametric methods is accepted by the EU EMEA.

For another example, there is tremendous debate regarding whether the Bayesian approach should be used in drug evaluation. Pros and cons have been discussed in the literature and major statistical conferences in the past decade. However, no universal agreement is reached. At this time, the use of the Bayesian approach for the evaluation of clinical efficacy and safety of medical devices is accepted by the CDRH of the FDA. However, it is not well accepted by CDER and CBER of the FDA. One of the fundamental differences between medical devices and drug/biologic products is that a medical device is less variable and the variation is controllable. These characteristics justify the use of the Bayesian approach. On the other hand, since the Bayesian approach is sensitive to the selection of prior distribution of the study parameters, it is not widely accepted by the CDER and CBER of the FDA for the evaluation of drug/biologic products.

26.4.4 Adaptive Design Methods in Clinical Trials

In recent years, the use of adaptive design methods in clinical trials has received much attention from both the pharmaceutical/biotechnology industry and the regulatory agencies due to its flexibility and efficiency for identifying the best clinical benefit of a test treatment under investigation.

An adaptive design is referred to as a clinical trial design that uses accumulating data to decide on how to modify aspects of the study as it continues, without undermining the validity and integrity of the trial (Gallo et al., 2006). Based on the natural adaptation, Chow and Chang (2006) identified several commonly considered adaptive designs in clinical research and development.

26.4.4.1 Adaptive Randomization Design

A design that allows modification of randomization schedules (e.g., unequal probabilities of treatment assignment) for increasing the probability of success. Basically, there are three types of adaptive randomization, namely, treatment adaptive, covariate adaptive, and response adaptive. Note that for an adaptive randomization design, the randomization schedule may not be available prior to the conduct of the study. Thus, it may not be feasible for a large trial or a trial with relatively long treatment duration. Furthermore, a statistical inference on the treatment effect is often difficult to obtain if not impossible.

26.4.4.2 Adaptive Group Sequential Design

An adaptive group sequential design refers to a group sequential design that allows for prematurely stopping a trial due to safety, efficacy/futility, or both based on interim analysis results. At interim analysis, blinded sample size reestimation may be performed. In addition, some adaptations such as adaptive randomization or dropping the losers may be applied. One of the major concerns in practice is that the overall type I error rate may not be preserved when (1) there are additional adaptations (e.g., changes in hypotheses and/or study endpoints) and (2) there is a shift in target patient population after adaptations are made.

26.4.4.3 Flexible Sample Size Reestimation Trial Design

An adaptive design that allows for sample size adjustment or reestimation based on the observed data at interim either with blinding or unblinding. Sample size adjustment or reestimation is usually performed based on the criteria of (1) variability, (2) conditional power, or (3) reproducibility probability. It should be noted that sample size reestimation is performed based on estimates from the interim analysis. Note that a flexible sample size reestimation design is also known as an N-adjustable design. In clinical trials using flexible sample size reestimation trial design, a commonly asked question is that *Can we always start with a small number and perform sample size reestimation at interim?*

26.4.4.4 Drop-the-Loser Design

Drop-the-loser design or pick-the-winner design is a multiple-stage adaptive design that allows dropping the inferior treatment groups. There are several general principles to follow when applying drop-the-loser design. These principles include (1) drop the inferior arms, (2) retain the control arm, and (3) may modify or add additional arms. The drop-the-loser design is useful when there are uncertainties regarding the dose levels. The selection criteria and decision rules play important roles for drop-the-loser designs. Note that dose groups that are dropped may contain valuable information regarding dose response of the treatment under study.

26.4.4.5 Adaptive Dose Escalation Design

Dose escalation design is often used in early-phase clinical development to identify the maximum tolerable dose, which is usually considered the optimal dose for the later phase clinical trials. In practice, two types of designs are commonly employed. They are the traditional "3 + 3" dose escalation rule, which is an algorithm-based trial design, and the continual reassessment method in conjunction with the Bayesian approach, which is a model-based trial design. When applying the adaptive dose escalation trial design, the following questions are necessarily asked: How to select the initial dose? How to select the dose range under study? How to achieve statistical significance with a desired power with a limited number of subjects? What are the selection criteria and decision rules? What is the probability of achieving the optimal dose?

26.4.4.6 Biomarker-Adaptive Design

A design that allows for adaptation based on the responses of biomarkers such as genomic markers for the assessment of treatment effect. It involves qualification and standard, optimal screening design, establishment of predictive model, and validation of the established predictive model. A classifier marker usually does not change over the course of study and can be used to identify the patient population who would benefit from the treatment from those who will not (e.g., DNA marker and other baseline marker for population selection). A prognostic marker informs the clinical outcomes, which is independent of treatment. A predictive marker informs the treatment effect on the clinical endpoint. A predictive marker can be population-specific. That is, a marker can be predictive for a population A but not population B. It should be noted that the correlation between a biomarker and true endpoint makes a prognostic marker. In addition, the correlation between a biomarker and true endpoint does not make a predictive biomarker. In practice, there is a gap between identifying genes that are associated with clinical outcomes and establishing a predictive model between relevant genes and clinical outcomes.

26.4.4.7 Adaptive Treatment-Switching Design

A design that allows the investigator to switch a patient's treatment from an initial assignment to an alternative treatment if there is evidence of lack of efficacy or safety of the initial treatment. Adaptive treatment-switching design is commonly employed in cancer trials. In practice, a high percentage of patients may switch due to disease progression. Thus, estimation of survival is a challenge to the biostatistician. A high percentage of subjects who switched could lead to a change in hypotheses to be tested. Sample size adjustment for achieving a desired power is critical to the success of the study.

26.4.4.8 Adaptive-Hypotheses Design

A design that allows change in hypotheses based on interim analysis results often considered before database lock and/or prior to data unblinding. Some examples include (1) switch from a superiority hypothesis to a non-inferiority hypothesis change in study endpoints (e.g., switch primary and secondary endpoints) and (2) switch between non-inferiority and superiority. When switching from a superiority hypothesis to a non-inferiority hypothesis, the selection of the non-inferiority margin is very critical. Note that for switching between the primary endpoint and the secondary endpoints, it is suggested that the switch from the primary endpoint to a co-primary endpoint or a composite endpoint be considered.

26.4.4.9 Adaptive Seamless Phase II/III Trial Design

An adaptive seamless phase II/III trial design is a trial design that combines two separate trials (i.e., a phase IIb and a phase III trial) into one trial and uses data from patients enrolled before and after the adaptation in the final analysis. An adaptive seamless phase II/III trial design is a two-stage design which consists of a learning phase (phase IIb) at the first stage and a confirmatory phase (phase III) at the second stage. In practice, an adaptive seamless trial design may combine two separate studies which may have similar but different study objectives and may use different study endpoints (e.g., a biomarker or a surrogate endpoint versus a clinical endpoint). In this case, it is a concern how the overall type I error rate can be controlled.

26.4.4.10 Multiple Adaptive Design

A design that is any combination of the above adaptive designs. A multiple adaptive design is very flexible and yet complicated. In practice, if it is not impossible, statistical inference is often difficult to obtain.

Adaptive design is attractive due to its flexibility. However, it is a concern that so-called operational bias and variation may be introduced after the application of adaptations. Thus, the regulatory agency requires that a strategy for preventing operational bias and variation be provided when utilizing adaptive design methods in clinical trials. In addition, it is a great concern whether the overall type I error rate is preserved after the adaptations are made during the conduct of the clinical trials. The FDA does not discourage the use of adaptive design methods in clinical trials. However, their slow response to the rapid development has been criticized. Regulatory guideline/guidance on adaptive design methods in clinical trials should be necessarily developed in order to assist the sponsor to adopt this new concept of adaptive design in clinical research and development.

26.5 Concluding Remarks

For evaluation of clinical efficacy and safety of a test treatment under investigation, it is always an obstacle and challenge to review staff, especially when there exist no gold standards although regulatory drug-specific guidelines/guidances are available. GRP is necessary to ensure quality, scientific validity, and consistency for drug evaluation. GSP is a key to the success of GRP. If one fails to follow GSP, GRP could lead to some controversial issues with no resolutions.

Although it should not be so, statistical/clinical research is driven by regulatory requirement or preference. A typical example is the development of criteria and statistical methods for the assessment of population and individual bioequivalence in bioequivalence trials in the 1990s. The developed criteria and statistical methods lacked input from scientists from clinical pharmacology and biostatisticians from the pharmaceutical/generic industry and consequently were dropped by the FDA in early 2000. For another example, missing data imputation was not accepted by the FDA in the 1980s. In recent years, however, the FDA requires that statistical methods for missing value imputation be provided for handling missing data.

In recent years, the FDA has led advances in statistical research and development for drug evaluation since the *Opportunity List* as the result of the *Critical Path Initiative* published in 2006. For example, the FDA has established working groups on critical topics such as adaptive design, non-inferiority trials, and QT/QTc studies with recording replicates. These research works will not only enhance GRP but also improve the quality, validity, and consistency of GRP.

27

Probability of Success

27.1 Introduction

In the past several decades, it has been recognized that increasing spending of biomedical research does not reflect an increase in the success rate of pharmaceutical/clinical research and development. Woodcock (2005) indicated that the low success rate of pharmaceutical/clinical development could be because (1) there is a diminished margin for improvement that escalates the level of difficulty in proving drug benefits, (2) genomics and other new sciences have not yet reached their full potential, (3) mergers and other business arrangements have decreased candidates, (4) easy targets are the focus as chronic diseases are harder to study, (5) failure rates have not improved, and (6) rapidly escalating costs and complexity decreases willingness/ability to bring many candidates forward into the clinic.

As indicated in Chapters 1 and 3, the United States Food and Drug Administration (FDA) kicked off the *Critical Path Initiative* to assist the sponsors in (1) identifying possible causes, (2) providing resolutions, and (3) increasing the efficiency and the probability of success (POS) in clinical development. The POS in clinical development is usually referred to as (1) the POS of a given clinical trial, (2) the POS of a given phase of clinical development (e.g., phase II or phase III clinical development), and (3) the POS of the overall clinical development process. For a given clinical trial, the POS can be assessed using the idea/concept of generalizability and reproducibility probabilities described in Chapter 25. A similar idea/concept can be applied to assess the POS for a given phase of clinical development. Thus, in this chapter, we will focus on the assessment of the POS of the entire clinical development process.

As indicated in Chapter 1, clinical development is an important stage of pharmaceutical development. The success of clinical development is the key to the success of pharmaceutical development of a promising compound which consists of multiple stages of development such as translational research from preclinical studies to first-in-man clinical trials, phases I–III of clinical development, and phase IV of post-approval safety surveillance. The development process of a promising compound is a lengthy and costly process. At the screening stage, many candidate compounds may be dropped due to intolerable toxicity/safety or lack of efficacy based on preclinical data. In practice, it is most likely that only a handful of promising compounds can make it to the stage of clinical development. As a result, how to select the most promising compound among this handful of compounds for continued clinical development has become a challenge to the clinical development team under possible resources/budget constraints. A wrong decision could lead to a total disaster to the sponsor (all of the efforts and investments would be wasted).

In the next section, traditional approaches for the assessment of POS that are commonly adopted by the pharmaceutical industry are described. Section 27.3 studies the POS assuming that the pharmaceutical development process is a multiple-stage process. A brief concluding remark is given in the last section of the chapter.

27.2 Go/No-Go Decision in Development Process

As indicated in Chapter 1, the pharmaceutical/clinical development process of a promising compound is a sequential process which consists of a preclinical development phase and phases I–III of the clinical development phase. The sponsor of the promising compound usually will not move from one phase to the next unless significantly positive results are observed at the current phase of development. At the early phase of pharmaceutical/clinical development of a promising compound, it is crucial for the sponsor to make a go/no-go decision because the budget and/or resources available may be very limited. This is critical especially when there are several promising compounds under development. The go/no-go decision at a given phase of development is usually made based on the evaluation of the POS at the phase of development. The objective for assessment of the POS is multifold. First, it is to obtain accurate and reliable individual estimates of the POS at each stage and an overall estimate of the overall POS. Second, it is to obtain a lower confidence bound of the overall POS. Third, it is to perform a sensitivity analysis of the POS with respect to relative changes at the early phase of clinical development versus the later phase of clinical development assuming a fixed total resources/budget.

27.2.1 Simple Approach for Decision Making

At the early phase of pharmaceutical/clinical development of a promising compound, a go/no-go decision is usually made based on limited information available from preclinical animal studies, first-in-man proof-of-concept, and/or early efficacy studies. For example, at the preclinical phase of development, the sponsor may conduct a number of small-scale animal studies for the evaluation of toxicity, tolerability, and/or efficacy of the test compound in animals. A simple approach for making a go/no-go decision is to consider the proportion of studies with positive results. If the percentage is greater than a prespecified threshold, we then make a "go" decision. For a promising compound, suppose that five studies are conducted and four show positive results. In this case, the simple approach gives an 80% (four out of five) success rate. If the sponsor considers 75% being the threshold of the promising compound, a "go" decision is made because 80% is greater than the threshold value of 75%.

The approach for making a go/no-go decision based on the proportion of positive studies described above is simple and easy to implement. However, this simple approach ignores (1) differences in study objectives and/or hypotheses, (2) heterogeneity in variability associated with the study endpoint across studies, and (3) sample sizes across studies. As a result, this simple approach for making a go/no-go decision has the following limitations. First, it is not applicable when there is only a study or a couple of studies available. In this case, a commonly considered alternative approach is to make a go/no-go decision based on *subjective* evaluation, which is usually made by the researchers or scientists according to their best knowledge and experience of the compound under investigation. Second, the simple approach assumes that all studies are equally important. In practice, it is very likely that (1) different studies may be evaluated by different study endpoints and (2) different studies may exhibit different variabilities in observing the responses of the study endpoints. Third, the sizes of different studies may vary from study to study. It is a concern that all of the smaller studies show positive results while the largest study fails.

One of the controversial issues regarding making a go/no-go decision based on either the simple approach using the proportion of positive studies or the subjective evaluation from the perspectives of the researchers or scientists in the subject area is that the accuracy and reliability of the decision is questionable. Consider the same example described above. A small-scale study does not provide satisfactory inference regarding the treatment of the test compound under study. In other words, the false positive of the small-scale study could be high; consequently, the decision made based on the proportion of positive results may not be reliable. As a result, the decision regarding go/no-go may be biased and hence misleading. A wrong decision may kill a promising (an effective) compound and hence lead the sponsor to a disastrous situation. The method of subjective evaluation based on the perspectives of the researchers and scientists suffers from the same difficulty and dilemma.

27.2.2 Decision-Tree Approach

Another commonly considered approach for making a go/no-go decision in the pharmaceutical/clinical development process is the use of a simple decision tree. A simple decision tree is a classifier (go or no-go) in the form

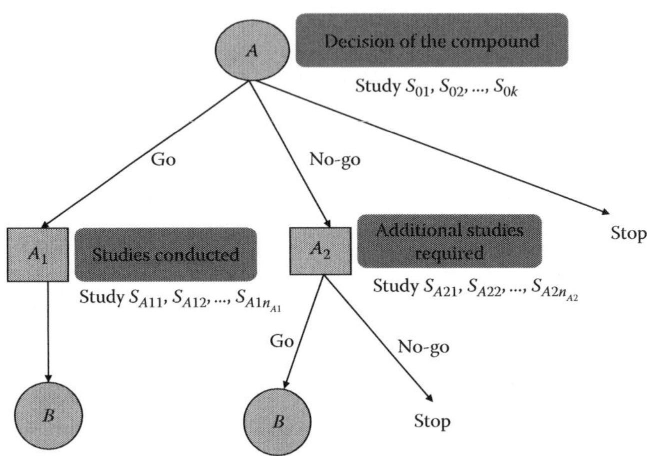

FIGURE 27.1
An example of decision tree.

of a tree structure (see Figure 27.1), where each node is either a leaf node (squares) or a decision node (circles). A leaf node indicates the tests or studies to be conducted at a specific phase of development, while a decision node specifies whether the test results meet prespecified thresholds. For example, as can be seen from Figure 27.1, suppose the first node (node A) is to make a decision whether the sponsor will take the compound through the development process based on some studies conducted prior to node A. If the decision is "go," then we move to leaf node A_1, which identifies the studies to be conducted prior to moving to node B. If the decision is "no-go with current information," then we move to leaf node A_2, which outlines additional studies required for further evaluation of the compound prior to moving to node B or node S (stop). If the decision is "no-go," we move to node S and stop the development of the compound.

More specifically, let S_{ijk} denote the kth study to be conducted under the jth decision made at the ith decision node, $i = A, B, ...; j = 1, 2, 3$ (stop); $k = 1, ..., n_{ij}$. Note that at decision node A, there are K studies that have been conducted, denoted by $S_{0k}, k = 1, ..., K$. Based on the results of $S_{0k}, k = 1, ..., K$, the sponsor will then make a decision at decision node A by checking the results with some prespecified thresholds at the leaf nodes. Let T_{ijk} and C_{ijk} be the test statistic and the corresponding threshold for the kth study under the jth decision made at the ith decision node, $i = A, B, ...; j = 1, 2, 3$ (stop); $k = 1, ..., n_{ij}$, respectively. A study is considered to have a positive result if $P\{T > C\} < 0.05$, where T and C are the test statistic and the corresponding critical value. As indicated in the previous subsection, a simple approach is to consider the criterion based on the proportion of positive results.

For example, if the proportion of studies with positive results is greater than 75%, we then proceed to the next phase of clinical development. On the other hand, if the proportion of positive studies is between 50% and 75%, we may conduct additional studies for further evaluation before proceeding to the next phase of development. If the proportion of studies with positive results is less than 50%, the sponsor may consider stopping the development of the test treatment under investigation. Thus, if the test results from the K studies, i.e., S_{0k}, $k = 1, \ldots, K$, meet the criterion for proceeding, we will move to leaf node A_1, which outlines the studies (i.e., S_{A1k}, $k = 1, \ldots, n_{A1}$) to be conducted before proceeding to the next decision node. If the test results from the K studies, i.e., S_{0k}, $k = 1, \ldots, K$, do not meet the criterion for proceeding but meet the criterion for further evaluation, we will move to leaf node A_2, which outlines additional studies (i.e., S_{A2k}, $k = 1, \ldots, n_{A2}$) required to be conducted for further evaluation. In the case where the test results of the K studies, i.e., S_{0k}, $k = 1, \ldots, K$, fail to meet both the criteria for proceeding and opportunity for further evaluation, the development of the test treatment will stop.

As can be seen, the decision-tree approach for making a go/no-go decision is simple and easy to understand. However, there are relative advantages and limitations of the decision-tree method. The advantages of the decision-tree method include the following: (1) it is able to generate understandable rules, (2) it allows the sponsor to make decisions without requiring much computation, (3) it is able to handle both continuous and categorical variables, and (4) it provides a clear indication of which nodes (decision points) are most important for assuring the success of the development process. On the other hand, the limitations of the decision-tree method include the following: (1) it is less appropriate for estimation tasks where the goal is to predict the value of a continuous attribute, (2) it is prone to errors in decision making with many classes and relatively small number of studies to be conducted at specific phases of development, and (3) the process of growing a decision tree can be computationally expensive.

It should be noted that a more complicated decision-tree approach with different rules and criteria may be applied at the leaf nodes and the decision nodes. Statistical properties of the decisions made in the decision tree should be carefully evaluated for providing accurate and reliable decisions when making go/no-go decisions.

27.2.3 An Example

A pharmaceutical company would like to assess the POS of one of their promising compounds under development. The company first determines the target product profile of the compound, which will be based on the POS assessment. The promising compound is intended for the treatment of choice for the reduction of psychotic symptoms in dementia (e.g., Lewy body, Alzheimer's, Parkinson's, and other dementias). For this promising compound, it is expected that (1) the efficacy of the compound is at least equal to that of competitors for the treatment of psychotic symptoms, (2) the compound is superior to that of competitors for treating the constellation of associated symptoms such as depression, and (3) the compound is superior to that of competitors in safety and tolerability such as rapid titration and risk of tardive dyskinesia. The ultimate goal is not only to restore functionality and delay progression to associated living (e.g., nursing home) but also to reduce caregiver burden and overall health-care cost.

Under the prespecified target product profile and ultimate goal of the promising compound, the company then identifies the critical phases of clinical development as phase Ia, phase Ib, phase IIa, phase IIb, and phase III. Under each identified critical phase of clinical development, the POS is assessed using the methods described above. As an example, consider the POS assessment of phase IIa. The company first identifies the sub-hurdles for the assessment of the overall POS of phase IIa. For example, the sub-hurdles at phase IIa of clinical development may involve the conduct of clinical studies for the assessment of (1) clinical efficacy and (2) safety and tolerability. Thus, in order to assess the overall POS, the company would assess the POS of the sub-hurdles based on some prespecified minimum success and target outcome. The minimum success is usually defined as the minimum criteria necessary to continue to the next phase of clinical development. For example, for the sub-hurdle of clinical efficacy (or demonstration of proof of concept), the minimum success could be the statistically significant efficacy in the control of delusions and hallucinations as compared to the placebo. The target outcomes could be the same as the minimum, with superior efficacy in the control of sleep, depression, and cognitive symptoms.

Now based on the predefined minimum of success and target outcome, the sub-hurdles POS can be assessed. Suppose that the sub-hurdles for (1) clinical efficacy and (2) safety and tolerability are determined to be 80% and 90%, respectively. Then the overall POS for phase IIa can be determined as 72% (see Figure 27.2).

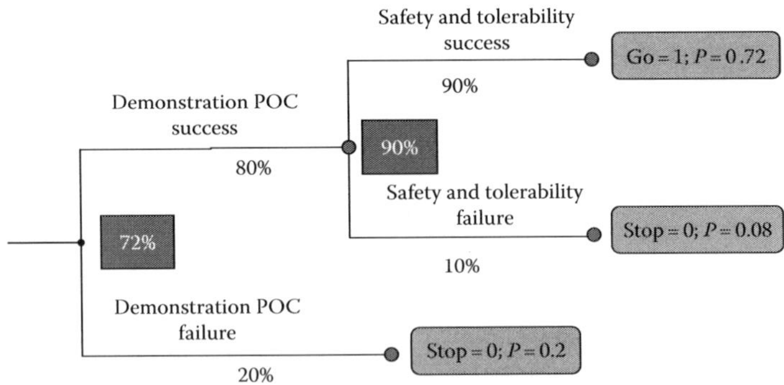

FIGURE 27.2
The assessment of overall POS at phase IIa. Note that Go = 1 means proceed to the next phase, while Stop = 1 indicates stop the development.

27.3 POS Assessment

As indicated earlier, the pharmaceutical/clinical development process of a compound is a sequential process that consists of several phases of development such as preclinical phase and phases I–III of clinical development. At each phase of development, a go/no-go decision is necessarily made. As discussed in the previous section, the go/no-go decision is usually made at each phase of development either based on a subjective evaluation, the simple approach, or a decision-tree approach. In this section, we attempt to study the assessment of the POS of the development process.

Let S_1, S_2, ..., and S_K denote Stage 1, 2, ..., and K of the development process of a pharmaceutical compound, respectively. Also, let p_1, p_2, ..., and p_K be the POS at Stage 1, 2, ..., and K, respectively. Thus, the POS can be obtained as

$$P(\text{Success}) = P(S_1)P(S_2 \mid S_1) \cdots P(S_K \mid S_{K-1}),$$

where $P(S_i)$ is defined as the probability of observing a positive result at the ith stage. That is,

$$P(S_i) = P(\text{positive} \mid T_i, n_i),$$

where a positive result is referred to as the rejection of the null hypothesis of no treatment difference at the α level of significance and there is an 80% power for correctly detecting a clinically important difference δ, in which n_i and T_i are the corresponding sample size and test statistic of the study conducted at the ith stage, where $i = 1, 2, ..., K$. It should be noted that in practice, there may be more than one study conducted at the same stage. In other words, $n_i = n_{ij}$ and $T_i = T_{ij}$, where $j = 1, 2, ..., J_i$. In this chapter, for simplicity, we will consider the case where $J_i = 1$ for all i. For illustration purpose, Table 27.1 summarizes the POS of pharmaceutical development of a promising compound with various scenarios of success at the early stage of pharmaceutical development.

TABLE 27.1

POS for Pharmaceutical Development

$P(S_1)$	$P(S_2\mid S_1)$	$P(S_3\mid S_2)$	$P(S_4\mid S_3)$	$P(\text{Success})$
0.5	0.9	0.9	0.9	0.365
0.6	0.9	0.9	0.9	0.437
0.7	0.9	0.9	0.9	0.510
0.8	0.9	0.9	0.9	0.583
0.9	0.9	0.9	0.9	0.656
0.95	0.9	0.9	0.9	0.693

Note: S_i indicates the ith stage of pharmaceutical development.

As can be seen from Table 27.1, the POS at the early stage of clinical development is critical. If the POS at the early stage is less than 70%, we may have an overall POS less than 50% even if the POS at subsequent stages of clinical development are as high as 90%.

In practice, it is of interest to perform a cost-effective analysis under the constraints of a fixed budget

$$C = C_1 + C_2 + \cdots + C_K,$$

where $C_1, C_2, \ldots,$ and C_K are the associated cost at $S_1, S_2, \ldots,$ and S_K, respectively. It is of interest to know what is the impact on the POS if we increase the sample size n_1 (i.e., put more investment in the early stage). It is also of interest to study the impact of an inaccurate assessment of p_i at the ith stage. Note that we may kill the project if the POS falls below a prespecified confidence level (say 70%) at any stage.

Suppose that at each Stage i, hypothesis testing is relevant with the significance level fixed as α, $i = 1, \ldots, K$. Let p_i be the observed conditional p-value at Stage i. Let n_i be the sample size needed at Stage i to detect a preselected known relative effective size d_i at Stage i with a conditional power of γ_i (conditional on rejecting the null hypotheses at all the previous stages, that is, $p_k \leq \alpha$ for all $k = 1, \ldots, i-1$). Write $\gamma = (\gamma_1, \gamma_2, \ldots, \gamma_K)'$, then n_i is a function of γ, denoted by $n_i = \xi_i(\gamma)$. The POS (when the compound is indeed effective with the prespecified effective size d_i), as a function of γ, is

$$P(\text{Success}) = H(\gamma) = \prod_{i=1}^{K} \gamma_i.$$

Let C_i be the cost for each experiment unit at Stage i, then the total cost is

$$C(\gamma) = \sum_{i=1}^{K} C_i n_i = \sum_{i=1}^{K} C_i \xi_i(\gamma).$$

The problem of maximizing the chance of success under a fixed total budget C is mathematically equivalent to maximizing $H(\gamma)$ in the above equation under the constraints of $C(\gamma) = C$. This can be done by applying, for example, the Lagrange multiplier method.

27.4 Concluding Remarks

In the pharmaceutical industry, it is common practice to conduct an exercise to prospectively assess the POS of promising compounds under investigation. Although the assessment of the POS does provide a *guide* as to which promising compound should be developed first, it may lead to a compound which is shown to be ineffective at a later phase of clinical development. Consequently, all of the efforts and investments are wasted. To improve accuracy and reliability of the assessment of the POS, it is suggested that more information or studies (i.e., more investment) should be conducted to obtain a more accurate and reliable estimate of the POS at the early stages of development. In practice, since the investment at the early phase of clinical development is usually very limited as compared to that of the later phase of clinical development, it is of interest to study the impact of an increasing investment (i.e., conduct more studies or increase sample size for obtaining more information) on the assessment of the POS.

In the pharmaceutical industry, the sponsor always puts less emphasis on the early phase of clinical development of a promising compound. However, a go/no-go decision is usually made at the early phase of clinical development. As a result, the establishment of criteria for making a go/no-go decision early and the assessment of POS have become common practice in the pharmaceutical industry. The studies conducted at the early phase of clinical development are usually small. Thus, one of the controversial issues regarding making a go/no-go decision at the early phase of development concerns the accuracy and reliability of the go/no-go decision based on the limited information available. In the pharmaceutical/clinical development, although it is desirable to shorten the development process by making a go/no-go decision as early as possible, it is not desirable to either wrongly bring an ineffective compound through the development process or kill a promising compound early. In practice, it is very likely that we may kill a promising compound before it shows positive results if we make a go/no-go decision early. There is always a risk to make a decision early based on the limited information available.

Another controversial issue is related to the false-positive rate and/or false-negative rate of the go/no-go decision. What is the chance that the observed positive result is by chance alone and hence is not reproducible? Note that false-positive rate and false-negative rate represent patient's risk and sponsor's risk, respectively. In addition, it is of particular interest to the sponsor as to "how the go/no-go decision relates to the POS of the development process?" In practice, only a handful of promising compounds will make it to the early phase of clinical development. Among these promising compounds, only one or two will gain regulatory approval and reach the marketplace. It is of interest to the investigator as to what the POS of a promising compound with a "go" decision at the early phase of development is.

As can be seen from Table 27.1, the POS at the early phase of development has a significant impact on the POS of the development process. Thus, it is suggested that more resources be put at the early phase of development in order to (1) increase the accuracy and reliability when making a go/no-go decision, (2) decrease both the false-positive rate and the false-negative rate, and (3) increase the POS of the development process.

References

Afonja, B. (1972). The moments of the maximum of correlated normal and t-variates. *Journal of the Royal Statistical Society, B,* 34, 251–262.

Agin, M.A., Aronstein, W.S., Ferber, G., Geraldes, M.C., Locke, C., and Sager, P. (2008). QT/QTc prolongation in placebo-treated subjects: A PhRMA collaborative data analysis. *Journal of Biopharmaceutical Statistics,* 18, 408–426.

Akaike, H. (1974). A new look at statistical model identification. *IEEE Transactions on Automatic Control,* 19, 716–723.

Alosh, M. (2009). The impact of missing data in a generalized integer-valued autoregression model for count data. *Journal of Biopharmaceutical Statistics,* 19, 1039–1054.

Babb, J., Rogatko, A., and Zacks, S. (1998). Cancer phase I clinical trials: Efficient dose escalation with overdose control. *Statistics in Medicine,* 17, 1103–1120.

Babb, J.S. and Rogatko, A. (2004). Bayesian methods for cancer phase I clinical trials. In: *Advances in Clinical Trial Biostatistics,* N.L. Geller (ed.). Marcel Dekker, Inc., New York.

Barry, M.J., Fowler, F.J. Jr., O'Leary, M.P., Bruskewitz, R.C., Holtgrewe, H.L., Mebust, W.K., and Cockett, A.T. (1992). The American Urological Association Symptom Index for benign prostatic hyperplasia. *Journal of Urology,* 148, 1549–1557.

Basford, K.E., Greenway, D.R., McLachlan, G.J., and Peel, D. (1997). Standard errors of fitted component means of normal mixtures. *Computational Statistics,* 12(1), 1–17.

Bauer, P. and Kieser, M. (1999). Combining different phases in the development of medical treatments within a single trial. *Statistics in Medicine,* 14, 1595–1607.

Bauer, P. and Kohne, K. (1994). Evaluation of experiments with adaptive interim analysis. *Biometrics,* 50, 1029–1041.

Bauer, P. and Rohmel, J. (1995). An adaptive method for establishing a dose-response relationship. *Statistics in Medicine,* 14, 1595–1607.

Benjamini, Y. and Hochberg, Y. (1995). Controlling the false discovery rate: A practical and powerful approach to multiple testing. *Journal of Royal Statistical Society, B,* 57, 289–300.

Berger, J.O. (1985). *Statistical Decision Theory and Bayesian Analysis,* 2nd edn. Springer-Verlag, New York.

Berger, R.L. (1982). Multiparametric hypothesis testing and acceptance sampling. *Technometrics,* 24, 295–300.

Bergner, M., Bobbitt, R.A., Carter, W.B., and Gilson, B.S. (1981). The sickness impact profile: Development and final revision of a health status measure. *Medical Care,* 19, 787–805.

Bergum, J.S. (1988). Constructing acceptance limits for multiple stage USP tests. In: *Proceedings of the Biopharmaceutical Section of the American Statistical Association,* Alexandria, VA, pp. 197–201.

Branson, M. and Whitehead, W. (2002). Estimating a treatment effect in survival studies in which patients switch treatment. *Statistics in Medicine,* 21, 2449–2463.

Breunig, R. (2001). An almost unbiased estimator of the coefficient of variation. *Economics Letters*, 70, 15–19.

Brookmeyer, R. and Crowley, J. (1982). A confidence interval for the median survival time. *Biometrics*, 38, 29–41.

Brownell, K.D. and Stunkard, A.J. (1982). The double-blind in danger untoward consequences of informed consent. *American Journal of Psychiatry*, 139, 1487–1489.

Canales, R.D., Luo, Y., Willey, J.C. et al. (2006). Evaluation of DNA microarray results with quantitative gene expression platforms. *Nature Biotechnology*, 24, 1115–1122.

Caraco, Y. (2004). Genes and the response to drugs. *The New England Journal of Medicine*, 351, 2867–2869.

Casciano, D.A. and Woodcock, J. (2006). Empowering microarrays in the regulatory setting. *Nature Biotechnology*, 24, 1103.

CBER/FDA. (1999). CBER/FDA Memorandum. Summary of CBER considerations on selected aspects of active controlled trial design and analysis for the evaluation of thrombolytics in acute MI. Center for Biological Evaluation and Research/Food and Drug Administration, Rockville, MD, June 1999.

Chang, M. (2005a). Bayesian adaptive design with biomarkers. Invited presentation at the *IBC's Second Annual Conference: Implementing Adaptive Designs for Drug Development*, Princeton, NJ, November 7–8, 2005.

Chang, M. (2005b). Adaptive clinical trial design. Presented at *International Conference for Stochastic Process and Data Analysis*, Brest, France, May, 2005.

Chang, M. (2007). Adaptive design method based on sum of p-values. *Statistics in Medicine*, 26, 2772–2784.

Chang, M. (2008). *Adaptive Design Theory and Implementation Using SAS and R*. Chapman and Hall/CRC Press, Taylor & Francis, New York.

Chang, M. (2011). *Monte Carlo Simulation for the Pharmaceutical Industry*. Chapman and Hall/CRC, Taylor & Francis, New York.

Chang, M. and Chow, S.C. (2005). A hybrid Bayesian adaptive design for dose response trials. *Journal of Biopharmaceutical Statistics*, 15, 677–691.

Chen, J. and Chen, C. (2003). Microarray gene expression. In: *Encyclopedia of Biopharmaceutical Statistics*, S.C. Chow (ed.). Marcel Dekker, Inc., New York, pp. 599–613.

Chen, M.L. (1995). Individual bioequivalence. Invited presentation at *International Workshop: Statistical and Regulatory Issues on the Assessment of Bioequivalence*. Dusseldorf, Germany, October 19–20, 1995.

Chen, M.L. (1997). Individual bioequivalence—A regulatory update. *Journal of Biopharmaceutical Statistics*, 7, 5–11.

Chen, M.L., Shah, V., Patinaik, R., Adams, W., Hussain, A., Conner, D., Mehta, M., Malinowski, H., Lazor, J., Huang, S.M., Hare, D., Lesko, L., Spom, D., and Williams, R. (2001). Bioavailability and bioequivalence: An FDA regulatory overview. *Pharmaceutical Research*, 18, 1645–1650.

Chen, X., Luo, X., and Capizzi, T. (2005). The application of enhanced parallel gatekeeping strategies. *Statistics in Medicine*, 24, 1385–1397.

Cheng, B. and Chow, S.C. (2010). Statistical inference for a multiple-stage transitional seamless trials designs with different study objectives and endpoints. Submitted.

Cheng, B. and Shao, J. (2007). Exact tests for negligible interaction in two-way linear models. *Statistica Sinica*, 17, 1441–1455.

Cheng, B., Chow, S.C., Burt, D., and Cosmatos, D. (2008). Statistical assessment of QT/QTc prolongation based on maximum of correlated normal random variables. *Journal of Biopharmaceutical Statistics*, 18, 494–501.

Chirino, A.J. and Mire-Sluis, A. (2004). Characterizing biological products and assessing comparability following manufacturing changes. *Nature Biotechnology*, 22, 1383–1391.

Chow, S.C. (1997). Good statistics practice in the drug development and regulatory approval process. *Drug Information Journal*, 31, 1157–1166.

Chow, S.C. (2000). Significant digits in basic research. Presented at Amgen, Inc., Thousand Oaks, CA, March, 2000.

Chow, S.C. (2007a). Statistics in translational medicine. Presented at *Current Advances in Evaluation of Research & Development of Translational Medicine*, National Health Research Institutes, Taipei, Taiwan, October 19, 2007.

Chow, S.C. (2007b). *Statistical Design and Analysis of Stability Studies*. Chapman and Hall/CRC, Taylor & Francis, New York.

Chow, S.C. and Chang, M. (2005). Statistical consideration of adaptive methods in clinical development. *Journal of Biopharmaceutical Statistics*, 15, 575–591.

Chow, S.C. and Chang, M. (2006). *Adaptive Design Methods in Clinical Trials*. Chapman and Hall/CRC Press, Taylor & Francis, New York.

Chow, S.C. and Ki, F. (1994). On statistical characteristics of quality of life assessment. *Journal of Biopharmaceutical Statistics*, 4, 1–17.

Chow, S.C. and Ki, F. (1996). Statistical issues in quality of life assessment. *Journal of Biopharmaceutical Statistics*, 6, 37–48.

Chow, S.C. and Liu, J.P. (1995). *Statistical Design and Analysis in Pharmaceutical Science: Validation, Process Control, and Stability*. Marcel Dekker, Inc., New York.

Chow, S.C. and Liu, J.P. (1998a). *Design and Analysis of Animal Studies in Pharmaceutical Development*. Marcel Dekker, Inc., New York.

Chow, S.C. and Liu, J.P. (1998b). *Design and Analysis of Clinical Trials*. John Wiley & Sons, New York.

Chow, S.C. and Liu, J.P. (2000a). *Design and Analysis of Bioavailability and Bioequivalence Studies—Revised and Expanded*, 2nd edn. Marcel Dekker, Inc., New York.

Chow, S.C. and Liu, J.P. (2000b). *Design and Analysis of Clinical Trials*. John Wiley & Sons, New York.

Chow, S.C. and Liu, J.P. (2004). *Design and Analysis of Clinical Trials*, 2nd edn. John Wiley & Sons, New York.

Chow, S.C. and Liu, J.P. (2008). *Design and Analysis of Bioavailability and Bioequivalence Studies*, 3rd edn. Chapman Hall/CRC Press, Taylor & Francis, New York.

Chow, S.C. and Lu, Q. (2011). Statistical methods for testing composite hypotheses of efficacy and safety in clinical trials. Unpublished manuscript (to be submitted).

Chow, S.C. and Shao, J. (2002). *Statistics in Drug Research*. Marcel Dekker, Inc., New York.

Chow, S.C. and Shao, J. (2003). Randomization. In: *Encyclopedia of Biopharmaceutical Statistics*, S.C. Chow (ed.), 2nd edn. Taylor & Francis, New York, pp. 828–832.

Chow, S.C. and Shao, J. (2004). Analysis of clinical data with breached blindness. *Statistics in Medicine*, 23, 1185–1193.

Chow, S.C. and Shao, J. (2005). Inference for clinical trials with some protocol amendments. *Journal of Biopharmaceutical Statistics*, 15, 659–666.

Chow, S.C. and Shao, J. (2006). On non-inferiority margin and statistical tests in active control trials. *Statistics in Medicine*, 25, 1101–1113.

Chow, S.C. and Shao, J. (2007). Stability analysis for drugs with multiple ingredients. *Statistics in Medicine*, 26, 1512–1517.

Chow, S.C. and Tse, S.K. (1991). On the estimation of total variability in assay validation. *Statistics in Medicine*, 10, 1543–1553.

Chow, S.C. and Tu, Y.H. (2009). On two-stage seamless adaptive design in clinical trials. *Journal of Formosan Medical Association*, 107 (12), S51–S59.

Chow, S.C. and Wang, H. (2001). On sample size calculation in bioequivalence trials. *Journal of Pharmacokinetics and Pharmacodynamics*, 28, 155–169.

Chow, S.C., Chang, M., and Pong, A. (2005). Statistical consideration of adaptive methods in clinical development. *Journal of Biopharmaceutical Statistics*, 15, 575–591.

Chow, S.C., Cheng, B., and Cosmatos, D. (2008a). On power and sample size calculation for QT studies with recording replicates at given time point. *Journal of Biopharmaceutical Statistics*, 18, 483–493.

Chow, S.C., Lu, Q., and Tse, S.K. (2007). Statistical analysis for two-stage adaptive design with different study endpoints. *Journal of Biopharmaceutical Statistics*, 17, 1163–1176.

Chow, S.C., Pong, A., and Chang, Y.W. (2006). On traditional Chinese medicine clinical trials. *Drug Information Journal*, 40, 395–406.

Chow, S.C., Shao, J., and Hu, Y.P. (2002a). Assessing sensitivity and similarity in bridging studies. *Journal of Biopharmaceutical Statistics*, 12, 385–400.

Chow, S.C., Shao, J., and Li, L. (2004). Assessing bioequivalence using genomic data. *Journal of Biopharmaceutical Statistics*, 14, 869–880.

Chow, S.C., Shao, J., and Wang, H. (2002b). A note on sample size calculation for mean comparisons based on non-central t-statistics. *Journal of Biopharmaceutical Statistics*, 12, 441–456.

Chow, S.C., Shao, J., and Wang, H. (2003). Statistical tests for population bioequivalence. *Statistica Sinica*, 13, 539–554.

Chow, S.C., Shao, J., and Wang, H. (2008b). *Sample Size Calculation in Clinical Research*. Chapman and Hall/CRC Press, Taylor & Francis, New York.

Chung, W.H., Hung, S.I., Hong, H.S., Hsih, M.S., Yang, L.C., Ho, H.C., Wu, J.Y., and Chen, Y.T. (2004). Medical genetics: A marker for Stevens-Johnson syndrome. *Nature*, 428 (6982), 486.

Chung-Stein, C. (1996). Summarizing laboratory data with different reference ranges in multi-center clinical trials. *Drug Information Journal*, 26, 77–84.

Church, J.D. and Harris, B. (1970). The estimation of reliability from stress-strength relationships. *Technometrics*, 12, 49–54.

Cochran, W.G. (1977). *Sampling Techniques*, 3rd edn. Wiley, New York.

Colton, T. (1974). *Statistics in Medicine*. Little, Brown and Company, Boston, MA.

Cosmatos, D. and Chow, S.C. (2008). *Translational Medicine*. Chapman and Hall/CRC Press, Taylor & Francis, New York.

Cox, D.R. and Snell, E.J. (1968). A general definition of residuals (with discussion). *Journal of the Royal Statistical Society B*, 30, 248–275.

CPMP. (1990). The Committee for Proprietary Medicinal Products Working Party on Efficacy of Medicinal Products. Note for Guidance; Good Clinical Practice for Trials on Medicinal Products in the European Community. Commission of European Communities, Brussels, Belgium, 1990—111/396/88-EN Final.

CPMP. (1997). Points to consider: The assessment of the potential for QT interval prolongation by non-cardiovascular products. Available at: www.coresearch.biz/regulations/cpmp.pdf

Crommelin, D., Bermejo, T., Bissig, M., Damianns, J., Kramer, I., Rambourg, P., Scroccaro, G., Strukelj, B., Tredree, R., and Ronco, C. (2005). Biosimilars, generic versions of the first generation of therapeutic proteins: Do they exist? *Contributions to Nephrology*, 149, 287–294.

Crowley, J. (2001). *Handbook of Statistics in Clinical Oncology*. Marcel Dekker, Inc., New York.

CTriSoft Intl. (2002). *Clinical Trial Design with ExpDesign Studio*. CTriSoft Intl., Lexington, MA. Available at: www.ctrisoft.net

Cui, L., Hung, H.M.J., and Wang, S.J. (1999). Modification of sample size in group sequential trials. *Biometrics*, 55, 853–857.

Dalton, W.S. and Friend, S.H. (2006). Cancer biomarkers—An invitation to the table. *Science*, 312, 1165–1168.

DeMets, D.L., Furberg, C.D., and Friedman, L.M. (2006). *Data Monitoring in Clinical Trials: A Case Studies Approach*. Springer, New York.

Demidenko, E. (2007). Sample size determination for logistic regression revisited. *Statistics in Medicine*, 26, 3385–3397.

Dempster, A.P., Laird, N.M., and Rubin, D.B. (1977). Maximum likelihood from incomplete data via the EM algorithm. *Journal of the Royal Statistical Society*, 39, 1–38.

Dent, S.F. and Eisenhauer, E.A. (1996). Phase I trial design: Are new methodologies being put into practice? *Annals of Oncology*, 7, 561–566.

DeSouza, C.M., Legedza, T.R., and Sankoh, A.J. (2009). An overview of practical approaches for handling missing data in clinical trials. *Journal of Biopharmaceutical Statistics*, 19, 1055–1073.

Diggle, P. and Kenward, M.G. (1994). Informative dropout in longitudinal data analysis (with discussion). *Applied Statistics*, 43, 49–94.

Dixon, D.O., Freedman, R.S., Herson, J., Hughes, M., Kim, K., Silerman, M.H., and Tangen, C.M. (2006). Guidelines for data and safety monitoring for clinical trials not requiring traditional data monitoring committees. *Clinical Trials*, 3, 314–319.

Dmitrienko, A., Offen, W., and Westfall, P.H. (2003). Gatekeeping strategies for clinical trials that do not require all primary effects to be significant. *Statistics in Medicine*, 22, 2387–2400.

Dmitrienko, A., Offen, W., Wang, O., and Xiao, D. (2006). Gatekeeping procedures in dose-response clinical trials based on the Dunnett test. *Pharmaceutical Statistics*, 5, 19–28.

Dobbin, K.K., Beer, D.G., Meyerson, M. et al. (2005). Interlaboratory comparability study of cancer gene expression analysis using oligonucleotide microarrays. *Clinical Cancer Research*, 11, 565–573.

DoH. (2004a). *Draft Guidance for IND of Traditional Chinese Medicine*. The Department of Health, Taipei, Taiwan.

DoH. (2004b). *Draft Guidance for NDA of Traditional Chinese Medicine*. The Department of Health, Taipei, Taiwan.

Dubey, S.D. (1991). Some thought on the one-sided and two-sided tests. *Journal of Biopharmaceutical Statistics*, 1, 139–150.

Dudoit, S., Yang, Y.H., Callow, M.J., and Speed, T.P. (2002). Statistical methods for identifying differentially expressed genes in replicated cDNA microarray experiments. *Statistica Sinica*, 12, 111–139.

Dunnett, C.W. (1955). Multivariate normal probability integrals with product correlation structure, Algorithm AS251. *Journal of the American Statistical Association*, 50, 1096–1121.

D'Agostino, R.B., Massaro, J.M., and Sullivan, L.M. (2003). Non-inferiority trials: Design concepts and issues—The encounters of academic consultants in statistics. *Statistics in Medicine*, 22, 169–186.

Eaton, M.L., Muirhead, R.J., Mancuso, J.Y., and Kolluri, S. (2006). A confidence interval for the maximal mean QT interval change caused by drug effect. *Drug Information Journal*, 40, 267–271.

Efron, B. (1982). *The Jackknife, the Bootstrap, and Other Resampling Plans*. Society of Industrial and Applied Mathematics CBMS-NSF Monographs, 38, Philadelphia, PA.

Efron, B. (1983). Estimating the error rate of a prediction rule: Improvement on cross-validation. *Journal of the American Statistical Association*, 78, 316–331.

Efron, B. (1986). How biased is the apparent error rate of a prediction rule? *Journal of the American Statistical Association*, 81, 461–470.

Efron, B. and Tibshirani, R.J. (1993). *An Introduction to the Bootstrap*. Chapman and Hall, New York.

Eisenhauer, E.A., O'Dwyer, P.J., Christian, M., and Humphrey, J.S. (2000). Phase I clinical trial design in cancer drug development. *Journal of Clinical Oncology*, 18, 684–692.

Ellenberg, S.S., Fleming, T.R., and DeMets, D.L. (2002). *Data Monitoring Committees in Clinical Trials: A Practical Perspective*. John Wiley & Sons, New York.

EMEA. (2001). Note for guidance on the investigation of bioavailability and bioequivalence. The European Medicines Agency Evaluation of Medicines for Human Use. EMEA/EWP/QWP/1401/98, London, U.K.

EMEA. (2003a). Note for guidance on comparability of medicinal products containing biotechnology-derived proteins as drug substance—Non clinical and clinical issues. The European Medicines Agency Evaluation of Medicines for Human Use. EMEA/CHMP/3097/02, London, U.K.

EMEA. (2003b). Rev. 1 Guideline on comparability of medicinal products containing biotechnology-derived proteins as drug substance—Quality issues. The European Medicines Agency Evaluation of Medicines for Human Use. EMEA/CHMP/BWP/3207/00/Rev 1, London, U.K.

EMEA. (2005a). Guideline on similar biological medicinal products. The European Medicines Agency Evaluation of Medicines for Human Use. EMEA/CHMP/437/04, London, U.K.

EMEA. (2005b). Draft guideline on similar biological medicinal products containing biotechnology-derived proteins as drug substance: Quality issues. The European Medicines Agency Evaluation of Medicines for Human Use. EMEA/CHMP/49348/05, London, U.K.

EMEA. (2005c). Draft annex guideline on similar biological medicinal products containing biotechnology-derived proteins as drug substance—Non clinical and clinical issues—Guidance on biosimilar medicinal products containing recombinant erythropoietins. The European Medicines Agency Evaluation of Medicines for Human Use. EMEA/CHMP/94526/05, London, U.K.

EMEA. (2005d). Draft annex guideline on similar biological medicinal products containing biotechnology-derived proteins as drug substance—Non clinical and clinical issues—Guidance on biosimilar medicinal products containing recombinant granulocyte-colony stimulating factor. The European Medicines Agency Evaluation of Medicines for Human Use. EMEA/CHMP /31329/05, London, U.K.

EMEA. (2005e). Draft annex guideline on similar biological medicinal products containing biotechnology-derived proteins as drug substance—Non-clinical and clinical issues—Guidance on biosimilar medicinal products containing somatropin. The European Medicines Agency Evaluation of Medicines for Human Use. EMEA/CHMP/94528/05, London, U.K.

EMEA. (2005f). Draft annex guideline on similar biological medicinal products containing biotechnology-derived proteins as drug substance—Non clinical and clinical issues—Guidance on biosimilar medicinal products containing recombinant human insulin. The European Medicines Agency Evaluation of Medicines for Human Use. EMEA/CHMP/32775/05, London, U.K.

EMEA. (2005g). Guideline on the clinical investigating of the pharmacokinetics of therapeutic proteins. The European Medicines Agency Evaluation of Medicines for Human Use. EMEA/CHMP/89249/04, London, U.K.

EMEA. (2007). Reflection paper on methodological issues in confirmatory clinical trials planned with an adaptive design. EMEA Doc. Ref. CHMP/EWP/2459/02, October 20. Available at: http://www.emea.europa.eu/pdfs/human/ewp/245902enadopted.pdf

Emerson, J.D. (1982). Nonparametric confidence intervals for the median in the presence of right censoring. *Biometrics*, 38, 17–27.

Endrenyi, L., Fritsch, S., and Yan, W. (1991). (Cmax)/AUC is a clearer measure than (Cmax) for absorption rates in investigations of bioequivalence. *International Journal of Clinical Pharmacology, Therapy and Toxicology*, 29, 394–399.

Enis, P. and Geisser, S. (1971). Estimation of the probability that Y < X. *Journal of American Statistical Association*, 66, 162–168.

Fairweather, W.R. (1994). Statisticians, the FDA and a time of transition. Presented at *Pharmaceutical Manufacturers Association Education and Research Institute Training Course in Non-Clinical Statistics*, Georgetown University Conference Center, Washington, DC, February 6–8, 1994.

FDA. (1987a). *Guideline for Submitting Documentation for the Stability of Human Drugs and Biologics*. Center for Drugs and Biologics, Office of Drug Research and Review, Food and Drug Administration, Rockville, MD.

FDA. (1987b). *Guideline on General Principles of Process Validation*. Center for Drug and Biologics and Center for Devices and Radiological Health, Food and Drug Administration, Rockville, MD.

FDA. (1988). *Guideline for Format and Content of the Clinical and Statistical Sections of New Drug Applications*. Center for Drug Evaluation and Research, Food and Drug Administration, Rockville, MD.

FDA. (1991). *Guidance for In Vivo Bioequivalence and In Vitro Drug Release*. Center for Drug Evaluation and Research, Food and Drug Administration, Rockville, MD.

FDA. (1997). *Guidance for Industry: Dissolution Testing of Immediate Release Solid Oral Dosage Forms*. The United States Food and Drug Administration, Rockville, MD.

FDA. (2001). *Guidance on Statistical Approaches to Establishing Bioequivalence*. Center for Drug Evaluation and Research, The US Food and Drug Administration, Rockville, MD.

FDA. (2003a). *Draft Guidance for Industry—Multiplex Tests for Heritable DNA Markers, Mutations and Expression Patterns*. The United States Food and Drug Administration, Rockville, MD.

FDA. (2003b). *Guidance on Bioavailability and Bioequivalence Studies for Orally Administrated Drug Products—General Considerations.* Center for Drug Evaluation and Research, The US Food and Drug Administration, Rockville, MD.

FDA. (2004). *Guidance for Industry—Botanical Drug Products.* The United States Food and Drug Administration, Rockville, MD.

FDA. (2005). *Draft Concept Paper on Drug-Diagnostic Co-Development.* Food and Drug Administration, Rockville, MD.

FDA. (2006a). *Draft Guidance on In Vitro Diagnostic Multivariate Index Assays.* Food and Drug Administration, Rockville, MD.

FDA. (2006b). *Guidance for Clinical Trial Sponsors: Establishment and Operation of Clinical Trial Data Monitoring Committees.* CBER/CDER/CDRH, The United States Food and Drug Administration, Rockville, MD. Available at: http://www.fda.gov/cber/gdlns/clintrialdmc.pdf

FDA. (2010a). *Guidance for Industry—Non-inferiority Clinical Trials.* The United States Food and Drug Administration, Rockville, MD.

FDA. (2010b). *Draft Guidance for Industry—Adaptive Design Clinical Trials for Drugs and Biologics.* The United States Food and Drug Administration, Rockville, MD.

FDA/TPD. (2003). *Preliminary Concept Paper: The Clinical Evaluation of QT/QTc Interval Prolongation and Proarrythmic Potential for Non-arrythmic Drug Products.* Released on November 15, 2002. Revised on February 6, 2003.

Feeny, D.H. and Torrance, G.W. (1989). Incorporating utility-based quality-of-life assessment measures in clinical trials. *Medical Care,* 27, S198–S204.

Fleiss, J.L., Levin, B. and Paik, M.C. (2003). *Statistical Methods for Rates and Proportions.* John Wiley & Sons, New York.

Fontanarosa, P.B., Flanagin, A., and DeAngelis, C.D. (2005). Reporting conflicts of interest, financial aspects of research and role of sponsors in funded studies. *Journal of the American Medical Association,* 294, 110–111.

Frank, R.G. (2007). Regulation of follow-on biologics. *New England Journal of Medicine,* 357, 841–843.

Frueh, F.W. (2006). Impact of microarray data quality on genomic data submissions to the FDA. *Nature Biotechnology,* 24, 1105–1107.

Gail, M.H. and Simon, R. (1985). Testing for qualitative interactions between treatment effects and patient subsets. *Biometrics,* 41, 361–372.

Gallo, P., Chuang-Stein, C., Dragalin, V., Gaydos, B., Krams, M., and Pinheiro, J. (2006). Adaptive design in clinical drug development—An executive summary of the PhRMA Working Group (with discussions). *Journal of Biopharmaceutical Statistics,* 16 (3), 75–283.

Gentle, J.E. (1998), *Random Number Generator and Monte Carlo Methods.* Springer-Verlag, New York.

Genz, A. and Bretz, F. (2002). Methods for the computation of multivariate t-probabilities. *Journal of Computational and Graphical Statistics,* 11, 950–971.

Goldberg, J.D. and Kury, K.J. (1990). Design and analysis of multicenter trials. In: *Statistical Methodology in the Pharmaceutical Industry,* D. Berry (ed.). Marcel Dekker, Inc., New York, pp. 201–237.

Goodman, S.N. (1992). A comment on replication, p-values and evidence. *Statistics in Medicine,* 11, 875–879.

Goodman, S.N., Zahurak, M.L., and Piantadosi, S. (1995). Some practical improvements in the continual reassessment method for phase I studies. *Statistics in Medicine,* 14, 1149–1161.

Guilford, J.P. (1954). *Psychometric Methods*, 2nd edn. McGraw-Hill, New York.

Gunst, G.F. and Mason, R.L. (1980). *Regression Analysis and Its Application*. Marcel Dekker, Inc., New York.

Guyatt, G.H., Veldhuyen Van Zanten, S.J.O., Feeny, D.H., and Patric, D.L. (1989). Measuring quality of life in clinical trials: A taxonomy and review. *Canadian Medical Association Journal*, 140, 1441–1448.

Haidar, S.H., Davit, B., Chen, M.-L., Conner, D., Lee, L., Li, Q.H., Lionberger, R., Makhlouf, F., Patel, D., Schuirmann, D.J., and Yu, L.X. (2008). Bioequivalence approaches for highly variable drugs and drug products. *Pharmaceutical Research*, 25(1), 237–241.

Hall, P. (1992). *The Bootstrap and Edgeworth Expansion*. Springer-Verlag, New York.

Hauck, W.W. and Anderson, S. (1984). A new statistical procedure for testing equivalence in two-group comparative bioavailability trials. *Journal of Pharmacokinetics and Biopharmaceutics*, 12, 83–91.

Hemmings, R. and Day, S. (2004). Regulatory perspectives on data safety monitoring boards: Protecting the integrity of data. *Drug Safety*, 27, 1–6.

Herson, J. (2009). *Data and Safety Monitoring Committees in Clinical Trials*. Chapman and Hall/CRC Press, Taylor & Francis, New York.

Heyd, J.M. and Carlin, B.P. (1999). Adaptive design improvements in the continual reassessment method for phase I studies. *Statistics in Medicine*, 18, 1307–1321.

Ho, H. and Chow, S.C. (1998). Design and analysis of multinational clinical trials. *Drug Information Journal*, 32, 1309S–1316S.

Hochberg, Y. (1988). A sharper Bonferroni procedure for multiple tests of significance. *Biometrika*, 75, 800–803.

Hochberg, Y. and Benjamini, Y. (1990). More powerful procedures for multiple significance testing. *Statistics in Medicine*, 9, 811–818.

Hochberg, Y. and Tamhane, A.C. (1987). *Multiple Comparison Procedures*. Wiley, New York.

Hollenberg, N.K., Testa, M., and Williams, G.H. (1991). Quality of life as a therapeutic end-point—An analysis of therapeutic trials in hypertension. *Drug Safety*, 6, 83–93.

Holm, S. (1979). A simple sequentially rejective multiple test procedure. *Scandinavian Journal of Statistics*, 6, 65–70.

Hommel, G. (1988). A stagewise rejective multiple test procedure based on a modified Bonferroni test. *Biometrika*, 75, 383–386.

Hommel, G., Lindig, V., and Faldum, A. (2005). Two stage adaptive designs with correlated test statistics. *Journal of Biopharmaceutical Statistics*, 15, 613–623.

Hosmane, B. and Locke, C. (2005). A simulation study of power in thorough QT/QTc studies and a normal approximation for planning purposes. *Drug Information Journal*, 39, 447–455.

Hsiao, C.F., Tsou, H.H., Pong, A., Liu, J.P., Lin, C.H., Chang, Y.J., and Chow, S.C. (2009). Statistical validation of traditional Chinese diagnostic procedure. *Drug Information Journal*, 43, 83–95.

Hsieh, E., Lu, Y., Yang, L.Y., and Chow, S.C. (2010). The impact of the block sizes on randomized controlled trials. Submitted.

Hsu, J.C. (1996). *Multiple Comparisons—Theory and Methods*. Chapman and Hall, London, U.K.

Hung, H.M.J. (2003). Statistical issues with design and analysis of bridging clinical trial. Presented at *the 2003 Symposium on Statistical Methodology for Evaluation of Bridging Evidence*, Taipei, Taiwan.

Hung, H.M.J. and Wang, S.J. (2009). Some controversial multiple testing problems in regulatory applications. *Journal of Biopharmaceutical Statistics*, 19, 1–11.

Hung, H.M.J., Wang, S.J., and O'Neil, R. (2007). Issues with statistical risks for testing methods in noninferiority trial without a placebo arm. *Journal of Biopharmaceutical Statistics*, 17, 201–213.

Hung, H.M.J., Wang, S.J., Tsong, Y., Lawrence, J., and O'Neil, R.T. (2003). Some fundamental issues with non-inferiority testing in active controlled trials. *Statistics in Medicine*, 22, 213–225.

ICH. (1993). Q1A stability testing of new drug substances and products. In: *Tripartite International Conference on Harmonization Guideline*. Geneva, Switzerland.

ICH. (1995). Guideline for structure and content of clinical study report. *International Conference on Harmonization*, Yokohama, Japan.

ICH. (1996a). E6 Good clinical practice. *Tripartite International Conference on Harmonization Guideline*. Available at: http://www/ich.org/LOB/media/MEDIA482.pdf

ICH. (1996b). Q2B Validation of analytical procedures: Methodology. *Tripartite International Conference on Harmonization Guideline*. Geneva, Switzerland, November, 1996.

ICH. (1996c). Q5C Guideline on quality of biotechnological products: Stability testing of biotechnological/biological products. Center for Drug Evaluation and Research, Center for Biologics Evaluation and Research, The U.S. Food and Drug Administration, Rockville, MD.

ICH. (1997). E5 Guideline on ethnic factors in the acceptability of foreign data. The U.S. Federal Register, Vol. 83, 31790–31796.

ICH. (1998). E9 Guideline for statistical principles for clinical trials. *Tripartite International Conference on Harmonization Guideline*. Centre for Drug Evaluation and Research, center for Biologics Evaluation and Research. The U.S. Food and Drug Administration, Rockville, MD.

ICH. (1999). Q6B Guideline on test procedures and acceptance criteria for biotechnological/biological products. Center for Drug Evaluation and Research, Center for Biologics Evaluation and Research, The U.S. Food and Drug Administration, Rockville, MD.

ICH. (2000). *E10 International Conference on Harmonization Guideline: Guidance on Choice of Control Group and Related Design and Conduct Issues in Clinical Trials*. Food and Drug Administration, DHHS, July 2000.

ICH. (2005a). E14 The clinical evaluation of QT/QTc interval prolongation and proarrythmic potential for non-antiarrythmic drugs. *Tripartite International Conference on Harmonization Guideline*, Geneva, Switzerland, May 2005.

ICH. (2005b). Q5E Guideline on comparability of biotechnological/biological products subject to changes in their manufacturing process. Center for Drug Evaluation and Research, Center for Biologics Evaluation and Research, The U.S. Food and Drug Administration, Rockville, MD.

Irizarry, R.A., Warren, D., Spencer, F. et al. (2005). Multi-laboratory comparison of microarray platforms. *Nature Methods*, 2, 345–349.

Jachuck, S.J., Brierley, H., and Wilcox, P.M. (1982). The effect of hypotensive drugs on the quality of life. *Journal of Royal College* of *General Practitioners*, 32, 103–105.

Jennison, C. and Turnbull, B.W. (2000). *Group Sequential Tests with Applications to Clinical Trials*. Chapman & Hall, London/Boca Raton, FL.

Ji, H. and Davis, R.W. (2006). Data quality in genomics and microarray. *Nature Biotechnology*, 24, 1112–1113.

Johnson, J., Williams, G., and Pazdur, R. (2003). End points and United States Food and Drug Administration approval of oncology drugs. *Journal of Clinical Oncology*, 21, 1404–1411.

Johnson, N.L. and Kotz, S. (1970). *Distributions in Statistics—Continuous Univariate Distributions—1*. John Wiley & Sons, New York.

Johnson, N.L. and Kotz, S. (1972). *Distributions in Statistics: Continuous Multivariate Distributions*. John Wiley & Sons, New York.

Julious, S.A., Tan, S.B., and Machin, D. (2009). *An Introduction to Statistics in Early Phase Clinical Trials*. Wiley-Blackwell, Chichester, U.K.

Kalton, G. and Kasprzyk, D. (1986). The treatment of missing data. *Survey Methodology*, 12, 1–16.

Kamp, B., Bretz, F., Dmitrienko, A. et al. (2007). Innovative approaches for designing and analyzing adaptive dose-ranging trials. *Journal of Biopharmaceutical Statistics*, 17, 965–995.

Kaplan, R.M., Bush, J.W., and Berry, C.C. (1976). Health status: Types of validity and index of well-being. *Health Services Research*, 4, 478–507.

Karlowski, T.R., Chalmers, T.C., Frenkel, L.D., Kapikian, A.Z., Lewis, T.L., and Lynch, J.M. (1975). Ascorbic acid for the common cold: A prophylactic and therapeutic trial. *Journal of the American Medical Association*, 231, 1038–1042.

Kawai, N., Stein, C., Komiyama, O., and Li, Y. (2008). An approach to rationalize partitioning sample size into individual regions in a multiregional trial. *Drug Information Journal*, 42, 139–147.

Keith, O.W. (2007). Biosimilars: Are we there yet? Presented at *Biosimilars 2007*, George Washington University, Washington, DC.

Kelly, P.J., Sooriyarachchi, M.R., Stallard, N., and Todd, S. (2005a). A practical comparison of group-sequential and adaptive designs. *Journal of Biopharmaceutical Statistics*, 15, 719–738.

Kelly, P.J., Stallard, N., and Todd, S. (2005b). An adaptive group sequential design for phase II/III clinical trials that select a single treatment from several. *Journal of Biopharmaceutical Statistics*, 15, 641–658.

Kessler, D.A. (1989). The regulation of investigational drugs. *New England Journal of Medicine*, 320, 281–288.

Khatri, C.G. and Shah, K.R. (1974). Estimation of location of parameters from two linear models under normality. *Communications in Statistics—Theory and Methods*, 3, 647–663.

Khongphatthanayothin, A., Lane, J., Thomas, D., Yen, L., Chang, D., and Bubolz, B. (1998). Effects of cisapride on QT interval in children. *Journal of Pediatrics*, 133, 51–56.

Ki, F.Y.C. and Chow, S.C. (1994). Analysis of quality of life with parallel questionnaires. *Drug Information Journal*, 28, 69–80.

Ki, F.Y.C. and Chow, S.C. (1995). Statistical justification for the use of composite score in quality assessment. *Drug Information Journal*, 29, 715–727.

Kimko, H.C. and Duffull, S.B. (eds.). (2003). *Simulation for Designing Clinical Trials*. Marcel Dekker, Inc., New York.

Ko, F.S., Tsou, H.H., Liu, J.P., and Hsiao, C.F. (2010). Sample size determination for a specific region in a multi-regional trial. *Journal of Biopharmaceutical Statistics*, 20 (4), 870–875.

Kong, F., Chen, Y.F., and Jin, K. (2009). A bias correction in testing treatment effect under informative dropout in clinical trials. *Journal of Biopharmaceutical Statistics*, 19, 980–1000.

Korn, E.L. and Simon, R. (1996). Data monitoring committees and problems of lower-than-expected accural or event rates. *Controlled Clinical Trials*, 17, 527–536.

Korn, E.L., Midthune, D., Chen, T.T., Rubinstein, L.V., Christian, M.C., and Simon, R. (1999). Commentary. *Statistics in Medicine*, 18, 2691–2692.

Korteweg, M. (2002). Benchmarking of GRP—Quality management system in the framework of PERF. *The Regulatory Affairs Journals*, Ltd, 109–113 (February, 2002).

Koti, K.M. (2007a). Use of the Fieller-Hinkley distribution of the ratio of random variables in testing for noninferiority. *Journal of Biopharmaceutical Statistics*, 17, 215–228.

Koti, K.M. (2007b). New tests for null hypothesis of non unity ratio of proportions. *Journal of Biopharmaceutical Statistics*, 17, 229–245.

Koyfman, S.A., Agrawal, M., Garrett-Mayer, E., Krohmal, B., Wolf, E., Emanuel, E.J., and Gross, C.P. (2007). Risks and benefits associated with novel phase 1 oncology trial designs. *Cancer*, 110, 1115–1124.

Kozlowski, S. (2007). FDA Policy on follow on biologics. Presented at *Biosimilars 2007*, George Washington University, Washington, DC.

Krams, M., Burman, C.F., Dragalin, V., Gaydos, B., Grieve, A.P., Pinheiro, J., and Maurer, W. (2007). Adaptive designs in clinical drug development: Opportunities challenges, and scope reflections following PhRMA's November 2006 Workshop. *Journal of Biopharmaceutical Statistics*, 17, 957–964.

Kuhlmann, M. and Covic, A. (2006). The protein science of biosimilars. *Nephrology Dialysis Transplantation*, 21(Suppl. 5), v4–v8.

Lachin, J.M. (1988). Statistical properties of randomization in clinical trials. *Controlled Clinical Trials*, 9, 289–311.

Lachin, J.M. and Foulkes, M.A. (1986). Evaluation of sample size and power for analysis of survival with allowance for nonuniform patient entry, losses to follow-up, noncompliance, and stratification. *Biometrics*, 42, 507–519.

Lakatos, E. (1986). Sample size determination in clinical trials with time dependent rates of losses and noncompliance. *Controlled Clinical Trials*, 7, 189–199.

Lakshminarayanan, M.Y. (2010). Multiple comparisons. In: *Encyclopedia of Biopharmaceutical Statistics*, S.C. Chow (ed.). Taylor & Francis, New York.

Larkin, J.E., Frank, B.C., Gavras, H., Sultana, R., and Quackenbush, J. (2005). Independence and reproducibility across microarray platforms. *Nature Methods*, 2, 337–343.

Lasser, K.E., Allen, P.D., Woolhandler, S.J., Himmelstein, D.U., Wolfe, S.M., and Bor, D.H. (2002). Timing of new black box warnings and withdrawals for prescription medications. *Journal of the American Medical Association*, 287, 2215–2220.

Laster, L.L. and Johnson, M.F. (2003). Non-inferiority trials: The 'at least as good as' criterion. *Statistics in Medicine*, 22, 187–200.

Lee, Y., Shao, J., Chow, S.C., and Wang, H. (2002a). Test for inter-subject and total variabilities under crossover design. *Journal of Biopharmaceutical Statistics*, 12, 503–534.

Lee, Y., Wang, H., and Chow, S.C. (2002b). Comparing variabilities in clinical research. In: *Encyclopedia of Biopharmaceutical Statistics*, S.C. Chow (ed.). Marcel Dekker, Inc., New York.

Lee, Y., Wang, H., and Chow, S.C. (2008). A bootstrap-median approach for stable sample size determination when the specification parameters are estimated from a small pilot study. Unpublished manuscript.

Leeson, L.J. (1995). In Vitro/in vivo correlation. *Drug Information Journal*, 29, 903–915.

Lehmacher, W. and Wassmer, G. (1999). Adaptive sample size calculations in group sequential trials. *Biometrics*, 55, 1286–1290.

Le Tourneau, C., Lee, J.J., and Siu, L.L. (2009). Dose escalation methods in phase I cancer clinical trials. *Journal of the National Cancer Institute*, 101, 708–720.

Lewis, J.A. (1995). Statistical issues in the regulation of medicine. *Statistics in Medicine*, 14, 127–136.

Li, C.R., Liao, C.T., and Liu, J.P. (2008). On the exact interval estimation for the difference in paired areas under the ROC curves. *Statistics in Medicine*, 27, 224–242.

Li, L., Chow, S.C., and Smith, W. (2004). Cross-validation for linear model with unequal variances in genomic analysis. *Journal of Biopharmaceutical Statistics*, 14, 723–739.

Li, W.J., Shih, W.J., and Wang, Y. (2005). Two-stage adaptive design for clinical trials with survival data. *Journal of Biopharmaceutical Statistics*, 15, 707–718.

Liang, B.A. (2007). Regulating follow-on biologics. *Harvard Journal on Legislation*, 44, 363–373.

Liao, C.T., Lin, C.Y., and Liu, J.P. (2007). Noninferiority tests based on concordance correlation coefficient for assessment of the agreement for gene expression data from microarray experiments. *Journal of Biopharmaceutical Statistics*, 17, 309–327.

Liao, C.T., Lin, C.Y., and Liu, J.P. (2007). Noninferiority tests based on concordance correlation coefficient for assessment of the agreement for gene expression data from microarray experiments. *Journal of Biopharmaceutical Statistics*, 17, 309–327.

Lin, L.I. (1989). A concordance correlation coefficient to evaluate reproducibility. *Biometrics*, 45, 255–268.

Lin, L.I. (1992). Assay validation using the concordance correlation coefficient. *Biometrics*, 48, 599–604.

Lin, L.I., Hedayat, A.S., Sinha, B., and Yang, M. (2002). Statistical methods in assessing agreement: Models, issues, and tools. *Journal of the American Statistical Association*, 97, 257–270.

Lin, M., Chu, C.C., Chang, S.L. et al. (2001). The origin of Minnan and Hakka, the so-called "Taiwanese," inferred by HLA study. *Tissue Antigen*, 57, 192–199.

Lin, M., Hsieh, E., Yang, L.Y., and Chow, S.C. (2010). On center grouping in multi-center clinical trials. Submitted.

Little, R.J. (1994). A class of pattern-mixture models for normal missing data. *Biometrika*, 81, 471–483.

Little, R.J. and Rubin, D.B. (1987). *Statistical Analysis with Missing Data*. Wiley, New York.

Little, R.J. and Rubin, D.B. (2002). *Statistical Analysis with Missing Data*, 2nd edn. Wiley, New York.

Liu, J.P. (1998). Statistical evaluation of individual bioequivalence. *Communications in Statistics, Theory and Methods*, 27, 1433–1451.

Liu, J.P. and Chow, S.C. (1996). Statistical issues on FDA conjugated estrogen tablets guideline. *Drug Information Journal*, 30, 881–889.

Liu, J.P. and Chow, S.C. (2008). Statistical issues on the diagnostic multivariate index assay and targeted clinical trials. *Journal of Biopharmaceutical Statistics*, 18, 167–182.

Liu, J.P. and Lin, J.R. (2008). Statistical methods for targeted clinical trials under enrichment design. *Journal of the Formosan Medical Association*, 107, S34–S41.

Liu, J.P. and Weng, C.S. (1992). Estimation of direct formulation effect under log-normal distribution in bioavailability/bioequivalence studies. *Statistics in Medicine*, 11, 881–896.

Liu, J.P. and Weng, C.S. (1995). Bias and two one-sided tests procedure in assessment of bioequivalence. *Statistics in Medicine*, 14, 853–861.

Liu, J.P., Dai, J.Y., Lee, T.C., and Liao, C.T. (2007). A new hypothesis to test minimal fold changes of gene expression levels. In: *The 5th International Conference on Multiple Comparison Procedures*, Vienna, Austria, July 9–11.

Liu, J.P., Hsueh, H.M., and Hsiao, C.F. (2002a). Bayesian approach to evaluation of the bridging studies. *Journal of Biopharmaceutical Statistics*, 12, 401–408.

Liu, J.P., Hsueh, H.M., Hsieh, E., and Chen, J.J. (2002b). Tests for equivalence or non-inferiority for paired binary data. *Statistics in Medicine*, 21, 231–245.

Liu, J.P., Lin, J.R., and Chow, S.C. (2009). Inference on treatment effects for targeted clinical trials under enrichment design. *Pharmaceutical Statistics*, 8, 356–370.

Liu, J.P., Ma, M.C., Wu, C.Y., and Tai, J.Y. (2006). Tests of equivalence and non-inferiority for diagnostic accuracy based on the paired areas under ROC curves. *Statistics in Medicine*, 25, 1219–1238.

Liu, Q. and Chi, G.Y.H. (2001). On sample size and inference for two-stage adaptive designs. *Biometrics*, 57, 172–177.

Liu, Q., Proschan, M.A., and Pledger, G.W. (2002). A unified theory of two-stage adaptive designs. *Journal of American Statistical Association*, 97, 1034–1041.

Lohr, S.L. (1999). *Sampling Design and Analysis*. Duxbury Press, Pacific Grove, CA.

Loke, Y.C., Tan, S.B., Cai, Y., and Machin, D. (2006). A Bayesian dose finding design for dual endpoint phase I trials. *Statistics in Medicine*, 25, 3–22.

Longford, N.T. (1993). *Random Coefficient Models*. Oxford University Press, Inc., New York.

Lu, Q., Chow, S.C., and Tse, S.K. (2007). Statistical quality control process for traditional Chinese medicine with multiple correlative components. *Journal of Biopharmaceutical Statistics*, 17, 791–808.

Lu, Y., Chow, S.C., and Zhang, Z. (2010). Statistical inference for clinical trials with random shift in scale parameter of target patient population. Submitted.

Ma, H., Smith, B., and Dmitrienko, A. (2008). Statistical analysis methods for QT/QTc prolongation. *Journal of Biopharmaceutical Statistics*, 18, 553–563.

Maca, J., Bhattacharya, S., Dragalin, V., Gallo, P., and Krams, M. (2006). Adaptive seamless phase II/III designs—Background, operational aspects, and examples. *Drug Information Journal*, 40, 463–474.

Maitournam, A. and Simon, R. (2005). On the efficiency of targeted clinical trials. *Statistics in Medicine*, 24, 329–339.

Malik, M. and Camm, A.J. (2001). Evaluation of drug-induced QT interval prolongation. *Drug Safety*, 24, 323–351.

Mallows, C.L. (1973). Some comments on Cp. *Technometrics*, 15, 661–675.

Mankoff, S.P., Brander, C., Ferrone, S., and Marincola, F.M. (2004). Lost in translation: Obstacles to translational medicine. *Journal of Translational Medicine*, 2, 14.

MAQC Consortium. (2006). The MAQC project shows inter- and intraplatform reproducibility of gene expression measurements. *Nature Biotechnology*, 24, 1151–1161.

Marcus, R., Peritz, E., and Gabriel, K.B. (1976). On closed testing procedures with special reference to ordered analysis of variance. *Biometrika*, 63, 655–660.

Margolies, M.E. (1994). Regulations of combination products. *Applied Clinical Trials*, 3, 50–65.

Maxwell, C., Domenet, J.G., and Joyce, C.R.R. (1971). Instant experience in clinical trials: A novel aid to teaching by simulation. *Journal of Clinical Pharmacology*, 11, 323–331.

McLachlan, G.J. and Krishnan, T. (1997). *The EM Algorithm and Extensions*. Wiley, New York.

McLachlan, G.J. and Peel, D. (2000). *Finite Mixture Models*. Wiley, New York.

Meier, P. (1953). Variance of a weighted mean. *Biometrics*, 9, 59–73.

Members of the Toxicogenomic Research Consortium. (2005). Standardization of global gene expression analysis between laboratories and across platforms. *Nature Methods*, 2, 351–356.

MHLW. (2007). *Guidance on Basic Principles on Global Clinical Trials*. The Ministry of Health, Labor, and Welfare of Japan, Tokyo, Japan.

MINDACT Design and MINDACT trial overview. Available at: http://www.breast internationalgroup.org/transbig.html (accessed on June 5, 2006).

Moore, J.W. and Flanner, H.H. (1996). Mathematical comparison of curves with an emphasis on dissolution profiles. *Pharmaceutical Technology*, 20, 64–74.

Moore, K.L. and van der Laan, M.J. (2009). Increasing power in randomized trials with right censored outcomes through covariate adjustment. *Journal of Biopharmaceutical Statistics*, 19, 1099–1131.

MOPH. (2002). *Guidance for Drug Registration*. Ministry of Public Health, Beijing, China.

Moss, A.J. (1993). Measurement of the QT interval and the risk associated with QT interval prolongation. *American Journal of Cardiology*, 72, 23B–25B.

Muller, H.H. and Schafer, H. (2001). Adaptive group sequential designs for clinical trials: Combining the advantages of adaptive and classical group sequential approaches. *Biometrics*, 57, 886–891.

Myrand, S.P., Sekiguchi, K., Man, M.Z. et al. (2008). Pharmacokinetics/genotype associations for major cytochrome P450 enzymes in native and first- and third-generation Japanese populations: Comparison with Korean, Chinese, and Caucasian populations. *Clinical Pharmacology & Therapeutics*, 84 (3), 347–361.

NCCLS. (2001). *User Demonstration of Performance for Precision and Accuracy*. Approved Guidance, NCCLS document EP15-A, National Committee for Clinical Laboratory Standards, Wayne, PA.

Ng, T.H. (2007). Simultaneous testing of noninferiority and superiority increases the false discover rate. *Journal of Biopharmaceutical Statistics*, 17, 259–264.

Nie, L., Chu, H., Cheng, Y., Spurney, C., Nagaraju, K., and Chen, J. (2009). Marginal and conditional approaches to multivariate variables subject to limit of detection. *Journal of Biopharmaceutical Statistics*, 19, 1151–1161.

NIH. (1998). NIH policy for data and safety monitoring. The United States National Institutes of Health, June 1998. Available at: http://grants1.nih.gov/grants/guide/notice-files/not98-084.html

NIH. (2000). Further guidance on data and safety monitoring for phase I and II trials. The United States National Institutes of Health, OD-00-038, June 2000.

Nityasuddhi, D. and Böhning, D. (2003). Asymptotic properties of the EM algorithm estimate for normal mixture models with component specific variances. *Computational Statistics & Data Analysis*, 41, 591–601.

NRC. (2010). *The Prevention and Treatment of Missing Data in Clinical Trials*. Panel on Handling Missing Data in Clinical Trials, Committee on National Statistics, Division of Behavioral and Social Sciences and Education, National Research Council of the National Academies. The National Academies Press, Washington, DC.

Olschewski, M. and Schumacher, M. (1990). Statistical analysis of quality of life data in cancer clinical trials. *Statistics in Medicine*, 9, 749–763.

Ott, L. (1984). *An Introduction to Statistical Method and Data Analysis*, 2nd edn. Duxbury Press, Boston, MA.

O'Brien, P.C. and Fleming, T.R. (1979). A multiple testing procedure for clinical trials. *Biometrics*, 35, 549–556.

O'Quigley, J. (1999). Another look at two phase I clinical trial designs. *Statistics in Medicine*, 18, 2683–2690.

O'Quigley, J. (2001). Dose-finding designs using continual reassessment method. In: *Handbook of Statistics in Clinical Oncology*, J. Crowley (ed.). Marcel Dekker, Inc., New York, pp. 35–72.

O'Quigley, J. and Chevret, S. (1991). Methods for dose finding studies in cancer clinical trials: A review and results of a Monte Carlo study. *Statistics in Medicine*, 10, 1647–1664.

O'Quigley, J. and Shen, L. (1996). Continual reassessment method: A likelihood approach. *Biometrics*, 52, 673–684.

O'Quigley, J., Hughes, M.D., and Fenton, T. (2001). Dose finding designs for HIV studies. *Biometrics*, 57, 1018–1029.

O'Quigley, J., Pepe, M., and Fisher, L. (1990). Continual reassessment method: A practical design for phase I clinical trials in cancer. *Biometrics*, 46, 33–48.

Paik, S., Shak, S., Tang, G. et al. (2004). A multigene assay to predict recurrence of tamoxifen-treated, node-negative breast cancer. *New England Journal of Medicine*, 351, 2817–2826.

Paik, S., Tang, G., Shak, S. et al. (2006). Gene expression and benefit of chemotherapy in women with node-negative, estrogen receptor-positive breast cancer. *Journal of Clinical Oncology*, 24, 1–12.

Paoletti, X. and Kramar, A. (2009). A comparison of model choices for the continual reassessment method in phase I cancer trials. *Statistics in Medicine*, 28, 3012–3028.

Parmigiani, G. (2002). *Modeling in Medical Decision Making*. John Wiley & Sons, West Sussex, England.

Patel, H.I. (1994). Dose–response in pharmacokinetics. *Communications in Statistics—Theory and Methods*, 23, 451–465.

Patnaik, R.N., Lesko, L.J., Chen, M.L., Williams, R. and FDA Individual Bioequivalence Working Group (1997). Individual bioequivalence—New concepts in the statistical assessment of bioequivalence metrics. *Clinical Pharmacokinetics*, 33, 1–6.

Patterson, S., Agin, M., Anziano, R. et al. (2005a). Investigating drug-induced QT and QTc prolongation in the clinic: A review of statistical design and analysis considerations: Report from the Pharmaceutical Research and Manufacturers of America QT Statistics Expert Team. *Drug Information Journal*, 39, 243–266.

Patterson, S.D., Jones, B., and Zariffa, N. (2005b). Modelling and interpreting QTc prolongation in clinical pharmacology studies. *Drug Information Journal*, 39, 437–445.

Patterson, T.A., Lobenhofer, E.K., Fulmer-Smentek, S.B. et al. (2006). Performance comparison of one-color and two-color platforms with the MAQC project. *Nature Biotechnology*, 24, 1140–1150.

PDR. (1998). *Physicians' Desk Reference for Herbal Medicines*. Medical Economics Company, Montvale, NJ.

Peabody, F. (1927). The care of the patient. *JAMA*, 88, 877.

Philipp, E. and Weihrauch, T.R. (1993). Multinational drug development and clinical research: A bird's eye view of principles and practice. *Drug Information Journal*, 27, 1121–1132.

Phillips, K.F. (2003). A new test of non-inferiority for anti-infective trials. *Statistics in Medicine*, 22, 201–212.

PhRMA. (2003). Investigating drug-induced QT and QTc prolongation in the clinic: Statistical design and analysis considerations. Report from the Pharmaceutical Research on Manufacturers of America QT Statistics Expert Team, August 14, 2003.

Piantadosi, S. and Liu, G. (1996). Improved designs for dose escalation studies using pharmacokinetic measurements. *Statistics in Medicine*, 15, 1605–1618.

Pizzo, P.A. (2006). *The Dean's Newsletter*. Stanford University School of Medicine. Stanford, CA.

Pong, A. and Chow, S.C. (2010). Handbook of Adaptive Designs in Pharmaceutical and Clinical Development. Chapman and Hall/CRC Press, Taylor & Francis, New York.

Pong, A. and Raghavarao, D. (2002). Comparing distributions of drug shelf lives for two components in stability analysis for different designs. *Journal of Biopharmaceutical Statistics*, 12, 277–293.

Posch, M. and Bauer, P. (2000). Interim analysis and sample size reassessment. *Biometrics*, 56, 1170–1176.

Pratt, C.M., Hertz, R.P., Ellis, B.E., Crowell, S.P., Louv, W., and Moye, L. (1994). Risk of developing life-threatening ventricular arrhythmia associated with terfenadine in comparison with over-the-counter antihistamines, ibuprofen and clemastine. *American Journal of Cardiology*, 73, 346–352.

Proschan, M.A. and Hunsberger, S.A. (1995). Designed extension of studies based on conditional power. *Biometrics*, 51, 1315–1324.

Proschan, M.A. and Wittes, J. (2000). An improved double sampling procedure based on the variance. *Biometrics*, 56, 1183–1187.

Quan, H., Zhao, P.L., Zhang, J., Roessner, M., and Aizawa, K. (2010). Sample size considerations for Japanese patients based on MHLW guidance. *Pharmaceutical Statistics*, 9 (2), 100–112.

Rao, J.N.K. and Scott, A.J. (1981). The analysis of categorical data from complex sample surveys: Chi-square tests for goodness-of-fit and independence in two-way tables. *Journal of American Statistical Association*, 76 (374), 221–230.

Rao, J.N.K. and Scott, A.J. (1987). On simple adjustments to chi-square tests with sample survey data. *Journal of Annals of Statistics*, 15, 1–12.

Rao, J.N.K. and Shao, J. (1992). Jackknife variance estimation with survey data under hot deck imputation. *Biometrika*, 79, 811–822.

Reiser, B., and Faraggi, D. (1997). Confidence intervals for the general ROC criterion. *Biometrics*, 53, 644–652.

Roger, S.D. (2006). Biosimilars: How similar or dissimilar are they? *Nephrology*, 11, 341–346.

Roger, S.D. and Mikhail, A. (2007). Biosimilars: Opportunity or cause for concern? *Journal of Pharmaceutical Science*, 10, 405–410.

Rom, D.M. (1990). A sequentially rejective test procedure based on a modified Bonferroni inequality. *Biometrika*, 77, 663–665.

Rosenberger, W.F. and Lachin, J.M. (2003). *Randomization in Clinical Trials*. John Wiley & Sons, Inc., New York.

Rothmann, M.D., Koti, K., Lee, K.Y., Lu, H.L., and Shen, Y.L. (2009). Missing data in biologic oncology products. *Journal of Biopharmaceutical Statistics*, 19, 1074–1084.

Rotnitzky, A., Robins, J.M., and Scharfstein, D.O. (1998). Semiparametric regression for repeated measures outcomes with non-ignorable non-response. *Journal of the American Statistical Association*, 93, 1321–1339.

Sampson, A. and Sill, M.W. (2005). Drop-the-losers design: Normal case. *Biometrical Journal*, 47 (3), 257–268.

Sarkar, S. and Chang, C.K. (1997). Simes method for multiple hypothesis testing with positively dependent test statistics. *Journal of American Statistical Association*, 91, 1601–1608.

Saul, S. (2007). More generics slow rise in drug prices. *The New York Times*, August 8, 2007.

Schafer, J.L. (1997). *Analysis of Incomplete Multivariate Data*. Chapman and Hall, London, U.K.

Schall, R. and Luus, H.G. (1993). On population and individual bioequivalence. *Statistics in Medicine*, 12, 1109–1124.

Scheffé, H. (1959). *The Analysis of Variance*. Wiley, New York.

Schellekens, H. (2004). How similar do 'biosimilar' need to be? *Nature Biotechnology*, 22, 1357–1359.

Schuirmann, D.J. (1987). A comparison of the two one-sided tests procedure and the power approach for assessing the equivalence of average bioavailability. *Journal of Pharmacokinetics and Biopharmaceutics*, 15, 657–680.

Searle, S.R. (1971). *Linear Models*. John Wiley, New York.

Serfling, R.J. (1980). *Approximation Theorems of Mathematical Statistics*. Wiley, New York.

Shao, J. (1993). Linear model selection by cross-validation. *Journal of American Statistical Association*, 88, 486–494.

Shao, J. (1999). *Mathematical Statistics*. Springer-Verlag, New York.

Shao, J. and Chow, S.C. (1993). Two-stage sampling with pharmaceutical applications. *Statistics in Medicine*, 12, 1999–2008.

Shao, J. and Chow, S.C. (2002). Reproducibility probability in clinical trials. *Statistics in Medicine*, 21, 1727–1742.

Shao, J. and Chow, S.C. (2007). Variable screening in predicting clinical outcome with high-dimensional microarrays. *Journal of Multivariate Analysis*, 98, 1529–1538.

Shao, J. and Tu, D. (1995). *The Jackknife and Bootstrap*. Springer, New York.

Shao, J. and Wang, H. (2002). Sample correlation coefficients based on survey data under regression imputation. *Journal of American Statistical Association*, 97, 544–552.

Shao, J. and Zhong, B. (2003). Last observation carry-forward and last observation analysis. *Statistics in Medicine*, 23, 3241–3244.

Shao, J., Chang, M., and Chow, S.C. (2005). Statistical inference for cancer trials with treatment switching. *Statistics in Medicine*, 24, 1783–1790.

Shardell, M. and El-Kamary, S. (2009). Sensitivity analysis of informatively coarsened data using pattern mixture models. *Journal of Biopharmaceutical Statistics*, 19, 1018–1038.

Shibata, R. (1981). An optimal selection of regression variables. *Biometrika*, 68, 45–54.

Shih, W.J. (2001). Clinical trials for drug registrations in Asian Pacific countries: Proposal for a new paradigm from a statistical perspective. *Controlled Clinical Trials*, 22, 357–366.

Shippy, R., Fulmer-Smentek, S., Jensen, R.V. et al. (2006). Using RNA sample titrations to assess microarray platform performance and normalization techniques. *Nature Biotechnology*, 24, 1123–1131.

Simes, R.J. (1986). An improved Bonferroni procedure for multiple test procedures. *Journal of the American Statistical Association*, 81, 826–831.

Simon, R. (1989). Optimal two-stage designs for phase II clinical trials. *Controlled Clinical Trials*, 10, 1–10.

Simon, R. (2006). Validation of pharmacogenomics biomarker classifier for treatment selection. *Cancer Biomarkers*, 2, 89–96.

Simon, R. (2008). Development and validation of biomarker classifier for treatment selection. *Journal of Statistical Planning and Inference*. 138, 308–320.

Simon, R. and Maitournam, A. (2004). Evaluating the efficiency of targeted designs for randomized clinical trials. *Clinical Cancer Research*, 10, 6759–6763.

Simon, R.M., Korn, E.L., McShane, L.M., Radmacher, M.D., Wright, G.W., and Zhao, Y. (2003). *Design and Analysis of DNA Microarray Investigations.* Springer, New York.

Smith, N. (1992). FDA perspectives on quality of life studies. Presented at *DIA Workshop*. Hilton Head, SC.

Soon, G. (2009). Editorial: Missing data—Prevention and analysis. *Journal of Biopharmaceutical Statistics*, 19, 941–944.

Sprarano, J., Hayes, D., Dees, E. et al. (2006). Phase III randomized study of adjuvant combination chemotherapy and hormonal therapy versus adjuvant hormonal therapy alone in women with previously resected axillary node-negative breast cancer with various levels of risk for recurrence (TAILORX Trial). Available at: http://www.cancer.gov/clinicaltrials/ECOG- PACCT-1 (accessed on June 5, 2006).

Spriet, A. and Dupin-Spriet, T. (1992). Good biometrics practice proposals for a set of procedures. *Drug Information Journal*, 26, 405–409.

Srivastava, M.S. and Carter, E.M. (1986). The maximum likelihood method for non-response in sample surveys. *Survey Methodology*, 12, 61–72.

Storer, B.E. (1989). Design and analysis of phase I trials. *Biometrics*, 45, 925–937.

Storer, B.E. (1993). Small-sample confidence sets for the MTD in a phase I clinical trial. *Biometrics*, 49, 1117–1125.

Storer, B.E. (2001). An evaluation of phase I clinical trial designs in the continuous dose-response setting. *Statistics in Medicine*, 20, 2399–2408.

Strieter, D., Wu, W., and Agin, M. (2003). Assessing the effects of replicate ECGs on QT variability in healthy subjects. Presented at *Midwest Biopharmaceutical Workshop*, Muncie, Indiana, May 21, 2003.

Su, J.Q. and Liu, J.S. (1993). Linear combination of multiple diagnostic markers. *Journal of the American Statistical Association*, 88, 1350–1355.

SUPAC-IR. (1995). The United States Food and Drug Administration Guideline: Immediate release solid oral dosage forms: Scale-up and postapproval changes: Chemistry, manufacturing, and controls, in vitro dissolution testing, and in vivo bioequivalence documentation, Rockville, MD.

Suwelack, D. and Weihrauch, T.R. (1992). Practical issues in design and management of multinational trials. *Drug Information Journal*, 26, 371–378.

Swain, S.M. (2006). A step in the right direction. *Journal of Clinical Oncology*, 24 (23), 1–2.

Tandon, P.K. (1990). Applications of global statistics in analyzing quality of life. *Statistics in Medicine*, 9, 819–827.

Temple, R. (2003). Overview of the concept paper, history of the QT/TdP concern; regulatory implications of QT prolongation. Presentations at *Drug Information Association/FDA Workshop on QT Prolongation,* January 13, Rockville, MD.

Testa, M.A. (1987). Interpreting quality of life clinical trial data for use in clinical practices of antihypertensive therapy. *Journal of Hypertension*, 5, S9–S13.

Testa, M.A., Anderson, R.B., Nackley, J.F., and Hollenberg, N.K. (1993). Quality of life and antihypertensive therapy in men: A comparison of Captopril with Enalapril. *New England Journal of Medicine*, 328, 907–913.

Thall, P.F. and Russel, K.E. (1998). A strategy for dose-finding and safety monitoring based on efficacy and adverse outcomes in phase I/II clinical trials. *Biometrics*, 54, 251–264.

Tian, H. and Natarajan, J. (2008). Effect of baseline measurement on the change from baseline in QTc intervals. *Journal of Biopharmaceutical Statistics*, 18, 542–552.

Tibshirani, R. (1996). Regression shrinkage and selection via the Lasso. *Journal of Royal Statistical Society, B*, 58, 267–288.

Todd, S. (2003). An adaptive approach to implementing bivariate group sequential clinical trial designs. *Journal of Biopharmaceutical Statistics*, 13, 605–619.

Tong, W., Lucas, A.B., Shippy, R. et al. (2006). Evaluation of external RNA controls for the assessment of microarray performance. *Nature Technology*, 24, 1132–1139.

Torrance, G.W. (1976). Toward a utility theory foundation for health status index models. *Health Services Research*, 4, 349–369.

Torrance, G.W. (1987). Utility approach to measuring health-related quality of life. *Journal of Chronic Diseases*, 40, 593–600.

Torrance, G.W. and Feeny, D.H. (1989). Utilities and quality-adjusted life years. *Journal of Technology Assessment* in *Health Care*, 5, 559–575.

Tse, S.K. and Chow, S.C. (2011). Clinical strategy for endpoint selection. Unpublished manuscript (to be submitted).

Tse, S.K., Chang, J.Y., Su, W.L., Chow, S.C., Hsiung, C., and Lu, Q. (2006). Statistical quality control process for traditional Chinese medicine. *Journal of Biopharmaceutical Statistics*, 16, 861–874.

Tsiatis, A.A. and Mehta, C. (2003). On the inefficiency of the adaptive design for monitoring clinical trials. *Biometrika*, 90, 367–378.

Tsong, Y. and Shen, M. (2007). An alternative approach to assess exchangeability of a test treatment and the standard treatment with normally distributed response. *Journal of Biopharmaceutical Statistics*, 17, 329–338.

Tsong, Y. and Zhang, J. (2005). Testing superiority and noninferiority hypotheses in active controlled clinical trials. *Biometrical Journal*, 47, 62–74.

Tsong, Y. and Zhang, J. (2007). Simultaneous test for superiority and noninferiority hypotheses in active-controlled clinical trials. *Journal of Biopharmaceutical Statistics*, 17, 247–257.

Tsong, Y. and Zhang, J. (2008). Guest editors' notes on statistical issues in design and analysis of thorough QTc studies. *Journal of Biopharmaceutical Statistics*, 18, 405–407.

Tsong, Y., Higgins, K., Wang, S.J., and Hung, H.M.J. (1999). An overview of equivalence testing—CDER reviewers' perspective. In: *Proceedings of the Biopharmaceutical Section of American Statistical Association*, Alexandria, VA, pp. 214–219.

Tsong, Y., Shen, M., Zhong, J., and Zhang, J. (2008). Statistical issues of QT prolongation assessment based on linear concentration modeling. *Journal of Biopharmaceutical Statistics*, 18, 564–584.

Tsong, Y., Zhang, J., and Levenson, M. (2007). Choice of δ noninferiority margin and dependency of the noninferiority trials. *Journal of Biopharmaceutical Statistics*, 17, 279–288.

Tsong, Y., Zhong, J., and Chen, W.J. (2008). Validation testing in thorough QT/QTc clinical trials. *Journal of Biopharmaceutical Statistics*, 18, 529–541.

Tsou, H.H., Chow, S.C., Wang, S.J., Hung, H.M., Lan, K.K., Liu, J.P., Wang, M., Chen, H.D., Ho, L.T., Hsiung, C.A., and Hsiao, C.F. (2011). Proposal of statistical consideration to evaluation of results for a specific region in multi-regional trials—Asian perspective. *Pharmaceutical Statistics* (in press).

Tsou, H.H., Hsiao, C.F., Chow, S.C., Yue, L., Xu, Y., and Lee, S. (2007). Mixed non-inferiority margin and statistical tests in active controlled trials. *Journal of Biopharmaceutical Statistics*, 17, 339–357.

Tusher, V.G., Tibshirani, R., and Chu, G. (2001). Significance analysis of microarrays applied to ionizing radiation response, *Proceedings of National Academy of Sciences*, 98, 5116–5121.

U.S. FDA (2007) Draft guidance on *In Vitro Diagnostic Multivariate Index Assays*. The U.S. Food and Drug Administration, Rockville, Maryland.

Uesaka, H. (2009). Sample size allocation to regions in multiregional trial. *Journal of Biopharmaceutical Statistics*, 19, 580–594.

Ueta, M., Sotozono, C., Tokunaga, K., Yabe, T., and Kinoshita, S. (2007). Strong association between HLA-A*0206 and Stevens-Johnson syndrome in the Japanese. *American Journal of Ophthalmology*, 143 (2), 367–368.

USP/NF. (2000). *United States Pharmacopeia 24 and National Formulary 19*, United States Pharmacopeial Convention, Inc., Rockville, MD.

Van de Vijver, M.J., He, Y.D., van't Veer, L.J. et al. (2002). A gene-expression signature as a predictor of survival in breast cancer. *New England Journal of Medicine*, 347, 1999–2009.

van't Veer, L.J., Dai, H., van de Vijver, M.J. et al. (2002). Gene expression profiling predicts clinical outcome of breast cancer. *Nature*, 415, 530–536.

Varmus, H. (2006). The new era in cancer research. *Science*, 312, 1162–1165.

Wang, H. (2001). Two-way contingency tables with marginally and conditionally imputed nonrespondents. PhD thesis, Department of Statistics, University of Wisconsin, Madison, WI.

Wang, H. and Chow, S.C. (2002). On statistical power for average bioequivalence testing under replicated crossover design. *Journal of Biopharmaceutical Statistics*, 12, 295–309.

Wang, H., Chow, S.C., and Li, G. (2002a). On sample size calculation based on odds ratio in clinical trials. *Journal of Biopharmaceutical Statistics*, 12, 471–483.

Wang, S. and Ethier, S. (2004). A generalized likelihood ratio test to identify differentially expressed genes from microarray data. *Bioinformatics*, 20, 100–104.

Wang, S.J., Hung, H.M.J., and Tsong, Y. (2002b). Utility and pitfall of some statistical methods in active controlled clinical trials. *Controlled Clinical Trials*, 23, 15–28.

Wang, X., Wu, Y., and Zhou, H. (2009). Outcome- and auxiliary-dependent subsampling and its statistical inference. *Journal of Biopharmaceutical Statistics*, 19, 1132–1150.

Wang, Y., Pan, G., and Balch, A. (2008). Bias and variance evaluation of QT interval correction methods. *Journal of Biopharmaceutical Statistics*, 18, 427–450.

Ware, J.E. (1987). Standards for validating health measures definition and content. *Journal of Chronic Diseases*, 40, 473–480.

Ware, J.H., Mosteller, F., and Ingelfinger, J.A. (1986). P-values. In: *Medical Use of Statistics*, J.C. Bailar and F. Mosteller (eds.). NEJM Books, Waltham, MA, Chap. 8.

Webber, K.O. (2007). Biosimilars: Are we there yet? Presented at *Biosimilars 2007*, George Washington University, Washington, DC.

Wei, X. and Chappel, R. (2005). A test for non-inferiority with a mixed multiplicative/additive null hypothesis, Presentation in *2005 ENAR Spring Meeting*. Austin, Texas.

Westfall, P. and Bretz, F. (2010). Multiplicity in clinical trials. In: *Encyclopedia of Biopharmaceutical Statistics*, S.C. Chow (ed.), 3rd edn. Taylor & Francis, New York.

Westfall, P.H., Tobias, R.D., Rom, D., Wolfinger, R.D., and Hochberg, Y. (1999). *Multiple Comparisons and Multiple Tests Using the SAS System*. SAS Institute, Cary, NC.

Westlake, W.J. (1976). Symmetrical confidence intervals for bioequivalence trials. *Biometrics*, 32, 741–744.

Whitehead, J. and Williamson, D. (1998). An evaluation of Bayesian decision procedures for dose-finding studies. *Journal of Biopharmaceutical Medicine*, 8, 445–467.

WHO. (2005). *World Health Organization Draft Revision on Multisource (Generic) Pharmaceutical Products: Guidelines on Registration Requirements to Establish Interchangeability*. WHO, Geneva, Switzerland.

Wiles, A., Atkinson, G., Huson, L., Morse, P., and Struthers, L. (1994). Good statistical practices in clinical research: Guideline standard operating procedures. *Drug Information Journal*, 28, 615–627.

Williams, G., Pazdur, R., and Temple, R. (2004). Assessing tumor-related signs and symptoms to support cancer drug approval. *Journal of Biopharmaceutical Statistics*, 14, 5–21.

Williams, G.H. (1987). Quality of life and its impact on hypertensive patients. *American Journal of Medicine*, 82, 98–105.

Woodcock, J. (2004). *FDA's Critical Path Initiative*. Available at: FDA website: http://www.fda.gov/oc/initiatives/criticalpath/woodcock0602/woodcock0602.html

Woodcock, J. (2005). FDA introduction comments: Clinical studies design and evaluation issues. *Clinical Trials*, 2, 273–275.

Woodcock, J., Griffin, J., Behrman, R. et al. (2007). The FDA's assessment of follow-on protein products: A historical perspective. *Nature Reviews Drug Discovery*, 6, 437–442.

Wu, J.C.F. (1983). On the convergence properties of the EM algorithm. *Annals of Statistics*, 11, 95–103.

Wysowski, D.K., Corken, A., Gallo-Torres, H., Talarico, L., and Rodriguez, E.M. (2001). Postmarketing reports of QT prolongation and ventricular arrhythmia in association with cisapride and Food and Drug Administration regulatory actions. *American Journal of Gastroenterology*, 96, 1698–1703.

Yan, X., Lee, S., and Li, N. (2009). Missing data handling methods in medical device clinical trials. *Journal of Biopharmaceutical Statistics*, 19, 1085–1098.

Yan, X., Wang, M.C., and Su, X. (2007). Test for the consistency of noninferiority from multiple clinical trials. *Journal of Biopharmaceutical Statistics*, 17, 265–278.

Yang, L.Y., Chi, Y.C., and Chow, S.C. (2011). Statistical inference for clinical trials with binary responses when there is a shift in patient population. *Journal of Biopharmaceutical Statistics*, 21 (in press).

Zhang, H. and Paik, M.C. (2009). Handling missing responses in generalized linear mixed model without specifying missing mechanism. *Journal of Biopharmaceutical Statistics*, 19, 1001–1017.

Zhang, J. (2008). Testing for positive control activity in a thorough QTc study. *Journal of Biopharmaceutical Statistics*, 18, 517–528.

Zhang, J. and Machado, S.G. (2008). Statistical issues including design and sample size calculation in thorough QT/QTc studies. *Journal of Biopharmaceutical Statistics*, 18, 451–467.

Zhang, L., Dmitrienko, A., and Luta, G. (2008). Sample size calculations in thorough QT studies. *Journal of Biopharmaceutical Statistics*, 18, 468–482.

Zhou, Y., Whitehead, J., Bonvini, E., and Stevens, J.W. (2006). Bayesian decision procedures for binary and continuous bivariate dose-escalation studies. *Pharmaceutical Statistics*, 5, 125–133.

Index

Made in the USA
Monee, IL
07 July 2026

56550256R00337